Numerical Constants (k, m, n, r = positive integers)

Designation	Symbol	Representation	Section
Archimedes constant	π	$4\sum_{k=1}^{\infty}\left[\dfrac{(-1)^{k-1}}{2k-1}\left(\dfrac{4}{5^{2k-1}}-\dfrac{1}{239^{2k-1}}\right)\right]$	2.17
Base of nat. logarithm	e	$2\left(\dfrac{1}{1!}+\dfrac{2}{3!}+\dfrac{3}{5!}+\cdots\right)$	2.18
Bernoulli number	B_r	$2\dfrac{(2r)!}{(2\pi)^{2r}}\sum_{k=1}^{\infty}\left(\dfrac{1}{k}\right)^{2r}$	2.22
Beta function	$\mathrm{B}(m, n)$	$\dfrac{(m-1)!(n-1)!}{(m+n-1)!}$	2.31
Binomial coefficient	$\binom{n}{k}$	$\dfrac{n(n-1)(n-2)(n-3)\cdots(n-k+1)}{k(k-1)(k-2)\cdots3\cdot2\cdot1}$	2.09
Catalan constant	G	$\bar{Z}(2)=\dfrac{1}{1^2}-\dfrac{1}{3^2}+\dfrac{1}{5^2}-\cdots$	2.27
Digamma function	$\psi(0)$	$-\gamma$	2.32
Digamma function	$\psi(n)$	$1+\dfrac{1}{2}+\dfrac{1}{3}+\cdots+\dfrac{1}{n}-\gamma$	2.32
Double factorial	$(2n)!!$	$2n(2n-2)(2n-4)\cdots6\cdot4\cdot2$	2.30
Double factorial	$(2n-1)!!$	$(2n-1)(2n-3)(2n-5)\cdots5\cdot3\cdot1$	2.30
Euler's constant	γ	$\lim_{n\to\infty}\left(\dfrac{1}{1}+\dfrac{1}{2}+\dfrac{1}{3}+\cdots+\dfrac{1}{n}-\ln n\right)$	2.19
Euler number	E_r	$2\left(\dfrac{2}{\pi}\right)^{2r+1}(2r)!\sum_{k=1}^{\infty}(-1)^{k+1}\left(\dfrac{1}{2k-1}\right)^{2r+1}$	2.24
Factorial	$n!$	$n(n-1)(n-2)\cdots3\cdot2\cdot1$	2.08
Factorial polynomial	$X_m^{(n)}$	$x(x-m)(x-2m)(x-3m)\cdots(x-mn+m)$	2.11
Fibonacci number	F_r	$\dfrac{1}{\sqrt{5}}\left[\left(\dfrac{1+\sqrt{5}}{2}\right)^r-\left(\dfrac{1-\sqrt{5}}{2}\right)^r\right]$	2.21
Gamma function	$\Gamma(n)$	$(n-1)(n-2)(n-3)\cdots3\cdot2\cdot1$	2.28
Polygamma function	$\psi^{(m)}(n)$	$(-1)^{m+1}m!\left[Z(m+1)-\sum_{k=1}^{n}\dfrac{1}{k^{m+1}}\right]$	2.34
Zeta function	$Z(n)$	$\dfrac{1}{1^n}+\dfrac{1}{2^n}+\dfrac{1}{3^n}+\cdots$	2.26
Zeta function	$\bar{Z}(n)$	$\dfrac{1}{1^n}-\dfrac{1}{3^n}+\dfrac{1}{5^n}-\cdots$	2.27

HANDBOOK OF NUMERICAL CALCULATIONS IN ENGINEERING

OTHER McGRAW-HILL REFERENCE BOOKS OF INTEREST

HANDBOOKS

Avallone and Baumeister · MARKS' STANDARD HANDBOOK FOR
  MECHANICAL ENGINEERS
Brady and Clauser · MATERIALS HANDBOOK
Callender · TIME-SAVER STANDARDS FOR ARCHITECTURAL DESIGN
  DATA
Chopey and Hicks · HANDBOOK OF CHEMICAL ENGINEERING
  CALCULATIONS
Dean · LANGE'S HANDBOOK OF CHEMISTRY
Fink and Beaty · STANDARD HANDBOOK FOR ELECTRICAL ENGINEERS
Fink and Christiansen · ELECTRONICS ENGINEERS' HANDBOOK
Gaylord and Gaylord · STRUCTURAL ENGINEERING HANDBOOK
Gieck · ENGINEERING FORMULAS
Harris · SHOCK AND VIBRATION HANDBOOK
Juran and Gryna · JURAN'S QUALITY CONTROL HANDBOOK
Maynard · INDUSTRIAL ENGINEERING HANDBOOK
Merritt · STANDARD HANDBOOK FOR CIVIL ENGINEERS
Perry and Green · PERRY'S CHEMICAL ENGINEERS' HANDBOOK
Rohsenow, Hartnell, and Ganić · HANDBOOK OF HEAT TRANSFER
  FUNDAMENTALS
Rohsenow, Hartnett, and Ganić · HANDBOOK OF HEAT TRANSFER
  APPLICATIONS
Rosaler and Rice · STANDARD HANDBOOK OF PLANT ENGINEERING
Seidman, Mahrous, and Hicks · HANDBOOK OF ELECTRONIC POWER
  CALCULATIONS
Shigley and Misckhe · STANDARD HANDBOOK OF MACHINE DESIGN
Tuma · ENGINEERING MATHEMATICS HANDBOOK
Tuma · HANDBOOK OF STRUCTURAL AND MECHANICAL MATRICES

ENCYCLOPEDIAS

CONCISE ENCYCLOPEDIA OF SCIENCE AND TECHNOLOGY
ENCYCLOPEDIA OF ELECTRONICS AND COMPUTERS
ENCYCLOPEDIA OF ENERGY
ENCYCLOPEDIA OF ENGINEERING
ENCYCLOPEDIA OF PHYSICS

DICTIONARIES

DICTIONARY OF SCIENTIFIC AND TECHNICAL TERMS
DICTIONARY OF MECHANICAL AND DESIGN ENGINEERING
DICTIONARY OF COMPUTERS

For more information about other McGraw-Hill materials,
call 1-800-2-MCGRAW in the United States. In other
countries, call your nearest McGraw-Hill office.

HANDBOOK OF NUMERICAL CALCULATIONS IN ENGINEERING

DEFINITIONS • THEOREMS • COMPUTER MODELS • NUMERICAL EXAMPLES • TABLES OF FORMULAS • TABLES OF FUNCTIONS

Jan J. Tuma, Ph.D.

**Professor of Engineering
Arizona State University**

McGraw-Hill Publishing Company

New York   St. Louis   San Francisco   Auckland
Bogotá   Hamburg   London   Madrid   Mexico
Milan   Montreal   New Delhi   Panama
Paris   São Paulo   Singapore
Sydney   Tokyo   Toronto

Library of Congress Cataloging-in-Publication Data

Tuma, Jan J.
   Handbook of numerical calculations in engineering: definitions,
theorems, computer models, numerical examples, tables of formulas,
tables of functions/Jan J. Tuma.
      p.   cm.
  Bibliography: p.
  Includes index.
  ISBN 0-07-065446-8
  1. Engineering mathematics—Handbooks, manuals, etc.   2. Numerical
calculations—Handbooks, manuals, etc.   I. Title.
TA335.T85   1989   88-25944
620′.0042--dc 19   CIP

1234567890   HAL/HAL   895432109

ISBN 0-07-065446-8

The editors for this book were Harold B. Crawford and David E.
Fogarty, and the production supervisor was Dianne Walber. It was set
in Baskerville by The Universities Press (Belfast) Ltd.

Printed and bound by Arcata Graphics/Halliday.

For more information about other McGraw-Hill materials,
call 1-800-2-MCGRAW in the United States. In other
countries, call your nearest McGraw-Hill office.

To my friend
George C. Beakley
Dean of Engineering
Arizona State University

Some thirty years ago, as a young man, he
conceived a model of an Engineering College in
which undergraduate classes are taught by full
professors, and teaching excellence is placed at
par with research accomplishments. After thirty
years, as Dean of Engineering, he still requires
full professors to teach undergraduate classes and
rewards their teaching excellence.
J.J.T.

About the Author

Jan J. Tuma, Ph.D., is Professor of Engineering at Arizona State university. He has extensive experience as an engineering consultant and has solved many problems on frame, plate, and space structures, some of which have become landmarks of the American Southwest. Dr. Tuma is the author of numerous research papers and several volumes in McGraw-Hill's Schaum Outline Series. Among his many other works are the *Engineering Mathematics Handbook, Technology Mathematics Handbook, Handbook of Physical Calculations,* and the *Handbook of Structural and Mechanical Matrices,* all published by McGraw-Hill.

Contents

Preface

This handbook presents in one volume a concise summary of the major tools of numerical calculations encountered in the preparation of micro-, mini-, and mainframe computer programs in engineering and applied sciences. It was prepared to serve as a *professional, users-oriented, desktop reference book* for engineers, architects, and scientists.

Since the publication of his *Engineering Mathematics Handbook,* which now appears in its third edition, this author has received a large number of requests to supplement his mathematical handbook by a companion volume dealing exclusively with numerical methods, and to summarize these methods in the telescopic form which was found so useful by many users.

The nature of the subject of numerical calculations discouraged any prior efforts to write a handbook such as this one. Numerical calculations, as the name indicates, can be explained only by numerical examples, and defy the effort to reduce them to simple formulas.

In early stages of planning of this handbook, it became apparent that the material would require tripart presentation consisting of

(1) *Governing relations,* supported by theoretical reasoning
(2) *Particular cases,* expressed in symbolic form and serving as micro-numerical models
(3) *Numerical examples,* providing the illustrations of the analytical background and serving as flowcharts of calculations

This effort required derivations of many new formulas, large numbers of which appear in the literature for the first time, and a very careful selection of examples, which would provide the inside understanding, but would not extend in length over one page of the book.

The results of this effort, which spans a period of five years, are 1400 analytical relations, 7000 micronumerical models, and over 400 numerical examples, kept in a volume of reasonable size that will allow desktop handling.

The subject material is divided into sixteen chapters covering the following:

(*a*) Evaluation of numerical constants
(*b*) Approximations of elementary and advanced functions
(*c*) Numerical differentiation and integration
(*d*) Solutions of algebraic and transcendental equations
(*e*) Solutions of systems of equations
(*f*) Applications of Fourier series and Laplace transforms
(*g*) Solutions of ordinary and partial differential equations

Finally, ten-digit tables of numerical values of the most important functions, which cannot be displayed by hand-held calculators, are assembled in the Appendixes.

The form of presentation has the same special features as the mathematical handbook mentioned before.

(1) *Each page is a table*, designated by a title and section number.
(2) *Left and right pages present related or similar material*, and important models are placed in blocks allowing a rapid location of the desired information.
(3) *Numerical examples are located below the respective formulas*, providing a direct illustration of the application on the same page or on the opposite page.

Although every effort was made to avoid errors, it would be presumptuous to assume that none had escaped detection in a work of this scope. The author earnestly solicits comments and recommendations for improvements and future additions.

In closing, gratitude is expressed to Mrs. Aileene Sparling who typed the final draft of the manuscript and to my wife Hana, for her patience, understanding and encouragement during the preparation of this book.

Tempe, Arizona   Jan J. Tuma

1
NUMERICAL CALCULATIONS

(a) Methods of calculation fall into two major cateogries, designated as *analytical methods* and *numerical methods*. Analytical methods use algebraic and transcendental functions in the solution of problems, whereas the numerical methods use arithmetic operations only.

(b) Numerical methods which form the major part of this book are classified as:

(1) Approximations of constants	(6) Operations with series and products
(2) Approximations of functions	(7) Solutions of algebraic equations
(3) Approximations of derivatives	(8) Solutions of transcendental equations
(4) Approximations of integrals	(9) Solutions of systems of equations
(5) Summations of series and products	(10) Solutions of differential equations

Two frequently used abbreviations are CA = *computer application* and TF = *telescopic form*.

1.02 APPROXIMATIONS OF CONSTANTS

(a) Three types of constants occur in the solution of engineering and applied science problems. They are the fundamental physical constants, the basic numerical constants, and the derived numerical constants.

(b) Fundamental physical constants are products of natural laws, and as such their values can only be obtained by measurements. A complete list of these constants is given in Appendix B, and in the back endpapers.

(c) Basic numerical constants are special numbers given by their definitions and frequently occurring in numerical calcuations. They can be evaluated to a desired degree of accuracy as shown in sections indicated below. They are:

(1) Stirling's numbers $\mathcal{S}_k^{(p)}$, $\bar{\mathcal{S}}_k^{(p)}$ (2.11)	(5) Natural logarithm of 2 (2.20)
(2) Archimedes constant π (2.17)	(6) Fibonacci numbers F_r (2.21)
(3) Euler's constant e (2.18)	(7) Bernoulli numbers B_r, $\bar{B}_r$ (2.22)
(4) Euler's constant γ (2.19)	(8) Euler numbers E_r, $\bar{E}_r$ (2.24)

(d) Derived numerical constants are special numbers obtained as particular values of certain functions and again occurring frequently in numerical calculations. They can also be calculated to a desired degree of accuracy as shown in sections indicated below. They are:

(1) Gamma functions of integer argument (factorials) $\Gamma(n + 1) = n!$ (2.09), (2.28)
(2) Double factorials $(2n)!!$, $(2n + 1)!!$ (2.30)
(3) Binomial coefficients $\binom{n}{k}$ (2.10)
(4) Riemann zeta functions of integer argument $Z(n)$, $\bar{Z}(n)$ (2.26), (2.27)
(5) Digamma functions of integer argument $\psi(n)$ (2.32)
(6) Polygamma functions of integer argument $\psi^{(m)}(n)$ (2.34)

In these functions, $k, n = 0, 1, 2, \ldots$ and $m = 1, 2, 3, \ldots$.

(a) Algebraic polynomials are the only functions which can be evaluated at a given point and/or in a given interval in a direct and closed form.

(b) All other functions can only be evaluated for a given argument and/or in a given interval by some approximations. The most important approximations of functions are:

```
(1)  Collocation polynomials (Chap. 3)
(2)  Power series expansions (Chaps. 9, 10)
(3)  Continued fractions (Chap. 9)
(4)  Rational fractions (Chap. 10)
(5)  Orthogonal polynomials (Chap. 11)
(6)  Least-squares polynomials (Chap. 12)
(7)  Trigonometric series expansions (Chap. 13)
```

(c) Example. The evaluation of $f(x) = 5x^2 + 3x - 2$ for $x = 0.5$ is obtained by a direct substitution as

$$f(0.5) = 5(0.5)^2 + 3(0.5) - 2 = 0.75$$

whereas $g(x) = x \cos x$ for $x = 0.5$ must be evaluated by some approximate formula such as

$$g(0.5) = 0.5 - (0.5)^3/2! + (0.5)^5/4! - (0.5)^7/6! + (0.5)^9/8! - (0.5)^{11}/10!$$

$$= 0.438\ 791\ 280\ 9$$

The argument that the same value can be obtained by the calculator does not change the fact that the calculator evaluation is based on some kind of approximation built into the calculator.

1.04   APPROXIMATIONS OF DERIVATIVES

(a) Algebraic polynomials. The first- and higher-order derivatives of algebraic polynomials are the only derivatives which can be evaluated at a given point and/or in a given interval in a direct and closed form.

(b) First and higher derivatives of all other functions can only be evaluated for a particular argument and/or in a given interval by some approximations. Three classes of approximations are available for this purpose. They are:

```
(1)  Finite-difference methods (Chap. 3)
(2)  Direct methods (Chaps. 9 to 11, 13, 14)
(3)  Indirect methods (Chaps. 9 to 11, 13, 14)
```

(c) Finite-difference methods are utilized in cases of tabulated values of the respective function or in cases when the analytical form of the function is too involved for direct differentiation. All finite-difference approximations must be approached with caution.

(d) Direct methods approximate the analytical form of the derivative by one of the methods listed in (1.03). If applicable, these methods offer the best approximations.

(e) Indirect methods approximate the function (of which derivative is required) by one of the methods listed in (1.03) and then take the derivatives of this approximation. If applicable, this approach offers a good and workable approximation, particularly in cases where derivatives of different order are required.

(a) Algebraic polynomials are the only functions which can be integrated in given limits in a direct and closed form. Integrals of all other functions must be evaluated by some approximations. They are classified as:

> (1) Closed-form integrals evaluated by functional approximations (1.03)
> (2) Numerical quadratures in terms of difference polynomials (4.05) to (4.09)
> (3) Numerical quadratures in terms of asymptotic series (4.10)
> (4) Numerical quadratures in terms of orthogonal polynomials (4.11) to (4.16)
> (5) Integration by power series expansion (11.01)
> (6) Integration by trigonometric expansion (13.14)
> (7) Goursat's formulas (4.17)
> (8) Filon's formulas (4.18)
> (9) Multiple-integral reduction formulas (14.03)

(b) Closed-form integrals expressed in terms of elementary transcendental functions or in terms of combinations of algebraic and elementary transcendental functions are listed in standard integral tables (Refs. 1.01, 1.20, 4.02, 4.03, 4.06) and as such must be evaluated by the approximations given in (1.03).

(c) Numerical quadrature formulas (4.05) to (4.09) use either values of the integrand at equidistant spacing or use the difference interpolation polynomial as the substitute function. Their power is their simplicity. The trapezoidal rule and Simpson's rule are the most commonly used methods in this class.

(d) Euler–MacLaurin formula adds correction terms of asymptotic series to the trapezoidal rule and provides the numerical analysis with one of the most important relationships, which is used inversely in the summation of series (4.10).

(e) Gauss' integration formulas, known as Gauss–Chebyshev, Gauss–Legendre, Radau and Lobatto formulas, use orthogonal polynomials as the basis for the evaluation (4.11) to (4.16). They gained recent popularity in connection with the finite-element methods.

(f) Power series expansion of the integrand allows the evaluation of integrals of functions which cannot be evaluated in terms of elementary functions. The advantage of this method is its application in the evaluation of multiple integrals and its simplicity in handling in general.

(g) Trigonometric series expansion of the integrand offers the same advantages as the power series method. In addition the trigonometric expansion of the integrand gives a symbolic expression for the integral evaluation.

(h) Goursat's formulas allow the evaluation of product functions by a finite series and are particularly useful in products of algebraic and elementary transcendental functions.

(i) Filon's formulas offer the approximate evaluation of definite integrals of $f(x) \cos tx$ and $f(x) \sin tx$, where t is a constant.

(j) Multiple-integral reduction formulas reduce the multiple integral to a single integral by means of the convolution theorem (14.03).

(a) Series and products are finite or infinite, and they are classified as:

> (1) Series and products of constants (Chap. 5)
> (2) Series and products of functions (Chap. 9)

(b) Sums of series and products of constants are calculated by:

> (1) Fundamental theorem of sum calculus (5.03)
> (2) Difference series formula (5.04)
> (3) Power series formula (5.04)
> (4) Euler–MacLaurin formula (5.05)
> (5) General and special transformations (5.06), (5.07)

(c) Sums of series of special constants are available for:

  (1) Arithmetic and geometric series (5.08)
  (2) Arithmogeometric series (5.09), (5.10)
  (3) Series of powers of integers (5.11), (5.12)
  (4) Harmonic series of integers (5.13), (5.14)
  (5) Series of powers of reciprocal integers (5.15), (5.16)
  (6) Series of factorial polynomials (5.17), (5.18)
  (7) Series of binomial coefficients (2.15), (2.16), (5.17), (5.18)
  (8) Series of powers of numbers (5.19), (5.20)
  (9) Harmonic series of decimal numbers (5.21)
 (10) Series of powers of reciprocal numbers (5.22)

(d) Sums of series of products of special constants are available for:

  (1) Series of products of numbers (5.25), (5.26)
  (2) Series of products of fractions (5.27), (5.28)

(e) Series of functions are the power series and the transcendental series. The most important series in this group are:

> (1) Algebraic power series (Chaps. 9, 10)
> (2) Finite-difference power series (Chap. 9)
> (3) Fourier series (Chap. 13)

(f) Algebraic power series are either the result of evaluation of a function in terms of

  (1) Taylor's series expansion (9.03)
  (2) MacLaurin's series expansion (9.03)

or they are the solutions of differential equations designated by special names such as Bessel functions (10.07) to (10.17), Legendre functions (11.10), Chebyshev functions (11.14), Laguerre functions (11.16), Hermite functions (11.17), and many more.

(g) Finite-difference power series (9.03) may also be used for the evaluation of functions, but their applications yield only a particular value of the function at a point.

(h) Fourier series are used for the same purpose as the algebraic power series. They can be used in the continuous range and also over equally spaced values in that range. They are particularly useful in the solution of partial differential equations (13.01), (13.17).

(a) Nesting of power series for the purpose of summation is the most commonly used operation, in which the given series is replaced by a nested product (Ref. 1.20, p. 102) as

$$
\sum_{k=0}^{n} (\pm 1)^k a_k x^k = a_0 \pm a_1 x \pm a_2 x^2 \pm a_3 x^3 \pm \cdots + (\pm 1)^n a_n x^n
$$

$$
= a_0(1 \pm b_1 x(1 \pm b_2 x(1 \pm b_3 x(1 \pm \cdots \pm b_n)))) = a_0 \bigwedge_{k=1}^{n} [1 \pm b_k x]
$$

where a_k = constant, $b_k = a_k/a_{k-1}$, x = real or complex variable, $n + 1$ = number of terms, and

$\bigwedge_{}^{n} [\ ]$ = symbol of nested sum

(b) Alternative form of nesting is

$$
\sum_{k=0}^{n} (\pm 1)^k a_k x^k = (a_0 \pm x(a_1 + x(a_2 \pm x(a_3 \pm x(a_4 \pm \cdots \pm a_n x))))) = \bigwedge_{k=0}^{n} [a_k \pm x]
$$

where the last expression in the product is $a_{n-1} \pm a_n x$.

(c) Example. In (1.03c) with $x = 0.5$, the truncated series of $n = 6$ is

$$
x \cos x = x - x^3/2! + x^5/4! - x^7/6! + x^9/8! - x^{11}/10! = 0.438\ 791\ 280\ 9
$$

The same series in nested form is

$$
x \cos x = x\left(1 - \frac{x^2}{(1)(2)}\left(1 - \frac{x^2}{(3)(4)}\left(1 - \frac{x^2}{(5)(6)}\left(1 - \frac{x^2}{(7)(8)}\left(1 - \frac{x^2}{(9)(10)}\right)\right)\right)\right)\right)
$$

$$
= 0.438\ 791\ 280\ 9
$$

As apparent from the nested structure, x^2 is a constant used recurrently.

(d) Algebraic operations with standard power series are performed by:

(1) Summation theorem (9.14a)	(3) Division theorem (9.14c)
(2) Multiplication theorem (9.14b)	(4) Power theorem (9.18a)

All these operations can be performed in matrix form.

(e) Algebraic operations with Fourier series are possible and are given in series form in (13.14). The resulting series are again a Fourier series.

(f) Derivatives and integrals of standard power series are obtained by differentiating and integrating the series term by the term as an algebraic polynomial. These operations can be performed in a closed interval if and only if this interval lies entirely within the interval of uniform convergence of the series. Derivatives and integrals of Fourier series are given by relationships described in (13.14).

(g) Derivatives and integrals of nested series introduced in (a) above are given symbolically as

$$
\frac{d}{dx}\left[\sum_{k=0}^{n} (\pm 1)^k a_k x^k \right] = \pm a_1 \bigwedge_{k=2}^{n} \left[1 \pm \frac{kb_k}{k-1} x \right]
$$

$$
\int_{a}^{b}\left[\sum_{k=0}^{n} (\pm 1)^k a_k x^k \right] dx = \left[a_0 x \bigwedge_{k=1}^{n} \left[1 \pm \frac{kb_k}{k+1} x \right] \right]_{a}^{b}
$$

(a) Algebraic equations of *n*th degree have closed-form solution if $n \leq 4$. If $n > 4$, only special cases can be solved in closed form. In general, the higher-order algebraic equations can be solved only by approximations.

(b) General closed-form solutions are available for the following algebraic equations:

(1) Quadratic equations (6.02*a*)	(4) Cubic equations (6.03*a*)
(2) Binomial equations (6.02*b*)	(5) Quartic equations (6.04*a*)
(3) Trinomial equations (6.02*f*)	

(c) Special closed-form solutions are available for the following algebraic equations:

    (1) Symmetrical equations of fifth degree (6.06*d*)
    (2) Antisymmetrical equations of fifth degree (6.06*e*)

(d) General methods used in the approximate solution of algebraic and transcendental equations are based on the concept of iteration and interpolation.

(e) Iteration methods are applicable in solutions of equations of all types, and the most important methods in this group are:

(1) Bisection method (6.08)	(3) Tangent method (6.10)
(2) Secant method (6.09)	(4) General iteration method (6.11), (6.12)

(f) Polynomial methods of practical importance used in the search of real and complex roots of algebraic equations of higher degree are:

(1) Newton–Raphson's method (6.16)	(2) Bairstow's method (6.17)

Methods developed in the era of hand calculations, such as Horner's method, Graeffe's method, Bernoulli's method, and Laguerre's method, are of limited value in computer applications and are not covered in this book.

1.09 SOLUTIONS OF SYSTEMS OF EQUATIONS

(a) Systems of linear algebraic equations are classified as nonhomogeneous and homogeneous. Methods of solution of nonhomogeneous equations fall into two categories:

    (1) *Direct methods*, producing an exact solution by using a finite number of arithmetic operations
    (2) *Iterative methods*, producing an approximate solution of desired accuracy by yielding a sequence of solutions which converges to the exact solution as the number of iterations tends to infinity

(b) Direct methods introduced in this book are:

(1) Cramer's rule method (7.03)	(5) Cholesky method (7.09)
(2) Matrix inversion method (7.04), (7.05)	(6) Square-root method (7.11)
(3) Gauss elimination method (7.06)	(7) Inversion by partitioning (7.12)
(4) Successive transformation method (7.07)	

(c) Iterative methods introduced in this book are:

(1) Gauss–Seidel iteration method (7.13)	(2) Carryover method (7.14)

(a) Homogeneous system of *n* algebraic equations given in matrix form as

$$[\mathbf{A} - \lambda\mathbf{I}]\mathbf{X} = \mathbf{0}$$

where $\mathbf{A}$ = given matrix of constant coefficients, $\mathbf{I}$ = unit matrix, $\mathbf{0}$ = zero vector, is said to be an eigenvalue matrix equation if the determinant

$$|A - \lambda I| = 0$$

Then $\lambda_1, \lambda_2, \ldots, \lambda_n$ are the *eigenvalues* and $\mathbf{X}$ is the *eigenvector*.

(b) Methods of calculation of eigenvalues and eigenvectors fall into two major groups:

(1) Direct methods (8.07), (8.08)
(2) Iterative methods (8.05), (8.06)

1.11 SOLUTIONS OF ORDINARY DIFFERENTIAL EQUATIONS

(a) Ordinary differential equations are classified by the order of the highest derivative and by the degree of the highest derivative, and their coefficients are constants or functions in the independent variable.

(b) Four types of solutions are used in this book:

(1) Direct method (15.01)
(2) Fourier series method (15.02)
(3) Laplace transform method (15.03)
(4) Numerical methods (15.25)

(c) Numerical methods are designated as:

(1) One-step methods (15.26) to (15.28)
(2) Multistep methods (15.29) to (15.30)
(3) Finite-difference methods (15.34) to (15.40)

1.12 SOLUTIONS OF PARTIAL DIFFERENTIAL EQUATIONS

(a) Partial differential equations are again designated by the order and degree, and their coefficients are constants or functions in the independent variables.

(b) Three types of solutions are used in this book:

(1) Separation-of-variables method (16.02*b*)
(2) Laplace transform method (16.02*c*)
(3) Finite-difference method (16.02*d*)

The most recent addition is the *finite-element method* (Refs. 16.12, 16.22). Since this method and its applications are described comprehensively in a new handbook (Ref. 16.09) dealing exclusively with the finite-elements, the coverage of this topic is not included in Chap. 16.

2
NUMERICAL
CONSTANTS

(a) Sequence is a set of n numbers $a_1, a_2, a_3, \ldots, a_n$ arranged in a prescribed order and formed according to a definite rule. Each number a_k $(k = 1, 2, 3, \ldots, n)$ is called a term, and the sequence is defined by the number of terms as finite $(n < \infty)$ or infinite $(n = \infty)$.

(b) Series is the sum of a sequence.

$$\sum_{k=1}^{n} a_k = a_1 + a_2 + a_3 + \cdots + a_n = S_n \qquad (-\infty < S_n < \infty)$$

$$\sum_{k=1}^{\infty} a_k = a_1 + a_2 + a_3 + \cdots + a_\infty = S_\infty \qquad (-\infty \le S_\infty \le \infty)$$

General term a_k (kth term) defines the law of formation of the sequence.

(c) Monotonic and alternating series are, respectively,

$$\sum_{k=1}^{n} a_k = a_1 + a_2 + a_3 + \cdots + a_n \qquad \sum_{k=1}^{n} \beta a_k = a_1 - a_2 + a_3 - \cdots \alpha a_n$$

where $\beta = (-1)^{k+1}$ and $\alpha = (-1)^{n+1}$

(d) Convergent series approaches the limit S as the number of terms approaches infinity $(n \to \infty)$; that is, if

$$\sum_{k=1}^{n} a_k = S_n \qquad \text{and} \qquad \lim_{n \to \infty} S_n = S$$

exists, the series is said to be convergent and S is the sum $(|S| \ne \infty)$.

(e) Divergent series increases without bound as the number of terms increases.

(f) Examples

$$\sum_{k=1}^{\infty} \frac{1}{3^{k-1}} = 1 + \frac{1}{3} + \frac{1}{9} + \frac{1}{27} + \cdots \qquad \text{is convergent, since } S = \frac{3}{2}$$

$$\sum_{k=1}^{\infty} (1 + k) = 2 + 3 + 4 + \cdots \qquad \text{is divergent, since } S = \infty$$

(g) Absolutely convergent series is an infinite series whose absolute terms form a convergent series

$$\sum_{k=1}^{\infty} |a_k| = |S|$$

(h) Conditionlly convergent series is a series which is not absolutely convergent.

(i) Examples

$$\frac{1}{1 \cdot 3} - \frac{1}{3 \cdot 5} + \frac{1}{5 \cdot 7} - \cdots \qquad \text{is absolutely convergent, since}$$

$$\frac{1}{1 \cdot 3} + \frac{1}{3 \cdot 5} + \frac{1}{5 \cdot 7} + \cdots \qquad \text{is a convergent series } \left(|S| = \frac{1}{2}\right).$$

$$1 - \frac{1}{2} + \frac{1}{3} - \frac{1}{4} + \cdots = \ln 2 \qquad \text{is conditionally convergent, since}$$

$$1 + \frac{1}{2} + \frac{1}{3} + \frac{1}{4} + \cdots \qquad \text{is a divergent series } (|S| = \infty)$$

(a) Comparison test. When a_k is the kth term of a series to be tested, b_k is the kth term of another series which is known to be absolutely convergent, and c is a constant independent of k, then if

$$|a_k| < c\,|b_k|$$

the series of a terms is also absolutely convergent.

(b) Example. If two series are given as

$$a_1 + a_2 + a_3 + \cdots = \frac{1}{1\cdot 2} - \frac{1}{2\cdot 3} + \frac{1}{3\cdot 4} - \cdots \qquad |a_k| = \frac{1}{k(k+1)}$$

$$b_1 + b_2 + b_3 + \cdots = 2 - \frac{2}{3} + \frac{2}{9} - \cdots \qquad c\,|b_k| = \frac{1}{2}\left|\frac{2}{3^{k-1}}\right|$$

then $|a_k| < c\,|b_k|$ for any value of k and the a series is absolutely convergent.

(c) Cauchy's test. If a_k is the kth term of a series to be tested and

$$\lim_{k\to\infty} \sqrt[k]{a_k} = L \qquad \text{(test fails for } L = 1\text{)}$$

then the series is absolutely convergent for $L < 1$, divergent for $L > 1$.

(d) Ratio test. If a_k and a_{k+1} are two successive terms of a series to be tested and

$$\lim_{k\to\infty}\left|\frac{a_{k+1}}{a_k}\right| = L \qquad \text{(test fails for } L = 1\text{)}$$

then the given series is absolutely convergent for $L < 1$, divergent for $L > 1$.

(e) Example. If the given series is

$$a_1 + a_2 + a_3 + \cdots = \frac{1}{1!} + \frac{1}{2!} + \frac{1}{3!} + \cdots \qquad \text{then} \qquad \lim_{k\to\infty}\left|\frac{1/(k+1)!}{1/k!}\right| = \lim_{k\to\infty}\left|\frac{1}{k+1}\right| = 0$$

and since $L < 1$, the series is absolutely convergent.

2.03 OPERATIONS WITH SERIES OF NUMBERS

(a) Grouping of terms of an absolutely or conditionally convergent series may be changed without affecting the sum.

(b) Example

$$1 - \frac{1}{2} + \frac{1}{3} - \frac{1}{4} + \cdots = \sum_{k=1}^{\infty} \frac{(-1)^{k+1}}{k} = \ln 2 \qquad \frac{1}{1\cdot 2} + \frac{1}{3\cdot 4} + \cdots = \sum_{k=1}^{\infty} \frac{1}{k(k+1)} = \ln 2$$

(c) Order of terms of an absolutely convergent series can be rearranged without affecting the sum. However, a particular change in order of terms of a conditionally convergent series may change the sum.

(d) Example. The conditionally convergent series $1 - \frac{1}{2} + \frac{1}{3} - \frac{1}{4} + \cdots = \ln 2$, but the rearranged series, $1 + \frac{1}{3} - \frac{1}{2} + \frac{1}{5} + \frac{1}{7} - \frac{1}{4} + \cdots = \ln \sqrt{8}$.

(e) Sum, difference, and product of two or more absolutely convergent series is an absolutely convergent series.

(a) Product of a set of n factors $(1 + a_1)$, $(1 + a_2)$, ..., $(1 + a_n)$ is defined by the number of factors as a finite product $(n < \infty)$

$$\prod_{k=1}^{n} (1 + a_k) = (1 + a_1)(1 + a_2)(1 + a_3) \cdots (1 + a_n) = U_n$$

where $-\infty < U_n < \infty$, or as an infinite product $(n = \infty)$

$$\prod_{k=1}^{\infty} (1 + a_k) = (1 + a_1)(1 + a_2)(1 + a_3) \cdots (1 + a_\infty) = U_\infty$$

where $-\infty \le U_\infty \le \infty$. General term a_k (kth term) defines the law of formation of the product.

(b) Examples

$$\prod_{k=1}^{n} (1 + k) = 2 \cdot 3 \cdot 4 \cdots (1 + n) \qquad \prod_{k=1}^{n} [1 + (2k - 2)] = 1 \cdot 3 \cdot 5 \cdots (2n - 1)$$

$$\prod_{k=1}^{n} (1 + k \sin k\theta) = (1 + \sin \theta)(1 + 2 \sin 2\theta)(1 + 3 \sin 3\theta) \cdots (1 + n \sin n\theta)$$

$$\prod_{k=1}^{\infty} \left(1 - \frac{1}{k + 1}\right) = \underbrace{(1 - \tfrac{1}{2})}_{\frac{1}{2}}\underbrace{(1 - \tfrac{1}{3})}_{\frac{2}{3}}\underbrace{(1 - \tfrac{1}{4})}_{\frac{3}{4}} \cdots = 1$$

$$\prod_{k=1}^{\infty} \left(1 - \frac{k + 1}{k + 2}\right) = \underbrace{(1 - \tfrac{2}{3})}_{\frac{1}{3}}\underbrace{(1 - \tfrac{3}{4})}_{\frac{1}{4}}\underbrace{(1 - \tfrac{4}{5})}_{\frac{1}{5}} \cdots = 0$$

(c) Monotonic and alternating products are, respectively,

$$\prod_{k=1}^{n} (1 + a_k) = (1 + a_1)(1 + a_2)(1 + a_3) \cdots (1 + a_n)$$

$$\prod_{k=1}^{n} (1 + \beta a_k) = (1 + a_1)(1 - a_2)(1 + a_3) \cdots (1 + \alpha a_n)$$

where $\beta = (-1)^{k+1}$ and $\alpha = (-1)^{n+1}$.

(d) Logarithmic conversion. Every product of numbers can be converted into a series by taking the logarithm of the product

$$\ln \left[\prod_{k=1}^{n} (1 + a_k)\right] = \ln [1 + a_1] + \ln [1 + a_2] + \ln [1 + a_3] + \cdots + \ln [1 + a_n]$$

and the properties of this series are the properties of the respective product.

(e) Typical products used in the numerical analysis are

$$n! = n \text{ factorial (2.09)} \qquad n!! = \text{double factorial (2.30)}$$

$$\binom{n}{k} = \text{binomial coefficient (2.10)}$$

and the nested product

$$a_1 + a_2 + a_3 + \cdots + a_n = a_1(1 + a_2/a_1(1 + a_3/a_2(1 + \cdots + a_n/a_{n-1})))$$

introduced in (1.07).

(a) Convergent product is an infinite product that approaches the limit U as the number of factors approaches infinity ($n \to \infty$); that is, if

$$\prod_{k=1}^{n} (1 + a_k) = U_n \qquad \text{and} \qquad \lim_{n \to \infty} U_n = U$$

exists, the product is said to be convergent and $|U| \neq \infty$.

(b) Example. Infinite product $1 \cdot \frac{1}{2} \cdot \frac{1}{3} \cdot \frac{1}{4} \cdots$ governed by $a_k = 1/k$ is convergent, since $U = 0$.

(c) Divergent product is an infinite product that increases without bound as the number of factors increases.

(d) Example

$$\prod_{k=1}^{\infty} (1 + k) = 2 \cdot 3 \cdot 4 \cdots = \infty$$

(e) Absolutely convergent product is an infinite product whose absolute terms form a convergent series.

$$\prod_{k=1}^{\infty} (1 + |a_k|) = |U|$$

(f) Example. Infinite product

$$\left(1 - \frac{1}{1 \cdot 3}\right)\left(1 + \frac{1}{3 \cdot 5}\right)\left(1 - \frac{1}{5 \cdot 7}\right) \cdots \qquad \text{is absolutely convergent, since}$$

$$\left(1 + \frac{1}{1 \cdot 3}\right)\left(1 + \frac{1}{3 \cdot 5}\right)\left(1 + \frac{1}{5 \cdot 7}\right) \cdots \qquad \text{is a convergent product} \left(U = \frac{\pi}{2}\right)$$

(g) Conditionally convergent product is an infinite product which is not absolutely convergent.

(h) Example. Infinite product

$$\prod_{k=1}^{\infty} \left(1 - \frac{\beta}{k+1}\right) = \left(1 - \frac{1}{2}\right)\left(1 + \frac{1}{3}\right)\left(1 - \frac{1}{4}\right) \cdots = \frac{1}{2}$$

is conditionally convergent, since

$$\prod_{k=1}^{\infty} \left(1 + \left|\frac{\beta}{k+1}\right|\right) = \left(1 + \frac{1}{2}\right)\left(1 + \frac{1}{3}\right)\left(1 + \frac{1}{4}\right) \cdots = \infty$$

(i) Tests of convergence introduced in (2.02) are directly applicable in testing the convergence of infinite products.

2.06 OPERATIONS WITH PRODUCTS OF NUMBERS

(a) Grouping of factors in an absolutely or conditionally convergent product may be changed without affecting the product.

(b) Example. Conditionally convergent product

$$(1 - \tfrac{1}{2})(1 + \tfrac{1}{3})(1 - \tfrac{1}{4})(1 + \tfrac{1}{5})(1 - \tfrac{1}{6}) \cdots = \tfrac{1}{2} \qquad\qquad \tfrac{1}{2} \cdot \tfrac{4}{3} \cdot \tfrac{3}{4} \cdot \tfrac{6}{5} \cdot \tfrac{5}{6} \cdots = \tfrac{1}{2}$$

(c) Order of factors in an absolutely convergent product can be rearranged without affecting the product.

(a) Continued fraction is of the form

$$F_n = b_0 + \cfrac{a_1}{b_1 + \cfrac{a_2}{b_2 + \cfrac{a_3}{b_3 + \cdots}}} = \left[b_0, \frac{a_1}{b_1 +}, \frac{a_2}{b_2 +}, \frac{a_3}{b_3 +}, \cdots \right]$$

where the right-side expression is a symbolic representation.

(b) Number of terms $a_1, b_1, a_2, b_2, \ldots, a_n, b_n$ defines the continued fraction as terminating ($n < \infty$) or infinite ($n = \infty$).

(c) Simple continued fraction has all partial numerators equal to 1.

$$F_n = b_0 + \cfrac{1}{b_1 + \cfrac{1}{b_2 + \cfrac{1}{b_3 + \cdots}}} = \left[b_0, \frac{1}{b_1 +}, \frac{1}{b_2 +}, \frac{1}{b_3 +}, \cdots \right]$$

(d) Successive convergents for $a_k > 0$, $b_k > 0$ are defined as

$$F_0 = \frac{P_0}{Q_0} = b_0 \qquad\qquad F_1 = \frac{P_1}{Q_1} = b_0 + \frac{a_1}{b_1} = \frac{b_0 b_1 + a_1}{b_1}$$

$$F_2 = \frac{P_2}{Q_2} = b_0 + \cfrac{a_1}{b_1 + \cfrac{a_2}{b_2}} = \frac{b_0(b_1 b_2 + a_2) + a_1 b_1}{b_1 b_2 + a_2}$$

. . . .

$$F_k = \frac{P_k}{Q_k} = \frac{a_k P_{k-2} + b_k P_{k-1}}{a_k Q_{k-2} + b_k Q_{k-1}} \qquad \text{or} \qquad \begin{bmatrix} P_k \\ Q_k \end{bmatrix} = \begin{bmatrix} P_{k-2} & P_{k-1} \\ Q_{k-2} & Q_{k-1} \end{bmatrix} \begin{bmatrix} a_k \\ b_k \end{bmatrix}$$

and

$$P_n Q_{n-1} - P_{n-1} Q_n = (-1)^{n-1} \prod_{k=1}^{n} a_k$$

where $P_{-1} = 1$, $P_0 = b_0$, $Q_{-1} = 0$, $Q_0 = 1$.

(e) Convergent continued fraction. If

$$\lim_{n \to \infty} F_n = \lim_{n \to \infty} \frac{P_n}{Q_n} = F \qquad (-\infty < F < \infty)$$

exists, the fraction is convergent. If $a_k = 1$ and b_k are integers, the continued fraction is always convergent. If a_k, b_k are positive integers or fractions and

$$a_k \leq b_k \qquad b_k > 0$$

then the continued fraction is convergent. If

$$F_n = \left[0, \frac{a_1}{b_1 +}, \frac{a_2}{b_2 +}, \frac{a_3}{b_3 +}, \cdots \right]$$

and $\quad |a_k| + 1 \leq |b_k| \qquad (k = 1, 2, 3, \ldots)$

then this continued fraction converges and its absolute value does not exceed unity.

(a) Conversion of continued fraction to a simple fraction is obtained by performing all the operations indicated by the continued fraction.

(b) Example

$$\left[4, \frac{1}{5+}, \frac{1}{6+}, \frac{1}{7}\right] = 4 + \cfrac{1}{5 + \cfrac{1}{6 + \frac{1}{7}}} = 4 + \cfrac{1}{5 + \frac{7}{43}} = 4 + \frac{43}{222} = \frac{931}{222}$$

(c) Any positive rational number may be converted to a finite continued fraction by a reversed process.

(d) Example

$$\frac{47}{13} = 3 + \frac{8}{13} = 3 + \frac{1}{\frac{13}{8}} = 3 + \cfrac{1}{1 + \frac{5}{8}} = 3 + \cfrac{1}{1 + \cfrac{1}{1 + \frac{3}{5}}} = \left[3, \frac{1}{1+}, \frac{1}{1+}, \frac{3}{5+}\right]$$

(e) Any positive irrational number may be converted to an infinite continued fraction by the same process, or it may be approximated by the convergent F_k (2.07d).

(1) F_k of odd order is greater than F_n but decreasing.
(2) F_k of even order is less than F_n but increasing.

(f) Table of successive convergents constructed by relations (2.07d) is shown below.

k	-1	0	1	2	3	$\cdots$	$k-2$	$k-1$	k	$\cdots$
a_k		0	a_1	a_2	a_3	$\cdots$	a_{k-2}	a_{k-1}	a_k	$\cdots$
b_k		b_0	b_1	b_2	b_3	$\cdots$	b_{k-2}	b_{k-1}	b_k	$\cdots$
P_k	1	b_0	P_1	P_2	P_3	$\cdots$	P_{k-2}	P_{k-1}	P_k	$\cdots$
Q_k	0	1	Q_1	Q_2	Q_3	$\cdots$	Q_{k-2}	Q_{k-1}	Q_k	$\cdots$

From this table, $P_{-1} = 1$, $Q_{-1} = 0$, $P_0 = b_0$, $Q_0 = 1$, and recurrently,

$$\begin{bmatrix} P_1 \\ Q_1 \end{bmatrix} = \begin{bmatrix} 1 & b_0 \\ 0 & 1 \end{bmatrix}\begin{bmatrix} a_1 \\ b_1 \end{bmatrix}, \qquad \begin{bmatrix} P_2 \\ Q_2 \end{bmatrix} = \begin{bmatrix} b_0 & P_1 \\ 1 & Q_1 \end{bmatrix}\begin{bmatrix} a_2 \\ b_2 \end{bmatrix}, \qquad \begin{bmatrix} P_k \\ Q_k \end{bmatrix} = \begin{bmatrix} P_{k-2} & P_{k-1} \\ Q_{k-2} & Q_{k-1} \end{bmatrix}\begin{bmatrix} a_k \\ b_k \end{bmatrix}$$

(g) Example. If $\pi = 3.141\,59$, the corresponding truncated continued fraction is

$$3 + \cfrac{1}{7 + \cfrac{1}{15 + \cfrac{1}{1 + \cfrac{1}{25 + \cfrac{1}{1 + \frac{1}{7}}}}}}$$

$b_0 = 3$

$1/0.141\,59 - 7.062\,65$	$b_1 = 7$
$1/0.062\,65 = 15.962\,80$	$b_2 = 15$
$1/0.962\,80 = 1.038\,64$	$b_3 = 1$
$1/0.038\,64 = 25.878\,79$	$b_4 = 25$
$1/0.878\,79 = 1.137\,93$	$b_4 = 1$
$1/0.137\,93 = 7.250\,08$	$b_5 = 7$

and with $a_1 = a_2 = a_3 = \cdots = a_k = 1$, the convergents calculated by (f) are

$$F_1 = P_1/Q_1 = 22/7 = 3.142\,86 \qquad F_4 = P_4/Q_4 = 9208/2831 = 3.141\,59$$

$$F_2 = P_2/Q_2 = 333/113 = 3.141\,51 \qquad F_5 = P_5/Q_5 = 9563/3044 = 3.141\,59$$

$$F_3 = P_3/Q_3 = 355/113 = 3.141\,59$$

(a) Factorial of positive integer n is defined as

$$n! = n(n - 1)(n - 2) \cdots 3 \cdot 2 \cdot 1$$

$$= n(n - 1)!$$

$$= n(n - 1)(n - 2)!$$

$$\cdot\cdot\cdot\cdot\cdot\cdot\cdot\cdot\cdot\cdot\cdot\cdot\cdot\cdot\cdot\cdot\cdot$$

and by definition, $0! = 1$. Numerical values of $n!$ for $n = 1, 2, 3, \ldots, 100$ are tabulated in (A.03).

(b) Factorial of proper fraction u **(0 < u < 1)** is calculated by the procedure described in (2.29), and numerical values of $u!$ for $u = 0.005, 0.010, \ldots, 0.995$ are tabulated in (A.04).

(c) Factorial of any number N can be expressed as

$$N! = N(N - 1)(N - 2) \cdots (N - n + 1)u! \qquad (N > 0)$$

where $N - n = u$ and n is a positive integer nearest to N.

(d) Example. For $N = 3.785$, $n = 3$, $u = 0.785$,

$$N! = (3.785)(2.785)(1.785)(0.785)! = 17.410\,416$$

where from (A.04), $u! \doteq 0.927\,488$.

(e) Factorial of $N > 4$ can be approximated by Stirling's expansion

$$\boxed{N! = e^{\lambda}\left(\frac{N}{e}\right)^{N}\sqrt{2N\pi} + \epsilon \qquad |\epsilon| < 5 \times 10^{-10}}$$

where π = Archimedes constant (2.17), e = Euler's constant (2.18), and in terms of Bernoulli numbers $\bar{B}_2, \bar{B}_4, \bar{B}_6, \bar{B}_8$ (2.22),

$$\lambda = \sum_{r=1}^{4}\frac{\bar{B}_{2r}}{2r(2r - 1)N^{2r-1}} = \frac{\bar{B}_2 N^{-1}}{1 \cdot 2} + \frac{\bar{B}_4 N^{-3}}{3 \cdot 4} + \frac{\bar{B}_6 N^{-5}}{5 \cdot 6} + \frac{\bar{B}_8 N^{-7}}{7 \cdot 8}$$

$$= \frac{N^{-1}}{12} - \frac{N^{-3}}{360} + \frac{N^{-5}}{1260} - \frac{N^{-7}}{1680}$$

In nested form, $\quad \lambda = \dfrac{1}{12N}\left(1 - \dfrac{S}{30}\left(1 - \dfrac{2S}{7}\left(1 - \dfrac{3S}{4}\right)\right)\right) \qquad$ with $S = N^{-2}$

(f) Example. For $N = 7.3$, with $S = (7.3)^{-2}$,

$$\lambda = \frac{1}{12(7.3)}\left(1 - \frac{S}{30}\left(1 - \frac{2S}{7}\left(1 - \frac{3S}{4}\right)\right)\right) = 0.011\,408$$

By (e), $\quad N! = e^{0.011\,408}\left(\dfrac{7.3}{e}\right)^{7.3}\sqrt{14.6\pi} = 9\,281.39$

By (c), $\quad N! = (7.3)(6.3)(5.3)(4.3)(3.3)(2.3)(1.3)(0.3)! = 9\,281.39$

where from (A.04), $(0.3)! \doteq 0.897\,471$.

(g) Factorial of $N > 100$ can be approximated by Stirling's formula

$$N! = \left(\frac{N}{e}\right)^{N}\sqrt{2N\pi} + \epsilon \qquad |\epsilon| < 5 \times 10^{-10}$$

This formula applies particularly well in computing ratios of two factorials.

(a) Notation. $x = $ signed number $k, m, n = $ positive integers

(b) Binomial coefficient in x and k is by definition

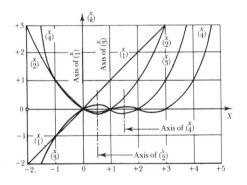

$$\binom{x}{k} = \begin{cases} \dfrac{x(x-1)(x-2)\cdots(x-k+1)}{k!} & (k > 1) \\[2mm] 1 & (k = 0) \\[2mm] x & (k = 1) \end{cases}$$

and for $k > 0$,

$$\binom{x+1}{k} = \binom{x}{k} + \binom{x}{k-1} \qquad \binom{x+1}{k+1} = \binom{x}{k} + \binom{x}{k+1}$$

$$\frac{\binom{x}{k+1}}{\binom{x}{k}} = \frac{n-k}{k+1} \qquad\qquad \frac{\binom{x}{k-1}}{\binom{x}{k}} = \frac{k}{x-k+1}$$

(c) Example

$$\binom{6.4}{3} = \frac{(6.4)(5.4)(4.4)}{(3)(2)(1)} = 25.344 \qquad \binom{0.5}{4} = \frac{(0.5)(-1.5)(-2.5)(-3.5)}{(4)(3)(2)(1)} = -0.273\,44$$

$$\binom{-6.4}{3} = \frac{(-6.4)(-7.4)(-8.4)}{(3)(2)(1)} = -66.304$$

(d) Binomial coefficient in n and k is by definition

$$\binom{n}{k} = \begin{cases} \dfrac{n!}{(n-k)!\,k!} & (n > k) \\[3mm] \dfrac{n(n-1)(n-2)\cdots(n-k+1)}{k!} & (|n| > k) \\[3mm] n & (k = 1 \text{ or } k = n-1) \\[1mm] 1 & (k = n \text{ or } k = 0) \\[1mm] 0 & (k < 0) \end{cases}$$

and for $0 < k \le n$,

$$\binom{n}{k} = \binom{n}{n-k} = (-1)^k\binom{k-n-1}{k} \qquad \binom{n}{k+1} = \frac{n-k}{k+1}\binom{n}{k} = (-1)^{k+1}\binom{k-n}{k+1}$$

$$\binom{2n}{n} = \frac{(2n)!}{(n!)^2} = (-1)^n\binom{-n-1}{n} \qquad \binom{2n-1}{n} = \frac{(2n-1)!}{n(n!)^2} = (-1)^n\binom{-n}{n}$$

Numerical values of $\binom{n}{k}$ for $n = 0, 1, 2, \ldots, 25$ and $k = 0, 1, 2, \ldots, 25$ are shown in (A.06).

(e) Examples

$$\binom{8}{3} = \frac{(8)(7)(6)}{(3)(2)(1)} = 56 \qquad \binom{8}{4} = \frac{(8)(7)(6)(5)}{(4)(3)(2)(1)} = \frac{8!}{(4!)^2} = 70$$

but $\displaystyle \binom{3}{8} = \frac{(3)(2)(1)(0)(-1)(-2)(-3)(-4)}{(8)(7)(6)(5)(4)(3)(2)(1)} = 0$

(a) Factorial polynomials $X_1^{(k)}$ in real variable x and $k = 1, 2, 3, \ldots, p$ are, by definition,

$$X_1^{(1)} = \binom{x}{1}1! = x = \mathscr{S}_1^{(1)}x$$

$$X_1^{(2)} = \binom{x}{2}2! = x(x - 1) = \mathscr{S}_1^{(2)}x + \mathscr{S}_2^{(2)}x^2$$

$$X_1^{(3)} = \binom{x}{3}3! = x(x - 1)(x - 2) = \mathscr{S}_1^{(3)}x + \mathscr{S}_2^{(3)}x^2 + \mathscr{S}_3^{(3)}x^3$$

. . . .

$$X_1^{(p)} = \binom{x}{p}p! = x(x - 1)(x - 2) \cdots [x - (p - 1)] = \mathscr{S}_1^{(p)}x + \mathscr{S}_2^{(p)}x^2 + \cdots + \mathscr{S}_p^{(p)}x^p$$

where $\mathscr{S}_k^{(p)}$ is called a *Stirling number of the first kind* (A.07).

(b) First recursion formula. If $\mathscr{S}_k^{(m)}$ and $\mathscr{S}_{k-1}^{(m)}$ are known, then for $m \geq k > 0$,

$$\mathscr{S}_k^{(m+1)} = \mathscr{S}_{k-1}^{(m)} - m\mathscr{S}_k^{(m)} \qquad (m = 0, 1, 2, \ldots)$$

where for $k = m$, $\mathscr{S}_m^{(m)} = 1$, and for $k \leq 0$ or for $k \geq m + 1$, $\mathscr{S}_k^{(m)} = 0$.

(c) Example. If $\mathscr{S}_1^{(2)} = -1$ and $\mathscr{S}_2^{(2)} = +1$, then by definition, $\mathscr{S}_0^{(2)} = 0$ and

$$\mathscr{S}_1^{(3)} = \mathscr{S}_0^{(2)} - 2\mathscr{S}_1^{(2)} = +2 \qquad\qquad \mathscr{S}_2^{(3)} = \mathscr{S}_1^{(2)} - 2\mathscr{S}_2^{(2)} = -3 \qquad\qquad \mathscr{S}_3^{(3)} = 1$$

(d) Powers x^k in terms of factorial polynomials $X^{(1)}, X^{(2)}, \ldots, X^{(p)}$ are

$$x = \bar{\mathscr{S}}_1^{(1)}X_1^{(1)}$$

$$x^2 = \bar{\mathscr{S}}_1^{(2)}X_1^{(1)} + \bar{\mathscr{S}}_2^{(2)}X_1^{(2)}$$

$$x^3 = \bar{\mathscr{S}}_1^{(3)}X_1^{(1)} + \bar{\mathscr{S}}_2^{(3)}X_1^{(2)} + \bar{\mathscr{S}}_3^{(3)}X_1^{(3)}$$

. . . .

$$x^p = \bar{\mathscr{S}}_1^{(p)}X_1^{(1)} + \bar{\mathscr{S}}_2^{(p)}X_1^{(2)} + \cdots + \bar{\mathscr{S}}_p^{(p)}X_1^{(p)}$$

where $\bar{\mathscr{S}}_k^{(p)}$ is called a *Stirling number of the second kind* (A.08).

(e) Second recursion formula. If $\bar{\mathscr{S}}_k^{(m)}$ and $\bar{\mathscr{S}}_{k-1}^{(m)}$ are known, then for $m \geq k > 0$,

$$\bar{\mathscr{S}}_k^{(m+1)} = \bar{\mathscr{S}}_{k-1}^{(m)} + k\bar{\mathscr{S}}_k^{(m)} \qquad (m = 0, 1, 2, \ldots)$$

where for $k = m$, $\bar{\mathscr{S}}_m^{(m)} = 1$, and for $k \leq 0$ or $k \geq m + 1$, $\bar{\mathscr{S}}_k^{(m)} = 0$.

(f) Example. If $\bar{\mathscr{S}}_1^{(2)} = +1$ and $\bar{\mathscr{S}}_2^{(2)} = +1$, then by definition, $\bar{\mathscr{S}}_0^{(2)} = 0$ and

$$\bar{\mathscr{S}}_1^{(3)} = \bar{\mathscr{S}}_0^{(2)} + \bar{\mathscr{S}}_1^{(2)} = +1 \qquad\qquad \bar{\mathscr{S}}_2^{(3)} = \bar{\mathscr{S}}_1^{(2)} + 2\bar{\mathscr{S}}_2^{(2)} = +3 \qquad\qquad \bar{\mathscr{S}}_3^{(3)} = +1$$

(a) Sum of binomial functions. If $\binom{x}{p}$ and $\binom{x}{q}$ are binomial functions in a real variable x and p, q are positive integers $(p > q)$, then

$$\binom{x}{p} + \binom{x}{q} = \frac{X_1^{(p)}}{p!} + \frac{X_1^{(q)}}{q!} = \sum_{k=1}^{p} \left[\frac{\mathscr{S}_k^{(p)}}{p!} + \frac{\mathscr{S}_k^{(q)}}{q!} \right] x^k$$

where $\mathscr{S}_k^{(q)}/q!$ is zero for $k > q$.

(b) Derivatives of $\binom{x}{p}$ are

$$\frac{d\left[\binom{x}{p} \right]}{dx} = \frac{1}{p!} \sum_{k=1}^{p} \binom{k}{1} \mathscr{S}_k^{(p)} x^{k-1} \qquad \frac{d^2 \left[\binom{x}{p} \right]}{dx^2} = \frac{2}{p!} \sum_{k=2}^{p} \binom{k}{2} \mathscr{S}_k^{(p)} x^{k-2}$$

and in general,

$$D_n \left[\binom{x}{p} \right] = \begin{cases} \dfrac{n!}{p!} \displaystyle\sum_{k=n}^{p} \binom{k}{n} \mathscr{S}_k^{(p)} x^{k-n} & (n < p) \\ 1 & (n = p) \\ 0 & (n > p) \end{cases}$$

where $D^n = d^n/dx^n$ and $n = 1, 2, 3, \ldots$.

(c) Indefinite integrals of $\binom{x}{p}$ are

$$\int \binom{x}{p} dx = \frac{1}{p!} \sum_{k=1}^{p} \frac{\mathscr{S}_k^{(p)} x^{k+1}}{k+1} + C_1$$

$$\int\int \binom{x}{p} dx\, dx = \frac{1}{p!} \sum_{k=1}^{p} \frac{\mathscr{S}_k^{(p)} x^{k+2}}{(k+1)(k+2)} + C_1 x + C_2$$

and in general,

$$I_n^n \left[\binom{x}{p} \right] = \frac{1}{p! \, n!} \sum_{k=1}^{p} \left[\mathscr{S}_k^{(p)} \frac{x^{k+n}}{\binom{k+n}{n}} \right] + C_1 x^{n-1} + C_2 x^{n-2} + \cdots + C_n \qquad \text{for } n \leqslant p$$

where $I^n[\ ] = \int\int\int \cdots (dx)^n$, $n = 1, 2, 3, \ldots$, and $C_1, C_2, \ldots, C_n$ are the constants of integration.

(d) Particular integrals involving $\binom{x}{p}$ are

$$\int \binom{x+a}{p} dx = \frac{1}{p!} \sum_{k=1}^{p} \frac{\mathscr{S}_k^{(p)}(x+a)^{k+1}}{k+1} + C \qquad \int \binom{ax+b}{p} dx = \frac{1}{p!} \sum_{k=1}^{p} \frac{\mathscr{S}_k^{(p)}(ax+b)^{k+1}}{a(k+1)} + C$$

$$\int x^n \binom{x}{p} dx = \frac{1}{n!} \sum_{k=1}^{p} \frac{\mathscr{S}_k^{(p)} x^{n+k+1}}{n+k+1} + C \qquad \int \frac{\binom{x}{p}}{x^n} dx = \frac{1}{p!} \left(\ln x - \frac{\mathscr{S}_{n+1}^{(p)}}{x} \right) + \frac{1}{p!} \sum_{k=1}^{p} \frac{\mathscr{S}_k^{(p)} x^{k+1-n}}{k+1-n} + C$$

where a, b are signed numbers and $n + 1 \leq p$.

(a) General case. The nth power of $a \pm x$ can be expanded by Newton's formula in a power series called a *binomial series*.

$$(a \pm x)^n = a^n(1 \pm u)^n = a^n\left[1 \pm \binom{n}{1}u + \binom{n}{2}u^2 \pm \cdots \alpha\binom{n}{n}u^n\right] = a^n \sum_{k=0}^{n} \beta\binom{n}{k}u^k$$

where n = signed number, $\binom{n}{k}$ = binomial coefficient, $\alpha = (\pm 1)^n$, $\beta = (\pm 1)^k$, and $u = x/a$.

(b) Classification. If in the expansion above

$n = 0, 1, 2, 3, \ldots, p$	the series is finite
$n \neq 0, 1, 2, 3, \ldots, p$ and $u^2 < 1$	the series is convergent
$n \neq 0, 1, 2, 3, \ldots, p$ and $u^2 > 1$	the series is divergent

(c) CA model. The most convenient model of this series is the nested series introduced in (1.07), which for $u = x/a$ is

$$(a \pm x)^n = a^n \bigwedge_{k=1}^{n} (1 \pm c_k u) = a^n\left(1 \pm nu\left(1 \pm \frac{n-1}{2}u\left(1 \pm \frac{n-2}{3}u(1 \pm \cdots)\right)\right)\right)$$

Five distinct forms of this model are given below in (d), and selected particular cases are tabulated in (2.14).

(d) Distinct CA models. If $u^2 < 1$ and $X = u/p$,

$n = -m, p = 1$	$\dfrac{1}{(1 \pm u)^m} = 1 \mp mu\left(1 \mp \dfrac{m+1}{2}u\left(1 \mp \dfrac{m+2}{3}u\left(1 \mp \dfrac{m+3}{4}u(\mp \cdots)\right)\right)\right)$
$n = \dfrac{1}{p}, m = 1$	$\sqrt[p]{1 \pm u} = 1 \pm X\left(1 \mp \dfrac{p-1}{2}X\left(1 \mp \dfrac{2p-1}{3}X\left(1 \mp \dfrac{3p-1}{4}X(1 \mp \cdots)\right)\right)\right)$
$n = -\dfrac{1}{p}, m = 1$	$\dfrac{1}{\sqrt[p]{1 \pm u}} = 1 \mp X\left(1 \mp \dfrac{p+1}{2}X\left(1 \mp \dfrac{2p+1}{3}X\left(1 \mp \dfrac{3p+1}{4}X(1 \mp \cdots)\right)\right)\right)$
$n = \dfrac{m}{p}$	$\sqrt[p]{(1 \pm u)^m} = 1 \pm mX\left(1 \mp \dfrac{p-m}{2}X\left(1 \mp \dfrac{2p-m}{3}X\left(1 \mp \dfrac{3p-m}{4}X(1 \mp \cdots)\right)\right)\right)$
$n = -\dfrac{m}{p}$	$\dfrac{1}{\sqrt[p]{(1 \pm u)^m}} = 1 \mp mX\left(1 \mp \dfrac{p+m}{2}X\left(1 \mp \dfrac{2p+m}{3}X\left(1 \mp \dfrac{3p+m}{4}X(1 \pm \cdots)\right)\right)\right)$

where m, p are positive integers or positive decimal numbers.

(e) Absolute truncation error $|\epsilon|$ in these series is less than the absolute value of the first deleted term.

In series (a), $\quad |\epsilon_{r+1}| \leq \left|\binom{n}{r+1}u^{r+1}\right| a^n$ $\qquad\qquad$ In series (c), $\quad |\epsilon_{r+1}| \leq |c_1 c_2 c_3 \cdots c_r c_{r+1} u^{r+1}| a^n$

(f) Example. By series (c).

$$(1 + 0.234)^{0.567} = 1 + 0.567(0.234)\left(1 - \frac{0.433}{2}(0.234)\left(1 - \frac{1.433}{3}(0.234)\left(1 - \frac{2.433}{4}(0.234)\right)\right)\right)$$

$$= 1.126\,601 \qquad \text{(correct value } 1.126\,615\text{)}$$

and $\quad |\epsilon_5| \leq (0.567)\left(\frac{0.433}{2}\right)\left(\frac{1.433}{3}\right)\left(\frac{2.433}{4}\right)\left(\frac{3.433}{5}\right)(0.234)^5 = 0.000\,017$

(a) Exponent $n = -m$ $(m = 1, 2, 3, 4; u = x/a)$

$$\frac{1}{a \pm x} = \frac{1}{a}(1 \mp u(1 \mp u(1 \mp u(1 \mp u(1 \mp u(1 \mp u(1 \mp \cdots)))))))$$

$$\frac{1}{(a \pm x)^2} = \frac{1}{a^2}\left(1 \mp \frac{2}{1}u\left(1 \mp \frac{3}{2}u\left(1 \mp \frac{4}{3}u\left(1 \mp \frac{5}{4}u\left(1 \mp \frac{6}{5}u\left(1 \mp \frac{7}{6}(1 \mp \cdots)\right)\right)\right)\right)\right)\right)$$

$$\frac{1}{(a \pm x)^3} = \frac{1}{a^3}\left(1 \mp \frac{3}{1}u\left(1 \mp \frac{4}{2}u\left(1 \mp \frac{5}{3}u\left(1 \mp \frac{6}{4}u\left(1 + \frac{7}{5}u\left(1 \mp \frac{8}{6}u(1 \mp \cdots)\right)\right)\right)\right)\right)\right)$$

$$\frac{1}{(a \pm x)^4} = \frac{1}{a^4}\left(1 \mp \frac{4}{1}u\left(1 \mp \frac{5}{2}u\left(1 \mp \frac{6}{3}u\left(1 \mp \frac{7}{4}u\left(1 \mp \frac{8}{5}u\left(1 \mp \frac{9}{6}u(1 \mp \cdots)\right)\right)\right)\right)\right)\right)$$

(b) Exponent $n = 1/m$ $(m = 2, 3, 4; u = x/am)$

$$\sqrt{a \pm x} = \sqrt{a}\left(1 \pm u\left(1 \mp \frac{1}{2}u\left(1 \mp \frac{3}{3}u\left(1 \mp \frac{5}{4}u\left(1 \mp \frac{7}{5}u\left(1 \mp \frac{9}{6}u(1 \mp \cdots)\right)\right)\right)\right)\right)\right)$$

$$\sqrt[3]{a \pm x} = \sqrt[3]{a}\left(1 \pm u\left(1 \mp \frac{2}{2}u\left(1 \mp \frac{5}{3}u\left(1 \mp \frac{8}{4}u\left(1 \mp \frac{11}{5}u\left(1 \mp \frac{14}{6}u(1 \mp \cdots)\right)\right)\right)\right)\right)\right)$$

$$\sqrt[4]{a \pm x} = \sqrt[4]{a}\left(1 \pm u\left(1 \mp \frac{3}{2}u\left(1 \mp \frac{7}{3}u\left(1 \mp \frac{11}{4}u\left(1 \mp \frac{15}{5}u\left(1 \mp \frac{19}{6}u(1 \mp \cdots)\right)\right)\right)\right)\right)\right)$$

(c) Exponent $n = -1/m$ $(m = 2, 3, 4; u = x/am)$

$$\frac{1}{\sqrt{a \pm x}} = \frac{1}{\sqrt{a}}\left(1 + u\left(1 \mp \frac{3}{2}u\left(1 \mp \frac{5}{3}u\left(u \mp \frac{7}{4}u\left(1 \mp \frac{9}{5}u\left(1 \mp \frac{11}{6}u(1 \mp \cdots)\right)\right)\right)\right)\right)\right)$$

$$\frac{1}{\sqrt[3]{a \pm x}} = \frac{1}{\sqrt[3]{a}}\left(1 \mp u\left(1 \mp \frac{4}{2}u\left(1 \mp \frac{7}{3}u\left(1 \mp \frac{10}{4}u\left(1 \mp \frac{13}{5}u\left(1 \mp \frac{16}{6}u(1 \mp \cdots)\right)\right)\right)\right)\right)\right)$$

$$\frac{1}{\sqrt[4]{a \pm x}} - \frac{1}{\sqrt[4]{a}}\left(1 \mp u\left(1 \mp \frac{5}{2}u\left(1 \mp \frac{9}{3}u\left(1 \mp \frac{13}{4}u\left(1 \mp \frac{17}{5}u\left(1 \mp \frac{21}{6}u(1 \mp \cdots)\right)\right)\right)\right)\right)\right)$$

(d) Exponent $n = p/q$ $(p = 2, 3, 4; q = 3, 4, 5; u = x/aq)$

$$\sqrt[3]{(a \pm x)^2} = a^2\left(1 \pm 2u\left(1 \mp \frac{1}{2}u\left(1 \mp \frac{4}{3}u\left(1 \mp \frac{7}{4}u\left(1 \mp \frac{10}{5}u\left(1 \mp \frac{13}{6}u(1 \mp \cdots)\right)\right)\right)\right)\right)\right)$$

$$\sqrt[4]{(a \pm x)^3} = a^3\left(1 \pm 3u\left(1 \mp \frac{1}{2}u\left(1 \mp \frac{5}{3}u\left(1 \mp \frac{9}{4}u\left(1 \mp \frac{13}{5}u\left(1 \mp \frac{17}{6}u(1 \mp \cdots)\right)\right)\right)\right)\right)\right)$$

$$\sqrt[5]{(a \pm x)^4} = a^4\left(1 \pm 4u\left(1 \mp \frac{1}{2}u\left(1 \mp \frac{6}{3}u\left(1 \mp \frac{11}{4}u\left(1 \mp \frac{16}{5}u\left(1 \mp \frac{21}{6}u(1 \mp \cdots)\right)\right)\right)\right)\right)\right)$$

(e) Exponent $n = -p/q$ $(p = 2, 3, 4; q = 3, 4, 5; u = x/aq)$

$$\frac{1}{\sqrt[3]{(a \pm x)^2}} = \frac{1}{\sqrt[3]{a^2}}\left(1 \mp 2u\left(1 \mp \frac{5}{2}u\left(1 \mp \frac{8}{3}u\left(1 \mp \frac{11}{4}u\left(1 \mp \frac{14}{5}u\left(1 \mp \frac{17}{6}u(1 \mp \cdots)\right)\right)\right)\right)\right)\right)$$

$$\frac{1}{\sqrt[4]{(a \pm x)^3}} = \frac{1}{\sqrt[4]{a^3}}\left(1 \mp 3u\left(1 \mp \frac{7}{2}u\left(1 \mp \frac{11}{3}u\left(1 \mp \frac{15}{4}u\left(1 \mp \frac{19}{5}u\left(1 \mp \frac{23}{6}u(1 \mp \cdots)\right)\right)\right)\right)\right)\right)$$

$$\frac{1}{\sqrt[5]{(a \pm x)^4}} = \frac{1}{\sqrt[5]{a^4}}\left(1 \mp 4u\left(1 \mp \frac{9}{2}u\left(1 \mp \frac{14}{3}u\left(1 \mp \frac{19}{4}u\left(1 \mp \frac{24}{5}u\left(1 \mp \frac{29}{6}u(1 \mp \cdots)\right)\right)\right)\right)\right)\right)$$

2.15 GENERAL SERIES AND PRODUCTS OF BINOMIAL COEFFICIENTS

(a) General sums of complete, monotonic sequences $(n, r, N = 1, 2, 3, \ldots)$

$$\sum_{k=0}^{n} \binom{n}{k} = \binom{n}{0} + \binom{n}{1} + \binom{n}{2} + \cdots + \binom{n}{n} = 2^n$$

$$\sum_{k=0}^{n} 2^k \binom{n}{k} = \binom{n}{0} + 2\binom{n}{1} + 2^2\binom{n}{2} + \cdots + 2^n\binom{n}{n} = 3^n$$

$$\sum_{k=0}^{n} 3^k \binom{n}{k} = \binom{n}{0} + 3\binom{n}{1} + 3^2\binom{n}{2} + \cdots + 3^n\binom{n}{n} = 4^n$$

$$\sum_{k=0}^{n} r^k \binom{n}{k} = \binom{n}{0} + r\binom{n}{1} + r^2\binom{n}{2} + \cdots + r^n\binom{n}{n} = (1 + r)^n$$

$$\sum_{k=0}^{n/2} \binom{n}{2k} = \binom{n}{0} + \binom{n}{2} + \binom{n}{4} + \cdots + \binom{n}{n} = 2^{n-1} \qquad (n \neq 1, 3, 5, \ldots)$$

$$\sum_{k=0}^{(n-1)/2} \binom{n}{2k + 1} = \binom{n}{1} + \binom{n}{3} + \binom{n}{5} + \cdots + \binom{n}{n} = 2^{n-1} \qquad (n \neq 2, 4, 6, \ldots)$$

$$\sum_{k=n}^{N} \binom{k}{n} = \binom{n}{n} + \binom{n + 1}{n} + \binom{n + 2}{n} + \cdots + \binom{N}{n} = \binom{N + 1}{n + 1} \qquad (N > n)$$

(b) General sums of complete, alternating sequences $(n, r, N = 1, 2, 3, \ldots)$

$$\sum_{k=0}^{n} (-1)^k \binom{n}{k} = \binom{n}{0} - \binom{n}{1} + \binom{n}{2} - \cdots (-1)^n \binom{n}{n} = 0$$

$$\sum_{k=0}^{n} (-2)^k \binom{n}{k} = \binom{n}{0} - 2\binom{n}{1} + 2^2\binom{n}{2} - \cdots (-2)^n \binom{n}{n} = (-1)^n$$

$$\sum_{k=0}^{n} (-3)^k \binom{n}{k} = \binom{n}{0} - 3\binom{n}{1} + 3^2\binom{n}{2} - \cdots (-3)^n \binom{n}{n} = (-2)^n$$

$$\sum_{k=0}^{n} (-r)^k \binom{n}{k} = \binom{n}{0} - r\binom{n}{1} + r^2\binom{n}{2} - \cdots (-r)^n \binom{n}{n} = (1 - r)^n$$

$$\sum_{k=0}^{n/2} (-1)^k \binom{n}{2k} = \binom{n}{0} - \binom{n}{2} + \binom{n}{4} - \cdots (-1)^n \binom{n}{n} = 2^{n/2} \cos \frac{n\pi}{4} \qquad (n \neq 1, 3, 5, \ldots)$$

$$\sum_{k=0}^{(n-1)/2} (-1)^k \binom{n}{2k + 1} = \binom{n}{1} - \binom{n}{3} + \binom{n}{5} - \cdots (-1)^n \binom{n}{n} = 2^{n/2} \sin \frac{n\pi}{4} \qquad (n \neq 2, 4, 6, \ldots)$$

$$\sum_{k=n}^{N} (-1)^{k-n} \binom{N}{k} = \binom{N}{n} - \binom{N}{n + 1} + \binom{N}{n + 2} - \cdots (-1)^{N-n} \binom{N}{N} = \binom{N - 1}{n - 1} \quad (N > n)$$

(c) General sums of complete products $(n, r, s = 1, 2, 3, \ldots; r + s \geq n; s - n \geq 0)$

$$\sum_{k=0}^{n} \binom{r}{k}\binom{n}{k} = \binom{r}{0}\binom{n}{0} + \binom{r}{1}\binom{n}{1} + \binom{r}{2}\binom{n}{2} + \cdots + \binom{r}{n}\binom{n}{n} = \binom{r + n}{n}$$

$$\sum_{k=0}^{n} \binom{r}{k}\binom{s}{n + k} = \binom{r}{0}\binom{s}{n} + \binom{r}{1}\binom{s}{n + 1} + \binom{r}{2}\binom{s}{n + 2} + \cdots + \binom{r}{n}\binom{s}{2n} = \binom{r + s}{s - n}$$

$$\sum_{k=0}^{n} \binom{r}{k}\binom{s}{n - k} = \binom{r}{0}\binom{s}{n} + \binom{r}{1}\binom{s}{n - 1} + \binom{r}{2}\binom{s}{n - 2} + \cdots + \binom{r}{n}\binom{s}{0} = \binom{r + s}{n}$$

(a) Particular sums of complete, monotonic sequences $(n = 1, 2, 3, \ldots)$

$$\sum_{k=1}^{n} \binom{2n}{2k} = \binom{2n}{2} + \binom{2n}{4} + \binom{2n}{6} + \cdots + \binom{2n}{2n} = 2^{2n-1} - 1$$

$$\sum_{k=1}^{n} \binom{2n-1}{2k-1} = \binom{2n-1}{1} + \binom{2n-1}{3} + \binom{2n-1}{5} + \cdots + \binom{2n-1}{2n-1} = 2^{2n-2}$$

$$\sum_{k=0}^{n} (k+1)\binom{n}{k} = \binom{n}{0} + 2\binom{n}{1} + 3\binom{n}{2} + \cdots + (n+1)\binom{n}{n} = (1 + \tfrac{1}{2}n)2^n$$

$$\sum_{k=1}^{n} k\binom{n}{k} = \binom{n}{1} + 2\binom{n}{2} + 3\binom{n}{3} + \cdots + n\binom{n}{n} = n2^{n-1}$$

$$\sum_{k=0}^{n} (2k+1)\binom{n}{k} = \binom{n}{0} + 3\binom{n}{1} + 5\binom{n}{2} + \cdots + (2n+1)\binom{n}{n} = (1 + n)2^n$$

$$\sum_{k=1}^{n} (2k-1)\binom{n}{k} = \binom{n}{1} + 3\binom{n}{3} + 5\binom{n}{5} + \cdots + (2n-1)\binom{n}{n} = 1 + (n-1)2^n$$

$$\sum_{k=1}^{n} k^2\binom{n}{k} = \binom{n}{1} + 2^2\binom{n}{2} + 3^2\binom{n}{3} + \cdots + n^2\binom{n}{n} = n(1 + n)2^{n+1}$$

$$\sum_{k=0}^{n} \binom{n}{k} = \binom{n}{0}^2 + \binom{n}{1}^2 + \binom{n}{2}^2 + \cdots + \binom{n}{n}^2 = \binom{2n}{n}$$

$$\sum_{k=1}^{n} k\binom{n}{k}^2 = \binom{n}{1}^2 + 2\binom{n}{2}^2 + 3\binom{n}{3}^2 + \cdots + n\binom{n}{n}^2 = \tfrac{1}{2}n\binom{2n}{n}$$

$$\sum_{k=1}^{n} k^2\binom{n}{k}^2 = \binom{n}{1}^2 + 2^2\binom{n}{2}^2 + 3^2\binom{n}{3}^2 + \cdots + n^2\binom{n}{n}^2 = \tfrac{1}{2}n(n+1)\binom{2n}{n}$$

(b) Particular sums of complete, alternating sequences $(n = 1, 2, 3, \ldots)$

$$\sum_{k=1}^{n} (-1)^{k+1}\binom{n}{k} = \binom{n}{1} - \binom{n}{2} + \binom{n}{3} - \cdots (-1)^{n+1}\binom{n}{n} = 1$$

$$\sum_{k=1}^{n} (-1)^{k+1}k\binom{n}{k} = \binom{n}{1} - 2\binom{n}{2} + 3\binom{n}{3} - \cdots (-1)^{n+1}n\binom{n}{n} = 0$$

$$\sum_{k=1}^{n} (-1)^{k+1}k^n\binom{n}{k} = \binom{n}{1} - 2^n\binom{n}{2} + 3^n\binom{n}{3} - \cdots (-1)^{n+1}n^n\binom{n}{n} = (-1)^{n+1}n!$$

$$\sum_{k=0}^{2n} (-1)^k\binom{2n}{k}^2 = \binom{2n}{0}^2 - \binom{2n}{1}^2 + \binom{2n}{2}^2 - \cdots (-1)^n\binom{2n}{2n}^2 = (-1)^n\binom{2n}{n}$$

$$\sum_{k=0}^{2n+1} (-1)^k\binom{2n+1}{k}^2 = \binom{2n+1}{0}^2 - \binom{2n+1}{1}^2 + \binom{2n+1}{2}^2 - \cdots (-1)^n\binom{2n+1}{2n+1}^2 = 0$$

(c) Particular sums of incomplete sequences $(r \leq n = 1, 2, 3, \ldots)$

$$\sum_{k=0}^{r} \binom{n}{k} = 1 + n\left(1 + \frac{n-1}{2}\left(1 + \frac{n-2}{3}\left(1 + \frac{n-3}{4}\left(1 + \cdots + \frac{n-r}{1+r}\right)\right)\right)\right)$$

$$\sum_{k=0}^{r} (-1)^k\binom{n}{k} = 1 - n\left(1 - \frac{n-1}{2}\left(1 - \frac{n-2}{3}\left(1 - \frac{n-3}{4}\left(1 - \cdots - \frac{n-r}{1+r}\right)\right)\right)\right)$$

(a) Definition. The symbol π denotes the ratio of the circumference C of a circle to its diameter D.

$$\pi = \frac{C}{D} = 3.141\ 592\ 653\ 589\ 793\ 238\ 462\ 643 \cdots$$

In 1882 C. L. P. Lindermann proved that π is both an irrational and a transcendental number and thus showed that the problem of rectification and squaring a circle with ruler and compass alone is a mathematical impossibility.

(b) Calculation of π to a desired degree of accuracy can be accomplished by means of infinite products or infinite series as shown below. For simple and rapid approximation (2.08g),

$$\pi = \begin{cases} \dfrac{22}{7} - \epsilon = 3.142\ 857\ 143 - \epsilon & (\epsilon < 1.3 \times 10^{-3}) \\[3mm] \dfrac{355}{113} - \epsilon = 3.141\ 592\ 920 - \epsilon & (\epsilon < 2.7 \times 10^{-7}) \end{cases}$$

(c) Vieta's product given in explicit form as

$$\pi = \frac{2}{\sqrt{\frac{1}{2}}\ \sqrt{\frac{1}{2} + \frac{1}{2}\sqrt{\frac{1}{2}}}\ \sqrt{\frac{1}{2} + \frac{1}{2}\sqrt{\frac{1}{2} + \frac{1}{2}\sqrt{\frac{1}{2}}}}\cdots}$$

can be expressed as

$$\pi = \frac{4R_0}{R_1 \cdot R_2 \cdot R_3 \cdots} = 4R_0 \prod_{k=1}^{\infty} \frac{1}{R_k}$$

where $\quad R_0 = \sqrt{\dfrac{1}{2}},\ R_1 = \sqrt{\dfrac{1 + R_0}{2}},\ R_2 = \sqrt{\dfrac{1 + R_1}{2}},\ \ldots,\ R_n = \sqrt{\dfrac{1 + R_{n-1}}{2}}$

The convergence of this product is shown in (h) (for $n = 15$, $\epsilon < 5 \times 10^{-10}$).

(d) Wallis' product given in standard form as

$$\pi = 2\left(\frac{2 \cdot 2}{1 \cdot 3}\right)\left(\frac{4 \cdot 4}{3 \cdot 5}\right) \cdots = 2 \prod_{k=1}^{\infty} \frac{4k^2}{4k^2 - 1}$$

can be expressed in a limit form as

$$\pi = \lim_{n \to \infty} \left\{ \left[2\left(\frac{2 \cdot 4 \cdot 6 \cdots 2n}{1 \cdot 3 \cdot 5 \cdots 2n - 1} \right) \right]^2 \frac{1}{2n} \right\}$$

The very slow convergence of this product is shown in (h) (for $n = 40$, $\epsilon < 5 \times 10^{-4}$, for $n = 1253$, $\epsilon < 5 \times 10^{-7}$, and $n = 35\ 633$, $\epsilon < 5 \times 10^{-10}$).

(e) Standard Gregory–Leibnitz' series based on

$$\tan^{-1} x = x - \frac{x^3}{3} + \frac{x^5}{5} - \frac{x^7}{7} + \cdots$$

for $x = 1$ gives

$$\pi = 4\left(1 - \frac{1}{3} + \frac{1}{5} - \frac{1}{7} + \cdots\right) = 4 \sum_{k=1}^{\infty} \frac{(-1)^{k-1}}{2k - 1}$$

(f) Telescopic Gregory–Leibnitz' series is

$$\pi = 8\left(\frac{1}{1 \cdot 3} + \frac{1}{5 \cdot 7} + \frac{1}{9 \cdot 11} + \cdots\right) = 8 \sum_{k=1}^{\infty} \frac{1}{[2(2k-1)]^2 - 1}$$

The very slow convergence of this series is shown in (*h*) (for $n = 32$, $\epsilon < 5 \times 10^{-4}$, for $n = 1001$, $\epsilon < 5 \times 10^{-7}$, and for $n = 31\,623$, $\epsilon < 5 \times 10^{-10}$).

(g) Machin's series and Gauss' series are the most useful models, given, respectively, as

$$\pi = 4 \sum_{k=1}^{\infty} \left[\frac{(-1)^{k-1}}{2k-1} \left(\frac{4}{5^{2k-1}} - \frac{1}{239^{2k-1}} \right) \right]$$

$$\pi = 4 \sum_{k=1}^{\infty} \left[\frac{(-1)^{k-1}}{2k-1} \left(\frac{12}{18^{2k-1}} + \frac{8}{57^{2k-1}} - \frac{5}{239^{2k-1}} \right) \right]$$

The rapid convergence of these series is shown in (*h*) (in Machin's series of $n = 7$, $\epsilon < 5 \times 10^{-10}$; in Gauss' series of $n = 4$, $\epsilon < 5 \times 10^{-10}$). Considering that each term of Machin's series involves only two fractions, this series is the more convenient of the two models given above.

(h) Values of π based on n-term approximations

n	Vieta	Wallis	Leibnitz	Machin	Gauss
1	3.061 467 125	2.666 666 667	2.666 666 667	3.183 263 598	3.144 388 167
2	3.121 445 152	2.844 444 444	2.895 238 095	3.140 597 029	3.141 587 574
3	3.136 548 491	2.925 714 284	2.976 046 176	3.141 621 023	3.141 592 665
4	3.140 331 153	2.972 154 194	3.017 071 817	3.141 591 772	3.141 592 654
5	3.141 277 251	3.002 175 955	3.041 839 619	3.141 592 682	
6	3.141 513 801	3.023 170 192	3.058 102 766	3.141 592 652	
7	3.141 572 940	3.038 673 629	3.070 254 618	3.141 592 654	
8	3.141 587 725	3.050 588 266	3.079 153 394		
9	3.141 591 421	3.059 063 857	3.086 079 801		
10	3.141 592 345	3.066 730 687	3.091 623 807		
11	3.141 592 576	3.073 080 023	3.096 161 529		
12	3.141 592 634	3.078 424 510	3.099 944 033		
13	3.141 592 649	3.082 985 139	3.103 145 314		
14	3.141 592 652	3.086 922 540	3.105 889 726		
15	3.141 592 654	3.090 356 269	3.108 268 557		
16		3.093 377 145	3.110 360 274		
17		3.096 055 394	3.112 187 243		
18		3.098 446 170	3.113 820 229		
19		3.100 593 395	3.114 002 877		
20		3.102 532 478	3.114 167 270		

(a) Definition. The symbol e denotes the limit,

$$e = \lim_{m \to \infty} \left(1 + \frac{1}{m}\right)^m = \lim_{n \to 0} (1 + n)^{1/n} = 2.718\,281\,828\,459\,045\,235\,360\,287 \cdots$$

and is the base of the natural system of logarithms. In 1873, C. Hermite proved that e is both an irrational and a transcendental number.

(b) Calculation of e to a desired degree of accuracy can be accomplished by means of infinite series and continued fractions as shown below. For simple and rapid approximation,

$$e = \begin{cases} \dfrac{19}{7} + \epsilon = 2.714\,285\,714 + \epsilon & (\epsilon < 4 \times 10^{-3} \\[3mm] \dfrac{1264}{465} + \epsilon = 2.718\,279\,570 + \epsilon & (\epsilon < 2.3 \times 10^{-6}) \end{cases}$$

or directly by CA as

$$e = \begin{cases} \left(1 + \dfrac{1}{10^3}\right)^{10^3} + \epsilon = 2.716\,923\,932 + \epsilon & (\epsilon < 1.4 \times 10^{-3}) \\[3mm] \left(1 + \dfrac{1}{10^6}\right)^{10^6} + \epsilon = 2.718\,280\,469 + \epsilon & (\epsilon < 1.4 \times 10^{-6}) \\[3mm] \left(1 + \dfrac{1}{10^9}\right)^{10^9} + \epsilon = 2.718\,281\,828 + \epsilon & (\epsilon < 4.6 \times 10^{-10}) \end{cases}$$

(c) Euler's series based on

$$e = 1 + \frac{1}{1!} + \frac{1}{2!} + \frac{1}{3!} + \cdots = 1 + \sum_{k=1}^{\infty} \frac{1}{k!}$$

can be expressed in TF as

$$e = 2\left(\frac{1}{1!} + \frac{2}{3!} + \frac{3}{5!} + \cdots\right) = 2 \sum_{k=1}^{\infty} \frac{k}{(2k-1)!}$$

The rapid convergence of this series is shown in (e) (for $n = 7$, $\epsilon < 4.6 \times 10^{-10}$).

(d) Euler's continued fraction

$$e = 2 + \cfrac{1}{1 + \cfrac{1}{2 + \cfrac{2}{3 + \cfrac{3}{4 + \cfrac{4}{5 + \cfrac{5}{6 + \cdots}}}}}}$$

converges rapidly as shown in (e) (for $n = 11$, $\epsilon < 4.6 \times 10^{-10}$) and is also a good CA model.

(e) Values of e **based on** n**-term approximation**

m	Series	Fraction
1	2.000 000 000	2.000 000 000
2	2.666 666 667	2.666 666 667
3	2.716 666 667	2.727 272 727
4	2.718 253 968	2.718 981 132
5	2.718 281 526	2.718 263 332
6	2.718 281 826	2.718 283 694
7	2.718 281 828	2.718 281 658
8		2.718 281 843
9		2.718 281 827
10		2.718 281 829
11		2.718 281 828

(a) Definition. The symbol γ denotes the limit,

$$\gamma = \lim_{n \to \infty} \left(1 + \frac{1}{2} + \frac{1}{3} + \cdots + \frac{1}{n} - \ln n \right) = 0.577\ 215\ 664\ 901\ 532\ 860\ 606\ 512 \cdots$$

No proof is yet available whether γ is an irrational number.

(b) Calculation of γ to a desired degree of accuracy can be accomplished by the evaluation of the given limit for large n as shown below or by the continued fraction listed in (d).

(c) Example.

$$\gamma = \begin{cases} 1 + \dfrac{1}{2} + \dfrac{1}{3} + \cdots + \dfrac{1}{10^3} - \ln(10^3) - \epsilon = 0.577\ 715\ 664 - \epsilon & (\epsilon < 5 \times 10^{-4}) \\[2ex] 1 + \dfrac{1}{2} + \dfrac{1}{3} + \cdots + \dfrac{1}{10^6} - \ln(10^6) - \epsilon = 0.577\ 216\ 165 - \epsilon & (\epsilon < 5 \times 10^{-7}) \\[2ex] 1 + \dfrac{1}{2} + \dfrac{1}{3} + \cdots + \dfrac{1}{10^9} - \ln(10^9) - \epsilon = 0.577\ 215\ 665 - \epsilon & (\epsilon < 5 \times 10^{-10}) \end{cases}$$

(d) Euler's continued fraction

$$\gamma = 0 + \cfrac{1}{1 + \cfrac{1}{1 + \cfrac{1}{2 + \cfrac{1}{1 + \cfrac{1}{2 + \cfrac{1}{1 + \cfrac{1}{4 + \cfrac{1}{3 + \cfrac{1}{13 + \cfrac{1}{5 + \cdots}}}}}}}}}$$

converges very rapidly as shown in (e) (for $n = 11$, $|\epsilon| <$ 5×10^{-10}) and is also a good CA model.

(e) Values of γ **based on** n**-term approximation**

n	Fraction
1	1.000 000 000
2	0.500 000 000
3	0.600 000 000
4	0.571 428 571
5	0.578 947 368
6	0.576 923 077
7	0.577 235 772
8	0.577 215 190
9	0.577 215 671
10	0.577 215 664
11	0.577 215 665

(f) Simple and rapid approximations by simple fractions are the convergents of (d).

$$\gamma = \begin{cases} \dfrac{11}{19} - \epsilon = 0.578\ 947\ 368 - \epsilon & (\epsilon < 1.8 \times 10^{-3}) \\[2ex] \dfrac{71}{123} - \epsilon = 0.577\ 235\ 772 - \epsilon & (\epsilon < 2.1 \times 10^{-5}) \\[2ex] \dfrac{228}{395} + \epsilon = 0.577\ 215\ 190 + \epsilon & (\epsilon < 4.8 \times 10^{-7}) \end{cases}$$

and for routine calculations,

$$\gamma = \sqrt{\tfrac{1}{3}} - \epsilon = 0.577\ 350\ 269 - \epsilon \qquad (\varepsilon < 1.3 \times 10^{-4})$$

(g) Fibonacci's formula in terms of $R = \frac{1}{2}(1 + \sqrt{5})$ and $e = 2.718\ 281\ 828 \cdots$ $(2.18a)$,

$$\gamma \cong \tfrac{1}{2} \ln \left[\pi + \frac{1}{Re^3 + (\frac{5}{8})^4} \right] = 0.577\ 215\ 665$$

is the best approximation of all.

(a) Definition. The symbol ln 2 denotes the natural logarithm of 2, defined by the relationship

$$e^{\ln 2} = \left[\lim_{m \to \infty} \left(1 + \frac{1}{m}\right)^m\right]^{\ln 2} \qquad \text{where } \ln 2 = 0.693\ 147\ 180\ 559\ 945\ 309\ 417\ 232 \cdots$$

This constant is an important component in the construction of CA models of natural logarithms of large numbers (9.06).

(b) Calculation of ln 2 to a desired degree of accuracy can be accomplished by means of continued fractions and infinite series as shown below. For simple and rapid calculations,

$$\ln 2 = \sqrt{\tfrac{25}{52}} - \epsilon = 0.693\ 375\ 245 - \epsilon \qquad (\epsilon < 2.3 \times 10^{-4})$$

(c) Napier's continued fraction,

$$\ln 2 = 0 + \cfrac{1}{1 + \cfrac{1}{2 + \cfrac{1}{3 + \cfrac{1}{1 + \cfrac{1}{6 + \cfrac{1}{3 + \cfrac{1}{1 + \cfrac{1}{1 + \cfrac{1}{2 + \cdots}}}}}}}}}$$

converges as shown in (f) (for $n = 11$, $|\epsilon| < 2.8 \times 10^{-8}$). Simple and rapid approximations by simple fractions are the convergents of this fraction constructed by $(2.08f)$.

(d) Gauss' series based on

$$\ln \sqrt{\frac{x+1}{x-1}} = \frac{1}{x} + \frac{1}{3x^3} + \frac{1}{5x^5} + \cdots$$

for $x = \frac{5}{8}$ yields

$$\ln 2 = \frac{a}{1} + \frac{a^3}{3} + \frac{a^5}{5} + \cdots$$

where $a = \frac{3}{5}$. The relatively slow convergence of this series is shown in (f) (for $n = 11$, $\epsilon < 4.1 \times 10^{-7}$).

(e) Goursat's series based on

$$\ln \frac{N+1}{N} = 2\left[\frac{1}{2N+1} + \frac{1}{3(2N+1)^3}\right.$$

$$\left. + \frac{1}{5(2N+1)^5} + \cdots\right]$$

for $N = 1$ yields

$$\ln 2 = 2\left[\frac{b}{1} + \frac{b^3}{3} + \frac{b^5}{5} + \cdots\right]$$

where $b = \frac{1}{3}$. The better convergence of this series is shown in (f) (for $n = 8$, $\epsilon < 5.1 \times 10^{-10}$).

(f) Values of ln 2 based on *n*-term approximation

k	Fraction	Gauss	Goursat
1	1.000 000 000	0.600 000 000	0.666 666 667
2	0.666 666 667	0.672 000 000	0.691 358 025
3	0.700 000 000	0.687 552 000	0.693 004 115
4	0.692 307 692	0.691 551 016	0.693 134 757
5	0.693 181 818	0.692 670 830	0.693 146 047
6	0.693 140 794	0.693 000 646	0.693 147 074
7	0.693 150 685	0.693 011 129	0.693 147 170
8	0.693 146 417	0.693 132 459	0.693 147 180
9	0.693 147 362	0.693 142 154	
10	0.693 147 097	0.693 146 226	
11	0.693 147 208	0.693 146 672	

(a) Fibonacci numbers F_r $(r = 0, 1, 2, 3, \ldots)$ are defined by their generating relation

$$F_r = F_{r-1} + F_{r-2}$$

where $F_0 = 0$, $F_1 = 1$, $F_2 = 1$, $F_3 = 2$, $F_4 = 3, \ldots$. Numerical values of the first 50 Fibonacci numbers are given below in (*e*).

(b) Closed form of this number is

$$F_r = \frac{1}{\sqrt{5}}\left[\left(\frac{1 + \sqrt{5}}{2}\right)^r - \left(\frac{1 - \sqrt{5}}{2}\right)^r\right] = \frac{1}{\sqrt{5}}(\alpha^r - \beta^r)$$

where

$$\alpha = 1.618\,033\,988\,749\,894\,848\,204\cdots \qquad \beta = -0.618\,033\,988\,749\,894\,948\,204\cdots$$

are the *golden mean numbers*.

(c) Golden mean numbers are the roots of

$$x^2 - x - 1 = 0$$

which is the condition of the division of a line segment AB in the adjacent figure by the point C in the mean and exteme ratio, so that the greater segment AC is the geometric mean of AB and CB,

$$x = \sqrt{(1)(x + 1)}$$

and the *golden rectangle* is the rectangle whose sides have the ratio of $\frac{1}{2}(1 + \sqrt{5})$ and is supposed to have the most pleasing effect on the eye.

(d) Series representation of the $(r + 1)$th Fibonacci number is

$$F_{r+1} = \sum_{k=m}^{r}\binom{k}{r - k}$$

where m is the nearest higher integer above $\frac{1}{2}r$ or $\frac{1}{2}r$.

(f) Ratio of two consecutive Fibonacci numbers is

$$R_r = F_r / F_{r+1}$$

where

$$R_r = \cfrac{1}{1 + \cfrac{1}{1 + \cfrac{1}{1 + \cfrac{1}{1 + \cfrac{1}{1 + \cdots}}}}}$$

is a continued fraction with r divisions.

(e) Table of Fibonacci numbers

r	F_r	r	F_r	r	F_r
1	1	18	2 584	35	9 227 465
2	1	19	4 181	36	14 930 352
3	2	20	6 765	37	24 157 817
4	3	21	10 946	38	39 088 169
5	5	22	17 711	39	63 245 986
6	8	23	28 657	40	102 334 155
7	13	24	46 368	41	165 580 141
8	21	25	75 025	42	267 914 296
9	34	26	121 393	43	433 494 437
10	55	27	196 418	44	701 408 733
11	89	28	317 811	45	1 134 903 170
12	144	29	514 229	46	1 836 311 903
13	233	30	832 040	47	2 971 215 073
14	377	31	1 346 269	48	4 807 742 049
15	610	32	2 178 309	49	7 778 742 049
16	987	33	3 524 578	50	12 586 269 025
17	1 597	34	5 702 887		

(g) Examples

By (*b*), $\quad F_8 = \dfrac{1}{\sqrt{5}}(\alpha^8 - \beta^8) = 21$ $\qquad$ By (*d*), $\quad F_8 = \dbinom{4}{3} + \dbinom{5}{2} + \dbinom{6}{1} + \dbinom{7}{0} = 21$

(a) Generating function

$$\frac{x}{e^x - 1} = \sum_{r=0}^{\infty} \bar{B}_r \frac{x^r}{r!} = \frac{\bar{B}_0}{0!} + \frac{\bar{B}_1 x}{1!} + \frac{\bar{B}_2 x^2}{2!} + \frac{\bar{B}_3 x^3}{3!} + \cdots \qquad (|x| < 2\pi)$$

where

| $\bar{B}_0 = 1$ | $\bar{B}_2 = \frac{1}{6}$ | $\bar{B}_4 = -\frac{1}{30}$ | $\bar{B}_6 = \frac{1}{42}$ | $\bar{B}_8 = -\frac{1}{30}$ | $\bar{B}_{10} = \frac{5}{66}$ | $\cdots$ |
| $\bar{B}_1 = -\frac{1}{2}$ | $\bar{B}_3 = 0$ | $\bar{B}_5 = 0$ | $\bar{B}_7 = 0$ | $\bar{B}_9 = 0$ | $\bar{B}_{11} = 0$ | $\cdots$ |

are *Bernoulli numbers* of order $r = 0, 1, 2, \ldots$ (A.10).

(b) Auxiliary generating function

$$2 - \frac{x}{2} \cot \frac{x}{2} = \sum_{r=0}^{\infty} B_r \frac{x^{2r}}{(2r)!} = \frac{B_0}{0!} + \frac{B_1 x^2}{2!} + \frac{B_2 x^4}{4!} + \frac{B_3 x^6}{6!} + \cdots \qquad (|x| < \pi)$$

where

| $B_0 = 1$ | $B_1 = \frac{1}{6}$ | $B_2 = \frac{1}{30}$ | $B_3 = \frac{1}{42}$ | $B_4 = \frac{1}{30}$ | $B_6 = \frac{5}{66}$ | $\cdots$ |

are *auxiliary Bernoulli numbers* of order $r = 0, 1, 2, \ldots$ (A.10).

(c) Relationships.

Bernoulli numbers $\bar{B}_r$ and B_r of order r can be always computed in terms of their lower counterparts as

$$\bar{B}_r = \frac{r-1}{2(r+1)} - \frac{1}{r+1} \sum_{k=1}^{s} \binom{r+1}{2k} \bar{B}_{2k} \qquad (r = 4, 6, 8, \ldots)$$

$$B_r = \alpha \bar{B}_{2r} = \alpha \frac{2r-1}{2(2r+1)} - \frac{1}{2r+1} \sum_{k=1}^{t} \beta \binom{2r+1}{2k} B_k \qquad (r = 3, 4, 5, \ldots)$$

where $s = \frac{1}{2}r - 1$, $t = r - 1$, $\alpha = (-1)^{r+1}$, and $\beta = (-1)^{k+1}$.

(d) Examples.

If $\bar{B}_2 = \frac{1}{6}$, then by (c),

$$\bar{B}_4 = \frac{3}{10} - \frac{1}{5}\binom{5}{2}\bar{B}_2 = -\frac{1}{30} \qquad \qquad B_2 = (-1)^3 \bar{B}_4 = +\frac{1}{30}$$

Similarly, if $\bar{B}_2 = \frac{1}{6}$ and $\bar{B}_4 = -\frac{1}{30}$, then again by (c),

$$\bar{B}_6 = \frac{5}{14} - \frac{1}{7}\binom{7}{2}\bar{B}_2 - \frac{1}{7}\binom{7}{4}\bar{B}_4 = +\frac{1}{42} \qquad \qquad B_3 = (-1)^4 \bar{B}_6 = +\frac{1}{42}$$

(e) Series representation.

For $r = 2, 3, 4, \ldots$ and $\alpha = (-1)^{r+1}$,

$$B_r = \alpha \bar{B}_{2r} = 2\frac{(2r)!}{(2\pi)^{2r}}\left(\frac{1}{1^{2r}} + \frac{1}{2^{2r}} + \frac{1}{3^{2r}} + \cdots\right)$$

The series converges rapidly for $r > 3$.

(f) Example.

From (b), $B_5 = \frac{5}{66} = 0.075\,758$, and by a series of three terms,

$$B_5 = \bar{B}_{10} = 2\frac{10!}{(2\pi)^{10}}\left[1 + \frac{1}{2^{10}} + \frac{1}{3^{10}}\right] = 0.075\,758$$

(a) Definition. *Bernoulli polynomials* $\bar{B}_r(x)$ of order $r = 0, 1, 2, 3, \ldots$ are defined as

$$\bar{B}_r(x) = x^r \bar{B}_0 + \binom{r}{1} x^{r-1} \bar{B}_1 + \binom{r}{2} x^{r-2} \bar{B}_2 + \binom{r}{3} x^{r-3} \bar{B}_3 + \cdots + \binom{r}{r} \bar{B}_r$$

where $\bar{B}_0, \bar{B}_1, \bar{B}_2, \ldots, \bar{B}_r$ are the *Bernoulli numbers* introduced in (2.22a). Particular polynomials for $r = 0, 1, 2, 3, \ldots, 25$ are tabulated in (A.11).

(b) Properties. In general,

$$\bar{B}_r(x + 1) - \bar{B}_r(x) = rx^{r-1} \qquad\qquad \bar{B}_r(x) = -\alpha \bar{B}_r(1 - x)$$

For $x = 0$, $r = 1, 2, 3, \ldots$, the polynomials define the respective Bernoulli numbers as

$$\bar{B}_{2r}(0) = \bar{B}_{2r} = \alpha B_r \qquad\qquad \bar{B}_{2r+1}(0) = \bar{B}_{2r+1} = 0$$

where $\alpha = (-1)^{r+1}$.

(c) Derivatives of $\bar{B}_r(x)$ are

$$\frac{d\bar{B}_r(x)}{dx} = r\bar{B}_{r-1}(x)$$

$$\frac{d^2 \bar{B}_r(x)}{dx^2} = r(r - 1)\bar{B}_{r-2}(x)$$

and in general,

$$D_n[\bar{B}_r(x)] = \begin{cases} \binom{r}{n} n! \, \bar{B}_{R-n}(x) & (n < r) \\ n! & (n = r) \\ 0 & (n > r) \end{cases}$$

where $D_n = \dfrac{d^n}{dx^n}$ and $n = 1, 2, 3, \ldots$

(d) Graphs of the first four polynomials are shown below.

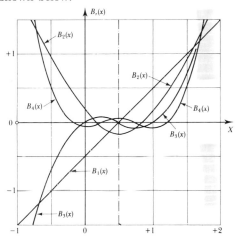

(e) Indefinite integrals of $\bar{B}_r(x)$ are

$$\int \bar{B}_r(x)\, dx = \frac{\bar{B}_{r+1}(x)}{r + 1} + C_1 \qquad\qquad \iint \bar{B}_r(x)\, dx\, dx = \frac{\bar{B}_{r+2}(x)}{(r + 1)(r + 2)} + C_1 x + C_2$$

and in general,

$$I^n[\bar{B}_r(x)] = \frac{\bar{B}_{r+n}(x)}{\binom{r+n}{n} n!} + C_1 \frac{x^{n-1}}{(n - 1)!} + C_2 \frac{x^{n-2}}{(n - 2)!} + \cdots + C_n \qquad \text{for } n \leq r$$

where $I^n[\;] = \iiint \cdots (dx)^n$, $n = 1, 2, 3, \ldots$, and $C_1, C_2, \ldots, C_n$ are the constants of integration.

(f) Fourier expansions of $\bar{B}_r(x)$ are

$$\bar{B}_r(x) = \begin{cases} +2\alpha \dfrac{r!}{(2\pi)^r} \displaystyle\sum_{k=1}^{\infty} \dfrac{\cos 2k\pi x}{k^r} & \text{if } r \text{ is even} \\[3ex] -2\alpha \dfrac{r!}{(2\pi)^r} \displaystyle\sum_{k=1}^{\infty} \dfrac{\sin 2k\pi x}{k^r} & \text{if } r \text{ is odd} \end{cases}$$

where $\alpha = (-1)^{r+1}$ and $0 \leq x \leq 1$.

(a) Generating function

$$\frac{2\sqrt{e^x}}{e^x + 1} = \sum_{r=0}^{\infty} \frac{\bar{E}_r x^r}{2^r r!} = \frac{\bar{E}_0}{0!} + \frac{\bar{E}_1 x}{2(1!)} + \frac{\bar{E}_1 x^2}{4(2!)} + \frac{\bar{E}_3 x^3}{8(3!)} + \cdots \qquad (|x| < \pi)$$

where

$\bar{E}_0 = 1$	$\bar{E}_2 = -1$	$\bar{E}_4 = 5$	$\bar{E}_6 = -61$	$\bar{E}_8 = 1\,385$	$\bar{E}_{10} = -50\,521$	$\cdots$
$\bar{E}_1 = 0$	$\bar{E}_3 = 0$	$\bar{E}_5 = 0$	$\bar{E}_7 = 0$	$\bar{E}_9 = 0$	$\bar{E}_{11} = 0$	$\cdots$

are *Euler numbers* of order $r = 0, 1, 2, \ldots$ (A.12).

(b) Auxiliary generating function

$$\sec x = \sum_{r=0}^{\infty} E_r \frac{x^{2r}}{(2r)!} = \frac{E_0}{0!} + \frac{E_1 x^2}{2!} + \frac{E_2 x^4}{4!} + \frac{E_3 x^6}{6!} + \cdots \qquad (|x|) < \tfrac{1}{2}\pi)$$

where

$E_0 = 1$	$E_1 = 1$	$E_2 = 5$	$E_3 = 61$	$E_4 = 1\,385$	$E_5 = 50\,521$	$\cdots$

are *auxiliary Euler numbers* of order $r = 0, 1, 2, \ldots$ (A.12)

(c) Relationships.

Euler numbers $\bar{E}_r$ and E_r of order r can be always computed in terms of their lower counterparts as

$$\bar{E}_r = -1 - \sum_{k=1}^{s} \binom{r}{2k} \bar{E}_{2k} \qquad (r = 4, 6, 8, \ldots)$$

$$E_r = -\alpha \bar{E}_{2r} = \alpha \left[1 - \sum_{k=1}^{t} \beta \binom{2r}{2k} E_k \right] \qquad (r = 3, 5, 7, \ldots)$$

where $s = \tfrac{1}{2}r - 1$, $t = r - 1$, $\alpha = (-1)^{r+1}$, and $\beta = (-1)^{k+1}$.

(d) Examples.

If $\bar{E}_2 = -1$, then by (c),

$$\bar{E}_4 = -1 - \binom{4}{2}\bar{E}_2 = +5 \qquad\qquad E_2 = -(-1)^3 \bar{E}_4 = +5$$

Similarly, if $E_2 = -1$ and $E_4 = 5$, then again by (c),

$$\bar{E}_6 = -1 - \binom{6}{2}\bar{E}_2 - \binom{6}{4}\bar{E}_4 = -61 \qquad\qquad E_3 = -(-1)^4 \bar{E}_6 = +61$$

(e) Series representation.

For $r = 2, 3, 4, \ldots$, $\alpha = (-1)^{r+1}$,

$$E_r = -\alpha \bar{E}_{2r} = 2\left(\frac{2}{\pi}\right)^{2r+1} (2r)! \left(\frac{1}{1^{2r+1}} - \frac{1}{3^{2r+1}} + \frac{1}{5^{2r+1}} - \cdots \right)$$

The series converges rapidly for $r > 3$.

(f) Example.

From (b), $E_5 = 50\,521$, and by a series of three terms,

$$E_5 = -\bar{E}_{10} = 2\left(\frac{2}{\pi}\right)^{11} (10!)\left(1 - \frac{1}{3^{11}} + \frac{1}{5^{11}} \right) = 50\,521$$

(a) Definition. *Euler polynomials* $\bar{E}_r(x)$ of order $r = 0, 1, 2, 3, \ldots$ are defined as

$$\bar{E}_r(x) = \frac{2}{r+1}\left[(1-2)\binom{r+1}{1}x^r\bar{B}_1 + (1-2^2)\binom{r+1}{2}x^{r-1}\bar{B}_2\right.$$
$$\left. + \cdots + (1-2^{r+1})\binom{r+1}{r+1}\bar{B}_{r+1}\right]$$

where $\bar{B}_1, \bar{B}_2, \bar{B}_3, \ldots, \bar{B}_{r+1}$ are *Bernoulli numbers* introduced in (2.22a). Particular polynomials for $r = 0, 1, 2, 3, \ldots, 25$ are tabulated in (A.13).

(b) Properties. In general,

$$\bar{E}_r(x+1) + \bar{E}_r(x) = 2x^r \qquad\qquad \bar{E}_r(x) = -\alpha\bar{E}_r(1-x)$$

For $x = \frac{1}{2}$, $r = 1, 2, 3, \ldots$, the polynomials define the respective Euler numbers as

$$\bar{E}_{2r}(\tfrac{1}{2}) = 2^{-2r}\bar{E}_{2r} = -2^{-2r}\alpha E_r \qquad\qquad \bar{E}_{2r+1}(\tfrac{1}{2}) = \bar{E}_{2r+1} = 0$$

where $\alpha = (-1)^{r+1}$.

(c) Derivatives of $\bar{E}_r(x)$ are

$$\frac{d\bar{E}_r(x)}{dx} = r\bar{E}_{r-1}(x) \qquad \frac{d^2\bar{E}_r}{dx^2} = r(r-1)\bar{E}_{r-2}(x)$$

and in general,

$$D_n[\bar{E}_r(x)] = \begin{cases} \binom{r}{n}n!\,\bar{E}_{r-n}(x) & (n < r) \\ n! & (n = r) \\ 0 & (n > r) \end{cases}$$

where $D_n = \dfrac{d^n}{dx^n}$ and $n = 1, 2, 3, \ldots$.

(d) Graphs of the first four polynomials are shown below.

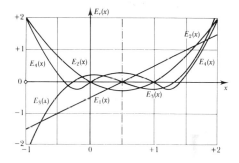

(e) Indefinite Integrals of $\bar{E}_r(x)$ are

$$\int \bar{E}_r(x)\,dx = \frac{\bar{E}_{r+1}(x)}{r+1} + C_1 \qquad\qquad \int\int \bar{E}_r(x)\,dx\,dx = \frac{\bar{E}_{r+2}(x)}{(r+1)(r+2)} + C_1 x + C_2$$

and in general,

$$I^n[\bar{E}_r(x)] = \frac{E_{r+n}(x)}{\binom{r+n}{n}n!} + C_1\frac{x^{n-1}}{(n-1)!} + C_2\frac{x^{n-2}}{(n-2)!} + \cdots + C_n \qquad \text{for } n \leqq r$$

where $I^n[\ ] = \int\int\int \cdots (dx)^n$, $n = 1, 2, 3, \ldots$, C_n and $C_1, C_2, \ldots$, are the constants of integration.

(f) Fourier expansions of $\bar{E}_r(x)$ are

$$\bar{E}_r(x) = \begin{cases} -\dfrac{4\alpha}{\pi^{r+1}}\displaystyle\sum_{k=0}^{\infty}\frac{\sin(2k+1)\pi x}{(2k+1)^{r+1}} & \text{if } r \text{ is even} \\[4mm] +\dfrac{4\alpha}{\pi^{r+1}}\displaystyle\sum_{k=0}^{\infty}\frac{\cos(2k+1)\pi x}{(2k+1)^{r+1}} & \text{if } r \text{ is odd} \end{cases}$$

where $\alpha = (-1)^{r+1}$ and $0 < x < 1$.

(a) Definition. *Rieman zeta functions* $Z(r)$ of order $r = 1, 2, 3, \ldots$ are defined by the series

$$Z(r) = \sum_{k=1}^{\infty} \frac{1}{k^r} = \frac{1}{1^r} + \frac{1}{2^r} + \frac{1}{3^r} + \cdots$$

or by the product

$$Z(r) = \prod_{p} \left(1 - \frac{1}{p^r}\right)^{-1} = \frac{1}{1 - \dfrac{1}{2^r}} \cdot \frac{1}{1 - \dfrac{1}{3^r}} \cdot \frac{1}{1 - \dfrac{1}{5^r}} \cdots$$

where the product is taken over all prime numbers $p = 2, 3, 5, 7, 11, 13, \ldots$.

(b) Alternative series defining $Z(r)$ is

$$Z(r) = \frac{2^r}{2^r - 1} \sum_{k=1}^{\infty} \frac{1}{(2k-1)^r} = \frac{2^r}{2^r - 1}\left(\frac{1}{1^r} + \frac{1}{3^r} + \frac{1}{5^r} + \cdots\right)$$

The series converges rapidly for $r > 5$.

(c) Special values of $Z(r)$ are

$$Z(1) = \infty, \; Z(2) = \frac{\pi^2}{6}, \; Z(4) = \frac{\pi^4}{90}, \; Z(6) = \frac{\pi^6}{945}, \ldots$$

and in general, in terms of Bernoulli numbers and polynomials (2.22), (2.23),

$$Z(r) = \sum_{k=1}^{\infty} \frac{1}{k^r} = \begin{cases} +\alpha\dfrac{(2\pi)^r}{r!}\left|\dfrac{\bar{B}_r}{2}\right| & \text{if } r \text{ is even} \\[2ex] -\alpha\dfrac{(2\pi)^r}{r!}\displaystyle\int_0^1 \dfrac{\bar{B}_r(x)}{2}(\cot \pi x)\,dx & \text{if } r \text{ is odd} \end{cases}$$

where $r > 1$ and $\alpha = (-1)^{r+1}$. Numerical values of $Z(r)$ for $r = 1, 2, 3, \ldots, 20$ are tabulated in (2.27c). Large tables of $Z(r)$, including decimal values of argument, can be found in Ref. 2.03.

(d) Sums of reciprocal powers

$$\sum_{k=1}^{\infty} \frac{1}{k^r} = \frac{1}{1^r} + \frac{1}{2^r} + \frac{1}{3^r} + \frac{1}{4^r} + \cdots = Z(r) \qquad\qquad (r = 2, 3, 4, \ldots)$$

$$\sum_{k=1}^{\infty} \frac{1}{(2k)^r} = \frac{1}{2^r} + \frac{1}{4^r} + \frac{1}{6^r} + \frac{1}{8^r} + \cdots = \left(\frac{1}{2}\right)^r Z(r) \qquad\qquad (r = 2, 3, 4, \ldots)$$

$$\sum_{k=1}^{\infty} \frac{1}{(2k-1)^r} = \frac{1}{1^r} + \frac{1}{3^r} + \frac{1}{5^r} + \frac{1}{7^r} + \cdots = \left[1 - \left(\frac{1}{2}\right)^r\right] Z(r) \qquad\qquad (r = 2, 3, 4, \ldots)$$

$$\sum_{k=1}^{\infty} \frac{\beta}{k^r} = \frac{1}{1^r} - \frac{1}{2^r} + \frac{1}{3^r} - \frac{1}{4^r} + \cdots = \left[1 - \left(\frac{1}{2}\right)^{r-1}\right] Z(r) \qquad\qquad (r = 2, 3, 4, \ldots)$$

$$\sum_{k=1}^{\infty} \frac{\beta}{(2k)^r} = \frac{1}{2^r} - \frac{1}{4^r} + \frac{1}{6^r} - \frac{1}{8^r} + \cdots = \left(\frac{1}{2}\right)^r\left[1 - \left(\frac{1}{2}\right)^{r-1}\right] Z(r) \qquad\qquad (r = 2, 3, 4, \ldots)$$

(a) Definition. Complementary Rieman zeta functions $\bar{Z}(r)$ of order $r = 1, 2, 3, \ldots$ are defined by the series as

$$\bar{Z}(r) = \sum_{k=1}^{\infty} \frac{\beta}{(2k-1)^r} = \frac{1}{1^r} - \frac{1}{3^r} + \frac{1}{5^r} - \cdots$$

The series converges rapidly for $r > 5$.

(b) Special values of $\bar{Z}(r)$ are

$$\bar{Z}(1) = \frac{\pi}{4}, \bar{Z}(3) = \frac{\pi^3}{32}, \bar{Z}(5) = \frac{5\pi^5}{1\,536}, \ldots$$

and in general in terms of Euler numbers and polynomials (2.24), (2.25),

$$\bar{Z}(r) = \sum_{k=1}^{\infty} \frac{\beta}{(2k-1)^r} = \begin{cases} \left| \dfrac{\pi^r}{(r-1)!} \displaystyle\int_0^1 \dfrac{\bar{E}_{r-1}(x)}{4\cos \pi x}\, dx \right| & \text{if } r \text{ is even} \\[3mm] \left| \dfrac{\bar{E}_{r-1}}{2^{r+1}} \dfrac{\pi^r}{(r-1)!} \right| & \text{if } r \text{ is odd} \end{cases}$$

where $r \geq 1$ and $\beta = (-1)^{k+1}$. Numerical values of $\bar{Z}(r)$ for $r = 1, 2, 3, \ldots, 20$ are tabulated in (c) below. Large tables of $\bar{Z}(r)$, including decimal values of argument, can be found in Ref. 2.03. A particular case, $\bar{Z}(2) = 0.915\,965\,594\,177\,219\,015 \cdots$, is called the *Catalan constant*.

(c) Numerical values*

r	$Z(r)$ $\sum_{k=1}^{\infty} \dfrac{1}{k^r}$	$[1 - (\frac{1}{2})^r]Z(r)$ $\sum_{k=1}^{\infty} \dfrac{1}{(2k-1)^r}$	$[1 - (\frac{1}{2})^{r-1}]Z(r)$ $\sum_{k=1}^{\infty} \dfrac{\beta}{k^r}$	$\bar{Z}(r)$ $\sum_{k=1}^{\infty} \dfrac{\beta}{(2k-1)^r}$
1	∞	∞	$0.693\,147\,181 = \ln 2$	$0.785\,398\,163 = \pi/4$
2	1.644 934 067	1.233 700 550	0.822 467 033	0.915 965 594
3	1.202 056 903	1.051 799 790	0.901 542 677	0.968 946 146
4	1.082 323 234	1.014 678 032	0.947 032 829	0.988 944 552
5	1.036 927 755	1.004 523 763	0.972 119 770	0.996 157 828
6	1.017 343 062	1.001 447 077	0.985 551 091	0.998 685 222
7	1.008 349 277	1.000 471 549	0.992 593 820	0.999 554 508
8	1.004 077 356	1.000 155 179	0.996 233 002	0.999 849 990
9	1.002 008 393	1.000 051 345	0.998 094 298	0.999 949 684
10	1.000 994 575	1.000 017 041	0.999 039 508	0.999 983 164
11	1.000 494 189	1.000 005 666	0.999 517 143	0.999 994 375
12	1.000 246 087	1.000 001 886	0.999 757 685	0.999 998 122
13	1.000 122 713	1.000 000 628	0.999 878 543	0.999 999 374
14	1.000 061 248	1.000 000 209	0.999 939 170	0.999 999 791
15	1.000 030 589	1.000 000 070	0.999 969 551	0.999 999 930
16	1.000 015 282	1.000 000 023	0.999 984 764	0.999 999 977
17	1.000 007 637	1.000 000 008	0.999 923 782	0.999 999 992
18	1.000 003 817	1.000 000 003	0.999 996 188	0.999 999 997
19	1.000 001 908	1.000 000 001	0.999 998 094	0.999 999 999
20	1.000 000 956	1.000 000 000	0.999 999 047	1.000 000 000

*Ref. 1.20, p. 461.

(a) Euler's integral. Gamma function $\Gamma(x + 1)$ is the generalization of the factorial (2.09a) and is defined as

$$\Gamma(x + 1) = \int_0^\infty e^{-t}t^x \, dt = x! \qquad (x > 0)$$

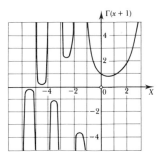

(b) Gauss' limit is a broader definition of the gamma function expressed as

$$\Gamma(x + 1) = \lim_{n\to\infty} \frac{n! \, n^{x+1}}{(x + n)(x + n - 1) \cdots (x + 2)(x + 1)}$$
$$= \prod_{k=1}^\infty \left[\frac{k}{x + k} \left(\frac{k + 1}{k} \right)^k \right] = x!$$

The graph shows that $\Gamma(x + 1)$ is single-valued except at $x = -1, -2, -3, \ldots$, and its alternating extremes are $\Gamma(1.462) = 0.886$, $\Gamma(-0.504) = -3.545$, $\Gamma(-1.573) = 2.302$, $\Gamma(-2.611) = -0.888$, $\Gamma(-3.635) = 0.245$, $\Gamma(-4.653) = -0.053$, $\Gamma(-5.667) = 0.009$, $\Gamma(-6.678) = -0.001$ (Ref. 2.08).

(c) Functional equations. With the restrictions placed upon x in (b),

$$\Gamma(x + n) = (x + n - 1)(x + n - 2) \cdots (x + 2)(x + 1)\Gamma(x + 1)$$
$$\Gamma(x - n) = \frac{\Gamma(x + 1)}{(x - n)(x - n + 1) \cdots (x - 2)(x - 1)x}$$

where n = positive integer and x = signed number (positive or negative, integer or fraction).

(d) Integer and fraction arguments $(n, p, q = 1, 2, 3, \ldots)$

$\Gamma(n + 1)$ $= n(n - 1)(n - 2) \cdots 3 \cdot 2 \cdot 1 = n!$	$\Gamma(n - 1)$ $= (n - 2)(n - 3) \cdots 3 \cdot 2 \cdot 1 = (n - 2)!$
$\Gamma(n + p + 1) = (n + p)!$	$\Gamma(n - p + 1) = (n - p)!$
$\Gamma\left(n \pm \frac{p}{q}\right) = \left(n \pm \frac{p}{q} - 1\right)\left(n \pm \frac{p}{q} - 2\right) \cdots \left(3 \pm \frac{p}{q}\right)\left(2 \pm \frac{p}{q}\right)\left(1 \pm \frac{p}{q}\right)\Gamma\left(1 \pm \frac{p}{q}\right)$	

(e) Reflection and duplication formulas

$$\Gamma(u)\Gamma(-u) = \frac{-\pi}{u \sin u\pi} \qquad \Gamma(2x) = \frac{4^x}{\sqrt{4\pi}}\Gamma(x)\Gamma(x + \tfrac{1}{2}) \qquad \Gamma(u)\Gamma(1 - u) = \frac{\pi}{\sin u\pi}$$

where x = signed number, $n = 1, 2, 3, \ldots$, and $u < 1$.

(f) Special values $(n = 1, 2, 3, \ldots)$

$\Gamma(-n) = \infty$	$F(0) = \infty$	$\Gamma(1) = 1$	$F(n) = (n - 1)!$
$\Gamma(-\tfrac{1}{2}) = -2\sqrt{\pi}$		$\Gamma(\tfrac{1}{2}) = \sqrt{\pi}$	
$\Gamma(n + \tfrac{1}{2}) = \dfrac{(2n)! \sqrt{\pi}}{n! \, 4^n}$		$\Gamma(-n + \tfrac{1}{2}) = (-4)^n \dfrac{n! \sqrt{\pi}}{(2n)!}$	

Numerical values of $\Gamma(n + 1)$ and $\Gamma(u)$ are tabulated in (A.03) and (A.04), respectively.

(a) Small argument. For $0 < u \leq \frac{1}{2}$, the gamma function can be evaluated to a desired degree of accuracy as

$$\Gamma(1 + u) = e^{\tau}\sqrt{\frac{(1 - u)u\pi}{(1 + u)\sin u\pi}} + \epsilon$$

$$\Gamma(1 - u) = e^{-\tau}\sqrt{\frac{(1 + u)u\pi}{(1 - u)\sin u\pi}} + \epsilon$$

where in terms of a_k given in (b),

$$\tau = a_1 u - a_3 u^3 - a_5 u^5 - \cdots - a_{13}u^{13}$$

and $|\epsilon| \leq 5 \times 10^{-10}$. Nested form in terms of A_k given in (b) is

$$\tau = A_1 u(1 - A_3 u^2(1 - A_5 u^2(1 - A_7 u^2(1 - \cdots - A_{13}u^2))))$$

The general expressions for a_k and A_k are $a_1 = 1 - \gamma$, $a_3 = Z(3) - 1$, $a_5 = Z(5) - 1, \ldots$ and $A_1 = a_1$, $A_3 = a_3/a_1$, $A_5 = a_5/a_3, \ldots$, where $\gamma =$ Euler's constant (2.19) and $Z(r) =$ Riemann zeta function (2.26).

(b) Factors a_k and A_k

k	a_k	A_k
1	0.422 784 335	0.422 784 335
3	0.067 352 301	0.159 306 581
5	0.007 385 551	0.109 655 511
7	0.001 192 754	0.161 498 309
9	0.000 223 155	0.187 092 229
11	0.000 044 926	0.201 321 951
13	0.000 009 439	0.284 490 050

(c) Example. If $\Gamma(1.25)$, $\Gamma(0.75)$, $\Gamma(3.25)$, and $\Gamma(3.75)$ are desired, e^{τ} is calculated first with $u = 0.25$.

$$a_1 u = 0.105\ 696\ 083\ 8 \qquad -a_5 u^5 = -0.000\ 007\ 212\ 4 \qquad -a_9 u^9 = -0.000\ 000\ 000\ 8$$

$$-a_3 u^3 = -0.001\ 032\ 379\ 7 \qquad -a_7 u^7 = -0.000\ 000\ 072\ 7$$

the sum of which is $\tau = 0.104\ 636\ 418\ 2$ and $e^{\tau} = 1.110\ 306\ 850$. Then by relations (a),

$$\Gamma(1 + 0.25) = e^{\tau}\sqrt{\frac{(0.75)(0.25)\pi}{1.25 \sin 0.25\pi}} \qquad \Gamma(1 - 0.25) = e^{-\tau}\sqrt{\frac{(1.25)(0.25)\pi}{0.75 \sin 0.25\pi}}$$

$$= 0.906\ 402\ 477\ 3 + \epsilon \qquad\qquad = 1.225\ 416\ 702\ 2 + \epsilon$$

where $\epsilon = -2 \times 10^{-10}$ based on comparison with the tabulated values in (A.04). The remaining values are calculated by (2.28d).

(d) Large argument. If $x > 4$ is an integer or decimal number, the Stirling's expansion (2.09e),

$$\Gamma(x + 1) = e^{\lambda}\left(\frac{x}{e}\right)^x \sqrt{2x\pi} + \epsilon = x!$$

yields good results as shown in (e) below.

(e) Errors of Stirling's expansion of $\Gamma(x + 1)$

$x + 1$	Correct value of $\Gamma(x + 1)$	Approximate value of $\Gamma(x + 1)$
3.00	2.000 000 000 (+00)	1.999 997 800 (+00)
3.25	2.549 256 967 (+00)	2.549 255 928 (+00)
3.50	3.323 350 971 (+00)	3.323 350 418 (+00)
3.75	4.422 988 410 (+00)	4.422 988 086 (+00)
4.00	6.000 000 000 (+00)	5.999 999 792 (+00)
5.00	2.400 000 000 (+01)	2.399 999 993 (+01)
6.00	1.200 000 000 (+02)	1.200 000 000 (+02)
7.00	7.200 000 000 (+02)	7.199 999 999 (+02)
8.00	5.040 000 000 (+03)	5.040 000 000 (+03)
10.00	3.628 800 000 (+05)	3.628 800 000 (+05)
20.00	1.216 451 004 (+17)	1.216 451 004 (+17)
50.00	6.082 818 640 (+62)	6.082 818 640 (+62)

(a) Binomial coefficient $\binom{x}{k}$ can be expressed in terms of gamma functions as

$$\binom{x}{k} = \frac{x!}{(x-k)!\,k!} = \frac{\Gamma(x+1)}{\Gamma(x-k+1)\Gamma(k+1)}$$

where $k = 0, 1, 2, \ldots$ and $x = $ signed number except $-1, -2, -3, \ldots$.

(b) Factorial polynomial of step 1 (2.11a) of order m in x is

$$X_1^{(m)} = x(x-1)(x-2)\cdots(x-m+1) = \frac{\Gamma(x+1)}{\Gamma(x-m+1)}$$

$$X_1^{(-m)} = \frac{1}{(x+1)(x+2)\cdots(x+m-1)(x+m)} = \frac{\Gamma(x+1)}{\Gamma(x+m+1)}$$

where $m = $ positive but $-m \neq -1$, and x is the signed number restricted by (2.28b).

(c) Factorial polynomial of step h of order m in x is

$$X_h^{(m)} = x(x-h)(x-2h)\cdots[x-(m-1)h] = \frac{h^m \Gamma\left(\dfrac{x}{h}+1\right)}{\Gamma\left(\dfrac{x}{h}-m+1\right)}$$

$$X_h^{(-m)} = \frac{1}{(x+h)(x+2h)\cdots(x+mh)} = \frac{\Gamma\left(\dfrac{x}{h}+1\right)}{h^m \Gamma\left(\dfrac{x}{h}+m+1\right)}$$

where m, x are those defined in (b) above and h is a positive constant.

(d) Double factorial denoted as $x!!$ is a special case of a factorial polynomial with $x = 2n$ or $x = 2n - 1$, $h = 2$, and $n = 1, 2, 3, \ldots$. In particular,

$$(2n)!! = 2n(2n-2)(2n-4)\cdots 6\cdot 4\cdot 2 = 2^n n! = 2^n \Gamma(n+1)$$

$$(2n-1)!! = (2n-1)(2n-3)(2n-5)\cdots 5\cdot 3\cdot 1 = \frac{2^n(n-\frac{1}{2})!}{\sqrt{\pi}} = \frac{2^n \Gamma(n+\frac{1}{2})}{\sqrt{\pi}}$$

(e) Natural logarithm of the gamma function for $0 < u \leq \frac{1}{2}$ is

$$\ln \Gamma(1+u) = \ln \sqrt{\frac{(1-u)u\pi}{(1+u)\sin u\pi}} + a_1 u - a_3 u^3 - a_5 u^5 - \cdots - a_{13}u^{13}$$

$$\ln \Gamma(1-u) = \ln \sqrt{\frac{(1+u)u\pi}{(1-u)\sin u\pi}} - a_1 u + a_3 u^3 + a_5 u^5 + \cdots + a_{13}u^{13}$$

where $a_1, a_3, a_5, \ldots$ are the factors listed in (2.29b). For $1 \leq (n+u) \leq 4$, $n = 2, 3, \ldots$,

$$\ln \Gamma(n+u) = \ln(n+u-1) + \ln(n+u-2) + \cdots + \ln(1+u) + \ln \Gamma(1+u)$$

where $\ln \Gamma(1+u)$ is the first relation above.

(f) Stirling's expansion of $\ln \Gamma(x+1)$ for $x \geq 4$ is $\ln \Gamma(x+1) = x(\ln x - 1) + \ln \sqrt{2x\pi} + \lambda$, where λ is the series in (2.09e), and for very large x, $\lambda \cong 0$.

(a) Pi function $\Pi(x)$ is the generalization of the factorial (2.09a) and is defined as

$$\Pi(x) = \Gamma(x + 1) = \int_0^\infty e^{-t}t^x \, dt = x! \qquad (x > 0)$$

where $x!$ is the general factorial. All relations introudced in (2.28) to (2.30) are applicable in operations with this function.

(b) Beta function $B(x, y)$ is defined as

$$B(x, y) = \int_0^1 t^{x-1}(1 - t)^{y-1} \, dt = \int_0^\infty \frac{t^{x-1}}{(1 + t)^{x+y}} \, dt \qquad (x > 0, y > 0)$$

and is related to the gamma function and pi function as

$$B(x, y) = \frac{\Gamma(x)\Gamma(y)}{\Gamma(x, y)} \qquad\qquad B(x, y) = \frac{\Pi(x - 1)\Pi(y - 1)}{\Pi(x + y - 1)}$$

(c) Reflection formulas $(n = 1, 2, 3, \ldots)$

$$B(x, y) = B(y, x) \qquad\qquad B(x, y, z) = B(z, x + y)$$

$$B(x, y)B(x + y, z) = B(x, z)B(y + z, x)$$

$$B(n + x, n - x) = \frac{\Gamma(n + x)\Gamma(n - x)}{\Gamma(2n)} = \frac{\pi P_{n-1}}{(2n - 1)! \sin \pi x}$$

$$B(n + \tfrac{1}{2} + x, n + \tfrac{1}{2} - x) = \frac{\Gamma(n + \tfrac{1}{2} + x)\Gamma(n + \tfrac{1}{2} - x)}{\Gamma(2n + 1)} = \frac{\pi Q_n}{(2n)! \cos \pi x}$$

where

$$P_0 = x, \qquad P_1 = x(1 - x^2), \qquad P_2 = x(1 - x^2)(2^2 - x^2),$$
$$P_3 = x(1 - x^2)(2^2 - x^2)(3^2 - x^2), \ldots$$

$$Q_0 = 1, \qquad Q_1 = [(\tfrac{1}{2})^2 - x^2], \qquad Q_2 = [(\tfrac{1}{2})^2 - x^2][(\tfrac{3}{2})^2 - x^2],$$
$$Q_3 = [(\tfrac{1}{2})^2 - x^2][(\tfrac{3}{2})^2 - x^2][(\tfrac{5}{2})^2 - x^2], \ldots$$

For $u < 1$, $\quad B(u, 1 - u) = \Gamma(u)\Gamma(1 - u) = \dfrac{\pi}{\sin u\pi}$

(d) Integer and fraction arguments $(p, q = 1, 2, 3, \ldots)$

$$B(p, q) = \frac{(p - 1)! \, (q - 1)!}{(p + q - 1)!} = \left[p\binom{p + q - 1}{q - 1} \right]^{-1} = \left[q\binom{p + q - 1}{p - 1} \right]^{-1}$$

$$B(p + 1, p + 1) = \frac{(p!)^2}{(2p + 1)!} = \frac{p!}{(2p + 1)2p(2p - 1)(2p - 2)\cdots(p + 1)} \qquad B(p, p) = \frac{B(p, \tfrac{1}{2})}{4^p - 1}$$

$$B(p + 1, p - 1) = \frac{p! \, (p - 2)!}{(2p - 1)!} = \frac{(p - 2)!}{(2p - 1)(2p - 2)(2p - 3)\cdots(p + 1)} \qquad B(\tfrac{1}{2}, \tfrac{1}{2}) = \pi$$

$$B(p, q) = \frac{p - 1}{p + q - 1} B(p - 1, q) = \frac{(p - 1)(p - 2)}{(p + q - 1)(p + q - 2)} B(p - 2, q) = \cdots$$

(a) Derivative of natural logarithm of $\Gamma(x + 1)$ in (2.28a) is called the *digamma function* $\psi(x)$.

$$\psi(x) = \frac{d}{dx}\{\ln[\Gamma(x + 1)]\} = \frac{\Gamma'(x + 1)}{\Gamma(x + 1)}$$

where $\Gamma'(x + 1)$ is the derivative of $\Gamma(x + 1)$ with respect to x.

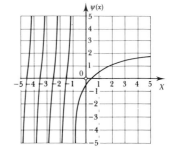

(b) Natural logarithm of Gauss' limit in (2.29b) can be differentiated term by term so that

$$\psi(x) = \lim_{n \to \infty}\left(\ln n - \sum_{k=1}^{n}\frac{1}{x + k}\right) = \sum_{k=1}^{\infty}\left(\frac{1}{k} - \frac{1}{x + k}\right) - \gamma$$

where γ = Euler's constant (2.19a). In some literatures $\psi(x)$ is called the *psi function*. The graph in (a) shows that $\psi(x)$ is single-valued except at $x = -1, -2, -3, \ldots$, and its zero points are at $x = 0.462, -1.504, -2.573, -3.611, -4.635, -5.653, -6.667, -7.678, \ldots$.

(c) Functional equations with restrictions placed upon x in (b) are

$$\psi(x + r) = \psi(x) + \sum_{k=1}^{r}\frac{1}{x + k} \qquad \psi(x - r) = \psi(x) - \sum_{k=1}^{r}\frac{1}{x + 1 - k}$$

where r is a positive integer and x is a signed number.

(d) Integer and fraction arguments $(n, p, q = 1, 2, 3, \ldots)$

$$\psi(n) = 1 + \frac{1}{2} + \frac{1}{3} + \cdots + \frac{1}{n - 1} + \frac{1}{n} - \gamma = \sum_{k=1}^{n}\frac{1}{k} - \gamma$$

$$\psi\left(n + \frac{p}{q}\right) = \psi\left(\frac{p}{q}\right) + q\sum_{k=1}^{n}\frac{1}{p + qk} \qquad \psi\left(n - \frac{p}{q}\right) = \psi\left(-\frac{p}{q}\right) + q\sum_{k=1}^{n}\frac{1}{qk - p}$$

(e) Reflection and duplication formulas

$$\psi(x + 1 - n) - \psi(x + 1) = \sum_{k=1}^{n}\frac{1}{x - k} \qquad \psi(x) - \psi(x + 1) = \frac{\pi}{\tan \pi x}$$

$$2\psi(2x - 2) = \psi(x - 1) + \psi(x - \tfrac{1}{2}) + \ln 4 \qquad \psi(u) - \psi(-u) = \frac{1}{u} - \frac{\pi}{\tan u\pi}$$

where x = signed number, $n = 1, 2, 3, \ldots$, and $|u| < 1$.

(f) Special values

$\psi(-n) = \pm\infty \quad \psi(0) = -\gamma \quad \psi(1) = 1 - \gamma \quad \psi(2) = 1 + \frac{1}{2} - \gamma \quad \psi(3) = 1 + \frac{1}{2} + \frac{1}{3} - \gamma$	
$\psi(-\frac{1}{2}) = -\gamma - \ln 4$	$\psi(\frac{1}{2}) = 2 - \gamma - \ln 4$
$\psi(-\frac{1}{3}) = -\gamma - \ln\sqrt{27} + \dfrac{\sqrt{3}\,\pi}{6}$	$\psi(\frac{1}{3}) = 3 - \gamma - \ln\sqrt{27} - \dfrac{\sqrt{3}\,\pi}{6}$
$\psi(-\frac{1}{4}) = -\gamma - \ln 16 + \dfrac{\pi}{2}$	$\psi(\frac{1}{4}) = 4 - \gamma - \ln 16 - \dfrac{\pi}{2}$

Numerical values of $\psi(u)$ are tabulated in (A.05). Numerical values of $\psi\left(n \pm \dfrac{p}{q}\right)$ are calculated by (d) or by (2.33d).

(a) Small argument. For $0 < u \leq \frac{1}{2}$, the digamma function can be evaluated to a desired degree of accuracy as

$$\psi(u) = 1 + \frac{1}{2u} - \frac{1}{(1+u)(1-u)} - \frac{\pi}{2 \tan u\pi} - \omega$$

$$\psi(-u) = 1 - \frac{1}{2u} - \frac{1}{(1+u)(1-u)} + \frac{1}{2 \tan u\pi} - \omega$$

where in terms of b_k given in (b),

$$\omega = b_0 + b_2 u^2 + b_4 u^4 + \cdots + b_{16} u^{16} + \epsilon$$

and $|\epsilon| = 5 \times 10^{-11}$.

(b) Factors b_k

k	b_k
0	0.577 215 664 9
2	0.202 056 903 1
4	0.036 927 755 1
6	0.008 349 277 4
8	0.002 008 392 8
10	0.000 494 188 6
12	0.000 122 713 3
14	0.000 030 588 2
16	0.000 007 637 2

(c) Example. If $\psi(\frac{1}{2})$, $\psi(\frac{3}{2})$ and $\psi(-\frac{5}{2})$ are desired, ω is calculated first with $u = \frac{1}{2}$ as

$b_0 = 0.577\ 215\ 664\ 9$	$b_6 u^6 = 0.000\ 130\ 457\ 5$	$b_{12} u^{12} = 0.000\ 000\ 030\ 0$
$b_2 u^2 = 0.050\ 514\ 225\ 8$	$b_8 u^8 = 0.000\ 007\ 845\ 3$	$b_{14} u^{14} = 0.000\ 000\ 001\ 9$
$b_4 u^4 = 0.002\ 307\ 984\ 7$	$b_{10} u^{10} = 0.000\ 000\ 482\ 6$	$b_{16} u^{16} = 0.000\ 000\ 000\ 1$

$\omega = \sum_{k=0}^{8} b_{2k} u^{2k} = 0.630\ 176\ 692\ 8$ and by the relation (a),

$$\psi(\tfrac{1}{2}) = 1 + 1 - \frac{1}{(1.5)(0.5)} - \frac{\pi}{2 \tan \frac{1}{2}u} - 0.630\ 176\ 692\ 8 - \epsilon = 0.036\ 489\ 973\ 9 - \epsilon$$

Since by $(2.32f)$, $\psi(\frac{1}{2}) = 2 - \gamma - \ln 4 = 0.036\ 489\ 073\ 9$, the error in this particular case is below the one indicated in (a). The remaining values are calculated by $(2.32d)$ in terms of $\psi(\frac{1}{2})$.

(d) Large argument. If $x > 5$ is an integer or decimal number, the Stirling's expansion

$$\psi(x) = \ln x + \frac{1}{2x} + \phi$$

yields good results as shown in (e) below, where in terms of $\lambda(2.09e)$,

$$\phi = \frac{d\lambda}{dx} = -\sum_{r=1}^{4} \frac{\bar{B}_{2r}}{2r x^{2r}} = -\frac{\bar{B}_2}{2x^2} - \frac{\bar{B}_4}{4x^4} - \frac{\bar{B}_6}{6x^6} - \frac{\bar{B}_8}{8x^8} = -\frac{1}{12x^2} + \frac{1}{120x^4} - \frac{1}{252x^6} + \frac{1}{240x^8}$$

(e) Errors of Stirling's expansion of $\psi(x)$

x	Correct value of $\psi(x)$		Approximate value of $\psi(x)$	
1	4.227 843 351	(-01)	4.251 198 413	(-01)
2	9.227 843 351	(-01)	9.227 889 526	(-01)
3	1.256 117 668	$(+00)$	1.256 117 768	$(+00)$
4	1.506 117 668	$(+00)$	1.506 117 675	$(+00)$
5	1.706 117 668	$(+00)$	1.706 117 779	$(+00)$
6	1.872 784 335	$(+00)$	1.872 784 335	$(+00)$
7	2.015 641 478	$(+00)$	2.015 641 478	$(+00)$
10	2.351 752 589	$(+00)$	2.351 751 589	$(+00)$
50	3.921 989 671	$(+00)$	3.921 989 671	$(+00)$

(a) Higher derivatives of the digamma function $\psi(x)$ in (2.32a), called the *polygamma functions* $\psi^{(1)}(x), \psi^{(2)}(x), \ldots, \psi^{(m)}(x)$, are defined symbolically as

$$\psi^{(m)}(x) = \frac{d^m}{dx^m}[\psi(x)] = \frac{d^{m+1}}{dx^{m+1}}\{\ln[\Gamma(x+1)]\}$$

where $\Gamma(x+1)$ = gamma function (2.28b) and x = signed number restricted in (2.32b).

(b) Series representation of the polygamma functions are

$$\psi^{(1)}(x) = \sum_{k=1}^{\infty} \frac{1}{(x+k)^2} \qquad \psi^{(2)}(x) = -2\sum_{k=1}^{\infty} \frac{1}{(x+k)^3}$$

and in general with $\alpha = (-1)^{m+1}$,

$$\psi^{(m)}(x) = \alpha m! \sum_{k=1}^{\infty} \frac{1}{(x+k)^{m+1}}$$

The series representing these functions converge very slowly and are of a very limited practical value. If, however, these series occur in the solution of a problem, they can be approximated by the rapidly converging *asymptotic series* introduced in (2.35a).

(c) Functional equations. With the restrictions placed upon x in (2.32b),

$$\psi^{(m)}(x) = -\frac{(-1)^{m+1}m!}{x^{m+1}} + \psi^{(m)}(x-1) \qquad \psi^{(m)}(x) = \frac{(-1)^{m+1}m!}{(x+1)^{m+1}} + \psi^{(m)}(x+1)$$

which are the mth derivatives of the functional equations in (2.32c).

(d) Integer and fraction arguments $(m, n = 1, 2, 3, \ldots)$

$\psi^{(1)}(n) = Z(2) - \sum_{k=1}^{n}\frac{1}{k^2}$	$\psi^{(m)}(n) = \alpha m!\left[Z(m+1) - \sum_{k=1}^{n}\frac{1}{k^{m+1}}\right]$
$\psi^{(1)}(\frac{1}{2}) = 3Z(2) - 4$	$\psi^{(m)}(\frac{1}{2}) = \alpha m!\left[(2^{m+1}-1)Z(m+1) - 2^{m+1}\right]$

where $Z(2), Z(3), \ldots, Z(m+1)$ are the Riemann zeta functions (2.26a).

(e) Reflection formulas

$$\psi^{(1)}(x) - \psi^{(1)}(x+1) = \frac{\pi^2}{\sin^2\pi x} \qquad \psi^{(m)}(x) - \psi^{(m)}(x+1) = \frac{d^m}{dx^m}\frac{\pi}{\tan\pi x}$$

$$\psi^{(1)}(u) + \psi^{(1)}(-u) = -\frac{1}{u^2} + \frac{\pi^2}{\sin^2\pi u} \qquad \psi^{(m)}(u) + \alpha\psi^{(m)}(-u) = -\frac{\alpha m!}{u^{m+1}} - \frac{d^m}{du^m}\frac{\pi}{\tan\pi u}$$

where x = signed number and $|u| < 1$.

(f) Special values

$\psi^{(1)}(-n) = \pm\infty$	$\psi^{(1)}(0) = Z(2)$	$\psi^{(1)}(1) = Z(2) - 1$
$\psi^{(m)}(-n) = \pm\infty$	$\psi^{(m)}(0) = Z(m+1)$	$\psi^{(m)}(1) = \alpha m![Z(m+1) - 1]$

Numerical values of $\psi^{(m)}(x \pm u)$ are calculated by the procedures described in (2.35).

(a) Small argument. For $0 < u \le \frac{1}{2}$, the polygamma functions can be evaluated in terms of the respective derivative of the formulas in $(2.33a)$ as

$$\frac{d\psi(u)}{du} = \psi^{(1)}(u) = -\frac{1}{2u^2} + \frac{1}{2(u+1)^2} - \frac{1}{2(1-u)^2} + \frac{\pi^2}{2\sin^2 u\pi} - \frac{d\omega}{du}$$

$$\frac{d^2\psi(u)}{du^2} = \psi^{(2)}(u) = \frac{1}{u^3} - \frac{1}{(1+u)^3} + \frac{1}{(1-u)^3} - \frac{\pi^3\cos u}{2\sin^3 u\pi} - \frac{d^2\omega}{du^2}$$

and in general with $\alpha = (-1)^{m+1}$,

$$\frac{d^m\psi(u)}{du^m} = \psi^{(m)}(u) = -\frac{\alpha m!}{2u^{m+1}} + \frac{\alpha m!}{2(1+u)^{m+1}} - \frac{m!}{2(1-u)^{m+1}} - \frac{\pi d^m}{2du^m}(\cot u\pi) - \frac{d^m\omega}{du^m}$$

(b) Examples

$$\frac{d\omega}{du} = 2ub_2 + 4u^3b_4 + 6u^5b_6 + 8u^7b_8 + \cdots + 20u^{19}b_{20} = \sum_{k=1}^{10} u^{2k-1}c_{2k}$$

$$\frac{d^2\omega}{du^2} = 2b_2 + 12u^2b_4 + 30u^4b_6 + 56u^6b_8 + \cdots + 180u^{18}b_{20} = \sum_{k=1}^{10} u^{2k}d_{2k}$$

and no error occurs in 10-digit display if b_k's are taken up to $k = 10$. In nested form,

$$\frac{d\omega}{du} = uC_2(1 + C_4u^2(1 + C_6u^2(1 + C_8u^2(1 + \cdots + C_{20}u^2))))$$

$$\frac{d^2\omega}{du^2} = D_2(1 + D_4u^2(1 + D_6u^2(1 + D_8u^2(1 + \cdots + D_{20}u^2))))$$

where $c_{2k}, d_{2k}, C_{2k}, D_{2k}$ are the factors tabulated in (c) below.

(c) Factors $c_{2k}, d_{2k}, C_{2k}, D_{2k}$

k	c_{2k}	d_{2k}	C_{2k}	D_{2k}
1	0.404 113 806 4	0.404 113 806 4	0.404 113 806 4	0.404 113 806 4
2	0.147 110 020 6	0.443 132 665 2	0.365 518 371 9	1.096 555 116 0
3	0.050 095 664 3	0.250 478 322 0	0.339 146 423 2	0.565 244 038 6
4	0.016 067 142 6	0.112 469 996 8	0.320 729 205 5	0.449 020 887 7
5	0.004 941 885 0	0.044 769 740 0	0.307 577 156 8	0.395 456 344 4
6	0.001 472 560 2	0.016 198 155 6	0.297 975 339 6	0.364 192 081 7
7	0.000 428 235 3	0.005 567 052 4	0.290 810 057 8	0.346 846 138 1
8	0.000 122 195 2	0.001 832 928 0	0.285 345 836 4	0.329 245 195 8
9	0.000 034 347 0	0.000 583 909 2	0.281 089 924 2	0.318 568 580 8
10	0.000 009 538 7	0.000 181 222 0	0.277 707 792 5	0.310 379 297 4

(d) Example. If the value of $\psi^{(1)}(\frac{1}{2})$ is desired, $d\omega/du$ is calculated first with $u = \frac{1}{2}$,

$$uc_2 = 0.202\ 056\ 903\ 1 \qquad u^7c_8 = 0.000\ 125\ 524\ 5 \qquad u^{13}c_{14} = 0.000\ 000\ 052\ 2$$

$$u^3c_4 = 0.018\ 463\ 877\ 6 \qquad u^9c_{10} = 0.000\ 009\ 652\ 1 \qquad u^{15}c_{16} = 0.000\ 000\ 003\ 7$$

$$u^5c_6 = 0.001\ 565\ 489\ 5 \qquad u^{11}c_{12} = 0.000\ 000\ 719\ 0 \qquad u^{17}c_{18} = 0.000\ 000\ 000\ 2$$

$$\frac{d\omega}{du} = \sum_{k=1}^{9} u^{2k-1}c_{2k} = 0.222\ 222\ 222\ 1, \text{ and by the relation } (a),$$

$$\psi^{(1)}(\tfrac{1}{2}) = -\frac{1}{2(0.5)^2} + \frac{1}{2(1.5)^2} - \frac{1}{2(0.5)^2} + \frac{\pi^2}{2\sin^2\frac{1}{2}\pi} - 0.222\ 222\ 222\ 1 = 0.934\ 802\ 200\ 7$$

The correct value is given by $(2.34d)$ as $\psi^{(1)}(\frac{1}{2}) = 3Z(2) - 4 = 0.934\ 802\ 200\ 5$, and the error is due to round-off only.

(e) Intermediate argument. For $\frac{1}{2} < x < 5$, the recurrent formulas in (2.34c) yield good results

$$\psi^{(1)}(r + u) = \psi^{(1)}(u) - \sum_{k=1}^{r} \frac{1}{(k + u)^2} \qquad \psi^{(1)}(r - u) = \psi^{(1)}(-u) + \sum_{k=1}^{r} \frac{1}{(k - u)^2}$$

$$\psi^{(2)}(r + u) = \psi^{(2)}(u) + 2\sum_{k=1}^{r} \frac{1}{(k + u)^3} \qquad \psi^{(2)}(r - u) = \psi^{(2)}(-u) + 2\sum_{k=1}^{r} \frac{1}{(k - u)^3}$$

where $r = 1, 2, 3, \ldots$, $\psi^{(1)}(u)$, $\psi^{(2)}(u)$ are calculated by (2.35a), and $\psi^{(1)}(-u)$, $\psi^{(2)}(-u)$ are obtained by (2.34e) in terms of the preceding values.

(f) Examples. If the values of $\psi^{(1)}(3.5)$ and of $\psi^{(1)}(-0.5)$ are desired, then by (e),

$$\psi^{(1)}(3 + 0.5) = \psi^{(1)}(0.5) - \frac{1}{1.5^2} - \frac{1}{2.5^2} - \frac{1}{3.5^2} = 0.248\,725\,103\,2$$

and by (2.34e),

$$\psi^{(1)}(-0.5) = -\frac{1}{(0.5)^2} + \frac{\pi^2}{\sin^2 \frac{1}{2}\pi} - \psi^{(1)}(0.5) = 4.934\,802\,200\,7$$

(g) Large argument If $x \geq 5$ is an integer or decimal number, the derivatives of Stirling's expansion in (2.33d) yield

$$\psi^{(1)}(x) = \frac{1}{x} - \frac{1}{2x^2} + \frac{d\phi}{dx} \qquad \psi^{(2)}(x) = -\frac{1}{x^2} + \frac{1}{x^3} + \frac{d^2\phi}{dx^2}$$

and in general with $\alpha = (-1)^{m+1}$,

$$\boxed{\psi^{(m)}(x) = \frac{\alpha(m - 1)!}{x^m}\left(1 - \frac{m}{2x}\right) + \frac{d^m\phi}{dx^m}}$$

where

$$\frac{d\phi}{dx} = \frac{1}{6x^3} - \frac{1}{30x^5} + \frac{1}{42x^7} - \frac{1}{30x^9} + \frac{5}{66x^{11}} \qquad \frac{d^2\phi}{dx^2} = -\frac{1}{2x^4} + \frac{1}{6x^6} - \frac{1}{6x^8} + \frac{3}{10x^{10}} - \frac{5}{6x^{12}}$$

$$\frac{d^m\phi}{dx^m} = \frac{\alpha}{x^m}\left[\frac{(m + 1)!}{12x^2} - \frac{(m + 3)!}{120(3!)x^4} + \frac{(m + 5)!}{252(5!)x^6} - \frac{(m + 7)!}{240(7!)x^8} + \frac{5(m + 9)!}{660(9!)x^{10}}\right]$$

If five terms of the asymptotic series are included, no error occurs on 10-digit display. Table (h) below shows the range of applications of this formula for $\psi^{(1)}(x)$.

(h) Errors of Stirling's expansion of $\psi^{(1)}(x)$

x	Correct value of $\psi^{(1)}(x)$		Approximate value of $\psi^{(1)}(x)$
−0.5	$\pi^2/2$	$= 4.934\,802\,201$ (+00)	
0	$\pi^2/6$	$= 1.644\,934\,067$ (+00)	
0.5	$\pi^2/2 - 4$	$= 9.348\,822\,005$ (−01)	Not applicable for $x \leq 1.5$
1	$\pi^2/6 - 1$	$= 6.449\,340\,673$ (−01)	
1.5	$\pi^2/2 - 4(1 + 1/3^2)$	$= 4.903\,577\,561$ (−01)	
2	$\pi^2/6 - (1 + 1/2^2)$	$= 3.949\,340\,668$ (−01)	3.949\,912\,574 (−01)
3	$\pi^2/6 - (1 + 1/2^2 + 1/3^2)$	$= 2.838\,229\,557$ (−01)	2.838\,230\,064 (−01)
4	$\pi^2/6 - (1 + 1/2^2 + 1/3^2 + 1/4^2)$	$= 2.213\,229\,557$ (−01)	2.213\,229\,587 (−01)
5	$\pi^2/6 - (1 + 1/2 + 1/3^2 + 1/4^2 + 1/5^2)$	$= 1.813\,229\,557$ (−01)	1.813\,229\,557 (−01)
10	$\pi^2/6 - (1 + 1/2^2 + 1/3^2 + \cdots + 1/10^2)$	$= 9.516\,633\,556$ (−02)	9.516\,633\,556 (−02)

3
NUMERICAL
DIFFERENCES

(a) Finite differences of a function $y = f(x)$ that is continuous and single-valued in a given interval (which can be infinite) are classified as *divided differences, forward differences, backward differences,* and *central differences.*

(b) Divided difference of first order defined as

$$f(x_k, x_{k+1}) = \frac{f(x_{k+1}) - f(x_k)}{x_{k+1} - x_k} = \frac{y_{k+1} - y_k}{x_{k+1} - x_k}$$

is the slope of the chord $k, k + 1$,

$$\tan \phi = \frac{f(x_{k+1}) - f(x_k)}{x_{k+1} - x_k}$$

where $k = 0, \pm 1, \pm 2, \pm 3, \dots$.

(c) Forward difference of first order, defined as

$$f(x_{k+1}) - f(x_k) = y_{k+1} - y_k$$

is the difference of the values of the function at x_{k+1}, x_k, denoted as

$$\Delta y_k = y_{k+1} - y_k$$

(d) Backward difference of first order, defined as

$$f(x_k) - f(x_{k-1}) = y_k - y_{k-1}$$

is the difference of the values of the function at x_k, x_{k-1}, denoted as

$$\nabla y_k = y_k - y_{k-1}$$

(e) Central difference of first order, defined as

$$f(x_{x+1/2}) - f(x_{k-1/2}) = y_{k+1/2} - y_{k-1/2}$$

is the difference of the values of the function at $x_{k+1/2}$, $x_{k-1/2}$, denoted as

$$\delta y_k = y_{k+1/2} - y_{k-1/2}$$

where $y_{k-1/2}$, $y_{k+1/2}$ are the vertical coordinates of midpoints between $k - 1, k$ and $k, k + 1$, respectively.

(f) Central means of first order, defined as

$$\frac{f(x_{k-1/2}) + f(x_{k+1/2})}{2} = \frac{y_{k-1/2} + y_{k+1/2}}{2}$$

is the arithmetic means of the values of the function at $x_{k-1/2}$ and $x_{k+1/2}$, denoted as

$$\mu y_k = \tfrac{1}{2}(y_{k-1/2} + y_{k+1/2})$$

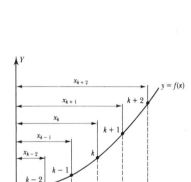

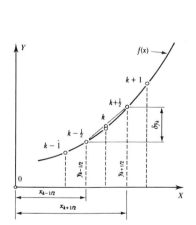

(a) Operator denoted by the general symbol $\mathbf{L}$ and preceding the function $f(a + h)$ or $f[a + (k + \frac{1}{2})h]$ called the operand indicates the nature of the operation to be performed on this function.

(b) Main properties of operators are

$$\mathbf{L}[f_1(x) + f_2(x)] = \mathbf{L}f_1(x) + \mathbf{L}f_2(x)$$

$$\mathbf{L}^m\mathbf{L}^n[f(x)] = \mathbf{L}^{m+n}[f(x)] \qquad \mathbf{L}^m[f(x)] = \mathbf{L}\{\mathbf{L}^{m-1}[f(x)]\}$$

where m, n are signed numbers.

(c) Classification of operators. In this chapter three classes of operators occur: *transformation operators* $\mathbf{E}$ *and* $\mathbf{1}$, *difference operators* Δ, ∇, δ, μ, and *derivative operator* D.

(d) Table of applications $(n = 1, 2, 3, \ldots; k = 0, \pm 1, \pm 2, \pm 3, \pm \ldots; r = 0, 1, 2, 3, \ldots)$

Unit operator $\mathbf{1}$	$\mathbf{1}y_k = y_k$	$\mathbf{1}^n y_k = y_k$
Shift operator $\mathbf{E}$	$\mathbf{E}y_k = y_{k+1}$	$\mathbf{E}^n y_k = y_{k+n}$
	$\mathbf{E}^{-1}y_k = y_{k-1}$	$\mathbf{E}^{-n}y_k = k_{k-n}$
Forward difference operator Δ	$\Delta y_k = y_{k+1} - y_k$	$\Delta^n y_k = \displaystyle\sum_{r=0}^{n} (-1)^r \binom{n}{r} y_{k+n-r}$
Backward difference operator ∇	$\nabla y_k = y_k - y_{k-1}$	$\nabla^n y_k = \displaystyle\sum_{r=0}^{n} (-1)^r \binom{n}{r} y_{k-r}$
Central difference operator δ	$\delta y_k = y_{k+1/2} - y_{k-1/2}$	$\delta^n y_k = \displaystyle\sum_{r=0}^{n} (-1)^r \binom{n}{r} y_{k+1/2-r}$
Central means operator μ	$\mu y_k = \frac{1}{2}(y_{k-1/2} + y_{k+1/2})$	$\mu^n y_k = \dfrac{1}{2^n} \displaystyle\sum_{r=0}^{n} \binom{n}{r} y_{k+1/2-r}$
Derivative operator D	$Dy_k = (dy/dx)_k$	$D^n y_k = (d^n y/dx^n)_k$

where the superscript n defines the order of the operation.

(e) Examples. If $y_k = f(x_k)$, then

$$\mathbf{1}^2 y_k = y_k \qquad\qquad \mathbf{1}^3 y_k = y_k \qquad\qquad \mathbf{E}^2 y_k = y_{k+2} \qquad\qquad \mathbf{E}^3 y_k = y_{k+3}$$

$$\Delta^2 y_k = y_{k+2} - 2y_{k+1} + y_k \qquad\qquad \Delta^3 y_k = y_{k+3} - 3y_{k+2} + 3y_{k+1} - y_k$$

$$\nabla^2 y_k = y_k - 2y_{k-1} + y_{k-2} \qquad\qquad \nabla^3 y_k = y_k - 3y_{k-1} + 3y_{k-2} - y_{k-3}$$

$$\delta^2 y_k = y_{k+1} - 2y_k + y_{k-1} \qquad\qquad \delta^3 y_k = y_{k+3/2} - 3y_{k+1/2} + 3y_{k-1/2} - y_{k-3/2}$$

$$\mu^2 y_k = \tfrac{1}{4}(y_{k-1} + 2y_k + y_{k+1}) \qquad\qquad \mu^3 y_k = \tfrac{1}{8}(y_{k-3/2} + y_{k-1/2} + 3y_{k+1/2} + y_{k+3/2})$$

$$D^2 y_k = (d^2y/dx^2)_k \qquad\qquad D^3 y_k = (d^n y/dx^n)_k$$

(f) Relations of differences $(m, n = 1, 2, 3, \ldots)$

$\Delta^n y_k = \nabla^n y_{k+n} = \delta^n y_{k+n/2}$	$\nabla^n y_k = \Delta^n y_{k-n} = \delta^n y_{k-n/2}$
$\Delta^m \nabla^n y_k = \Delta^{m+n} y_{k-n} = \delta^{m+n} y_{k+m/2-n/2}$	$\delta^n y_k = \Delta^n y_{k-n/2} = \nabla^n y_{k+n/2}$

(a) Notation. n, r = positive integers $\qquad$ $\mathbf{E}, \Delta, \nabla, \delta, \mu$ = operators (3.02)

(b) Relationships between operators

	$\mathbf{E}$	Δ	∇	δ	D
$\mathbf{E}$	$\mathbf{E}$	$\Delta + 1$	$(1 - \nabla)^{-1}$	$1 + \frac{1}{2}\delta^2 + \delta\sqrt{1 + \frac{1}{4}\delta^2}$	e^{hD}
Δ	$\mathbf{E} - 1$	Δ	$(1 - \nabla)^{-1} - 1$	$\frac{1}{2}\delta^2 + \delta\sqrt{1 + \frac{1}{4}\delta^2}$	$e^{hD} - 1$
∇	$1 - \mathbf{E}^{-1}$	$1 - (1 + \Delta)^{-1}$	∇	$-\frac{1}{2}\delta^2 + \delta\sqrt{1 + \frac{1}{4}\delta^2}$	$1 - e^{-hD}$
δ	$\mathbf{E}^{1/2} - \mathbf{E}^{-1/2}$	$\Delta(1 + \delta)^{-1/2}$	$\nabla(1 - \nabla)^{-1/2}$	δ	$2\sinh\frac{1}{2}hD$
μ	$\frac{1}{2}(\mathbf{E}^{1/2} + \mathbf{E}^{-1/2})$	$\frac{1}{2}(2 + \Delta)(1 + \Delta)^{1/2}$	$\frac{1}{2}(2 - \nabla)(1 - \nabla)^{-1/2}$	$\sqrt{1 + \frac{1}{4}\delta^2}$	$2\cosh\frac{1}{2}hD$
D	$h^{-1}\ln \mathbf{E}$	$h^{-1}\ln(1 + \Delta)$	$-h^{-1}\ln(1 - \nabla)$	$2h^{-1}\sinh^{-1}(\delta/2)$	D

(c) E series in terms of Δ, ∇, and D

$$\mathbf{E}^n = (1 + \Delta)^n = \sum_{r=0}^{n} \binom{n}{r}\Delta^r \qquad\qquad \mathbf{E}^{-n} = (1 + \Delta)^{-n} = \sum_{r=0}^{\infty} (-1)^r \binom{n}{r}\Delta^r$$

$$\mathbf{E}^n = (1 - \nabla)^{-n} = \sum_{r=0}^{\infty} \binom{n}{r}\nabla^r \qquad\qquad \mathbf{E}^{-n} = (1 - \nabla)^{-n} = \sum_{r=0}^{n} (-1)^r \binom{n}{r}\nabla^r$$

$$\mathbf{E}^n = e^{nhD} = \sum_{r=0}^{\infty} \frac{(nh)^r D^r}{r!} \qquad\qquad \mathbf{E}^{-n} = e^{-nhD} = \sum_{r=0}^{\infty} (-1)^r \frac{(nh)^r D^r}{r!}$$

(d) Δ series in terms of E and D

$$\Delta^n = (\mathbf{E} - 1)^n = \sum_{r=0}^{n} (-1)^r \mathbf{E}^{n-r} \qquad\qquad \Delta = e^{hD} - 1 = \sum_{r=0}^{\infty} \frac{h^r D^r}{r!}$$

$$\Delta^2 = (e^{hD} - 1)^2 = \sum_{r=0}^{\infty} (2^r - 2) \frac{h^r D^r}{r!} \qquad\qquad \Delta^3 = (e^{hD} - 1)^3 = \sum_{r=3}^{\infty} [3^r - (3)2^r + 3] \frac{h^r D^r}{r!}$$

$$\Delta^n = (e^{hD} - 1)^n = \sum_{r=n}^{\infty} \left[n^r - \binom{n}{r}(n - 1)^r + \binom{n}{2}(n - 2)^r - \cdots (-1)^{n+1}\binom{n}{n-1}1^r \right] \frac{h^r D^r}{r!}$$

(e) ∇ series in terms of E and D

$$\nabla^n = (1 - \mathbf{E}^{-1})^n = \sum_{r=0}^{n} (-1)^r \mathbf{E}^{-r} \qquad\qquad \nabla = 1 - e^{-hD} = \sum_{r=1}^{\infty} (-1)^{r+1} \frac{h^r D^r}{r!}$$

$$\nabla^2 = (1 - e^{-hD})^2 = \sum_{r=2}^{\infty} (-1)^r (2^r - 2) \frac{h^r D^r}{r!}$$

$$\nabla^3 = (1 - e^{-hD})^3 = \sum_{r=3}^{\infty} (-1)^{r+1}[3^r - (3)2^r + 3] \frac{h^r D^r}{r!}$$

$$\nabla^n = (1 - e^{-hD})^n = \sum_{r=n}^{\infty} (-1)^\lambda \left[n^r - \binom{n}{1}(n - 1)^r + \binom{n}{2}(n - 2)^2 - \cdots (-1)^{n+1}\binom{n}{n-1}1^r \right] \frac{h^r D^r}{r!}$$

where $\lambda = r$ if n is even and $\lambda = r + 1$ if n is odd.

(f) δ series in terms of E and D

$$\delta^n = (\mathbf{E}^{1/2} - \mathbf{E}^{-1/2})^n = \sum_{r=0}^{n} (-1)^{1/2} \binom{n}{r} \mathbf{E}^{n/2-r} \qquad\qquad \delta = 2\sinh\frac{1}{2}hD = 2\sum_{r=1}^{\infty} \frac{(\frac{1}{2}h)^r D^r}{r!}$$

$$\delta^n = 2^n \sinh^n \frac{1}{2}hD = 2\sum_{r=n}^{\infty} \left[n^r - \binom{n}{1}(n - 2)^r + \binom{n}{2}(n - 4)^r - \cdots (-1)^{n+1}\binom{n}{n}(-n)^r \right] \frac{(\frac{1}{2}h)^r D^r}{r!}$$

(a) Types of tables. The difference tables shown in (d) to (f) below are the flowcharts for the calculation of finite differences. In general, two formats of these tables are available: the diagonal format and the horizontal format.

(b) Construction of tables. First, the numerical values of x_k and y_k are recorded in the first two columns of the respective table. Next, the respective differences are successively calculated from the nearest column on the left. This process is repeated to the desired order.

(c) Three rules must be observed in the construction of the difference table:
   (1) The computation of the nth-order difference requires $n + 1$ values of x_k and y_k.
   (2) The nth-order differences of an nth-order polynomial in x are the same values, and the $(n + 1)$th-order differences of the same polynomial are zeros.
   (3) The sum of numbers in one column equals the difference of the top and bottom numbers in the preceding column

(d) Forward differences

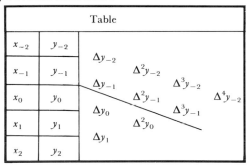

Table					
x_{-2}	y_{-2}				
		Δy_{-2}			
x_{-1}	y_{-1}		$\Delta^2 y_{-2}$		
		Δy_{-1}		$\Delta^3 y_{-2}$	
x_0	y_0		$\Delta^2 y_{-1}$		$\Delta^4 y_{-2}$
		Δy_0		$\Delta^3 y_{-1}$	
x_1	y_1		$\Delta^2 y_0$		
		Δy_1			
x_2	y_2				

Example					
-4	200				
		-180			
-2	20		162		
		-18		654	
0	2		816		-1068
		798		-414	
2	800		402		
		1200			
4	2000				

(e) Backward differences

Table					
x_{-2}	y_{-2}				
		∇y_{-1}			
x_1	y_1		$\nabla^2 y_0$		
		∇y_0		$\nabla^3 y$	
x_0	y_0		$\nabla^2 y_1$		$\nabla^4 y_2$
		∇y_1		$\nabla^3 y_2$	
x_1	y_1		$\nabla^2 y_2$		
		∇y_2			
x_2	y_2				

Example					
-4	200				
		-180			
-2	20		162		
		-18		654	
0	2		816		-1068
		798		-414	
2	800		402		
		1200			
4	2000				

(f) Central differences

Table					
x_{-2}	y_{-2}				
		$\delta y_{-3/2}$			
x_{-1}	y_{-1}		$\delta^2 y_{-1}$		
		$\delta y_{-1/2}$		$\delta^3 y_{-1/2}$	
x_0	y_0		$\delta^2 y_0$		$\delta^4 y_0$
		$\delta y_{1/2}$		$\delta^3 y_{1/2}$	
x_1	y_1		$\delta^2 y_1$		
		$\delta y_{3/2}$			
x_2	y_2				

Example					
-4	200				
		-180			
-2	20		162		
		-18		654	
0	2		816		-1068
		798		-414	
2	800		402		
		1200			
4	2000				

(a) Derivative operator. By definition, the first-order derivative of $y_k = f(x_k)$ is

$$Dy_k = \frac{dy_k}{dx} = \lim_{h \to 0} \frac{\Delta y_k}{h}$$

and the higher-order derivatives are

$$D^2 y_k = \frac{d^2 y_k}{dx^2} = \lim_{h \to 0} \frac{\Delta^2 y_k}{h^2}, \dots, D^n y_k = \frac{d^n y_k}{dx^n} = \lim_{h \to 0} \frac{\Delta^n y_k}{h^n}$$

where $\Delta y_k, \Delta^2 y_k, \dots, \Delta^n y_k$ are forward differences of y_k and D is the derivative operator used as

$$D = \frac{d}{dx}, D^2 = \frac{d^2}{dx^2}, \dots, D^n = \frac{d^n}{dx^n}$$

(b) Rules of difference and differential calculus bear close resemblance. If $u = f(x)$, $v = g(x)$, $w = g(x + h)$ and h, c are constants, then

$D[c]$	$=$	0	$\Delta[c]$	$=$	0
$D[x]$	$=$	1	$\Delta[x]$	$=$	h
$D\left[\dfrac{1}{x}\right]$	$=$	$-\dfrac{1}{x^2}$	$\Delta\left[\dfrac{1}{x}\right]$	$=$	$-\dfrac{h}{x(x+h)}$
$D[cu]$	$=$	$c\,Du$	$\Delta[cu]$	$=$	$c\,\Delta u$
$D[u+v]$	$=$	$Du + Dv$	$\Delta[u+v]$	$=$	$\Delta u + \Delta v$
$D[uv]$	$=$	$u\,Dv + v\,Du$	$\Delta[uv]$	$=$	$u\,\Delta v + v\,\Delta u + \Delta u\,\Delta v$
$D\left[\dfrac{u}{v}\right]$	$=$	$\dfrac{v\,Du - u\,Dv}{v^2}$	$\Delta\left[\dfrac{u}{v}\right]$	$=$	$\dfrac{v\,\Delta u - u\,\Delta v}{vw}$

(c) Examples

$$\Delta[x^2 + 1] = [(x + h)^2 + 1] - [x^2 + 1] = 2hx + h^2$$

$$\Delta\left[\frac{1}{x^2 + 1}\right] = \left[\frac{1}{(x + h)^2 + 1}\right] - \left[\frac{1}{x^2 + 1}\right] = -\frac{2hx + h^2}{(x^2 + 1)(x^2 + 2hx + h^2 + 1)}$$

(d) Products and quotients. If $u_k = f(x + kh)$ and $v_k = g(x + kh)$, then

$\Delta[u_k v_k] = u_k\,\Delta v_k + v_{k+1}\,\Delta u_k$	$\Delta[u_k^2] = (u_k + u_{k+1})\,\Delta u_k$
$\Delta\left[\dfrac{1}{u_k}\right] = -\dfrac{\Delta u_k}{u_k u_{k+1}}$	$\Delta\left[\dfrac{u_k}{v_k}\right] = \dfrac{v_k\,\Delta u_k - u_k\,\Delta v_k}{v_k v_{k+1}}$

where $u_k \neq 0$ and $v_k \neq 0$.

(e) Examples

$$\Delta\left[\frac{1}{x + 1}\right] = \frac{(x + h + 1) - (x + 1)}{(x + 1)(x + h + 1)} = -\frac{h}{(x + 1)(x + h + 1)}$$

$$\Delta\left[\frac{x + 1}{x - 1}\right] = \frac{(x - 1)h - (x + 1)h}{(x - 1)(x + h - 1)} = -\frac{2h}{(x - 1)(x + h - 1)}$$

(a) Notation. a, b, c = constants h = spacing m, n, r = positive integers

$$x^{(m)} = x(x - h)(x - 2h) \cdots [x - (m - 1)h] \qquad x^{(-m)} = \frac{1}{(x + h)(x + 2h)(x + 3h) \cdots (x + mh)}$$

(b) Algebraic functions $(m \geq n, r \geq\ )$

$$\Delta[x^{(m)}] = mhx^{(m-1)} \qquad \Delta^n[x^{(m)}] = \frac{h^n m!}{(m - n + 1)!} x^{(m-n)}$$

$$\Delta[x^{(-m)}] = - mhx^{(-m-1)} \qquad \Delta^n[x^{(-m)}] = - \frac{h^n m!}{(m - n + 1)!} x^{(-m-n)}$$

$$\Delta\left[\binom{x}{r}\right] = h\binom{x}{r - 1} \qquad \Delta^n\left[\binom{x}{r}\right] = h^n\binom{x}{r - n}$$

(c) Transcendental functions

$$\Delta[a^{bx}] = (a^{bh} - 1)a^{bx} \qquad \Delta[\sin(a + bx)] = 2 \sin(\tfrac{1}{2}bh) \sin[bx + a + \tfrac{1}{2}(bh + \pi)]$$

$$\Delta[e^{bx}] = (e^{bx} - 1)e^{bx} \qquad \Delta[\cos(a + bx)] = 2 \sin(\tfrac{1}{2}bh) \cos[bx + a + \tfrac{1}{2}(bh + \pi)]$$

$$\Delta^n[a^{bx}] = (a^{bh} - 1)^n a^{bx} \qquad \Delta^n[\sin(a + bx)] = 2^n \sin^n(\tfrac{1}{2}bh) \sin[bx + a + \tfrac{1}{2}(nbh + n\pi)]$$

$$\Delta^n[e^{bx}] = (e^{bx} - 1)^n e^{bx} \qquad \Delta^n[\cos(a + bx)] = 2^n \sin^n(\tfrac{1}{2}bh) \cos[bx + a + \tfrac{1}{2}(nbh + n\pi)]$$

(d) Leibnitz' rule. If $u_0 = f(x)$, $u_1 = f(x + h), \ldots, u_k = f(x + kh)$, $v_0 = g(x)$, $v_1 = g(x + h)$, $\ldots, v_k = g(x + kh)$, then

$$\Delta^n[u_0 v_0] = \sum_{k=0}^{n} (-1)^{n+k}\binom{n}{k}u_k v_k \qquad (k = 0, 1, 2, 3, \ldots)$$

(e) Example. If $u_0 = \sin x$ and $v_0 = \cos x$, then by (d),

$$\Delta^2[u_0 v_0] - \tfrac{1}{2}\sin 2x - \sin 2(x + h) + \tfrac{1}{2}\sin 2(x + 2h)$$

(f) Error ϵ in the difference tables used for the calculation of differences spreads as shown below.

k	ϵ	$\Delta\epsilon$	$\Delta^2\epsilon$	$\Delta^3\epsilon$	$\Delta^4\epsilon$	$\Delta^5\epsilon$	$\Delta^6\epsilon$
-3	0						ϵ
		0				ϵ	
-2	0		0		ϵ		-6ϵ
		0		ϵ		-5ϵ	
-1	0		ϵ		-4ϵ		15ϵ
		ϵ		-3ϵ		10ϵ	
0	ϵ		-2ϵ		6ϵ		-20ϵ
		$-\epsilon$		3ϵ		-10ϵ	
1	0		ϵ		-4ϵ		15ϵ
		0		$-\epsilon$		5ϵ	
2	0		0		ϵ		-6ϵ
		0				$-\epsilon$	
3	0						ϵ

(a) **Process of interpolation** is used for the development of a substitute function $\bar{y}(x)$ which closely approximates a more complicated function $y(x)$ in a given interval and coincides (collocates) with $y(x)$ at certain specific points. The substitute function of a polynomial form in terms of finite differences is called the *finite-difference collocation polynomial*. Collocation polynomials of finite-difference type are of general or equidistant spacing. All formulas in (3.08) to (3.13) are exact and yield the same values if the function to be represented is a polynomial. If the function is not a polynomial, the remainder R_{r+1} must be added.

(b) **Newton–Gregory formulas** (3.10) are used mainly for interpolations at the near or far end of the interval.

(c) **Central difference formulas** (3.11) to (3.13) are in general preferred over those involving forward and backward differences and are particularly useful for interpolations at intermediate points of the interval.

3.08   LAGRANGE'S INTERPOLATION FORMULAS, GENERAL SPACING

(a) **Lagrange's formula.** In terms of $r + 1$ values of x_k and y_k, which are not necessarily equally spaced,

$$\boxed{y(x) = y_0\lambda_0(x) + y_1\lambda_1(x) + \cdots + y_r\lambda_r(x) + R_{r+1} = \bar{y}(x) + R_{r+1}}$$

where $\lambda_k(x) = \displaystyle\prod_{\substack{j=0 \\ j\neq k}}^{r} \frac{x - x_j}{x_k - x_j} = \frac{(x - x_0)(x - x_1) \cdots (x - x_{k-1})(x - x_{k+1}) \cdots (x - x_r)}{(x_k - x_0)(x_k - x_1) \cdots (x_k - x_{k-1})(x_k - x_{k+1}) \cdots (x_k - x_r)}$

are the Lagrange multipliers and the remainder

$$\boxed{R_{r+1} = \frac{(x - x_0)(x - x_1)(x - x_2) \cdots (x - x_r)}{(r + 1)!} [\max D^{r+1}f(\eta)]}$$

The sum of Lagrange's multipliers is equal to 1 for all values of x in the range of interpolation, and η is the argument in $[x_0, x_r]$, which makes the value of the derivative a maximum.

(b) **Example.** If

x_k	0	1	2	4
y_k	-12	-12	-24	-60

$$\bar{y}(x) = \frac{(x - 1)(x - 2)(x - 4)}{(0 - 1)(0 - 2)(0 - 4)}(-12) + \frac{x(x - 2)(x - 4)}{1(1 - 2)(1 - 4)}(-12) + \frac{x(x - 1)(x - 4)}{2(2 - 1)(2 - 4)}(-24)$$

$$+ \frac{x(x - 1)(x - 2)}{4(4 - 1)(4 - 2)}(-60) = x^3 - 9x^2 + 8x - 12$$

which must check: $\bar{y}(0) = -12$, $\bar{y}(+1) = -12$, $\bar{y}(+2) = -24$, $\bar{y}(+4) = -60$.

(c) **Special formulas for equal spacing** h in terms of $u = (x - x_0)/h$ are

$$\bar{y}(x) = \frac{a}{2}\left(-\frac{y_{-1}}{1 + u} + \frac{2y_0}{u} + \frac{y_1}{1 - u}\right) \qquad a = (1 - u)u(1 + u)$$

$$\bar{y}(x) = \frac{b}{24}\left(\frac{y_{-2}}{2 + u} - \frac{4y_{-1}}{1 + u} + \frac{6y_0}{u} + \frac{4y_1}{1 - u} - \frac{y_2}{2 - u}\right) \qquad b = (2 - u)(1 - u)u(1 + u)(2 + u)$$

where the reaminders are $R_3 = 0.065\Delta^3 y_0$ and $R_5 = 0.012\Delta^5 y_0$ for $|u| \leq 1$.

(a) Newton's divided-difference formula in terms of x_k, y_k defined in (3.08a) is

$$\bar{y}(x) = y_0 + (x - x_0)\Delta_{11} + (x - x_0)(x - x_1)\Delta_{22} + \cdots [(x - x_0)(x - x_1) \cdots + (x - x_{r-1})]\Delta_{rr}$$

where $\quad \Delta_{11} = \dfrac{y_1 - y_0}{x_1 - x_0}, \Delta_{12} = \dfrac{y_2 - y_0}{x_2 - x_0}, \ldots, \Delta_{1r} = \dfrac{y_r - y_0}{x_r - x_0}$

$$\Delta_{22} = \frac{\Delta_{12} - \Delta_{11}}{x_2 - x_1}, \Delta_{23} = \frac{\Delta_{13} - \Delta_{11}}{x_3 - x_0}, \ldots, \Delta_{2r} = \frac{\Delta_{1r} - \Delta_{11}}{x_r - x_1}$$

$$\Delta_{33} = \frac{\Delta_{23} - \Delta_{22}}{x_3 - x_2}, \Delta_{34} = \frac{\Delta_{24} - \Delta_{22}}{x_4 - x_2}, \ldots, \Delta_{3r} = \frac{\Delta_{2r} - \Delta_{22}}{x_r - x_2}$$

. . . .

The flowchart of the calculation of divided differences is shown in (*b*) below.

(b) Divided-difference table

x_k	y_k	$x_k - x_0$	$y_k - y_0$	Δ_{1k}	$x_k - x_1$	$\Delta_{1k} - \Delta_{11}$	Δ_{2k}	$x_k - x_2$	$\Delta_{2k} - \Delta_{22}$	Δ_{3k}
x_0	y_0									
x_1	y_1	$x_1 - x_0$	$y_1 - y_0$	Δ_{11}						
x_2	y_2	$x_2 - x_0$	$y_2 - y_0$	Δ_{12}	$x_2 - x_1$	$\Delta_{12} - \Delta_{11}$	Δ_{22}			
x_3	y_3	$x_3 - x_0$	$y_3 - y_0$	Δ_{13}	$x_3 - x_1$	$\Delta_{13} - \Delta_{11}$	Δ_{23}	$x_3 - x_2$	$\Delta_{23} - \Delta_{22}$	Δ_{33}
x_4	y_4	$x_4 - x_0$	$y_4 - y_0$	Δ_{14}	$x_4 - x_1$	$\Delta_{14} - \Delta_{11}$	Δ_{24}	$x_4 - x_2$	$\Delta_{24} - \Delta_{22}$	Δ_{34}

(c) Example. If x_k, y_k are the same as in (3.08b), then the difference table prepared according to (*b*) above is

x_k	y_k	$x_k - x_0$	$y_k - y_0$	Δ_{1k}	$x_k - x_1$	$\Delta_{1k} - \Delta_{11}$	Δ_{2k}	$x_k - x_2$	$\Delta_{2k} - \Delta_{22}$	Δ_{3k}
0	−12									
1	−12	1	0	0						
2	−24	2	−12	−6	1	−6	−6			
4	−60	4	−48	−12	3	−12	−4	2	2	1

By (*a*) in terms of y_0, Δ_{11}, Δ_{22}, and Δ_{33},

$$\bar{y}(x) = -12 + (x - 0)(0) + (x - 0)(x - 1)(-6) + (x - 0)(x - 1)(x - 2)(+1)$$
$$= x^3 - 9x^2 + 8x - 12$$

which is the same result as in (3.08b).

(d) Comparison. The advantage of Lagrange's method (3.08) is the absence of the difference table. The disadvantages are the difficulty of error estimation if $y(x)$ is not given in analytical form and the difficulty in subtraction of almost equal numbers for closely spaced arguments. Newton's method requires the construction of a difference table, but the error can be estimated by checking the first neglected term. Also the addition of one or more points requires a complete recalculation of all multipliers in Lagrange's methods, whereas in Newton's method only the higher differences must be computed and their polynomial added to the previously computed polynomial.

(a) Argument. If the increment of x is a constant value h so that $x_k = a + kh$, where $a = $ constant and $k = 0, \pm1, \pm2, \pm3, \ldots, \pm r$, then the substitute function may be taken in a polynomial form (collocation polynomial) whose coefficients can be expressed in forward, backward, or central differences of ascending order as shown below and in (3.11) to (3.13), where $u = (x - a)/h$,

$$u^{(1)} = u, u^{(2)} = u(u - 1), \ldots, u^{(r)} = u(u - 1)(u - 2) \cdots (u + 1 - r)$$

and $y_0 = f(a), y_1 = f(a + h), \ldots, y_k = f(a + kh), \ldots, y_r = f(a + rh)$

(b) Newton–Gregory forward difference interpolation formula in standard form in terms of u and y_k defined in (a) as

$$y(x) = y_0 + \frac{u^{(1)}}{1!}\Delta y_0 + \frac{u^{(2)}}{2!}\Delta^2 y_0 + \cdots + \frac{u^{(r)}}{r!}\Delta^r y_0 + R_{r+1} = \bar{y}(x) + R_{r+1}$$

where $y_0 = f(a)$ and the remainder R_{r+1} can be estimated as

$$|R_{r+1}| \leq \left|\frac{u^{(r+1)}}{(r+1)!}h^{r+1}[\max D^{r+1}f(\eta)] \cong \frac{u^{(r+1)}}{(r+1)!}\Delta^{r+1}y_0\right|$$

where η is an argument in $[a, a + rh]$ which if inserted in the $(r + 1)$st derivative of $f(x)$ makes the value of this derivative a maximum.

For *CA*,
$$\bar{y}(x) = y_0 + u\left(\Delta y_0 - \frac{1-u}{2}\left(\Delta^2 y_0 - \frac{2-u}{3}\left(\Delta^3 y_0 - \cdots - \frac{r-1-u}{r}\Delta^r y_0\right)\right)\right)$$

where $\Delta y_0, \Delta^2 y_0, \Delta^3 y_0, \ldots, \Delta^r y_0$ are the forward differences of first, second, third, $\ldots, r$th order at x_0 [table (3.04d)].

(c) Newton–Gregory backward difference interpolation formula in standard form in terms of u and y_k defined in (a) is

$$y(x) = y_0 + \frac{u^{(1)}}{1!}\nabla y_0 + \frac{(u+1)^{(2)}}{2!}\nabla^2 y_0 + \cdots + \frac{(u+r-1)^{(r)}}{r!}\nabla^r y_0 + R_{r+1} = \bar{y}(x) + R_{r+1}$$

where $y_0 = f(a)$ and the remainder R_{r+1} can be estimated as

$$|R_{r+1}| \leq \left|\frac{(u+r)^{(r+1)}}{(r+1)!}h^{r+1}[\max D^{r+1}f(\eta)] \cong \frac{(u+r)^{(r+1)}}{(r+1)!}\nabla^{r+1}y_0\right|$$

where η and $\max D^{(r+1)}f(\eta)$ have the same meaning as in (b).

For *CA*,
$$\bar{y}(x) = y_0 + u\left(\nabla y_0 + \frac{u+1}{2}\left(\nabla^2 y_0 + \frac{u+2}{3}\left(\nabla^3 y_0 + \cdots + \frac{u+r-1}{r}\nabla^r y_0\right)\right)\right)$$

where $\nabla y_0, \nabla^2 y_0, \nabla^3 y_0, \ldots, \nabla^r y_0$ are the backward differences of first, second, $\ldots, r$th order at x_0 [table (3.04e)].

(a) Notation. a, h, k, y_k are defined in (3.10a)

$$u = (x - a)/h \qquad v = 1 - u \qquad\qquad \delta = \text{central difference operator} \quad (3.02d)$$

$$u^{(k)} = u(u - 1)(u - 2) \cdots (u - k + 1) \qquad v^{(k)} = v(v - 1)(v - 2) \cdots (v - k + 1)$$

(b) Open-form formula $(r = 2, 4, 6, \ldots)$

$$
y(x) = \frac{v}{1!} y_0 + \frac{u}{1!} y_1 + \frac{(v + 1)^{(3)}}{3!} \delta^2 y_0 + \frac{(u + 1)^{(3)}}{3!} \delta^2 y_1 + \frac{(v + 2)^{(5)}}{5!} \delta^4 y_0 + \frac{(u + 2)^{(5)}}{5!} \delta^4 y_1
$$

$$
+ \cdots + \frac{[v + \frac{1}{2}(r - 2)]^{(r-1)}}{(r - 1)!} \delta^{r-2} y_0 + \frac{[u + \frac{1}{2}(r - 2)]^{(r-1)}}{(r - 1)!} \delta^{r-2} y_1 + R_r
$$

where $y_0, \delta^2 y_0, \delta^4 y_0, \ldots, y_1, \delta^2 y_1, \delta^4 y_1, \ldots$ are the numerical values taken from the difference table (3.04f) according to the flowchart (e) shown below.

(c) Nested-form formula

$$
y(x) = vy_0 + uy_1 - C_0(\lambda_1 - C_1(\lambda_2 - C_2(\lambda_3 - C_3(\lambda_4 - \cdots)))) + R_r
$$

where

$$C_0 = \frac{uv}{2 \cdot 3} \qquad C_1 = \frac{(u + 1)(v + 1)}{4 \cdot 5} \qquad C_2 = \frac{(u + 2)(v + 2)}{6 \cdot 7}$$

$$C_3 = \frac{(u + 3)(v + 3)}{8 \cdot 9} \qquad\qquad C_k = \frac{(u + k)(v + k)}{(2k + 2)(2k + 3)}$$

$$\lambda_1 = (v + 1)\delta^2 y_0 + (u + 1)\delta^2 y_1 \qquad \lambda_2 = (v + 2)\delta^4 y_0 + (u + 2)\delta^4 y_1$$

$$\lambda_3 = (v + 3)\delta^6 y_0 + (u + 3)\delta^6 y_1 \qquad \lambda_k = (v + k)\delta^{2k} y_0 + (u + k)\delta^{2k} y_1$$

(d) Remainder in (b) and (c) is

$$|R_r| \leq \left| \frac{(u + \frac{1}{2}r)^{(r)}}{r!} h^r [\max D^r f(\eta)] \right|$$

where η is defined in (3.10b).

(e) Flowchart

k	y_k	δy_k	$\delta^2 y_k$	$\delta^3 y_k$	$\delta^4 y_k$	$\delta^5 y_k$	$\cdots$
$\cdots$	$\cdots$		$\cdots$		$\cdots$		$\cdots$
		$\cdots$		$\cdots$		$\cdots$	
-2	$\cdots$		$\cdots$		$\cdots$		$\cdots$
		$\cdots$		$\cdots$		$\cdots$	
-1	$\cdots$		$\cdots$		$\cdots$		$\cdots$
		$\cdots$		$\cdots$		$\cdots$	
0	y_0		$\delta^2 y_0$		$\delta^4 y_0$		$\cdots$
		$\cdots$		$\cdots$		$\cdots$	
$+1$	y_1		$\delta^2 y_1$		$\delta^4 y_1$		$\cdots$
		$\cdots$		$\cdots$		$\cdots$	
$+2$	$\cdots$		$\cdots$		$\cdots$		$\cdots$
		$\cdots$		$\cdots$		$\cdots$	
$\cdots$	$\cdots$		$\cdots$		$\cdots$		$\cdots$

(a) Notation. a, h, k, y_k are defined in $(3.10a)$

$$u = (x - a)/h \qquad v = u - \tfrac{1}{2} \qquad \delta = \text{central difference operator} \quad (3.02d)$$

$$u^{(k)} = u(u - 1)(u - 2) \cdots (u - k + 1)$$

(b) Open-form formula $(r = 2, 4, 6, \ldots)$

$$y(x) = \tfrac{1}{2}(y_0 + y_1) + v\delta y_{1/2} + \frac{u^{(2)}}{2!}\lambda_2 + \frac{u^{(2)}}{3!}v\delta^3 y_{1/2} + \frac{(u+1)^{(4)}}{4!}\lambda_4 + \frac{(u+1)^{(4)}}{5!}v\delta^5 y_{1/2}$$

$$+ \cdots + \frac{(u+r-5)^{(r-2)}}{(r-2)!}\lambda_{r-2} + \frac{(u+r-5)^{(r-2)}}{(r-1)!}v\delta^{r-1}y_{1/2} + R_r$$

where $y_0, \delta^2 y_0, \delta^4 y_0, \ldots, y_1, \delta^2 y_1, \delta^4 y_1, \ldots$ and $\delta y_{1/2}, \delta^3 y_{1/2}, \ldots$ are the numerical values taken from the difference table $(3.04f)$ according to the flowchart (e) and $\lambda_2, \lambda_4, \ldots$ are the equivalents defined in (c) below.

(c) Nested-form formula

$$y(x) = \tfrac{1}{2}(y_0 + y_1) + v\left(\delta y_{1/2} + C_1\left(\lambda_2 + \frac{v}{3}\left(\delta^3 y_{1/2} + C_2\left(\lambda_4 + \frac{v}{5}(\delta^5 y_{1/2} + \cdots)\right)\right)\right)\right) + R_r$$

where

$$C_1 = -\frac{u(1-u)}{2v} \qquad C_2 = -\frac{(1+u)(2-u)}{4v} \qquad C_3 = -\frac{(2+u)(3-u)}{6v}$$

$$C_4 = -\frac{(3+u)(4-u)}{8v} \qquad\qquad C_k = -\frac{(k-1+u)(k-u)}{2kv}$$

$$\lambda_2 = \tfrac{1}{2}(\delta^2 y_0 + \delta^2 y_1) \qquad\qquad \lambda_4 = \tfrac{1}{2}(\delta^4 y_0 + \delta^4 y_1)$$

$$\lambda_6 = \tfrac{1}{2}(\delta^6 y_0 + \delta^6 y_1) \qquad\qquad \lambda_{2k} = \tfrac{1}{2}(\delta^{2k} y_0 + \delta^{2k} y_1)$$

(d) Remainder in (b) and (c) is

$$|R_r| \le \left| \frac{(u + \tfrac{1}{2}r)^{(r)} + [u + \tfrac{1}{2}(r-2)]^{(r)}}{r!} h^r [\max D^r f(\eta)] \right|$$

where η is defined in $(3.10b)$.

(e) Flowchart

k	y_k	δy_k	$\delta^2 y_k$	$\delta^3 y_k$	$\delta^4 y_k$	$\delta^5 y_k$	$\cdots$
$\cdots$	$\cdots$		$\cdots$		$\cdots$		$\cdots$
		$\cdots$		$\cdots$		$\cdots$	
-2	$\cdots$		$\cdots$		$\cdots$		$\cdots$
		$\cdots$		$\cdots$		$\cdots$	
-1	$\cdots$		$\cdots$		$\cdots$		$\cdots$
		$\cdots$		$\cdots$		$\cdots$	
0	y_0		$\delta^2 y_0$		$\delta^4 y_0$		$\cdots$
		$\delta y_{1/2}$		$\delta^3 y_{1/2}$		$\delta^5 y_{1/2}$	
$+1$	y_1		$\delta^2 y_1$		$\delta^4 y_1$		$\cdots$
		$\cdots$		$\cdots$		$\cdots$	
$+2$	$\cdots$		$\cdots$		$\cdots$		$\cdots$
		$\cdots$		$\cdots$		$\cdots$	
$\cdots$	$\cdots$		$\cdots$		$\cdots$		$\cdots$

(a) Notation.

a, h, k, y_k are defined in $(3.10a)$

$$u = (x - a)/h \qquad\qquad \delta = \text{central difference operator} \quad (3.02d)$$

$$u^{(k)} = u(u - 1)(u - 2) \cdots (u - k + 1)$$

(b) Open-form formula $(r = 2, 4, 6, \ldots)$

$$y(x) = y_0 + \frac{u^{(1)}}{1!}\,\omega_1 + \frac{u^{(1)}}{2!}\,u\delta^2 y_0 + \frac{(u + 1)^{(3)}}{3!}\,\omega_3 + \frac{(u + 1)^{(3)}}{3!}\,u\delta^4 y_0 + \frac{(u + 2)^{(5)}}{5!}\,\omega_5$$

$$+ \cdots + \frac{(u + r - 5)^{(r-2)}}{(r - 2)!}\,u\delta^{r-1} y_0 + \frac{(u + r - 4)^{(r-1)}}{(r - 1)!}\,\omega_{r-1} + R_r$$

where $y_0, \delta^4 y_0, \ldots, \delta y_{1/2}, \delta^3 y_{1/2}, \delta^5 y_{1/2}, \ldots, \delta y_{-1/2}, \delta^3 y_{-1/2}, \delta^5 y_{-1/2}, \ldots$ are the numerical values taken from the difference table $(3.04f)$ according to the flowchart (e) and $\omega_1, \omega_3, \omega_5, \ldots$ are the equivalents defined in (c) below.

(c) Nested-form formula

$$y(x) = y_0 + u\left(\omega_1 + \frac{u}{2}\left(\delta^2 y_0 + C_2\left(\omega_3 + \frac{u}{4}\left(\delta^4 y_0 + C_4\left(\omega_5 + \frac{u}{6}\left(\delta^6 y_0 + \cdots\right)\right)\right)\right)\right)\right) + R_r$$

where

$$C_2 = \frac{u^2 - 1}{3u} \qquad C_4 = \frac{u^2 - 2^2}{5u} \qquad C_6 = \frac{u^2 - 3^2}{7u}$$

$$C_{2k} = \frac{u^2 - k^2}{(2k - 1)u}$$

$$\omega_1 = \tfrac{1}{2}(\delta y_{-1/2} + \delta y_{1/2}) \qquad\qquad \omega_3 = \tfrac{1}{2}(\delta^3 y_{-1/2} + \delta^3 y_{1/2})$$

$$\omega_5 = \tfrac{1}{2}(\delta^5 y_{-1/2} + \delta^5 y_{1/2}) \qquad\qquad \omega_{2k-1} = \tfrac{1}{2}(\delta^{2k-1} y_{-1/2} + \delta^{2k-1} y_{1/2})$$

(d) Remainder in (b) and (c) is

$$|R_r| \le \left|\frac{(u + \tfrac{1}{2}r)^{(r)}}{r!}\, h^r [\max D^r f(\eta)]\right|$$

where η is defined in $(3.10b)$.

(e) Flowchart

k	y_k	δy_k	$\delta^2 y_k$	$\delta^3 y_k$	$\delta^4 y_k$	$\delta^5 y_k$	$\cdots$
$\cdots$	$\cdots$		$\cdots$		$\cdots$		$\cdots$
		$\cdots$		$\cdots$		$\cdots$	
-2	$\cdots$		$\cdots$		$\cdots$		$\cdots$
		$\cdots$		$\cdots$		$\cdots$	
-1	$\cdots$		$\cdots$		$\cdots$		$\cdots$
		$\delta y_{-1/2}$		$\delta^3 y_{-1/2}$		$\delta^5 y_{-1/2}$	
0	y_0		$\delta^2 y_0$		$\delta^4 y_0$		$\cdots$
		$\delta y_{1/2}$		$\delta^3 y_{1/2}$		$\delta^5 y_{1/2}$	
$+1$	$\cdots$		$\cdots$		$\cdots$		$\cdots$
		$\cdots$		$\cdots$		$\cdots$	
$+2$	$\cdots$		$\cdots$		$\cdots$		$\cdots$
		$\cdots$		$\cdots$		$\cdots$	
$\cdots$	$\cdots$		$\cdots$		$\cdots$		

(a) Statement of problem. The numerical values of $y_{-2} = \sin 0.4$, $y_{-1} = \sin 0.5$, $y_0 = \sin 0.6$, $y_{+1} = \sin 0.7$, $y_{+2} = \sin 0.8$, and their differences are tabulated in (*b*) below. The values of $\sin 0.456$, $\sin 0.656$, and $\sin 0.856$ are desired.

(b) Difference table

k	x	$y_k = \sin x_k$	δy_k	$\delta^2 y_k$	$\delta^3 y_k$	$\delta^4 y_k$
-2	0.4	0.389 418				
			0.090 008			
-1	0.5	0.479 426		$-0.004\ 791$		
			0.085 217		$-0.000\ 852$	
0	0.6	0.564 643		$-0.005\ 642$		0.000 056
			0.079 575		$-0.000\ 795$	
$+1$	0.7	0.644 218		$-0.006\ 437$		
			0.073 138			
$+2$	0.8	0.717 356				

(c) Newton–Gregory forward formula. For $\sin 0.456$ by $(3.10b)$ with

$$u = (0.456 - 0.400)/0.1 = 0.56 \text{ and } h = 0.1$$

and in terms of the numerical values of (*b*),

$$\sin 0.456 = y_{-2} + 0.56\left(\Delta y_{-2} - \frac{0.44}{2}\left(\Delta^2 y_{-2} - \frac{1.44}{3}\left(\Delta^3 y_{-2} - \frac{2.44}{4}\Delta^4 y_{-2}\right)\right)\right) + R_5$$

$$= 0.440\ 360 + R_5 \qquad \text{(correct value } 0.440\ 360)$$

where the remainder by the formula given in $(3.10b)$ is

$$|R_5| \le \left|\binom{0.56}{5}0.1^5 \cos 0.4\right| = |2.3 \times 10^{-7}|$$

(d) Everett's formula. For $\sin 0.656$ by $(3.11c)$ with

$$u = (0.656 - 0.600)/0.1 = 0.56 \text{ and } h = 0.1$$

and in terms of the numerical values of (*b*),

$$\sin 0.656 = 0.44 y_0 + 0.56 y_{+1} - C_0(\lambda_1 - C_1\lambda_2) + R_5$$

$$= 0.609\ 951 + R_5 \qquad \text{(correct value } 0.609\ 952)$$

where the remainder by the formula given in $(3.11d)$ is

$$|R_5| \le \left|\binom{2.56}{4}0.1^4 \sin 0.656\right| = |3 \times 10^{-6}|$$

(e) Lagrange's three-point formula. For $\sin 0.656$ by $(3.8c)$ with u calculated in (*d*), $a = (0.44)(0.56)(1.56)$, and y_{-1}, y_0, y_{+1} taken from (*b*),

$$\sin 0.656 = \frac{a}{2}\left(-\frac{y_{-1}}{1.56} + \frac{2y_0}{0.56} + \frac{y_{+1}}{0.44}\right) + R_4$$

$$= 0.609\ 900 + R_4 \qquad \text{(correct value } 0.609\ 952)$$

where the remainder by the formula given in $(3.08c)$ is

$$|R_4| \le |0.065(0.000\ 795)| = |5.2 \times 10^{-5}|$$

(a) Tabular interpolation is the interpolation process in a narrow sense and consists of the computation of intermediate values between those given in a table. Since most tabular values are for equally spaced arguments, all formulas in (3.10) to (3.13) and Lagrange's three- and five-point formulas (3.08c) are directly applicable. Tabular interpolations are classified as near-end-value interpolation depicted by the diagram (b), intermediate-value interpolation depicted in (c), and far-end-value interpolation depicted in (d).

(b) Near-end-value interpolation diagram   **(c) Intermediate-value interpolation diagram**

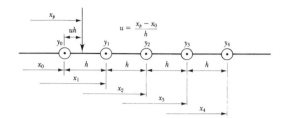   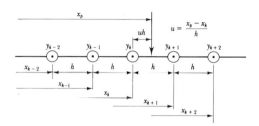

(d) Far-end-value interpolation diagram

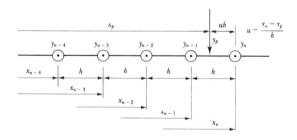

(e) Range of practicality of the interpolation formulas used in (3.16) and (3.17) is given in the table below. The u value is defined in (b) to (d).

Formula	Section	Range	Rapid convergence
Linear	3.16	$0 < u < 1$	
Lagrange's	3.08c	$0 < u < 1$	
Newton–Gregory	3.10	$0 < u < 1$	
Everett's	3.11	$0 < u < 1$	
Bessel's	3.12	$0 < u < 1$	$0.25 < u < 0.75$
Stirling's	3.13	$-\frac{1}{2} < u < \frac{1}{2}$	$-0.25 < u < 0.25$

(f) Subtabulation is the process of computation of values corresponding to smaller intervals than those listed in a table.

(a) Interpolation formula. If y_a and y_b are two consecutively tabulated values of $y = f(x)$, corresponding to $x = a$, $x = b$, respectively, as shown in the adjacent figure, then the approximate value of y_p at $x = p$ is

$$\bar{y}_p = y_a + \frac{y_b - y_a}{b - a}(p - a) = y_a + ud$$

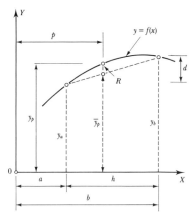

where $d = y_b - y_a$ is the tabular difference, $b - a = h$ is the interpolation interval, and $(p - a)/h = u$ is the interpolation factor $(u < 1)$.

(b) Remainder of this approximation can be estimated as

$$R_2 = y_p - \bar{y}_p = \frac{u(u - 1)}{2} h^2 [\max D^2 f(\eta)]$$

where η is so selected as to make the second derivative a maximum. If $f(x)$ is not known, its second derivative can be approximated as

$$\frac{\Delta^2 y_a}{h^2} \le [\max D^2 f(\eta)] \le \frac{\Delta^2 y_b}{h^2}$$

(c) n-place tables. If the values y_a, y_b are listed in n-decimal-place accuracy, then the condition of permissibility of linear interplation is

$$\frac{h^2}{8} |\max D^2 f(\eta)| < \frac{1}{2} \times 10^{-n}$$

where the second derivative can be approximated as in (b).

(d) Example. If $y_a = \sin 8° = 0.139\,173$, $y_b = \sin 9° = 0.156\,434$, and $y_p = \sin 8.345°$ is desired, then by (a), $d = 0.017\,261$, $h = 1$, $u = 0.345$, and

$$y_p = 0.139\,173 + 0.345(0.017\,261) + R_2 = 0.145\,128 + R_2$$

For the evaluation of R_2,

$$f(x) = \sin(\pi x/180) \qquad \text{and} \qquad D^2 f(x) = -(\pi/180)^2 \sin(\pi x/180)$$

$$\min D^2 f(8°) = -(\pi/180)^2 \sin(8\pi/180) = 0.000\,042$$

$$\max D^2 f(9°) = -(\pi/180)^2 \sin(9\pi/180) = 0.000\,047$$

$$R_2 = \tfrac{1}{2}(-0.345 \times 0.655) \times 1^2(-0.000\,047) = 5.3 \times 10^{-6}$$

The correct value is $y_p = 0.145\,133\,3$, which is satisfied by $\bar{y}_p + R_2$.

(e) Optional interval $\bar{h}$. If the question arises to determine the interval $\bar{h}$ at a to assure an accuracy of linear interpolation to n decimal places, then from (c) this interval can be estimated as

$$\bar{h} = \frac{2}{\sqrt{10^n [D^2 f(a)]}} \qquad [\text{in } (d), \bar{h} \le 0.31 \text{ for six-place accuracy}]$$

which shows that the linear interplation is suitable only for cases of very small $D^2 f(a)$.

(a) Greater precision of interpolation in tables can be obtained by using the difference formulas of (3.08) to (3.13).

(b) Near-end-values interpolation in $y_0, y_1, y_2, \ldots, y_n$ at $x_k < x_p < x_{k+1}$ is given conveniently by the Newton–Gregory formula (3.10b) as

$$y_p = y_{k+u} = y_k + A_1(\Delta y_k - A_2(\Delta^2 y_k - A_3(\Delta^3 y_k - A_4\Delta^4 y_k))) + R_5 \qquad\qquad R^5 = \binom{u}{5}\Delta^5 y_k$$

where k is 0 or 1 [figure (3.15b)], $u = (x_p - x_k)/(x_{k+1} - x_k)$, and $A_1 = u/1$, $A_2 = (1 - u)/2$, $A_3 = (2 - u)/3$, $A_r = (3 - u)/4, \ldots, A_r = (r - 1 - u)/v$

(c) Intermediate-value interpolation in y_{k-2}, y_{k-1}, y_k, y_{k+1}, y_{k+2} at $x_k < x_p < x_{k+1}$ is given conveniently by the five-point Lagrange's formula (3.08c) as

$$y_p = y_{k+u} = \frac{b}{24}\left(\frac{y_{k-2}}{2+u} - \frac{4y_{k-1}}{1+u} + \frac{6y_k}{u} + \frac{4y_{k+1}}{1-u} - \frac{y_{k+2}}{2-u}\right) + R_5 \qquad\qquad R_5 = 0.012\Delta^5 y_k$$

where $1 < k < n - 1$ [figure (3.15c)], $u = (x_p - x_k)/(x_{k+1} - x_k)$ and $b = (2 - u)(1 + u) \times u(1 + u)(2 + u)$.

(d) Far-end-values interpolation in $y_0, y_1, y_2, \ldots, y_{n-2}, y_{n-1}, y_n$ at $x_{k-1} < x_p < x_k$ is given conveniently by the Newton–Gregoy formula (3.10c) as

$$y_p = y_{k-u} = y_k + B_1(\nabla y_k + B_2(\nabla y_k + B_3(\nabla^3 y_k + B_4\nabla^4 y_k))) + R_5 \qquad\qquad R_5 = \binom{u+5}{5}\nabla^5 y_k$$

where k is n or $n - 1$ (figure (3.15d)], $u = (x_k - x_p)/(x_k - x_{k-1})$ and $B_1 = u/1$, $B_2 = (1 + u)/2$, $B_r = (r - 1 + u)/r$.

(e) Examples. The numerical values of $y = \sin\theta°$ are given for $\theta° = 8, 9, \ldots, 13°$ in the difference table shown below. The end value $\sin 8.345°$ and the intermediate value $\sin 10.345°$ are desired.

$u = 0.345$	A_r	+0.345 000	+0.327 500	+0.551 667	+0.405 000
x	$\sin\theta°$	Δ	Δ^2	Δ^3	Δ^4
8°	0.139 173 1				
		0.017 261 4			
9°	0.156 434 5		−0.000 047 7		
		0.017 213 7		−0.000 005 2	
10°	0.173 648 2		−0.000 052 9		0
		0.017 160 8		−0.000 005 2	
11°	0.190 809 0		−0.000 058 1		0
		0.017 102 7		−0.000 005 2	
12°	0.207 911 7		−0.000 063 3		
		0.017 039 4			
13°	0.224 951 1				
$u = 0.345$	B_r	+0.345 000	+0.327 500	+0.781 667	+0.836 250

By (b) in terms of y_8, differences, and A_r's given above, $\sin 8.345°$ is

$$y_{8.345} = 0.145\ 133\ 3 + R_5 \qquad \text{(correct value is 0.145 133 3)} \qquad R_5 = \binom{0.345}{5}\Delta^5 y_8 = 0$$

By (c) in terms of $y_8, y_9, y_{10}, y_{11}, y_{11}$ given above, $\sin 10.345°$ is

$$y_{10.345} = 0.179\ 574\ 9 + R_5 \qquad \text{(correct value is 0.179 574 9)} \qquad R_5 = 0.012\Delta^5 y_{10} = 0.$$

(a) Inverse process. *Interpolation* is the process of finding $f(x)$ for a particular x located between two known values. *Inverse interpolation* is the process of finding the argument x if the value of $f(x)$ is known.

(b) Inverse linear interpolation. If in (3.16a) a, b and y_a, y_p, y_b are given, then the desired p is calculated as

$$p = \bar{p} - R = a + \frac{y_p - y_a}{y_b - y_a}(b - a) - R \qquad\qquad R = \frac{(p - a)(p - b)}{2(b - a)(y_b - y_a)}[\max f''(\eta)] \qquad (a \le \eta \le b)$$

(c) Example. From (3.16d), $y_a = \sin a = 0.139\,173$, $y_p = \sin p° = 0.145\,133$, $y_b = \sin b = 0.156\,434$, and $a = 8°$, $b = 9°$. By (b) above,

$$p = 8 + \frac{0.145\,133 - 0.139\,173}{0.156\,434 - 0.139\,173}(9 - 8) - R = 8.345\,280° - R$$

$$R = \frac{(8.345\,280 - 8)(8.345\,280 - 9)}{2(0.156\,434 - 0.139\,173)}\left[-\left(\frac{\pi}{180}\right)^2 \sin 9°\right] = (3.12 \times 10^{-4})°$$

(d) Inverse Lagrange's interpolation. If in (3.08a) x and y are interchanged, then

$$\bar{p} = x_0\lambda_0(y_p) + x_1\lambda_1(y_p) = \cdots + x_r\lambda_r(y_p)$$

where

$$\lambda_k(y_0) = \prod_{\substack{j=0 \\ j \neq k}}^{r} \frac{y_p - y_j}{y_k - y_j} = \frac{(y_p - y_1)(y_p - y_2)\cdots(y_p - y_{k-1})(y_p - y_{k+1})\cdots(y_p - y_r)}{(y_k - y_0)(y_k - y_1)\cdots(y_k - y_{k-1})(y_k - y_{k-1})\cdots(y_k - y_r)}$$

No simple formula is available for the calculation of the remainder R.

(e) Five-point inverse Lagrange's interpolation formula. If in (3.08c) x and y are interchanged,

$$\bar{p} = c\left(\frac{x_{k-2}}{2+v} - \frac{4x_{k-1}}{1+v} + \frac{6x_k}{v} + \frac{4x_{k+1}}{1-v} - \frac{x_{k+2}}{2-v}\right)$$

where $v = (y_p - y_k)/(y_{k+1} - y_k)$ and $c = (2 + v)(1 + v)v(1 - v)(2 - v)/24$.

(f) Example. If $y_{-2} = \sin 6°$, $y_{-1} = \sin 7°$, $y_0 = 8°$, $y_1 = \sin 9°$, $y_2 = \sin 10°$, $y_p = \sin p° = 0.145\,133$, $v = (0.145\,133 - 0.139\,173)/(0.156\,434 - 0.139\,173) = 0.345\,274$, and

$$c = (2.345\,274)(1.345\,274)(0.345\,274)(0.654\,726)(1.654\,726)/24 = 0.049\,175$$

$$\bar{p} = (0.049\,175)\left[\frac{6}{2.345\,274} - \frac{4(7)}{1.345\,274} + \frac{6(8)}{0.345\,274} + \frac{4(9)}{0.654\,726} - \frac{10}{1.654\,726}\right] = 8.345\,307°$$

whereas the correct value is 8.344 981°, and the remainder R in (c) is also applicable here.

(g) Inverse interpolation by reversion of series. Any interpolation formula expressed in series form

$$y(x) = a_0 + a_1x + a_2x^2 + a_3x^3 + \cdots$$

can be reversed with $v = (y_p - a_0)/a_1$ as

$$x = b_1v + b_2v^2 + b_3v^3 + b_4v^4 + b_5v^6 + \cdots$$

where $b_1 = 1$ $\qquad\qquad\qquad b_4 = -a_4/a_1 + 5a_2a_3/a_1^2 - 5(a_2/a_1)^2$

$b_2 = -a_2/a_1$ $\qquad\qquad b_5 = -a_5/a_1 + 6a_2a_4/a_1^2 + 3(a_3/a_1)^2 21a_2^2a_3/a_1^3 + 14(a_2/a_1)^4$

$b_3 = -a_3/a_1 + 2b_2^2$ $\qquad\qquad$ .... (Ref. 1.20, p. 105)

(a) Functions of two variables can be interpolated in two ways. The simplest method is to first interpolate with respect to one variable and then with respect to the other, using any one of the standard formulas of (3.08) to (3.13). A more involved but more direct approach is by the bivariate interpolation formulas presented below.

(b) Newton's bivariate interpolation. With constant spacing p and q in the direction of x and y, respectively, the forward differences of $z = f(x,y)$ are

$$\Delta_{10}z = f(x + p,y) - f(x,y) \qquad = (E_x - 1)z$$

$$\Delta_{01}z = f(x,y + q) - f(x,y) \qquad = (E_y - 1)z$$

$$\Delta_{11}z = \Delta_{10}[f(x,y + q) - f(x,y)] = (E_x - 1)(E_y - 1)z$$

$$\qquad = \Delta_{01}[f(x + p,y) - f(x,y)] = (E_y - 1)(E_x - 1)z$$

$$\Delta_{20}z = \Delta_{10}[f(x + p,y) - f(x,y)] = (E_x - 1)^2 z$$

$$\Delta_{02}z = \Delta_{01}[f(x,y + p) - f(x,y)] = (E_y - 1)^2 z$$

$$\cdots\cdots\cdots\cdots\cdots\cdots\cdots\cdots\cdots\cdots\cdots\cdots\cdots$$

where $E_x z = f(x + p,y), E_x^2 z = f(x + 2p,y), \ldots, E_x^n z = f(x + np,y)$

    $E_y z = f(x,y + q), E_y^2 z = f(x,y + 2q), \ldots, E_y^n z = f(x,y + nq)$

With $u = (x - x_0)/p, \ldots, u^{(k)} = u(u - 1)(u - 2) \cdots (u - k + 1)$

   $v = (y - y_0)/q, \ldots, v^{(k)} = v(v - 1)(v - 2) \cdots (v - k + 1)$

the substitute function becomes

$$\bar{z}(x,y) = [1 + (u^{(1)}\Delta_{10} + v^{(1)}\Delta_{01}) + \frac{1}{2!}(u^{(2)}\Delta_{20} + 2u^{(1)}v^{(1)}\Delta_{11} + v^{(2)}\Delta_{02})$$

$$+ \frac{1}{3!}(u^{(3)}\Delta_{30} + 3u^{(2)}v^{(1)}\Delta_{21} + 3u^{(1)}v^{(2)}\Delta_{12} + v^{(3)}\Delta_{03})$$

$$+ \ldots + \frac{1}{n!}\sum_{k=0}^{n}\binom{n}{k}u^{(n-k)}v^{(k)}\Delta_{n-k,k}]f(x_0,y_0)$$

where $x = x_0 + up$ and $y = y_0 + vq$.

(c) Multipoint formulas. If the values of z are known at certain points of the grid shown in the adjacent diagram, then the value of z at $x = x_0 + uh, y = y_0 + vh$ can be approximated

 (1) *Three-point formula*

$$\bar{z}(x,y) = (1 - u - v)z_C + uz_T + vz_R + R_2$$

 (2) *Six-point formula*

$$\bar{z}(x,y) = \tfrac{1}{2}(u - 1)uz_L + \tfrac{1}{2}(v - 1)vz_B$$

$$+ (1 + uv - u^2 - v^2)z_C$$

$$+ \tfrac{1}{2}(v - 2u + 1)vz_R$$

$$+ \tfrac{1}{2}(u - 2v + 1)uz_T + uvz_{RT} + R_3$$

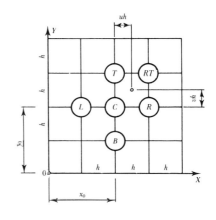

In these formulas the subscripts of z identify the respective point (L = left point, R = right point, B = bottom point, T = top point, RT = right top point) and R_2, R_3 are the remainders to be estimated as $0(h^2), 0(h^3)$, respectively.

3.20 ORDINARY DERIVATIVES IN TERMS OF DIFFERENCES

(a) Approximate derivatives of a tabulated function can be found at any point of its interval of continuity by representing this function by a selected interpolation polynomial $\bar{y}$ and differentiating it with respect to u. Thus the first derivative of $y = f(x)$ tabulated in constant intervals of length h is

$$\frac{dy}{dx} = Dy \cong \frac{d\bar{y}}{dx} = \frac{1}{h}\frac{d\bar{y}}{du} = D\bar{y}$$

and in general, its nth derivative is

$$\frac{d^n y}{dx^n} = D^n y \cong \frac{d^n \bar{y}}{dx^n} = \frac{1}{h^n}\frac{d^n \bar{y}}{du^n} = D^n \bar{y}$$

where $n = 1, 2, 3, \ldots$, D = derivative operator (3.02d), $u = (x - x_0)/h$, and $\bar{y} = \bar{f}(x)$ is one of the interpolation formulas (3.10) to (3.13).

(b) D operator in terms of Δ series

$$\begin{bmatrix} hD \\ h^2D \\ h^3D \\ h^4D \\ h^5D \end{bmatrix} = \begin{bmatrix} \ln(1-\Delta) \\ \ln^2(1+\Delta) \\ \ln^3(1+\Delta) \\ \ln^4(1+\Delta) \\ \ln^5(1+\Delta) \end{bmatrix} = \begin{bmatrix} 1 & -\frac{1}{2} & \frac{1}{3} & -\frac{1}{4} & \frac{1}{5} & -\frac{1}{6} & \frac{1}{7} & \cdots \\ 0 & 1 & -1 & \frac{11}{12} & -\frac{5}{6} & \frac{137}{180} & -\frac{7}{10} & \cdots \\ 0 & 0 & 1 & -\frac{3}{2} & \frac{7}{4} & -\frac{15}{8} & \frac{29}{15} & \cdots \\ 0 & 0 & 0 & 1 & -2 & \frac{17}{6} & -\frac{7}{2} & \cdots \\ 0 & 0 & 0 & 0 & 1 & -\frac{5}{2} & \frac{25}{6} & \cdots \end{bmatrix} \begin{bmatrix} \Delta \\ \Delta^2 \\ \Delta^3 \\ \Delta^4 \\ \Delta^5 \\ \cdots \end{bmatrix}$$

where Δ = forward difference operator (3.02d).

(c) D operator in terms of ∇ series

$$\begin{bmatrix} hD \\ h^2D \\ h^3D \\ h^4D \\ h^5D \end{bmatrix} = \begin{bmatrix} -\ln(1-\nabla) \\ \ln^2(1-\nabla) \\ -\ln^3(1-\nabla) \\ \ln^4(1-\nabla) \\ -\ln^5(1-\nabla) \end{bmatrix} = \begin{bmatrix} 1 & \frac{1}{2} & \frac{1}{3} & \frac{1}{4} & \frac{1}{5} & \frac{1}{6} & \frac{1}{7} & \cdots \\ 0 & 1 & 1 & \frac{11}{12} & \frac{5}{6} & \frac{137}{180} & \frac{7}{10} & \cdots \\ 0 & 0 & 1 & \frac{3}{2} & \frac{7}{4} & \frac{15}{8} & \frac{29}{15} & \cdots \\ 0 & 0 & 0 & 1 & 2 & \frac{17}{6} & \frac{7}{2} & \cdots \\ 0 & 0 & 0 & 0 & 1 & \frac{5}{2} & \frac{25}{6} & \cdots \end{bmatrix} \begin{bmatrix} \nabla \\ \nabla^2 \\ \nabla^3 \\ \nabla^4 \\ \nabla^5 \\ \cdots \end{bmatrix}$$

where ∇ = backward difference operator (3.02d).

(d) D operator in terms of δ series. The nth-order derivative operator is

$$D^n = \frac{2^n}{h^n}(\sinh^{-1}\lambda)^n$$

where $\lambda = \frac{1}{2}\delta$ and δ = central difference operator (3.02d). The series in terms of δ are then

$$\begin{bmatrix} hD \\ h^2D \\ h^3D \\ h^4D \\ h^5D \end{bmatrix} = \begin{bmatrix} 2\sinh^{-1}\lambda \\ (2\sinh^{-1}\lambda)^2 \\ (2\sinh^{-1}\lambda)^3 \\ (2\sinh^{-1}\lambda)^4 \\ (2\sinh^{-1}\lambda)^5 \end{bmatrix} = \begin{bmatrix} 2 & 0 & -\frac{1}{3} & 0 & \frac{3}{20} & 0 & -\frac{5}{28} & \cdots \\ 0 & 4 & 0 & -\frac{2}{3} & 0 & \frac{32}{45} & 0 & \cdots \\ 0 & 0 & 8 & 0 & -4 & 0 & \frac{37}{15} & \cdots \\ 0 & 0 & 0 & 16 & 0 & -\frac{32}{3} & 0 & \cdots \\ 0 & 0 & 0 & 0 & 32 & 0 & -\frac{80}{3} & \cdots \end{bmatrix} \begin{bmatrix} \lambda \\ \lambda^2 \\ \lambda^3 \\ \lambda^4 \\ \lambda^5 \\ \cdots \end{bmatrix}$$

3.21 ORDINARY DERIVATIVES IN TERMS OF ORDINATES

(a) Approximate derivatives of a tabulated function can be also found by expressing each difference in (3.20) in terms of the tabulated values $y_0, y_1, y_2, \ldots, y_{k-1}, y_k, y_{k+1}, \ldots, y_{n-2}, y_{n-1}, y_n$ as shown below in (*b*).

(b) Examples. From (3.20*b*) in terms of (3.02*d*), (3.02*e*), the first-order derivative approximations at 0 are

$$\left[\frac{dy}{dx}\right]_0 \cong (\Delta y_0)/h = -(y_0 - y_1)/h$$

$$\cong (\Delta y_0 - \tfrac{1}{2}\Delta^2 y_0)/h = -\tfrac{1}{2}(3y_0 - 4y_1 + y_2)/h$$

$$\cong (\Delta y_0 - \tfrac{1}{2}\Delta^2 y_0 + \tfrac{1}{3}\Delta^3 y_0)/h = -\tfrac{1}{6}(-11y_0 + 18y_1 - 9y_2 + 2y_3)/h$$

where the error is $0(h)$, $0(h^2)$, and $0(h^3)$, respectively.

(c) Forward difference approximations in terms of the first difference only are

$$\begin{bmatrix} hDy_0 \\ h^2D^2y_0 \\ h^3D^3y_0 \\ h^4D^4y_0 \\ h^5D^5y_0 \\ \cdots \end{bmatrix} = \begin{bmatrix} -1 & 1 & 0 & 0 & 0 & 0 \\ 1 & -2 & 1 & 0 & 0 & 0 \\ -1 & 3 & -3 & 1 & 0 & 0 \\ 1 & -4 & 6 & -4 & 1 & 0 \\ -1 & 5 & -10 & 10 & -5 & 1 \\ \cdots & \cdots & \cdots & \cdots & \cdots & \cdots \end{bmatrix} \begin{bmatrix} y_0 \\ y_1 \\ y_2 \\ y_3 \\ y_4 \\ y_5 \end{bmatrix} + 0(h)$$

(d) Backward difference approximations in terms of the first difference only are

$$\begin{bmatrix} hDy_n \\ h^2D^2y_n \\ h^3D^3y_n \\ h^4D^4y_n \\ h^5D^5y_n \\ \cdots \end{bmatrix} = \begin{bmatrix} 0 & 0 & 0 & 0 & -1 & 1 \\ 0 & 0 & 0 & 1 & -2 & 1 \\ 0 & 0 & -1 & 3 & -3 & 1 \\ 0 & 1 & -4 & 6 & -4 & 1 \\ -1 & 5 & -10 & 10 & -5 & 1 \\ \cdots & \cdots & \cdots & \cdots & \cdots & \cdots \end{bmatrix} \begin{bmatrix} y_{n-5} \\ y_{n-4} \\ y_{n-3} \\ y_{n-2} \\ y_{n-1} \\ y_n \end{bmatrix} + 0(h)$$

(e) Central difference approximations in terms of the first difference only are

$$\begin{bmatrix} hDy_k \\ h^2D^2y_k \\ h^3D^3y_k \\ h^4D^4y_k \\ h^5D^5y_k \\ h^6D^6y_k \\ \cdots \end{bmatrix} = \begin{bmatrix} 0 & 0 & -\frac{1}{2} & 0 & \frac{1}{2} & 0 & 0 \\ 0 & 0 & 1 & -2 & 1 & 0 & 0 \\ 0 & -\frac{1}{2} & 1 & 0 & -1 & \frac{1}{2} & 0 \\ 0 & 1 & -4 & 6 & -4 & 1 & 0 \\ -\frac{1}{2} & 2 & -2.5 & 0 & 2.5 & -2 & \frac{1}{2} \\ 1 & -6 & 15 & -20 & 15 & -6 & 1 \\ \cdots & \cdots & \cdots & \cdots & \cdots & \cdots & \cdots \end{bmatrix} \begin{bmatrix} y_{k-3} \\ y_{k-2} \\ y_{k-1} \\ y_k \\ y_{k+1} \\ y_{k+2} \\ y_{k+3} \end{bmatrix} + 0(h^2)$$

(f) Errors of these approximations are of order h in (*c*), (*d*) and of order h^2 in (*e*). Results should be used with caution, and even when very accurate input data are available, the accuracy diminishes with the increasing order of numerical differentiation.

(a) Forward differences of $y_k = f(x + kh)$ in terms of $D, D^2, D^3, \ldots$, are

$$
\begin{bmatrix} \Delta y_k \\ \Delta^2 y_k \\ \Delta^3 y_k \\ \Delta^4 y_k \\ \Delta^5 y_k \\ \Delta^6 y_k \end{bmatrix} = \begin{bmatrix} 1 & 1 & 1 & 1 & 1 & 1 & \cdots \\ 0 & 2 & 6 & 14 & 30 & 62 & \cdots \\ 0 & 0 & 6 & 36 & 182 & 540 & \cdots \\ 0 & 0 & 0 & 24 & 240 & 1560 & \cdots \\ 0 & 0 & 0 & 0 & 120 & 2440 & \cdots \\ 0 & 0 & 0 & 0 & 0 & 720 & \cdots \end{bmatrix} \begin{bmatrix} hD/1! \\ h^2 D^2/2! \\ h^3 D^3/3! \\ h^4 D^4/4! \\ h^5 D^5/5! \\ h^6 D^6/6! \end{bmatrix} y_k
$$

and the nth-order difference is

$$
\Delta^n y_k = (e^{hD} - 1)^n y_k = \left\{ \sum_{r=n}^{\infty} \frac{[n^r - \binom{n}{1}(n-1)^r + \binom{n}{2}(n-2)^r - \cdots]}{r!} h^r D^r \right\} y_k
$$

where $n = 1, 2, 3, \ldots$ and $r = n, n+1, n+2, \ldots, \infty$.

(b) Backward differences of the same function in terms of $D, D^2, D^3, \ldots$ are

$$
\begin{bmatrix} \nabla y_k \\ \nabla^2 y_k \\ \nabla^3 y_k \\ \nabla^4 y_k \\ \nabla^5 y_k \\ \nabla^6 y_k \end{bmatrix} = \begin{bmatrix} 1 & -1 & 1 & -1 & 1 & -1 & \cdots \\ 0 & 2 & -6 & 14 & -30 & 62 & \cdots \\ 0 & 0 & 6 & -36 & 182 & -540 & \cdots \\ 0 & 0 & 0 & 24 & -240 & 1560 & \cdots \\ 0 & 0 & 0 & 0 & 120 & -2440 & \cdots \\ 0 & 0 & 0 & 0 & 0 & 720 & \cdots \end{bmatrix} \begin{bmatrix} hD/1! \\ h^2 D^2/2! \\ h^3 D^3/3! \\ h^4 D^4/4! \\ h^5 D^5/5! \\ h^6 D^6/6! \end{bmatrix} y_k
$$

and the nth-order difference is

$$
\nabla^n y_k = (1 - e^{-hD})^n y_k = \left\{ \sum_{r=n}^{\infty} \frac{(-1)^\lambda [n^r - \binom{n}{1}(n-1)^r + \binom{n}{2}(n-2)^r \cdots]}{r!} h^r D^r \right\} y_k
$$

where n, r are the same as in (a) and $\lambda = r$ if n is even and $\lambda = r + 1$ if n is odd.

(c) Central differences of the same function in terms of $D, D^2, D^3, \ldots$ are

$$
\begin{bmatrix} \delta y_k \\ \delta^2 y_k \\ \delta^3 y_k \\ \delta^4 y_k \\ \delta^5 y_k \\ \delta^6 y_k \end{bmatrix} = \begin{bmatrix} 1 & 1 & 1 & 1 & 1 & 1 & \cdots \\ 0 & 4 & 8 & 16 & 32 & 64 & \cdots \\ 0 & 0 & 24 & 78 & 240 & 726 & \cdots \\ 0 & 0 & 0 & 192 & 896 & 3840 & \cdots \\ 0 & 0 & 0 & 0 & 1920 & 11990 & \cdots \\ 0 & 0 & 0 & 0 & 0 & 23040 & \cdots \end{bmatrix} \begin{bmatrix} \frac{1}{2}hD/1! \\ (\frac{1}{2}h)^2 D^2/2! \\ (\frac{1}{2}h)^3 D^3/3! \\ (\frac{1}{2}h)^4 D^4/4! \\ (\frac{1}{2}h)^5 D^5/5! \\ (\frac{1}{2}h)^6 D^6/6! \end{bmatrix} y_k
$$

and the nth-order difference is

$$
\delta^n y_k = \left(2^n \sinh^n \frac{hD}{2} \right) y_k = \left\{ \sum_{r=n}^{\infty} \frac{[n^r - \binom{n}{1}(n-2)^r + \binom{n}{2}(n-4)^r - \cdots]}{r!} (\tfrac{1}{2}h)^r D^r \right\} y_k
$$

where n, r are the same as in (a).

4
NUMERICAL
INTEGRALS

(a) Primitive function. The function $Y = F(x)$ is said to be a primitive function (indefinite integral) of the function $y = f(x)$ in the interval (a, b), if for all values of x in (a, b),

$$\frac{dF(x)}{dx} = D[F(x)] = f(x)$$

where $D = d/dx$ is the *derivative operator.*

(b) Indefinite integral. Since the derivative of $F(x) + C$, where C is a constant (or zero), is also equal to $f(x)$, all indefinite integrals of $f(x)$ are included in the expression

$$\int f(x)\, dx = I[f(x)] = F(x) + C$$

where $f(x)$ is called the integrand, C is the integration constant, and I is the *integral operator.*

(c) Definite integral. If $F(x)$ and its derivative $f(x)$ are single-valued and continuous in the closed interval $[a, b]$, then the definite integral of $f(x)$ between limits $x = a$ and $x = b$ is defined as

$$\int_a^b f(x)\, dx = \lim_{h \to 0} \{f(a) + f(a + h) + f(a + 2h) + \cdots + f[a + (n - 1)h]\}$$

$$= Y[a, b] = [F(x) + C]_a^b = F(b) - F(a)$$

where n = positive integer and $h = (b - a)/n$.

(d) Limits. The signed numbers a, b (which may be 0 or $\pm\infty$) are called the lower and upper limits of integration, and the closed interval $[a, b]$ is called the range of integration.

(e) Areas bounded by $f(x)$ may take one of the shapes shown below.

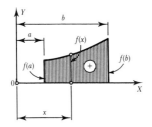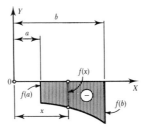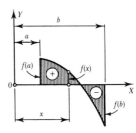

(f) Geometric interpretation. The definite integral introduced in (c) above equals numerically the area bounded by the graph of $f(x)$, the X axis, and the ordinates $f(a), f(b)$ as shown in (e) above. The sign of the area is governed by the sign of $f(x)$. If the graph intersects the X axis one or several times inside the interval $[a, b]$, then the definite integral equals the algebraic sum of all areas above the axis and of all areas below the axis. If the sum of areas above the axis (+ areas) and the sum of areas below the axis (− areas) are numerically equal, then the definite integral equals zero.

(g) Example. If $y = \sin x$ and the closed interval of integration is $[0, \pi]$, then

$$\int_0^\pi \sin x\, dx = [-\cos x]_0^\pi = 0$$

since the areas above and below the X axis are $+1$ and -1, respectively.

(a) Even function. If $f(x)$ is an even function so that $f(-x) = f(x)$, then $f(x)$ is symmetrical with respect to the Y axis and

$$\int_{-a}^{a} f(x)\,dx = 2\int_{0}^{a} f(x)\,dx = 2[F(a)]$$

where $a \neq 0$.

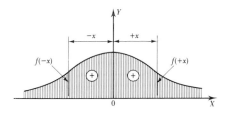

(b) Odd function. If $g(x)$ is an odd function so that $g(-x) = -g(x)$, then $g(x)$ is antisymmetrical with respect to the Y axis (odd function) and

$$\int_{-a}^{a} g(x)\,dx = 0$$

where $a \neq 0$.

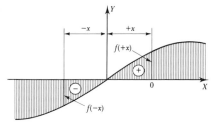

(c) Periodic function. If $n =$ integer, $T =$ positive number (period), and $f(x)$ is a periodic function so that $f(nT + x) = f(x)$, then for $b = nT + a$,

$$\int_{0}^{b} f(x)\,dx = nF(T) + F(a) - (n + 1)F(0)$$

where $dF(x)/dx = f(x)$.

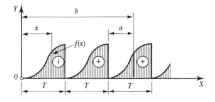

(d) Antiperiodic function. If n, T are the same as in (c) and $g(x)$ is an antiperiodic function so that $g[(2n - 1)T + x] = -g(x)$ and $g(2nT + x) = g(x)$, then for $b = (2n - 1)T + a$ and $c = 2nT + a$,

$$\int_{0}^{b} g(x)\,dx = G(T) - G(a)$$

$$\int_{0}^{c} g(x)\,dx = G(a) - G(0)$$

where $dG(x)/dx = g(x)$.

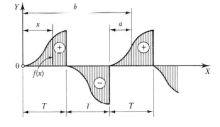

(e) Change in limits and variables. If $f(x)$ is integrable in $[a, b]$ and $x = g(t)$ and its derivative $dg(t)/dt$ are continuous in $\alpha \leq t \leq \beta$, then with $a = g(\alpha)$, $b = g(\beta)$,

$$\int_{x=a}^{x=b} f(x)\,dx = \int_{t=\alpha}^{t=\beta} f[g(t)]\frac{dg(t)}{dt}\,dt$$

Particular cases of this transformation are:

$$\int_{0}^{b} f(x)\,dx = \int_{0}^{b} f(b - x)\,dx \qquad\qquad \int_{a}^{b} f(x)\,dx = \int_{a}^{b} f(a + b - x)\,dx$$

$$\int_{a}^{b} f(x)\,dx = \frac{1}{\lambda}\int_{\lambda a}^{\lambda b} f\left(\frac{x}{\lambda}\right)\,dx \qquad\qquad \int_{a}^{b} f(x)\,dx = \int_{a\pm\lambda}^{b\pm\lambda} f(x \mp \lambda)\,dx$$

$$\int_{a}^{b} f(x)\,dx = \int_{0}^{b-a} f(x + a)\,dx = (b - a)\int_{0}^{1} f[(b - a)x + a]\,dx$$

(a) Numerical integration. Whenever the closed-form integration is too involved or is not feasible, the numerical value of a definite integral can be found (to any degree of accuracy) by means of any of several quadrature formulas which evaluate the definite integral as a linear combination of a selected set of integrands in the range of integration.

(b) Quadrature formulas fall into four basic categories: numerical integration in terms of ordinates (4.05), (4.06), in terms of finite differences (4.07) to (4.09), in terms of asymptotic series (4.10), and in terms of orthogonal polynomials (4.11) to (4.16).

(c) Special methods based on Taylor's series expansion (11.01), on Fourier series representation (13.14), and on integrodifferential representation of the integrand (14.17) are applicable in special cases.

4.04 LAGRANGE'S FORMULA

(a) Definite integral of the substitute function in (3.08a) is

$$\bar{Y}[a, b] = \int_a^b \bar{y}(x)\, dx = y_0 \mathscr{A}_0^{(r)} + y_1 \mathscr{A}_1^{(r)} + y_2 \mathscr{A}_2^{(r)} + \cdots + y_r \mathscr{A}_r^{(r)}$$

where $y_0, y_1, y_2, \ldots, y_r$ are the values of $y_k = f(x_k)$ at k and

$$\mathscr{A}_k^{(r)} = \int_a^b \lambda_k(x)\, dx = \int_a^b \left(\prod_{\substack{j=0 \\ j \neq k}}^{r} \frac{x - x_j}{x_k - x_j} \right) dx = \mathscr{A}_{r-k}^{(r)}$$

where $k, j = 0, 1, 2, \ldots, r$, $\lambda_k(x) =$ Lagrange's multiplier function defined in (3.08a), and $\mathscr{A}_k^{(r)} =$ weight which is symmetrical with respect to the center of the sequence.

(b) Equally spaced ordinates. If the spacing of $y_0, y_1, y_2, \ldots, y_r$ is constant so that $h = (b - a)/r$ and $u = (x - a)/h$, Lagrange's factor becomes

$$\mathscr{A}_k^{(r)} = \frac{(-1)^{r-k} h}{k!\,(r - k)!} \int_0^r \frac{u(u - 1)(u - 2) \cdots (u - r)}{u - k}\, du$$

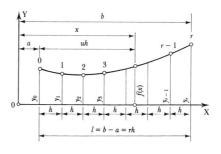

and it can be shown that

$$\mathscr{A}_k^{(r)} = \mathscr{A}_{r-k}^{(r)} \qquad \sum_{k=0}^{r} \mathscr{A}_k^{(r)} = rh$$

For $r = 1, 2, 3, 4, 5$, the factors are tabulated in (4.05b).

(c) Error of Lagrange's integration formula of constant spacing h can be estimated as

$$R_r \cong Y[a, b] - \bar{Y}[a, b] = \begin{cases} h^{r+2} D^{r+1} f(\eta) \displaystyle\int_0^r \binom{u}{r + 1}\, du & \text{if } r \text{ is odd} \\[4mm] \frac{1}{2} h^{r+3} D^{r+2} f(\eta) \displaystyle\int_0^r \frac{u - \frac{1}{2}r}{1 + \frac{1}{2}r} \binom{u}{r + 1}\, du & \text{if } r \text{ is even} \end{cases}$$

where $a \leq \eta \leq b$ and particular values of R_r are given in (4.05b).

(a) Closed-end formulas based on (4.04*b*) are in symbolic form

$$\int_a^b f(x)\,dx = \sum_{k=0}^{r} \mathscr{A}_k^{(r)} y_k + R_r$$

where $x_0 = a$, $l = b - a$, $h = l/r$, $x_r = a + rh$. The particular forms of $\mathscr{A}_k^{(r)}$ and R_r are shown in (*b*) below. If $r + 1$ is odd and $y(x)$ is a polynomial of $r + 1$ or less degree, the result is exact. But if $r + 1$ is even, only r-or-less-degree polynomials can be exactly evaluated. Consequently, odd-point formulas yield a higher degree of accuracy.

(b) Weight factors and remainders in closed-end formulas

r	$l = b - a$	$\mathscr{A}_0^{(r)}$	$\mathscr{A}_1^{(r)}$	$\mathscr{A}_2^{(r)}$	$\mathscr{A}_3^{(r)}$	$\mathscr{A}_4^{(r)}$	$\mathscr{A}_5^{(r)}$	R_r
1	h	$\dfrac{l}{2}$	$\dfrac{l}{2}$					$-\dfrac{h^3}{12}D^2 f(\eta)$
2	$2h$	$\dfrac{l}{6}$	$\dfrac{4l}{6}$	$\dfrac{l}{6}$				$-\dfrac{h^5}{90}D^4 f(\eta)$
3	$3h$	$\dfrac{l}{8}$	$\dfrac{3l}{8}$	$\dfrac{3l}{8}$	$\dfrac{l}{8}$			$-\dfrac{3h^5}{80}D^4 f(\eta)$
4	$4h$	$\dfrac{7l}{90}$	$\dfrac{32l}{90}$	$\dfrac{12l}{90}$	$\dfrac{32l}{90}$	$\dfrac{7l}{90}$		$-\dfrac{8h^7}{945}D^6 f(\eta)$
5	$5h$	$\dfrac{19l}{288}$	$\dfrac{75l}{288}$	$\dfrac{50l}{288}$	$\dfrac{50l}{288}$	$\dfrac{75l}{288}$	$\dfrac{19l}{288}$	$-\dfrac{275h^7}{12096}D^6 f(\eta)$

(c) Open-end formulas. If the end ordinates y_0 and y_r are undefined, unknown, or are singular points, the numerical evaluation of the definite integral can be expressed as

$$\int_a^b f(x)\,dx = \sum_{k=1}^{r-1} \mathscr{B}_k^{(r)} y_k + R_r$$

where a, b, h, l are the same as in (*a*) above and the weight factors $\mathscr{B}_k^{(r)}$ and the remainders R_r are tabulated in (*d*) below.

(d) Weight factors and remainders in open-end formulas

r	$l = b - a$	$\mathscr{B}_0^{(r)}$	$\mathscr{B}_1^{(r)}$	$\mathscr{B}_2^{(r)}$	$\mathscr{B}_3^{(r)}$	$\mathscr{B}_4^{(r)}$	$\mathscr{B}_5^{(r)}$	R_r
3	$3h$	0	$\dfrac{l}{2}$	$\dfrac{l}{2}$				$+\dfrac{3h^3}{8}D^2 f(\eta)$
4	$4h$	0	$\dfrac{2l}{3}$	$-\dfrac{2l}{3}$	$\dfrac{2l}{3}$			$+\dfrac{14h^5}{45}D^4 f(\eta)$
5	$5h$	0	$\dfrac{11l}{24}$	$\dfrac{l}{24}$	$\dfrac{l}{24}$	$\dfrac{11l}{24}$		$+\dfrac{95h^5}{144}D^4 f(\eta)$
6	$6h$	0	$\dfrac{11l}{20}$	$-\dfrac{14l}{20}$	$\dfrac{26l}{20}$	$-\dfrac{14l}{20}$	$\dfrac{11l}{20}$	$+\dfrac{41h^7}{170}D^6 f(\eta)$

(e) Accuracy of Newton–Cotes formulas based on r subdivisions of the interval $b - a$ does not necessarily increase as r increases and in some cases becomes asymptotic. For this reason, the application of composite formulas introduced in (4.06) to (4.10) is in many cases more efficient and more accurate.

(a) For large interval [a, b], the given definite integral can be replaced by the sums of the integrals of the same type but in shorter intervals, such as

$$\int_a^b f(x)\,dx = \int_a^{a+h} f(x)\,dx + \int_{a+h}^{a+2h} f(x)\,dx + \cdots + \int_{a+(n-1)h}^b f(x)\,dx = \sum_{k=1}^n \bar{Y}_k[a, a+kh] + R_n$$

where n = integer, $h = (b-a)/n = l/n$, $\bar{Y}[\ ]$ = approximation of the respective definite integral based on one of the formulas in (4.05), and

$$R_n \cong R[a, a+h] + R[a+h, a+2h] + \cdots + R[a+(n-1)h, b]$$

is the sum of all remainders.

(b) Trapezoidal rule approximates the area under $f(x)$ by a series of n trapezoids formed by ordinates y_k, y_{k+1} and the constant width $h = l/n$ as shown in the adjacent figure.

$$\boxed{\int_a^b f(x)\,dx = h(\tfrac{1}{2}y_0 + y_1 + y_2 + \cdots + y_{n-1} + \tfrac{1}{2}y_n) + R_n}$$

where $\quad R_n \cong -\dfrac{nh^3}{12} D^2 f(\eta) \qquad (a \le \eta \le b)$

A practical estimate of R_n is $\quad R_m \cong \tfrac{1}{3}(\bar{Y}_{2n} - \bar{Y}_n) \cong -\dfrac{l}{24}(\Delta^2 y_0 + \nabla^2 y_n)$

where $\bar{Y}_n$ = quadrature based on n subdivisions of l, $\bar{Y}_{2n}$ = quadrature based on $2n$ subdivisions of l, and $\Delta^2 y_0, \nabla^2 y_n$ = second-order differences at $0, n$, respectively.

(c) Example. The application of (b) is illustrated by

$$Y\left[0, \frac{\pi}{2}\right] = \int_0^{\pi/2} \cos x\,dx = \begin{cases} \bar{Y}_4 + R_4 \cong 0.987\,115\,801 + \dfrac{l}{12} h_4^2 = 1.007\,302\,179 \\[2mm] \bar{Y}_8 + R_8 \cong 0.996\,785\,172 + \dfrac{l}{12} h_8^2 = 1.001\,831\,767 \\[2mm] \bar{Y}_8 + R_m = 0.996\,785\,172 + \tfrac{1}{3}(\bar{Y}_8 - \bar{Y}_4) = 1.000\,008\,296 \end{cases}$$

where $\bar{Y}_4, \bar{Y}_8, R_4, R_8, R_m$ are tabulated below.

k	$h_4 = \pi/8$	For $\bar{Y}_4$	k	$h_8 = \pi/16$	For $\bar{Y}_8$
0	$\tfrac{1}{2}\cos 0$	0.500 000 000	0	$\tfrac{1}{2}\cos 0$	0.500 000 000
			1	$\cos h_8$	0.980 785 280
1	$\cos h_4$	0.923 879 533	2	$\cos 2h_8$	0.923 879 533
			3	$\cos 3h_8$	0.831 463 612
2	$\cos 2h_4$	0.707 106 781	4	$\cos 4h_8$	0.707 570 233
			5	$\cos 5h_8$	0.555 570 233
3	$\cos 3h_4$	0.382 683 432	6	$\cos 6h_8$	0.382 683 432
			7	$\cos 7h_8$	0.195 030 322
4	$\tfrac{1}{2}\cos 4h_4$	0.000 000 000	8	$\tfrac{1}{2}\cos 8h_8$	0.000 000 000
	$\sum = 2.513\,669\,746$			$\sum = 5.076\,585\,193$	
	$\bar{Y}_4 = h_4 \sum = 0.987\,115\,801$			$\bar{Y}_8 = h_8 \sum = 0.996\,785\,172$	
	$R_m = \tfrac{1}{3}(\bar{Y}_8 - \bar{Y}_4) = \tfrac{1}{3}(0.996\,785\,172 - 0.987\,115\,801) = 0.003\,223\,124$				

(d) Simpson's rule. If $f(x)$ is approximated by second-degree parabolas passing through equally spaced points $(0, 1, 2), (2, 3, 4), \ldots, (n - 2, n - 1, n)$ and n is an even integer, then with $h = (b - a)/n = l/n$,

$$\int_a^b f(x)\, dx = \frac{h}{3}\,(y_0 + 4y_1 + 2y_2 + 4y_3 + 2y_4 + \cdots + 2y_{n-2} + 4y_{n-1} + y_n) + R_n$$

where $R_n \cong -\dfrac{nh^5}{180}\, D^4 f(\eta)$ $(a \le \eta \le b)$

A practical estimate of R_n is $R_m \cong \frac{1}{15}(\bar{Y}_{2n} - \bar{Y}_n) \cong -\dfrac{l}{360}\,(\Delta^4 y_0 + \nabla^4 y_n)$

where the symbols have analogical meaning to those in (b) on the opposite page. This formula yields exact results if $f(x)$ is a polynomial of at most third degree.

(e) Example. The application of (d) is illustrated as a comparison with (c) by

$$Y\left(0, \frac{\pi}{2}\right) = \int_0^{\pi/2} \cos x\, dx = \begin{cases} \bar{Y}_4 + R_4 \cong 1.000\ 134\ 581 - \dfrac{l}{180} h_4^4 = 0.999\ 927\ 052 \\[2mm] \bar{Y}_8 + R_8 \cong 1.000\ 008\ 295 - \dfrac{l}{180} h_8^4 = 0.999\ 995\ 324 \\[2mm] \bar{Y}_8 + R_m \ \cong 1.000\ 008\ 295 + \frac{1}{15}(\bar{Y}_8 - \bar{Y}_4) = 0.999\ 999\ 864 \end{cases}$$

where $\bar{Y}_4$, Y_8, R_4, R_8, R_m are tabulated below.

k	$h_4 = \pi/8$	For $\bar{Y}$	k	$h_8 = \pi/16$	For $\bar{Y}_8$
0	$\cos 0$	1.000 000 000	0	$\cos 0$	1.000 000 000
			1	$4\cos h_8$	3.923 141 122
1	$4\cos h_4$	3.965 518 130	2	$2\cos 2h_8$	1.847 759 065
			3	$4\cos 3h_8$	3.325 878 449
2	$2\cos 2h_4$	1.414 213 562	4	$2\cos 4h_8$	1.414 213 562
			5	$4\cos 5h_8$	2.222 280 932
3	$4\cos 3h_4$	1.530 733 729	6	$2\cos 6h_8$	0.765 366 865
			7	$4\cos 7h_8$	0.780 361 288
4	$\cos 4h_4$	0.000 000 000	8	$\cos 8h_8$	0.000 000 000
$\sum = 7.640\ 465\ 421$			$\sum = 15.279\ 001\ 283$		
$\bar{Y}_4 = \frac{h_4}{3} \sum = 1.000\ 134\ 585$			$\bar{Y}_8 = \frac{h_8}{3} \sum = 1.000\ 008\ 295$		
$R_m = \frac{1}{15}(1.000\ 008\ 295 - 1.000\ 134\ 585) = -0.000\ 008\ 419$					

(f) Extrapolation formula. Because of its simplicity and good accuracy, Simpson's rule is the most-used method. If a greater accuracy is desired, the extrapolation formula introduced in (c) and (e) for R_m can be generalized for any composite rule based on weight factors of table $(4.05b)$ as

$$R_m \cong \frac{n^p \bar{Y}_{tn} - n^p \bar{Y}_n}{(tn)^p - n^p} = \frac{\bar{Y}_{tn} - \bar{Y}_n}{t^p - 1}$$

where p = order of derivative in R_n in $(4.05b)$, n = number of subdivisions used in calculation of quadrature $\bar{Y}_n$, and tn = number of subdivisions used in calculation of quadrature Y_{tn}.

(a) General case. As noted before in $(4.03b)$, any interpolation difference polynomial can be used for the development of numerical integration formulas. Symbolically, one of these polynomials is taken as

$$\bar{y}_s(u) = g_0(u)y_s + g_1(u)\omega y_s + g_2(u)\omega^2 y_s + \cdots + g_r(u)\omega^r y_s$$

where $0 < u < 1$, $\omega^k y_s = k$th-order difference of y_s, such as $\Delta^k y_s$, $\nabla^k y_s$, or $\delta^k y_s$, $g_k(u) =$ function factor in u, and s is a subscript specific for a particular point.

(b) Definite integral of $y = f(x)$ in $[a, b]$ in terms of $u = (x - a)/(b - a)$ can be approximated as

$$\int_a^b f(x)\,dx = (b - a)\int_0^1 \bar{y}(u)\,du + R_{r+1}^* = (b - a)\sum_{k=0}^r \mathscr{K}_k \omega^k y_s + R_{r+1}^*$$

where the change in limits is based on the relation $(4.02e)$,

$$\mathscr{K}_k = \int_0^1 g_k(u)\,du$$

is a specific constant for each formula, and R_{r+1}^* is the remainder shown in the subsequent sections in this chapter.

(c) Example. If the Newton–Gregory forward difference interpolation polynomial $(3.10b)$ is used,

$$\mathscr{K}_k = \mathscr{A}_k = \int_0^1 \binom{u}{k}\,du = \int_0^1 \frac{u(u-1)(u-2)\cdots(u-k+1)}{k!}\,du$$

and $\mathscr{A}_0 = 1$, $\mathscr{A}_1 = \frac{1}{2}, \ldots,$ $\mathscr{A}_k = \dfrac{1}{k!}\left(\dfrac{\mathscr{S}_1^{(k)}}{2} + \dfrac{\mathscr{S}_2^{(k)}}{3} + \cdots + \dfrac{\mathscr{S}_k^{(k)}}{k+1}\right)$

where $\mathscr{S}_1^{(k)}, \mathscr{S}_2^{(k)}, \ldots, \mathscr{S}_k^{(k)}$ are the Stirling's numbers of the first kind $(2.11a)$, $(A.07)$ and the particular values of $\mathscr{A}_k$ follow in $(4.08a)$. Similarly, if the Newton–Gregory backward difference polynomial $(3.10c)$ is used,

$$\mathscr{K}_k = \mathscr{B}_k = \int_{-1}^0 \binom{u}{k}\,du = -\int_0^{-1} \frac{u(u-1)(u-2)\cdots(u-k+1)}{k!}\,du$$

and $\mathscr{B}_0 = 1$, $\mathscr{B}_1 = -\frac{1}{2}, \ldots,$ $\mathscr{B}_k = \dfrac{1}{k!}\left(\dfrac{\mathscr{S}_1^{(k)}}{2} - \dfrac{\mathscr{S}_2^{(k)}}{3} + \cdots + (-1)^{k+1}\dfrac{\mathscr{S}_k^{(k)}}{k+1}\right)$

where the symbols in $\mathscr{B}_k$ are the same as in the calculation of $\mathscr{A}_k$.

4.08 NEWTON–GREGORY INTEGRATION FORMULAS

(a) Forward difference polynomial. If $y(u)$ in $(4.07b)$ is expressed by $(3.10b)$, then the definite integrals of $f(x)$ in $[a, a + h]$, $[a - h, a]$, and $[a - h, a + h]$ become, respectively,

$$\int_a^{a+h} f(x)\,dx = h\left[1 + \frac{\Delta}{2} - \frac{\Delta^2}{12} + \frac{\Delta^3}{24} - \cdots \mathscr{A}_r \Delta^r\right]y_0 + R_{r+1}^*[a, a + h]$$

$$\int_{a-h}^a f(x)\,dx = h\left[1 - \frac{\Delta}{2} + \frac{5\Delta^2}{12} - \frac{3\Delta^3}{8} + \cdots \mathscr{B}_r \Delta^r\right]y_0 + R_{r+1}^*[a - h, a]$$

$$\int_{a-h}^{a+h} f(x)\,dx = h\left[2 + \frac{\Delta^2}{3} - \frac{\Delta^3}{3} + \frac{59\Delta^4}{180} \rightarrow \cdots \mathscr{C}_r \Delta^r\right]y_0 + R_{r+1}^*[a - h, a + h]$$

where $\mathscr{A}_k$, $\mathscr{B}_k$, $\mathscr{C}_k$ for $k > 1$ and the remainders $R_{r+1}^*[\ ]$ are given in (b) on the opposite page.

(b) Numerical constants $(k = 1, 2, 3, \ldots, n)$

$$\mathcal{A}_k = \frac{1}{k!} \sum_{m=1}^{k} \frac{\mathscr{S}_m^{(k)}}{m+1} \qquad R_{r+1}^*[a, a+h] \cong h^{r+1}\mathcal{A}_{r+1}D^{r+1}f(\eta) \qquad (a \le \eta \le a+h)$$

$$\mathcal{B}_k = \frac{1}{k!} \sum_{m=1}^{k} \frac{(-1)^{m+1}\mathscr{S}_m^{(k)}}{m+1} \qquad R_{r+1}^*[a-h, a] \cong -h^{r+1}\mathcal{B}_{r+1}D^{r+1}f(\eta) \qquad (a-h \le \eta \le a)$$

$$\mathcal{C}_k = \mathcal{A}_k + \mathcal{B}_k \qquad R_{r+1}^*[a-h, a+h] \cong h^{r+1}\mathcal{C}_{r+1}D^{r+1}f(\eta) \qquad (a-h \le \eta \le a+h)$$

where $\mathscr{S}_m^{(k)}$ = Stirling's number of the first kind (2.11a), (A.07).

(c) Backward difference polynomial. If $\bar{y}(u)$ in (4.07b) is expressed by (3.10c), then the definite integrals of $f(x)$ in $[a, a+h]$, $[a-h, a]$, and $[a-h, a+h]$ become, respectively,

$$\int_a^{a+h} f(x)\,dx = h\left[1 + \frac{\nabla}{2} + \frac{5\nabla^2}{12} + \frac{3\nabla^3}{8} + \cdots + |\mathcal{B}_r|\,\nabla^r\right]y_0 + |R_{r+1}^*[a, a+h]|$$

$$\int_{a-h}^{a} f(x)\,dx = h\left[1 - \frac{\nabla}{2} - \frac{\nabla^2}{12} - \frac{\nabla^3}{24} - \cdots - |\mathcal{A}_r|\,\nabla^r\right]y_0 - |R_{r+1}^*[a-h, a]|$$

$$\int_{a-h}^{a+1} f(x)\,dx = h\left[1 + \frac{\nabla^2}{3} + \frac{\nabla^3}{3} + \frac{59\nabla^4}{180} + \cdots + |\mathcal{C}_r|\,\nabla^r\right]y_0 + |R_{r+1}^*[a-h, a+h]|$$

where $|\mathcal{A}_k|$, $|\mathcal{B}_k|$, $|\mathcal{C}_k|$ for $k \ge 1$ and the remainders $|R_{r+1}^*[\ ]|$ are the absolute values of their counterparts in (*a*) listed in (*b*) above.

(d) Composite formulas for $[x_0, x_0 + nh]$ are formally identical, but for $k \ge 1$, their constants take on the following general forms:

$$\mathcal{A}_k = \int_0^n \binom{u}{k}\,du = \frac{1}{k!}\sum_{m=1}^{k} \frac{n^{m+1}\mathscr{S}_m^{(k)}}{m+1} = \frac{1}{k!}\left[\frac{n^2\mathscr{S}_1^{(k)}}{2} + \frac{n^3\mathscr{S}_2^{(k)}}{3} + \cdots + \frac{n^{k+1}\mathscr{S}_k^{(k)}}{k+1}\right]$$

$$\mathcal{B} = \int_{-n}^{0} \binom{u}{k}\,du = -\frac{1}{k!}\sum_{m=1}^{k} \frac{(-n)^{m+1}\mathscr{S}_m^{(k)}}{m+1} = -\frac{1}{k!}\left[\frac{n^2\mathscr{S}_1^{(k)}}{2} - \frac{n^3\mathscr{S}_2^{(k)}}{3} + \cdots \frac{(-n)^{k+1}\mathscr{S}_k^{(k)}}{k+1}\right]$$

and $\mathcal{C}_k = \mathcal{A}_k + \mathcal{B}_k$. The remainders must be expressed in terms of $\mathcal{A}_{r+1}$, $\mathcal{B}_{r+1}$, and $\mathcal{C}_{r+1}$ of this section.

(e) Corrected Simpson's rule in terms of the relations stated in (*a*) is

$$\int_a^b f(x)\,dx = h\sum_{j=1}^{n-1}\left(2 + \frac{\Delta^2}{3}\right)y_j + h\sum_{j=1}^{n-1}\left(-\frac{\Delta^3}{3} + \frac{59\Delta^4}{180} - \cdots\right)y_j$$

$$= \frac{h}{3}\sum_{j=1}^{n-1}(y_{j-1} + 4y_j + y_{j+1}) + R_n^*[a, b] \qquad (j = 1, 3, 5, \ldots, n-1)$$

where n = even integer, $h = (b-a)/n$, and $R_n^*[a, b]$ is a better remainder than R_n in (4.06e). If the forward differences for the second half of $b - a$ are not available, $R_n^*[a, b]$ may be expressed as

$$R_n^*[a, b] \cong h\left[\sum_{j=1}^{p}\left(-\frac{\Delta^3}{3} + \frac{59\Delta^4}{180} - \cdots\right)y_j + \sum_{j=p}^{n-1}\left(\frac{\nabla^3}{3} + \frac{59\nabla^4}{180} + \cdots\right)y_j\right]$$

where $p = n/2$.

(a) Bessel's difference polynomial. If $\bar{y}(u)$ in (4.07b) is expressed by (3.12a), the definite integral of $f(x)$ in $[a, a + h]$ becomes

$$\int_a^{a+h} f(x)\, dx = h\left(\frac{y_0 + y_1}{2} - \frac{\lambda_2}{24} + \frac{11\lambda_4}{1440} - \frac{191\lambda_6}{120\,960} + \cdots \mathscr{C}_{2r}\lambda_{2r}\right) + R_{2r+2}^*[a, a + h]$$

where for $k \geq 1$, $\lambda_{2k} = \delta^{2k}y_0 + \delta^{2k}y_1$,

$$\mathscr{C}_{2k} = \int_0^1 \binom{u + k - 1}{2k}\, du = \frac{1}{(2k)!} \sum_{m=1}^{2k} \frac{k^{m+1} - (k-1)^{m+1}}{m + 1} \mathscr{S}_m^{(2k)}$$

$$R_{2r+2}^*[0, a + h] \cong h^{2r+2}\mathscr{C}_{2r+2}[D^{2r+2}f(a) + D^{2r+2}f(a + h)]$$

and $\mathscr{S}_m^{(2k)}$ = Stirling's number of the first kind (2.11a), (A.07).

(b) Stirling's difference polynomial. If $\bar{y}(u)$ in (4.07b) is expressed by (3.13a), the definite integral of $f(x)$ in $[a - h, a + h]$ becomes

$$\int_{a-h}^{a+h} f(x)\, dx = h\left(2 + \frac{\delta^2}{3} - \frac{\delta^4}{90} + \frac{\delta^6}{756} - \cdots \mathscr{C}_{2r}\delta^{2r}\right)y_0 + R_{2r+2}^*[a - h, a + h]$$

where for $k \geq 1$, $p = 2k - 1$,

$$\mathscr{C}_{2k} = \frac{1}{2k}\int_{-1}^1 \binom{u + k - 1}{p}u\, du = \frac{1}{(2k)!} \sum_{m=1}^{p} \frac{k^{m+2} - (k-2)^{m+2}}{m + 2} \mathscr{S}_m^{(p)}$$

$$R_{2r+2}^*[a - h, a + h] \cong h^{2r+2}D^{2r+2}f(\eta) \qquad (x_0 - h) \leq \eta \leq (x_0 + h)$$

and $\mathscr{S}_m^{(p)}$ = Stirling's number of the first kind (2.11a), (A.07).

(c) Corrected trapezoidal rule in terms of (a) above is

$$\int_a^b f(x)\, dx = \tfrac{1}{2}h \sum_{k=0}^{n-1} (y_k + y_{k+1}) + \underbrace{h \sum_{k=0}^{n-1} \left(-\frac{\lambda_2}{24} + \frac{11\lambda_4}{1440} - \frac{191\lambda}{120\,960} + \cdots\right)}_{R_n^*}$$

where $\lambda_2 = \delta^2 y_k + \delta^2 y_{k+1}$, $\lambda_4 = \delta^4 y_k + \delta^4 y_{k+1}$, $\lambda_6 = \delta^6 y_k + \delta^6 y_{k+1}, \ldots$, $h = (b - a)/n$, and n = integer. R_n^* is a better remainder than R_n in (4.06c).

(d) Corrected Simpson's rule in terms of (b) above is

$$\int_a^b f(x)\, dx = \frac{h}{3} \sum_{j=1}^{n-1} (y_{j-1} + 4y_j + y_{j+1}) + \underbrace{h \sum_{j=1}^{n-1} \left(-\frac{\delta^4}{90} + \frac{\delta^6}{756} - \cdots\right)y_j}_{R_n^*}$$

where $\delta^4 y_j = y_{j-3} - 4y_{j-2} + 6y_{j-1} - 4y_j + y_{j+1}$

$\delta^6 y_1 = y_{j-4} - 5y_{j-3} + 10y_{j-2} - 10y_{j-1} + 5y_j - y_{j+1}$

$\cdots \cdots \cdots \cdots \cdots \cdots \cdots \cdots \cdots \cdots \cdots$

and n = even integer. R_n^* is a better remainder than R_n in (4.06d). The spacing h is the same as in (c) above.

(a) General form. The Euler–MacLaurin formula is the most important relationship of numerical analysis. For equally spaced values $x_k = a + kh$, $y_k = f(a + kh)$ and $h = (b - a)/n$,

$$\int_a^b f(x)\,dx = \underbrace{\frac{h}{2}(y_0 + y_n) + h\sum_{k=1}^{n-1} y_k}_{I_1} - \underbrace{\sum_{k=1}^{r} G_{2k} + R_{2r+2}^*}_{I_2}$$

where the first part I_1 is the trapezoidal rule (4.06b) and the second part is an asymptotic series

$$I_2 = -\left[\frac{h^2 \bar{B}_2 \omega_1}{2!} + \frac{h^4 \bar{B}_4 \omega_3}{4!} + \frac{h^6 \bar{B}_6 \omega_5}{6!} + \frac{h^8 \bar{B}_8 \omega_7}{8!} + \cdots\right]$$

in which $\bar{B}_{2k}$ is the Bernoulli number (2.22a), (A.10) and

$$\omega_{2k-1} = \left[\frac{d^{2k-1} f(x)}{dx^{2n-1}}\right]_{x=b} - \left[\frac{d^{2k-1} f(x)}{dx^{2n-1}}\right]_{x=a}$$

The terms in I_2 first decrease and then begin to increase. Thus the most accurate result is obtained by stopping with the term G_{2r} preceding the smallest one G_{2r+2} which is left for the remainder.

$G_2 = \dfrac{h^2 \bar{B}_2 \omega_1}{2!} = \dfrac{h^2 \omega_1}{12}$	$G_4 = \dfrac{h^4 \bar{B}_4 \omega_3}{4!} = -\dfrac{h^4 \omega_3}{720}$	
$G_6 = \dfrac{h^6 \bar{B}_6 \omega_5}{6!} = \dfrac{h^6 \omega_5}{30\,240}$	$G_8 = \dfrac{h^8 \bar{B}_8 \omega_7}{8!} = -\dfrac{h^8 \omega_7}{1\,209\,600}$	

(b) CA model of part two is

$$I_2 = -\frac{h^2}{12}\left(\omega_1 - \frac{h^2}{60}\left(\omega_3 - \frac{h^2}{42}\left(\omega_5 - \frac{h^2}{40}(\omega_7 - \cdots)\right)\right)\right)$$

and the remainder can be estimated as

$$|R_{2r+2}^*| < \begin{cases} 2\,|G_{2r+2}| & \text{if } I_2 - \text{series is alternating} \\ |G_{2r+2}| & \text{if } I_2 - \text{series is monotonic} \end{cases}$$

(c) Example. To show the efficiency of this formula, the example (4.06c) is considered with $y = \cos x$, $a = 0$, $b = \frac{1}{2}\pi$, $n = 4$, $h = \pi/8$. Then

$$\omega_1 = -\sin\tfrac{1}{2}\pi + \sin 0 = -1 \qquad \omega_3 = \sin\tfrac{1}{2}\pi - \sin 0 = 1 \qquad \omega_5 = -\sin\tfrac{1}{2}\pi + \sin 0 = -1$$

$$\int_0^{\pi/2} \cos x\,dx = 0.987\,115\,801 + \frac{h}{12}\left(1 + \frac{h^2}{60}\left(1 + \frac{h^2}{42}\right)\right) + R_8^* = 0.999\,999\,999\,6 + R_8^*$$

where $0.987\,115\,801$ is the result of the trapezoidal rule given in table (4.06c). Since

$$G_2 = \frac{h^2(-1)}{12} = -0.018\,851\,047\,4 \qquad G_6 = \frac{h^6(-1)}{30\,240} = -0.000\,000\,121\,2$$

$$G_4 = -\frac{h^4(+1)}{720} = -0.000\,033\,029\,8 \qquad G_8 = -\frac{h^8(+1)}{1\,209\,600} = -0.000\,000\,000\,4$$

is monotonic,

$$R_8^* \leq |-G_8| = 0.000\,000\,000\,4$$

(a) Concept. In the development of formulas introduced in the preceding sections, the values of y_k were assumed to be known at constant spacing $h = (b - a)/n$ and the weights have been determined in such a way that the formula yielded an accuracy of highest possible order. It occurred to Gauss that some other spacing may produce a better approximation, which in turn posed the task of finding the coordinates of points at which the ordinates magnified by their weights would given the best approximation.

(b) General Gauss formula introduced below in symbolic form is the model of this type of approximation.

$$\int_a^b w(x)f(x)\,dx = C[A_1f(x_1) + \mathscr{A}_2f(x_2) + \cdots + \mathscr{A}_rf(x_r)] + R$$

where $w(x)$ = positive weight function to be selected in each particular case, $\mathscr{A}_k$ = weight factor (weight), x_k = coordinate, both given and tabulated in the following sections of this chapter, and C = scaling constant [for illustration refer to (4.12a)].

(c) Remainder R involves some high-order derivatives of $f(x)$ and is of limited practical value. A more direct and simpler approach is to compare the results of several approximations, each involving different number of terms ($r = 3, 4, 5, \ldots$). If this comparison leads to widely different results, the singularities or steep oscillation may be present.

4.12 GAUSS–LEGENDRE FORMULA

(a) Integration formula. With the transformation

$$x = \tfrac{1}{2}(b + a) + \tfrac{1}{2}(b - a)u \qquad dx = \tfrac{1}{2}(b - a)\,du$$

and $w(x) = 1$, the general formula (4.11b) reduces to

$$\int_a^b f(x)\,dx = \tfrac{1}{2}(b - a)\int_{-1}^1 f[\tfrac{1}{2}(b + a) + \tfrac{1}{2}(b - a)u]\,du$$
$$= \tfrac{1}{2}(b - a)[\mathscr{A}_1f(x_1) + \mathscr{A}_2f(x_2) + \cdots + \mathscr{A}_rf(x_r)] + R$$

where with $k = 1, 2, \ldots, r$, $P_r(u)$ = Legendre polynomial of order r (11.10), $u_1, u_2, \ldots, u_r$ = zeros of $P_r(u)$, and

$$\mathscr{A}_k = \frac{2(1 - u_k^2)}{(r + 1)^2 P_{r+1}^2(u_k)} \qquad R = \frac{(b - a)^{2r+1}(r!)^4}{(2r + 1)[(2r)!]^3} f^{(2r)}(\eta) \qquad (a \le \eta \le b)$$

The weight factors $\mathscr{A}_k$ are symmetrical and the coordinates u_k are antisymmetrical with respect to the center of $b - a$. The formula is exact for polynomials of degree up to $r + 1$.

(b) Particular case. For $r = 3$, the Legendre polynomial in (11.10d) of order three is

$$P_3(u) = \tfrac{1}{2}(5u^3 - 3u)$$

Its zeros are

$$u_1 = -\sqrt{\frac{3}{5}} \qquad u_2 = 0 \qquad u_3 = \sqrt{\frac{3}{5}}$$

and the respective weight factors are

$$\mathscr{A}_1 = 5/9 \qquad \mathscr{A}_2 = 8/9 \qquad \mathscr{A}_3 = 5/9$$

Table of u_k and $\mathscr{A}_k$ is shown on the opposite page for r up to 10.

(c) Numerical values of u_k and $\mathscr{A}_k$ are tabulated below for $r = 2, 3, \ldots, 10$. More elaborate tables are given in Ref. 1.01.

r	u_k	$\mathscr{A}_k$	r	u_k	$\mathscr{A}_k$
2	−0.577 350 269 189	1.000 000 000 000	8	−0.960 289 856 498	0.101 228 536 290
	0.577 350 269 189	1.000 000 000 000		−0.796 666 477 414	0.222 381 034 453
				−0.525 532 409 916	0.313 707 645 878
3	−0.774 596 669 241	0.555 555 555 556		−0.183 434 642 496	0.362 683 783 378
	0.000 000 000 000	0.888 888 888 889		0.183 434 642 496	0.362 683 783 328
	0.774 596 669 241	0.555 555 555 556		0.525 532 409 916	0.313 707 645 878
				0.796 666 477 414	0.222 381 034 453
4	−0.861 136 311 594	0.347 854 484 514		0.960 289 856 498	0.101 228 536 290
	−0.339 981 043 585	0.652 145 154 863			
	0.339 981 043 585	0.652 145 154 863	9	−0.968 160 239 508	0.081 274 388 362
	0.861 136 311 594	0.347 854 484 514		−0.836 031 107 327	0.180 648 160 695
				−0.613 371 432 701	0.260 610 696 403
5	−0.906 179 845 939	0.236 926 885 056		−0.324 253 423 404	0.312 347 077 040
	−0.538 469 310 106	0.478 628 670 050		0.000 000 000 000	0.330 239 355 500
	0.000 000 000 000	0.568 888 888 889		0.324 253 423 404	0.312 347 077 040
	0.538 469 310 106	0.478 628 670 050		0.613 371 432 701	0.260 610 696 403
	0.906 179 845 939	0.236 926 885 056		0.836 031 107 327	0.180 648 160 695
				0.968 160 239 508	0.081 274 388 362
6	−0.932 469 514 203	0.171 324 492 379			
	−0.661 209 386 466	0.360 761 573 048	10	−0.973 906 528 517	0.066 671 344 309
	−0.238 619 186 083	0.467 913 934 573		−0.865 063 366 689	0.149 451 349 151
	0.238 610 186 083	0.467 913 934 573		−0.679 409 568 299	0.219 086 362 516
	0.661 209 386 466	0.360 761 573 048		−0.433 295 394 129	0.269 266 719 310
	0.932 469 514 203	0.171 324 492 379		−0.148 874 338 982	0.295 524 224 715
				0.148 874 338 082	0.295 524 224 715
7	−0.949 107 912 343	0.129 484 966 169		0.433 395 394 129	0.269 266 919 310
	−0.741 531 185 599	0.279 705 391 489		0.679 409 568 299	0.219 086 362 516
	−0.405 845 151 377	0.381 830 050 505		0.865 063 366 689	0.149 451 349 151
	0.000 000 000 000	0.417 959 183 673		0.973 906 528 517	0.066 671 344 309
	0.405 845 151 377	0.381 830 050 505			
	0.741 531 185 599	0.279 705 391 489			
	0.949 107 912 343	0.129 484 966 169			

(d) Example. If $a = 0$, $b = \frac{1}{2}\pi$, $r = 3$, $f(x) = x \cos x$, $x_k = \frac{1}{2}[(b + a) + (b - a)u_k]$,

$$x_1 = \tfrac{1}{4}\pi(1 + u_1) = 0.177\,031 \qquad f(x_1) = 0.174\,264 \qquad \mathscr{A}_1 = 0.555\,556$$

$$x_2 = \tfrac{1}{4}\pi(1 + u_2) = 0.785\,398 \qquad f(x_2) = 0.555\,360 \qquad \mathscr{A}_2 = 0.888\,889$$

$$x_3 = \tfrac{1}{4}\pi(1 + u_3) = 1.393\,765 \qquad f(x_3) = 0.245\,453 \qquad \mathscr{A}_3 = 0.555\,556$$

and by (4.12a) on the opposite page

$$\int_0^{\pi/2} x \cos x \, dx = \tfrac{1}{4}\pi \sum_{k=1}^{3} \mathscr{A}_k f(x_k) + R = 0.570\,851 - 0.000\,070 = 0.570\,781$$

and the correct value is 0.570 796, which shows the efficiency of the formula even when a very small number of points is used.

(a) Integration formula. With $x = \frac{1}{2}[(b + a) + (b - a)u]$ and the weight $w(x) = (b - a)/r$, the formula (4.11b) reduces to

$$\int_a^b f(x)\,dx = \frac{1}{2}(b - a) \int_{-1}^1 f[\frac{1}{2}(b + a) + \frac{1}{2}(b - a)u]\,du \qquad R = \frac{[\frac{1}{2}(b - a)]^{2r-1}}{(2r)!} f^{(2r-2)}(\eta)$$

$$= \frac{b - a}{r}[f(x_1) + f(x_2) + \cdots + f(x_r)] + R \qquad a \le \eta \le b$$

where u_k are the zeros of the following polynomials:

$$G_2(u) = u^2 - \frac{1}{3} \qquad G_5(u) = u^5 - \frac{5}{6}u^3 + \frac{7}{72}u \qquad G_6(u) = u^6 - u^4 + \frac{1}{5}u^2 - \frac{1}{105}$$

$$G_3(u) = u^3 - \frac{1}{3}u \qquad G_7(u) = u^7 - \frac{7}{6}u^5 + \frac{119}{360}u^3 - \frac{1}{6480}u$$

$$G_4(u) = u^4 - \frac{2}{3}u^2 + \frac{1}{45} \qquad G_9(u) = u^9 - \frac{3}{2}u^7 + \frac{27}{40}u^5 - \frac{57}{560}u^3 + \frac{53}{22\,400}u$$

The arguments u_k are antisymmetrical with respect to the center of the interval $b - a$. $r = 2, 3, 4, 5, 6, 7, 9$, and for $r = 8$, $r \ge 10$, the method has no real solutions, which is the fundamental defect of the Chebshev method.

(b) Numerical values of u_k corresponding to particular values of r are tabulated below.

$r = 2$	$r = 3$	$r = 4$
$u_1 = -0.577\,350\,269$ $u_2 = 0.577\,350\,269$	$u_1 = -0.707\,106\,781$ $u_2 = 0.000\,000\,000$ $u_3 = 0.707\,106\,781$	$u_1 = -0.794\,654\,472$ $u_2 = -0.187\,592\,474$ $u_3 = 0.187\,592\,474$ $u_4 = 0.794\,654\,472$

$r = 5$	$r = 6$	$r = 7$
$u_1 = -0.832\,497\,487$ $u_2 = -0.374\,541\,410$ $u_3 = 0.000\,000\,000$ $u_4 = 0.374\,541\,410$ $u_5 = 0.832\,497\,487$	$u_1 = -0.866\,246\,818$ $u_2 = -0.422\,518\,654$ $u_3 = -0.266\,635\,402$ $u_4 = 0.266\,635\,402$ $u_5 = 0.422\,518\,654$ $u_6 = 0.866\,246\,818$	$u_1 = -0.883\,861\,701$ $u_2 = -0.529\,656\,775$ $u_3 = -0.323\,911\,811$ $u_4 = 0.000\,000\,000$ $u_5 = 0.323\,911\,811$ $u_6 = 0.529\,656\,775$ $u_7 = 0.883\,861\,701$

$r = 9$		
$u_1 = -0.911\,589\,308$ $u_2 = -0.601\,018\,655$ $u_3 = -0.528\,761\,783$	$u_4 = -0.167\,906\,184$ $u_5 = 0.000\,000\,000$ $u_6 = 0.167\,906\,184$	$u_7 = 0.528\,761\,783$ $u_8 = 0.601\,018\,655$ $u_9 = 0.911\,589\,308$

(c) Example. If $a = 0$, $b = \frac{1}{2}\pi$, $r = 4$, $f(x) = x \cos x$, $x_k = \frac{1}{4}\pi(1 + u_k)$,

$$x_1 = 0.161\,278 \qquad x_2 = 0.638\,063 \qquad x_3 = 0.932\,733 \qquad x_4 = 1.409\,519$$
$$f(x_1) = 0.159\,185 \qquad f(x_2) = 0.512\,524 \qquad f(x_3) = 0.555\,524 \qquad f(x_4) = 0.226\,340$$

and by (a) above,

$$\int_0^{\pi/2} x \cos x\,dx = \frac{\pi}{8}\sum_{k=1}^4 f(x_k) + R = 0.570\,836 - 0.000\,027 = 0.570\,809$$

and the correct value is 0.570 796, which again shows the efficiency of the formula even when a very small number of points is used.

(a) Integration formula. With $x = \frac{1}{2}[(b+a) + (b-a)u]$ and the weight function $w(u) = (1 - u^2)^{-1/2}$, the general form $(4.11b)$ reduces to

$$\int_a^b f(x)\,dx = \frac{1}{2}(b-a)\int_{-1}^{1}(1-u^2)^{-1/2}g(u)\,du \qquad\qquad R = 2\left(\frac{b-a}{2}\right)^{2r}\frac{\pi}{(2r)!}f^{(2r)}(\eta)$$

$$= \frac{(b-a)\pi}{2r}[g(u_1) + g(u_2) + \cdots + g(u_r)] + R \qquad\qquad (a \le \eta \le b)$$

where

$$u_k = \cos\frac{(2k-1)\pi}{2r} \qquad (k = 1, 2, 3, \ldots, r)$$

$$g(u_k) = (1 - u_k^2)^{1/2}f[\tfrac{1}{2}(b+a) + (b-a)u_k]$$

The weights are all equal, $\mathscr{A}_1 = \mathscr{A}_2 = \cdots = \mathscr{A}_r = \pi/r$, and the arguments u_k are the zeros of the Chebshev polynomial $(11.15d)$

$$T_r(u) = \cos\,(r\cos^{-1}u)$$

and are antisymmetrical with respect to the center of the interval $b - a$.

(b) Numerical values of u_k corresponding to particular values of r are tabulated below.

$r = 2$	$r = 3$	$r = 4$
$u_1 = -0.707\,106\,781$	$u_1 = -0.866\,025\,254$	$u_1 = -0.923\,879\,533$
$u_2 = 0.707\,106\,781$	$u_2 = 0.000\,000\,000$	$u_2 = -0.382\,683\,432$
	$u_3 = 0.866\,025\,254$	$u_3 = 0.382\,683\,132$
		$u_4 = 0.923\,879\,533$

$r = 5$	$r = 6$	$r = 7$
$u_1 = -0.951\,056\,516$	$u_1 = -0.965\,925\,826$	$u_1 = -0.974\,927\,912$
$u_2 = -0.587\,785\,252$	$u_2 = -0.707\,106\,781$	$u_2 = -0.781\,831\,483$
$u_3 = 0.000\,000\,000$	$u_3 = -0.258\,819\,045$	$u_3 = -0.433\,883\,739$
$u_4 = 0.587\,785\,252$	$u_4 = 0.258\,810\,045$	$u_4 = 0.000\,000\,000$
$u_5 = 0.951\,056\,516$	$u_5 = 0.707\,106\,781$	$u_5 = 0.433\,883\,739$
	$u_6 = 0.965\,025\,826$	$u_6 = 0.781\,831\,483$
		$u_7 = 0.974\,927\,912$

$r = 8$		
$u_1 = -0.980\,785\,280$	$u_4 = -0.195\,090\,322$	$u_7 = 0.831\,469\,612$
$u_2 = -0.831\,469\,612$	$u_5 = 0.195\,090\,322$	$u_8 = 0.980\,785\,280$
$u_3 = -0.555\,570\,023$	$u_6 = 0.555\,570\,023$	

(c) Two auxiliary cases in which the arguments u_k and the weight factors $\mathscr{A}_k$ are given in closed form are

$$\int_{-1}^{1}\sqrt{1 - u^2}\,g(u)\,du = \frac{\pi}{r+1}\sum_{k=1}^{r}\sin^2\frac{k\pi}{r+1}g\left(\cos\frac{k\pi}{r+1}\right) + \frac{\pi g^{(2r)}(\eta)}{2^{2r}(2r)!}$$

$$\int_{-1}^{1}\sqrt{\frac{1-u}{1+u}}\,g(u)\,du = \frac{4\pi}{2r+1}\sum_{k=1}^{r}\sin^2\frac{k\pi}{2r+1}g\left(\cos\frac{2k\pi}{2r+1}\right) + \frac{\pi g^{(2r)}(\eta)}{2^{2r}(2r)!}$$

The last term in each formula is the remainder R with $-1 \le \eta \le 1$.

(a) Notation, $r + 1 =$ number of terms $\qquad k = 1, 2, \ldots, r$

$P_r(u) =$ Legendre polynomial of order r (11.10d)

$u_1, u_2, \ldots, u_r =$ roots of $[P_r(u) + P_{r+1}(u)] = 0 \qquad v_k = \frac{1}{2}(1 + u_k)$

(b) Integration formula A. When in (4.11b) the interval of integration is transformed to $[-1, 1]$, $w(u) = 1$, and the lower limit is preassigned as $u_0 = -1$, then

$$\int_{-1}^{1} g(u)\, du = \frac{2}{(r+1)^2} g(-1) + \mathscr{A}_1 g(u_1) + \mathscr{A}_2 g(u_2) + \cdots + \mathscr{A}_r g(u_r) + R_r$$

where

$$\mathscr{A}_k = \frac{1 - u_k}{r^2 [P_r(u_k)]^2} \qquad R_r = \frac{(r+1) 2^{2r+1} (r!)^4}{[(2r+1)!]^3} g^{(2r+1)}(\eta) \qquad (-1 \le \eta \le 1)$$

This formula is exact for polynomials of degree up to $2r$.

(c) Integration formula B. When the same integral is transformed to $[0, 1]$, $w(u) = 1$, and the lower limit is preassigned as $u_0 = 0$, then

$$\int_{0}^{1} h(v)\, dv = \frac{1}{(r+1)^2} h(0) + \mathscr{B}_1 h(v_1) + \mathscr{B}_2 h(v_2) + \cdots + \mathscr{B}_r h(v_r) + R_r^*$$

where

$$\mathscr{B}_k = \frac{\frac{1}{2}(1 - u_k)}{r^2 [P_r(u_k)]^2} \qquad R_r^* = \frac{(r+1)(r!)^4}{[(2r+1)!]^3} h^{(2r+1)}(\eta) \qquad (0 \le \eta \le 1)$$

This formula is again exact for polynomials of degree up to $2r$.

(d) Numerical values of u_k, $\mathscr{A}_k$ and v_k, $\mathscr{B}_k$ with respective remainders are tabulated below.

$r = 1$	$u_1 = \;\;\; 0.333\,333$	$\mathscr{A}_1 = 1.500\,000$	$v_1 = 0.666\,667$	$\mathscr{B}_1 = 0.\;\;0\,000$
	$R_1 = 7.407(10)^{-2} g^{(3)}(\eta)$		$R_1^* = 9.259(10)^{-3} h^{(3)}(\eta)$	
$r = 2$	$u_1 = -0.289\,898$ $\quad$ $u_2 = \;\;\; 0.689\,897$	$\mathscr{A}_1 = 1.024\,972$ $\quad$ $\mathscr{A}_2 = 0.752\,806$	$v_1 = 0.355\,051$ $\quad$ $v_2 = 0.844\,949$	$\mathscr{B}_1 = 0.512\,486$ $\quad$ $\mathscr{B}_2 = 0.376\,403$
	$R_2 = 8.889(10)^{-4} g^{(5)}(\eta)$		$R_2^* = 2.778(10)^{-5} h^{(5)}(\eta)$	
$r = 3$	$u_1 = -0.575\,319$ $\quad$ $u_2 = \;\;\; 0.181\,066$ $\quad$ $u_3 = \;\;\; 0.822\,834$	$\mathscr{A}_1 = 0.657\,689$ $\quad$ $\mathscr{A}_2 = 0.776\,387$ $\quad$ $\mathscr{A}_3 = 0.440\,925$	$v_1 = 0.212\,341$ $\quad$ $v_2 = 0.590\,533$ $\quad$ $v_3 = 0.911\,417$	$\mathscr{B}_1 = 0.328\,835$ $\quad$ $\mathscr{B}_2 = 0.388\,194$ $\quad$ $\mathscr{B}_3 = 0.220\,463$
	$R_3 = 3.239(10)^{-7} g^{(7)}(\eta)$		$R_3^* = 2.530(10)^{-8} h^{(7)}(\eta)$	
$r = 4$	$u_1 = -0.720\,480$ $\quad$ $u_2 = -0.167\,181$ $\quad$ $u_3 = \;\;\; 0.446\,314$ $\quad$ $u_4 = \;\;\; 0.885\,792$	$\mathscr{A}_1 = 0.446\,207$ $\quad$ $\mathscr{A}_2 = 0.623\,653$ $\quad$ $\mathscr{A}_3 = 0.562\,712$ $\quad$ $\mathscr{A}_4 = 0.287\,427$	$v_1 = 0.139\,760$ $\quad$ $v_2 = 0.416\,410$ $\quad$ $v_3 = 0.723\,157$ $\quad$ $v_4 = 0.942\,896$	$\mathscr{B}_1 = 0.223\,104$ $\quad$ $\mathscr{B}_2 = 0.311\,827$ $\quad$ $\mathscr{B}_3 = 0.281\,356$ $\quad$ $\mathscr{B}_4 = 0.147\,214$
	$R_4 = 1.371(10)^{-11} g^{(9)}(\eta)$		$R_4^* = 2.670(10)^{-14} h^{(9)}(\eta)$	

(e) Example. If $g(u) = f[\frac{1}{2}(b + a) + \frac{1}{2}(b - a)u]$,

$$\int_{-1}^{1} g(u)\, du = \begin{cases} \frac{1}{2} g(-1) + \frac{3}{2} g(\frac{1}{3}) + \frac{2}{27} g^{(3)}(\eta) & \text{for } r = 1 \\[2ex] \frac{2}{9} g(-1) + \frac{16 + \sqrt{6}}{18} g\left(\frac{1 - \sqrt{6}}{5}\right) + \frac{16 - \sqrt{6}}{28} g\left(\frac{1 + \sqrt{6}}{5}\right) + \frac{1}{1125} g^{(5)}(\eta) & \text{for } r = 2 \end{cases}$$

(a) Notation. $r + 2$ = number of terms $k = 1, 2, \ldots, r$

$P_r(u)$ = Legendre polynomial of order r (11.10d)

$u_1, u_2, \ldots, u_r$ = roots of polynomial $[dP_{r+1}(u)/du = 0]$ $v_k = \tfrac{1}{2}(1 + u_k)$

(b) Integration formula A. When in (4.11b) the interval of integration is transformed to $[-1.1]$, $w(u) = 1$, and the lower limit and the upper limit are preassigned as $u_0 = -1$, $u_{r+1} = 1$, then

$$\int_{-1}^{1} g(u)\, du = \frac{2}{(r+1)(r+2)}\, [g(-1) + g(1)] + \mathscr{A}_1 g(u_1) + \mathscr{A}_2 g(u_2) + \cdots + \mathscr{A}_r g(u_r) + R_r$$

where

$$\mathscr{A}_k = \frac{2}{(r+1)(r+2)[P_{r+1}(u_k)]^2} \qquad R_r = -\frac{(r+2)(r+1)^3 2^{2r+3}(r!)^4}{(2r+3)[(2r+2)!]^3}\, g^{(2r+2)}(\eta) \qquad (-1 \le \eta \le 1)$$

This formula is exact for polynomials of degree up to $2r + 1$.

(c) Integration formula B. When the same integral is transformed to $[0, 1]$, $w(v) = 1$, and the lower limit and the upper limit are preassigned as $v_0 = 0$, $v_{r+1} = 1$, then

$$\int_{0}^{1} h(v)\, dv = \frac{1}{(r+1)(r+2)}\, [h(0) + h(1)] + \mathscr{B}_1 h(v_1) + \mathscr{B}_2 h(v_2) + \cdots + \mathscr{B}_r h(v_r) + R_r^{*}$$

where

$$\mathscr{B}_k = \frac{1}{(r+1)(r+2)[P_{r+1}(u_k)]^2} \qquad R_r^{*} = -\frac{2(r+2)(r+1)^2(r!)^4}{(2r+3)[(2r+2)!]^3}\, h^{(2r+2)}(\eta) \qquad (0 \le \eta \le 1)$$

This formula is exact for polynomials of degree up to $2r + 1$.

(d) Numerical values of u_k, $\mathscr{A}_k$ and v_k, $\mathscr{B}_k$ with the respective remainders are tabulated below.

$r = 1$	$u_1 =$ 0.000 000	$\mathscr{A}_1 =$ 1.333 333	$v_1 =$ 0.500 000	$\mathscr{B}_1 =$ 0.666 667
	$R_1 = 1.111(10)^{-2} g^{(4)}(\eta)$		$R_1^{*} = 6.944(10)^{-4} h^{(4)}(\eta)$	
$r = 2$	$u_1 = -0.447\,214$ $u_2 = 0.447\,214$	$\mathscr{A}_1 =$ 0.833 333 $\mathscr{A}_2 =$ 0.833 333	$v_1 =$ 0.276 393 $v_2 =$ 0.723 607	$\mathscr{B}_1 =$ 0.416 667 $\mathscr{B}_2 =$ 0.416 667
	$R_2 = 8.446(10)^{-5} g^{(6)}(\eta)$		$R_2^{*} = 1.323(10)^{-6} h^{(6)}(\eta)$	
$r = 3$	$u_1 = -0.654\,654$ $u_2 = 0.000\,000$ $u_3 = 0.654\,654$	$\mathscr{A}_1 =$ 0.544 444 $\mathscr{A}_2 =$ 0.711 111 $\mathscr{A}_3 =$ 0.544 444	$v_1 =$ 0.172 673 $v_2 =$ 0.500 000 $v_3 =$ 0.357 384	$\mathscr{B}_1 =$ 0.272 222 $\mathscr{B}_2 =$ 0.355 556 $\mathscr{B}_3 =$ 0.272 222
	$R_3 = 3.599(10)^{-7} g^{(8)}(\eta)$		$R_3^{*} = 1.406(10)^{-9} h^{(8)}(\eta)$	
$r = 4$	$u_1 = -0.285\,232$ $u_2 = -0.765\,055$ $u_3 = 0.765\,055$ $u_4 = 0.285\,232$	$\mathscr{A}_1 =$ 0.544 858 $\mathscr{A}_2 =$ 0.378 475 $\mathscr{A}_3 =$ 0.378 475 $\mathscr{A}_4 =$ 0.554 858	$v_1 =$ 0.357 384 $v_2 =$ 0.117 473 $v_3 =$ 0.882 528 $v_4 =$ 0.642 616	$\mathscr{B}_1 =$ 0.277 429 $\mathscr{B}_3 =$ 0.189 238 $\mathscr{B}_3 =$ 0.189 238 $\mathscr{B}_4 =$ 0.277 222
	$R_4 = 9.695(10)^{-10} g^{(10)}(\eta)$		$R_4^{*} = 9.468(10)^{-13} h^{(10)}(\eta)$	

(e) Example. If u_1, u_2 and $\mathscr{A}_1$, $\mathscr{A}_2$ are desired, the roots of $dP_3(u)/du = 0$ are calculated first as

$$\tfrac{1}{2}(15u^2 - 3) = 0 \qquad u_1 = -\sqrt{\tfrac{1}{5}} \qquad u_2 = \sqrt{\tfrac{1}{5}}$$

$$\mathscr{A}_1 = \frac{2}{(3)(4)[\tfrac{1}{2}(5u_1^3 - 3u_1)]^2} = \frac{5}{6} \qquad \mathscr{A}_2 = \frac{2}{(3)(4)[\tfrac{1}{2}(5u_2^3 - 3u_2)]^2} = \frac{5}{6}$$

(a) Notation. $u = u(x)$, $v = v(x)$ = differentiable functions of x

$\left. \begin{array}{l} u', u'', \ldots, u^{(n)} \\ v', v'', \ldots, v^{(n)} \end{array} \right\}$ = successive derivatives of u and v, respectively

$\left. \begin{array}{l} U_1, U_2, \ldots, U_n \\ V_1, V_2, \ldots, V_n \end{array} \right\}$ = successive integrals of u and v, *respectively*

a, b, λ = constants k, m, n = positive integers

(b) General integration formula

$$\int_a^b uv\, dx = \begin{cases} [U_1 v - U_2 v' + U_3 v'' - U_4 v''' + \cdots]_a^b \\ [uV_1 - u'V_2 + u''V_3 - u'''V_4 + \cdots]_a^b \end{cases}$$

(c) Particular cases

$$\int_a^b x^m v\, dx = m! \left[x^m \sum_{k=0}^m \frac{V_{k+1}}{(m-k)!\,(-x)^k} \right]_a^b$$

$$\int_a^b x^m \sin \lambda x\, dx = m! \left[x^m \sum_{k=0}^m \frac{\sin(\lambda x - \phi)}{(m-k)!\,(-\lambda x)^k} \right]_a^b \qquad [\phi = \tfrac{1}{2}(k+1)\pi]$$

$$\int_a^b x^m \cos \lambda x\, dx = m! \left[x^m \sum_{k=0}^m \frac{\cos(\lambda x - \phi)}{(m-k)!\,(-\lambda x)^k} \right]_a^b \qquad [\phi = \tfrac{1}{2}(k+1)\pi]$$

$$\int_a^b x^m e^{\lambda x}\, dx = m! \left[x^m \sum_{k=0}^m \frac{e^{\lambda x}}{(m-k)!(-\lambda x)^k} \right]_a^b$$

4.18 FILON'S FORMULAS

(a) Notation. n = positive integer a, b, λ = constants $k = 0, 1, 2, \ldots, 2n$

$h = (b-a)/2n$ = spacing $\theta = \lambda h = \lambda(b-a)/2n$ = parameter

$f_k = f(a + kh)$ $C_k = \cos \lambda(a + kh)$ $S_k = \sin \lambda(a + kh)$

$\mathscr{A}_{2n} = \tfrac{1}{2} f_0 C_0 + f_2 C_2 + f_4 C_4 + \cdots + \tfrac{1}{2} f_{2n} C_{2n}$

$\mathscr{A}_{2n-1} = f_1 C_1 + f_3 C_3 + f_5 C_5 + \cdots + f_{2n-1} C_{2n-1}$

$\mathscr{B}_{2n} = \tfrac{1}{2} f_0 S_0 + f_2 S_2 + f_4 S_4 + \cdots + \tfrac{1}{2} f_{2n} S_{2n}$

$\mathscr{B}_{2n-1} = f_1 S_1 + f_3 S_3 + f_5 S_5 + \cdots + f_{2n-1} S_{2n-1}$

$\gamma = \dfrac{\theta^2 + \theta \sin\theta \cos\theta - 2\sin^2\theta}{\theta^3}$ $\alpha = \dfrac{2[\theta(1 + \cos^2\theta) - 2\sin\theta\cos\theta]}{\theta^3}$

$\beta = \dfrac{4(\sin\theta - \theta\cos\theta)}{\theta^3}$

(b) Integration formulas

$$\int_a^b f(x)\cos\lambda x\, dx \simeq h[\alpha\mathscr{A}_{2n} + \beta\mathscr{A}_{2n-1} - \gamma f_0 S_0 + \gamma f_{2n} S_{2n}]$$

$$\int_a^b f(x)\sin\lambda x\, dx \cong h[\alpha\mathscr{B}_{2n} + \beta\mathscr{B}_{2n-1} + \gamma f_0 C_0 - \gamma f_{2n} C_{2n}]$$

5

SERIES AND PRODUCTS OF CONSTANTS

(a) Sum function. The function $Y = F(x)$ is said to be the sum function (indefinite sum) of the function $y = f(x)$, which is single-valued in the interval (a, b), if for all x in (a, b),

$$\Delta F(x) = hf(x)$$

where $\Delta F(x) = F(x + h) - F(x)$ is the forward difference of $F(x)$ and $h =$ constant difference interval introduced in (3.03).

(b) Indefinite sum. Since the forward difference of $F(x) + C(x)$, where $C(x + h) = C(x)$ is a periodic constant, is also equal to $hf(x)$, all indefinite sums of $f(x)$ are included in the expression

$$\Delta^{-1}[hf(x)] = \sum hf(x) + C(x) = F(x) + C(x)$$

where $f(x)$ is called the summand and $\Delta^{-1} = \sum$ is the sum operator so that $\Delta\sum = 1$.

(c) Limit theorem. The relationship of the indefinite sum to the indefinite integral is given by

$$\lim_{h \to 0} [\sum hf(x) + C(x)] = \int f(x)\, dx + C$$

which leads to the resemblances shown in (d) below.

(d) Rules of integral and sum calculus are summarized below, where

$a, b =$ constants	$h =$ spacing	$m =$ integers
$u = f(x)$	$u' = \dfrac{df(x)}{dx}$	$\Delta u = f(x + h) - f(x)$
$v = g(x)$	$v' = \dfrac{dg(x)}{dx}$	$\Delta v = g(x + h) - g(x)$
$w = g(x + h)$	$x^{(m)} = x(x - h)(x - 2h) \cdots [x - (m - 1)h]$	

and $C(x) =$ periodic constant, which is omitted but implied.

$\int a\, dx = ax$	$\Delta^{-1} a = \dfrac{ax}{h}$
$\int ax^m\, dx = \dfrac{ax^{m+1}}{m + 1}$	$\Delta^{-1} ax^{(m)} = \dfrac{ax^{(m+1)}}{(m + 1)h}$
$\int (a + bx)^m\, dx = \dfrac{(a + bx)^{m+1}}{(m + 1)b}$	$\Delta^{-1} (a + bx)^{(m)} = \dfrac{(a + bx)^{(m+1)}}{(m + 1)bh}$
$\int \dfrac{dx}{ax^m} = -\dfrac{1}{(m - 1)ax^{m-1}}$	$\Delta^{-1} \dfrac{1}{ax^{(m)}} = -\dfrac{1}{(m - 1)ahx^{(m-1)}}$
$\int \dfrac{dx}{(a + bx)^m} = -\dfrac{1}{(m - 1)b(a + bx)^{m-1}}$	$\Delta^{-1} \dfrac{1}{(a + bx)^{(m)}} = -\dfrac{1}{(m - 1)bh(a + bx)^{(m-1)}}$
$\int au\, dx = a\int u\, du$	$\Delta^{-1} au = a\sum u$
$\int (u + v)\, dx = uv + \int v\, dx$	$\Delta^{-1}(u + v) = \sum u + \sum v$
$\int uv'\, dx = uv - \int u'v\, dx$	$\Delta^{-1}(u\, \Delta v) = uv - \Delta u \sum w$

(a) Notation. In addition to the symbols defined in (5.01d),

$\bar{\mathcal{S}}_r^{(m)}$ = Stirling number of the second kind (2.11d)

e = Euler's constant (2.18a)

$N = [2 \sin (\frac{1}{2}bh)]^{-1}$

m, n, p = integers

(b) Power functions $(h = 1, x \neq 0)$

$$\Delta^{-1}x^m = \sum x^m = \bar{\mathcal{S}}_1^{(m)} \sum x^{(1)} + \bar{\mathcal{S}}_2^{(m)} \sum x^{(2)} + \cdots + \bar{\mathcal{S}}_m^{(m)} \sum x^{(m)}$$

$$= \bar{\mathcal{S}}_1^{(m)} \frac{x^{(2)}}{2} + \bar{\mathcal{S}}_2^{(m)} \frac{x^{(3)}}{3} + \cdots + \bar{\mathcal{S}}_m^{(m)} \frac{x^{(m+1)}}{m + 1}$$

(c) Binomial coefficients $(h = 1, x \neq 0)$

$$\Delta^{-1}\binom{x}{p} = \sum \frac{x(x - 1)(x - 2) \cdots (x - p + 1)}{p!} = \binom{x}{p + 1}$$

$$\Delta^{-1}\binom{a + bx}{p} = \sum \binom{a + bx}{p} = \binom{a + bx}{p + 1}$$

(d) Exponential functions $(x \neq 0)$

$$\Delta^{-1}a^{bx} = \frac{a^{bx}}{a^{bh-1}}$$

$$\Delta^{-1}e^{bx} = \frac{e^{bx}}{e^{bx} - 1}$$

$$\Delta^{-1}[xa^{bx}] = \frac{xa^{bx}}{a^{bh} - 1} - \frac{bha^b(x + h)}{(a^{bh} - 1)^2}$$

$$\Delta^{-1}[xe^{bx}] = \frac{xe^{bx}}{e^{bh} - 1} - \frac{bhe^b(x + h)}{(e^{bh} - 1)^2}$$

(e) Logarithmic functions $(x > 0)$

$$\Delta^{-1}\ln \frac{x + h}{x - h} = \ln \frac{x}{x - h}$$

$$\Delta^{-1}\log \frac{x + h}{x - h} = \log \frac{x}{x - h}$$

$$\Delta^{-1}\ln \frac{x + h}{x} = \ln x$$

$$\Delta^{-1}\log \frac{x + h}{x} = \log x$$

(f) Trigonometric functions

$$\Delta^{-1}\sin bx = -N \cos [b(x - \tfrac{1}{2}h)] \qquad \Delta^{-1}\cos bx = N \sin [b(x - \tfrac{1}{2}h)]$$

$$\Delta^{-1}x \sin bx = -Nx \cos [b(x - \tfrac{1}{2}h)] + N^2h \cos [b(x + \tfrac{1}{5}h)]$$

$$\Delta^{-1}x \cos bx = Nx \sin [b(x - \tfrac{1}{2}h)] - N^2h \sin [b(x + \tfrac{1}{2}h)]$$

$$\Delta^{-1}\sin (a + bx) = -N \sin (a + bx - \tfrac{1}{2}bh)$$

$$\Delta^{-1}\cos (a + bx) = N \cos (a + bx - \tfrac{1}{2}bh)$$

(g) General sum by parts. With

$$\Delta^{-1}f(x) = F_1(x), \Delta^{-1}F_1(x) = F_2(x), \ldots, \Delta^{-1}F_{n-1}(x) = F_n(x)$$

$$\Delta g(x) = g_1(x), \Delta g_1(x) = g_2(x), \ldots, \Delta g_{n-1}(x) = g_n(x)$$

$$\Delta^{-1}[f(x) \cdot g(x)] = F_1(x)g(x) - F_2(x + h)g_1(x) + F_3(x + 2h)g_2(x) - F_4(x + 3h)g_3(x) + \cdots$$

the series is finite only when $g_n(x) = 0$.

(a) Fundamental theorem. If $y = f(x)$ is single-valued and continuous in $[a, b]$, $n = 1, 2, 3, \ldots$, and $(b - a)/n = h$, then the definite sum of $f(x)$ in limits $x = a$, $x = b$ is

$$\sum_{x=a}^{x=b} f(x) = f(a) + f(a + h) + f(a + 2h) + \cdots + f(a + nh)$$

$$= \Delta^{-1}f(x)\Big|_a^{b+h} = F(x)\Big|_a^{b+h} = F(b + h) - F(a)$$

where $\Delta^{-1}f(x)$ is the indefinite sum given by the particular formula in (5.01), (5.02), which must be evaluated in terms of a, b, h, and n (given constants).

(b) Example. The sum of the sequence

$$\sum_{a=2}^{b=10} \sin 3x = \sin 6 + \sin 12 + \sin 18 + \sin 24 + \sin 30$$

where $h = 2$, $n = 4$, is by (5.03a), (5.02f),

$$\Delta^{-1}\sin 3x\Big|_2^{12} = -\frac{\cos 3(x - 1)}{2\sin 3}\Big|_2^{12} = -\frac{\cos 33 - \cos 3}{2\sin 3} = -3.460\,586$$

(c) Subscript notation. The sum of $n + 1$ terms given by the general expression $y_k = f(a + kh)$ is

$$\sum_{k=0}^{n} y_k = y_0 + y_1 + y_2 + \cdots + y_{n-1} + y_n = \Delta^{-1}y_k\Big|_0^{n+1} = Y(n + 1) - Y(0)$$

where $k = 0, 1, 2, \ldots, n$ and $\Delta^{-1}y_k$ is the indefinite sum given by the particular formula in (5.01), (5.02). For the evaluation of this form, y_k, a, n, and h must be given.

(d) Example. The sum of the sequel in (b),

$$\sum_{k=0}^{4} \sin 6(1 + k) = \sin 6 + \sin 12 + \sin 18 + \sin 24 + \sin 30$$

where $\sin(a + bk) = \sin(6 + 6k)$ and $n = 4$, is, by (5.03c), (5.02f),

$$\Delta^{-1}\sin(6 + 6k)\Big|_0^5 = -\frac{\cos(3 + 6k)}{2\sin 3}\Big|_0^5 = -\frac{\cos 33 - \cos 3}{2\sin 3} = -3.460\,586$$

(e) Change in limits. As in (4.02e),

$$\sum_a^b f(x) = \Delta^{-1}f(x)\Big|_a^{b+h} = \Delta^{-1}f\left(\frac{x}{\lambda}\right)\Big|_{\lambda a}^{\lambda(b+h)} = \Delta^{-1}f(x \mp c)\Big|_{a\pm c}^{b+h\pm c} = \Delta^{-1}f(x - a)\Big|_0^{b-a+h}$$

where a, b, c, λ = constants and n = positive integer.

(f) Example. The sum of the sequence

$$\sum_{a=2}^{b=10} \sin 3x = \Delta^{-1}\sin 3x\Big|_2^{12} = \Delta^{-1}\sin x\Big|_6^{36}$$

with $h = 6$, $n = 4$ is, by (5.03c), (5.02f),

$$\Delta^{-1}\sin x\Big|_6^{36} = -\frac{\cos 33 - \cos 3}{2\sin 3} = -3.460\,586$$

(a) Definite sum formula. When the kth term of a series can be expressed as the summand of one of the formulas in (5.01), (5.02), then the exact sum of n terms of the sequence is given by the fundamental theorem (5.03a).

(b) Example. For $h = 2$,

$$\sum_{x=1}^{109} x^{(-2)} = \frac{1}{(1)(3)} + \frac{1}{(3)(5)} + \frac{1}{(5)(7)} + \cdots + \frac{1}{(107)(109)} + \frac{1}{(109)(111)}$$

$$= \Delta^{-1}x^{(-2)}\Big|_1^{111} = -\frac{x^{(-1)}}{h}\Big|_1^{111} = -\frac{1}{(2)(111)} + \frac{1}{(2)(1)} = \frac{110}{222}$$

(c) Difference series formula. When $y_k = f(a + kh)$, then

$$\sum_{k=1}^{n} y_k = y_1 + y_2 + y_3 + \cdots + y_n = \left[\binom{n}{1} + \binom{n}{2}\Delta + \binom{n}{3}\Delta^2 + \binom{n}{4}\Delta^3 + \cdots\right]y_1$$

where the difference series terminates with $\binom{n}{r+1}\Delta^r y_1 = 0$. For $r < n$ the difference series involves less terms than the original series. For CA,

$$\boxed{\sum_{k=1}^{n} y_k = n\bigwedge_{k=1}^{n}\left(\Delta^{k-1} + \frac{n-k}{k+1}\right)y_1 = n\left(1 + \frac{n-1}{2}\left(\Delta + \frac{n-2}{3}(\Delta^2 + \cdots)\right)\right)y_1}$$

(d) Example. The sum of sequence given by $y_k = (5 + 2k)^2$ and $k = 1, 2, 3, \ldots, 100$ is

$$\sum_{k=1}^{100} (5 + 2k)^2 = 49 + 81 + 121 + 169 + \cdots + 42\,025$$

$$= 100(49 + \tfrac{99}{2}(32 + \tfrac{98}{3}(8))) = 1\,456\,900$$

where $y_1 = 49$, $\Delta y_1 = 32$, $\Delta^2 y_1 = 8$, and $\Delta^3 y_1 = 0$ are taken from the difference table given by the first four terms of the sequence.

y_k	Δy_k	$\Delta^2 y_k$	$\Delta^3 y_k$
49			
	32		
81		8	
	40		0
121		8	
	48		
169			

(e) Power series formula. If $y_k = u_k v_k x^k = a_k x^k$, u_k = polynomial in k, and

$$\sum_{k=1}^{n} v_k x^k = \phi(x) = \text{exact sum} \qquad \sum_{k=1}^{n} a_k x^k = \left[\phi(x) + \frac{x}{1!}\phi'(x)\Delta + \frac{x^2}{2!}\phi''(x)\Delta^2 + \cdots\right]u_1$$

where $\phi'(x), \phi''(x), \ldots$ are the derivatives of $\phi(x)$ and the difference series terminates with $\Delta^{r+1}u_1$ for $r < n$. For CA,

$$\boxed{\sum_{k=1}^{n} a_k x^k = \bigwedge_{k=0}^{n}\left[\phi^{(k)}(x)\Delta^k + \frac{x}{k+1}\right]u_1 = \left[\phi(x) + \frac{x}{1}\left[\phi'(x)\Delta + \frac{x}{2}[\phi''(x)\Delta^2 + \cdots]\right]\right]u_1}$$

(f) Example. The sum of the sequence given by $y_k = (1 + k^2)\dfrac{x^k}{k!}$ can be computed for $k = 0, 1, 2, \ldots, \infty$ in terms of

$$u_k = 1 + k^2 \qquad v_k x^k = \frac{1}{k!}x^k \qquad \text{and} \qquad \sum_{k=0}^{\infty} v_k x^k = e^x = \phi(x) \qquad \phi'(x) = e^x, \ldots$$

as $\displaystyle\sum_{k=0}^{\infty} (1 + k^2)\frac{x^k}{k!} = \left(1 + \frac{x}{1} + \frac{x^2}{2} + \frac{x^3}{3}\right)e^k$

(a) Notation

a, b = constants $\quad\quad k, m, n, r$ = integers $\quad\quad h = (h - a)/n$ = spacing

$\bar{B}_{2r}$ = Bernoulli number (2.22a), (A.10) $\quad\quad R_m$ = remainder

(b) Sum of finite series. If $y = f(x)$ is single-valued and continuous in $[a, b]$ and $y_k = f(a + kh)$,

$$\sum_{k=0}^{n} y_k \cong \frac{1}{h} \int_a^b f(x)\, dx + \frac{1}{2}(y_0 + y_n) + \Phi_m + R_m$$

where

$$\Phi_m = \frac{h}{12}(Dy_n - Dy_0) - \frac{h^3}{720}(D^3 y_n - D^3 y_0) + \cdots + \frac{h^{2m-1}}{(2m)!}\bar{B}_{2m}(D^{2m-1} y_n - D^{2m-1} y_0)$$

$$R_m = n\frac{\bar{B}_{2m+2}}{(2m+2)!} D^{2m+2} f(\eta) \quad\quad (a < \eta < b)$$

If $f(x)$ is a polynomial of m or $m + 1$ degree, Φ_m terminates with m, the remainder R_m is zero, and the sum is exact. If $f(x)$ is not a polynomial, the series is asymptotic and the sum is an approximation, called the *Euler–MacLaurin formula*.

(c) Sum of monotonic infinite series. If $b = \infty$ and all terms of the series are positive, then

$$\sum_{k=0}^{\infty} y_k \cong \frac{1}{h}\int_0^\infty f(x)\,dx + \begin{cases} \dfrac{1}{2}\left(1 - \dfrac{h}{6}D + \dfrac{h^3}{360}D^3 - \dfrac{h^4}{15\,120}D^5 + \cdots\right)y_0 \\[3mm] \dfrac{1}{2}\left(1 - \dfrac{h}{6}\Delta + \dfrac{1}{12}\Delta^2 - \dfrac{19}{360}\Delta^3 + \cdots\right)y_0 \end{cases}$$

where the remainder can be estimated by the second formula in (4.03) and (4.04), respectively.

(d) Sum of alternating infinite series. If $b = \infty$ and the signs of terms in the series alternate, then

$$\sum_{k=0}^{\infty} (-1)^k y_k \cong \begin{cases} \dfrac{1}{2}\left(1 - \dfrac{h}{2}D + \dfrac{h^3}{24}D^3 - \dfrac{h^5}{240}D^5 + \dfrac{17}{40\,320}D^7 - \cdots\right)y_0 \\[3mm] \dfrac{1}{2}\left(1 - \dfrac{1}{2}\Delta + \dfrac{1}{4}\Delta^3 - \dfrac{1}{8}\Delta^5 - \dfrac{1}{16}\Delta^7 - \cdots\right)y_0 \end{cases}$$

where the remainder is estimated as in (c) above.

(e) Example. If $y_k = (-1)^k \dfrac{1}{1+k}$, $h = 1$, then

$$\sum_{k=0}^{\infty} (-1)^k \frac{1}{1+k} = 1 - \frac{1}{2} + \frac{1}{3} - \frac{1}{4} + \frac{1}{5} - \frac{1}{6} + \cdots = \frac{1}{2} + \frac{1}{4} - \frac{3!}{48} + \frac{5!}{480} - \frac{17(7!)}{80\,620} + \cdots$$

which is divergent. To improve the sum of the asymptotic series, the first six terms are summed first and the asymptoic series (d) is added as

$$\sum_{k=0}^{\infty} (-1)\frac{1}{1+k} = 0.616\,666\,667 + \sum_{k=0}^{\infty} (-1)^k \frac{1}{7+k}$$

$$\cong 0.616\,666\,667 + \frac{1}{(2)7} + \frac{1}{(4)72} - \frac{3!}{(48)74} + \frac{5!}{(480)76} = 0.693\,147\,125$$

Since the correct value is $\ln 2 = 0.693\,147\,781$, the error is $\epsilon \leq |6(10)^{-7}|$.

(a) Superposition. Since for many series their sums are known or are available in collections of formulas, the relationships

$$\sum_{k=0}^{n} y_k = \sum_{k=0}^{\infty} y_k - \sum_{k=n+1}^{\infty} y_k = S_\infty - \sum_{k=n+1}^{\infty} y_k$$

$$\sum_{k=0}^{n} (-1)^k y_k = \sum_{k=0}^{\infty} (-1)^k y_k - \sum_{k=n+1}^{\infty} (-1)^k y_k = \bar{S}_\infty - \sum_{k=n+1}^{\infty} (-1) y_k$$

where S_∞ and $\bar{S}_\infty$ are the known sums.

(b) Example. If $b < 1$, the sum of the infinite series

$$\sum_{k=0}^{\infty} b^k = 1 + b + b^2 + \cdots = \frac{1}{1-b}$$

then the sum of the first 100 terms is

$$\sum_{k=0}^{99} b^k = \sum_{k=0}^{\infty} b^k - \sum_{k=100}^{\infty} b^k = \frac{1}{1-b} - \frac{b^{100}}{1-b} = \frac{1 - b^{100}}{1-b}$$

If the series alternates,

$$\sum_{k=0}^{99} (-1)^k b^k = \sum_{k=0}^{\infty} (-1)^k - \sum_{k=100}^{\infty} (-1)^k b^k = \frac{1}{1+b} - \frac{b^{100}}{1+b} = \frac{1 - b^{100}}{1+b}$$

(c) Resolution in partial sums. If the nth term of series A_k can be expressed as a sum of two or any number of terms, so that

$$A_k = B_{1k} + B_{2k} + \cdots + B_{rk}$$

then
$$\sum_{k=0}^{\infty} A_k = \sum_{k=0}^{\infty} B_{1k} + \sum_{k=0}^{\infty} B_{2k} + \cdots + \sum_{k=0}^{\infty} B_{rk}$$

where $r = 2, 3, \ldots$ and the partial sums of $B_{1k}, B_{2k}, \ldots, B_{rk}$ are either known values or may be evaluated by the relations of the preceding sections.

(d) Examples. If $n < \infty$,

$$\sum_{k=1}^{n} (a + kb)^3 = \sum_{k=1}^{n} (a^3 + 3a^2 bk + 3ab^2 k^2 + b^3 k^3)$$

$$= na^3 + 3a^2 b \mathscr{P}_1 + 3ab^2 \mathscr{P}_2 + b^3 \mathscr{P}_3$$

where $\mathscr{P}_1 = \sum_{k=1}^{n} k$, $\mathscr{P}_2 = \sum_{k=1}^{n} k^2$, $\mathscr{P}_3 = \sum_{k=1}^{n} k^3$ are the known sums (5.11).

(e) Resolution in finite sum and asymptotic series. If the lower limit in the integral of the sum formula in (5.05b) leads to an indeterminate expression or if the asymptotic series may be resolved in two series as

$$\sum_{k=0}^{\infty} y_k = \sum_{k=0}^{s} y_k + \sum_{k=s+1}^{\infty} y_k$$

where the first series in $k = 0, 1, 2, \ldots, s$ can be summed term by term and the second series can be evaluated by a selected formula.

(a) Kummer's transformation. Given two convergent series,

$$A_\infty = \sum_{k=0}^{\infty} a_k \qquad\qquad B_\infty = \sum_{k=0}^{\infty} b_k$$

where the sum A_∞ is desired and the sum B is known, and if $\displaystyle\lim_{k\to\infty} \frac{a_k}{b_k} = \lambda \neq 0$

then
$$A_\infty = \lambda B_\infty + \sum_{k=0}^{\infty}\left(1 - \lambda\frac{b_k}{a_k}\right)a_k = \lambda B_\infty + \sum_{k=0}^{\infty} c_k$$

where the series of c_k may be more convenient for numerical evaluation. This transformation is quite useful in the evaluation of slowly convergent series.

(b) Example. If

$$A_\infty = \sum_{k=1}^{\infty}\frac{1}{k^2} \qquad B_\infty = \sum_{k=1}^{\infty}\frac{1}{k(k+1)} = 1 \qquad \lim_{k\to\infty}\frac{k+1}{k} = 1$$

then $\quad A_\infty = 1 + \displaystyle\sum_{k=1}^{\infty}\frac{1}{k^2(k+1)}$

(c) Abel's transformation. When $u_k = f(k)$, $v_k = g(x)$, and $k = 0, 1, 2, \ldots$, then with $j = 1, 2, 3, \ldots$,

$$\sum_{k=0}^{n} u_k v_k = u_0 v_0 + u_{n+1}\sum_{k=1}^{n} v_k - \sum_{k=1}^{n}\left[\Delta u_k \sum_{j=1}^{k} v_j\right]$$

leads in some cases to a rapidly convergent series.

(d) Examples. In the series

$$\sum_{k=1}^{100} k e^k = e + 2e^2 + 3e^3 + \cdots + 100e^{100},$$

$$u_k = k \qquad v_k = e^k \qquad u_0 = 0 \qquad v_0 = 0 \qquad n = 100 \qquad u_{n+1} = 101$$

$$\sum_{k=1}^{100} e^k = \frac{e^{101} - e}{e - 1} \qquad \sum_{j=1}^{k} e^j = \frac{e^{k+1} - e}{e - 1} \qquad \Delta u_k = 1$$

and $\displaystyle\sum_{k=1}^{100} k e^k = 101\frac{e^{101} - e}{e - 1} - \frac{e^2 - e}{e - 1} - \frac{e^3 - e}{e - 1} - \cdots - \frac{e^{101} - e}{e - 1}$

$$= \frac{e(101e^{100} - 1)}{e - 1} - \frac{e^2(e^{100} - 1)}{(e - 1)^2} = \frac{100e^{102} - 101e^{101} + e}{(e - 1)^2}$$

(e) Difference series transformation. When a is a real number $(a \neq 0)$, $b = a - 1$, $\lambda = a/b$, and $k = 1, 2, \ldots, n$, then

$$\sum_{k=1}^{n} a^k f(k) = \frac{a^k}{b}[1 - \lambda\Delta + \lambda^2\Delta^2 - \lambda^3\Delta^3 + \cdots]f(k)\Big|_1^{n+1}$$

is the sum. If $f(k)$ is a polynomial in k of nth order, the sum terminates with the term containing $\Delta^n f(k)$. For CA,

$$\sum_{k=1}^{n} a^k f(k) = \frac{a^k}{b}(1 - \lambda(\Delta - \lambda(\Delta^2 - \lambda(\Delta^3 - \cdots))))f(x)\Big|_1^{n+1}$$

(a) Notation. a, b, c, h = signed numbers n = number of terms

$k = 1, 2, 3, \ldots, n$ A_k, B_k, C_k, D_k = kth terms $m = 1, 2, 3, \ldots$

$\alpha = (-1)^{n+1}$ $\beta = (-1)^{k+1}$

$$\theta = \begin{cases} 0 & \text{if } n \text{ is even} \\ 1 & \text{if } n \text{ is odd} \end{cases}$$

(b) Monotonic arithmetic series $(n < \infty)$

$$\sum_{k=1}^{n} A_k = a + (a + h) + (a + 2h) + \cdots + [a + (n - 1)h] = \tfrac{1}{2}n(A_1 + A_n)$$

$$A_k = a + (k - 1)h \qquad\qquad A_{k+1} - A_k = h \qquad\qquad n = 1 + \frac{A_n - A_1}{h}$$

(c) Alternating arithmetic series $(n < \infty)$

$$\sum_{k=1}^{n} B_k = a - (a + h) + (a + 2h) - \cdots \alpha[a + (n - 1)h] = -\frac{nh}{2} + \theta(V_n + \tfrac{1}{2}h)$$

$$B_k = \beta[a + (k - 1)h] \qquad\qquad |B_{k+1}| - |B_k| = h \qquad\qquad n = 1 + \frac{|B_n| - B_1}{h}$$

(d) Finite monotonic geometric series $(n < \infty)$

$$\sum_{k=1}^{n} A_k = a + ab^h + ab^{2h} + \cdots + ab^{(n-1)h} = \frac{cA_n - A_1}{c - 1}$$

$$A_k = ab^{(k-1)h} \qquad\qquad \frac{A_{k+1}}{A_k} = b^h = c \qquad\qquad n = 1 + \frac{\ln A_n - \ln A_1}{\ln c}$$

(e) Finite alternating geometric series $(n < \infty)$

$$\sum_{k=1}^{n} B_k = a - ab^h + ab^{2h} - \cdots \alpha ab^{(n-1)h} = \frac{B_1 - cB_n}{1 - c}$$

$$B_k = \beta ab^{(k-1)h} \qquad\qquad \frac{B_{k+1}}{B_k} = -b^h = c \qquad\qquad n = 1 + \frac{\ln |B_n| - \ln B_1}{\ln |c|}$$

(f) Example. In

$$3 - 12 + 48 - 192 + \cdots - 49\,152 \qquad\qquad c = -4 \qquad\qquad n = 1 + \frac{\ln 49\,152 - \ln 3}{\ln 4} = 8$$

$$\text{and } \sum_{k=1}^{8} B_k = \frac{3 - (-4)(-49\,152)}{1 - (-4)} = -39\,321$$

(g) Infinite geometric series $(n = \infty, -1 < b < 1)$

$$\sum_{k=1}^{\infty} C_k = a + ab^h + ab^{2h} + \cdots = \frac{a}{1 - c} \qquad\qquad \sum_{k=1}^{\infty} D_k = a^m + a^{m-1}b + a^{m-2}b^2 + \cdots = \frac{a^{m+1}}{a - b}$$

$$C_k = ab^{(k-1)h} \qquad \frac{C_{k+1}}{C_k} = b^h = c \qquad\qquad D_k = a^{m-k+1}b^{k-1} \qquad \frac{D_{k+1}}{D_k} = \frac{b}{a} = c$$

(a) Notation. a, b, c, h = positive numbers n = number of terms

$k = 1, 2, 3, \ldots, n$ A_k, B_k, C_k = kth term

(b) Finite series $(n < \infty, c \neq 1)$

$$\sum_{k=1}^{n} A_k = a + (a + b)c^h + (a + 2b)c^{2h} + \cdots + [a + (n - 1)b]c^{(n-1)h}$$

$$= \frac{nbc^{nh}}{c^h - 1} - \frac{[(a - c^h(a - b)](c^{nh} - 1)}{(c^h - 1)^2}$$

where

$$A_k = B_k C_k \qquad B_k = [a + (k - 1)b] \qquad C_k = c^{(k-1)h} \qquad n = \frac{\ln C_n - \ln C_1}{h \ln c} + 1$$

(c) Example. If $a = b = h = 1, c = 2$, then

$$\sum_{k=1}^{n} k(2^{k-1}) = 1 \cdot 2^0 + 2 \cdot 2^1 + 3 \cdot 2^2 + 4 \cdot 2^3 + \cdots + n2^{n-1} = 1 + (n - 1)2^n$$

(d) Infinite series. When in (b), $n = \infty$ and $c^h > 1$, the series is divergent. If however, $0 < c^h < 1$, the series converges and

$$\sum_{k=1}^{\infty} A_k = a + (a + b)c^h + (a + 2b)c^{2h} + \cdots = \frac{a - (a - b)c^h}{(1 - c^h)^2}$$

(e) Particular cases

$n < \infty$ $\qquad a \neq 1$	$n = \infty$ $\qquad 0 < a < 1$
$\sum_{k=1}^{n} ka^k = \frac{na^{n+1}}{a - 1} - \frac{a(a^n - 1)}{(a - 1)^2}$	$\sum_{k=1}^{\infty} ka^k = \frac{a}{(1 - a)^2}$
$\sum_{k=1}^{n} ka^{k-1} = \frac{na^n}{a - 1} - \frac{a^n - 1}{(a - 1)^2}$	$\sum_{k=1}^{\infty} ka^{k-1} = \frac{1}{(1 - a)^2}$
$\sum_{k=1}^{n} (2k - 1)a^k = \frac{2na^{n+1}}{a - 1} - \frac{a(a + 1)(a^n - 1)}{(a - 1)^2}$	$\sum_{k=1}^{\infty} (2k - 1)a^k = \frac{a(1 + a)}{(1 - a)^2}$
$\sum_{k=1}^{n} (2k - 1)a^{k-1} = \frac{2na^n}{a - 1} - \frac{(a + 1)(a^n - 1)}{(a - 1)^2}$	$\sum_{k=1}^{\infty} (2k - 1)a^{k-1} = \frac{1 + a}{(1 - a)^2}$
$n < \infty$ $\qquad b \neq 1$	$n = \infty$ $\qquad b > 1$
$\sum_{k=1}^{n} \frac{k}{b^k} = \frac{b(b^n - 1) - n(b - 1)}{b^n(b - 1)^2}$	$\sum_{k=1}^{\infty} \frac{k}{b^k} = \frac{b}{(b - 1)^2}$
$\sum_{k=1}^{n} \frac{k}{b^{k-1}} = \frac{b^2(b^n - 1) - nb(b - 1)}{b^n(b - 1)^2}$	$\sum_{k=1}^{\infty} \frac{k}{b^{k-1}} = \frac{1}{(b - 1)^2}$
$\sum_{k=1}^{n} \frac{2k - 1}{b^k} = \frac{(b + 1)(b^n - 1) - 2n(b - 1)}{b^n(b - 1)^2}$	$\sum_{k=1}^{\infty} \frac{2k - 1}{b^k} = \frac{1}{(b - 1)^2}$
$\sum_{k=1}^{n} \frac{2k - 1}{b^{k-1}} = \frac{b(b + 1)(b^n - 1) - 2nb(b - 1)}{b^n(b - 1)^2}$	$\sum_{k=1}^{\infty} \frac{2k - 1}{b^{k-1}} = \frac{b(b + 1)}{(b - 1)^2}$

(a) Notation. $a, b, c, h, k, n, A_k, B_k, C_k$ are defined in (5.09a)

$$\alpha = (-1)^{n+1} \qquad \beta = (-1)^{k+1}$$

(b) Finite series $(n < \infty, c > 0)$

$$\sum_{k=1}^{n} A_k = a - (a + b)c^h + (a + 2b)c^{2h} - \cdots \alpha[a + (n-1)b]c^{(n-1)h}$$

$$= \frac{\alpha nbc^{nh}}{1 + c^h} + \frac{[a + c^h(a - b)](\alpha c^{nh} + 1)}{(1 + c^h)^2}$$

where

$$A_k = B_k C_k \qquad B_k = \beta[a + (k-1)h] \qquad C_k = c^{(k-1)h} \qquad n = \frac{\ln C_n - \ln C_1}{h \ln c} + 1$$

(c) Example. If $a = b = h = 1$ and $c = 2$, then

$$\sum_{k=1}^{n} \beta k 2^{k-1} = 1 \cdot 2^0 - 2 \cdot 2^1 + 3 \cdot 2^2 - 4 \cdot 2^3 + \cdots \alpha n 2^{n-1} = \frac{1 + \alpha(3n + 1)2^n}{9}$$

(d) Infinite series. When in (b), $n = \infty$ and $c^h > 1$, the series is divergent. If however, $0 < c^h < 1$, the series converges and

$$\sum_{k=1}^{\infty} A_k = a - (a + b)c^h + (a + 2b)c^{2h} - \cdots = \frac{a}{1 + c^h} - \frac{bc^h}{(1 + c^h)^2}$$

(e) Particular cases

$n < \infty$ $\qquad\qquad a \neq -1$	$n = \infty$ $\qquad\qquad 0 < a < 1$
$\displaystyle\sum_{k=1}^{n} \beta k a^k = \frac{\alpha n a^{n+1}}{1 + a} - \frac{a(1 + \alpha a^n)}{(1 + a)^2}$	$\displaystyle\sum_{k=1}^{\infty} \beta k a^k = \frac{a}{(1 + a)^2}$
$\displaystyle\sum_{k=1}^{n} \beta k a^{k-1} = \frac{\alpha n a^n}{1 + a} + \frac{1 + \alpha a^n}{(1 + a)^2}$	$\displaystyle\sum_{k=1}^{\infty} \beta k a^{k-1} = \frac{1}{(1 + a)^2}$
$\displaystyle\sum_{k=1}^{n} \beta(2k - 1)a^k = \frac{2\alpha n a^{n+1}}{1 + a} + \frac{a(a - 1)(1 + \alpha a^n)}{(1 + a)^2}$	$\displaystyle\sum_{k=1}^{\infty} \beta(2k - 1)a^k = \frac{a(1 - a)}{(1 + a)^2}$
$\displaystyle\sum_{k=1}^{n} \beta(2k - 1)a^{k-1} = \frac{2\alpha n a^n}{1 + a} - \frac{(a - 1)(1 + \alpha a^n)}{(1 + a)^2}$	$\displaystyle\sum_{k=1}^{\infty} \beta(2k + 1)a^{k-1} = \frac{1 - a}{(1 + a)^2}$
$n < \infty$ $\qquad\qquad b > 0$	$n = \infty$ $\qquad\qquad b > 1$
$\displaystyle\sum_{k=1}^{n} \frac{k}{b^k} = \frac{\alpha n(1 + b) - b(b^n + \alpha)}{b^n(1 + b)^2}$	$\displaystyle\sum_{k=1}^{\infty} \frac{\beta k}{b^k} = \frac{b(b - 1)}{(1 + b)^2}$
$\displaystyle\sum_{k=1}^{n} \frac{\beta k}{b^{k-1}} = \frac{\alpha n(1 + b) + b^2(b^n + \alpha)}{b^n(1 + b)^2}$	$\displaystyle\sum_{k=1}^{\infty} \frac{\beta k}{b^{k-1}} = \left(\frac{b}{1 + b}\right)^2$
$\displaystyle\sum_{k=1}^{n} \frac{\beta(2k - 1)}{b^k} = \frac{2\alpha n(1 + b) - (b - 1)(b^n + \alpha)}{b^n(1 + b)^2}$	$\displaystyle\sum_{k=1}^{\infty} \frac{\beta(2k - 1)}{b^k} = \frac{b - 1}{(1 + b)^2}$
$\displaystyle\sum_{k=1}^{n} \frac{\beta(2k - 1)}{b^k} = \frac{2\alpha n(1 + b) + b(b - 1)(b^n + \alpha)}{b^n(1 + b)^2}$	$\displaystyle\sum_{k=1}^{\infty} \frac{\beta(2k - 1)}{b^k} = \frac{b}{(1 + b)^2}$

(a) Finite series of powers of integers are exactly summable for $k, m, n = 1, 2, 3, \ldots$ as

$$\sum_{k=1}^{n} k^m = 1^m + 2^m + 3^m + \cdots + n^m = \frac{\bar{B}_{m+1}(n+1) - \bar{B}_{m+1}(0)}{m+1} = \mathscr{P}_{m,n}$$

where $\bar{B}_{m+1}(\ ) = $ Bernoulli polynomial of order $m+1$ (2.22) and $\mathscr{P}_{m,n} = $ sum.

(b) Special series

$$\sum_{k=1}^{n} (2k)^m = 2^m \mathscr{P}_{m,n} \qquad \sum_{k=1}^{n} (2k-1) = \mathscr{P}_{m,n} - 2^{m+1}\mathscr{P}_{t,n} = \mathscr{P}^*_{m,n}$$

where $\mathscr{P}_{t,n} = \sum_{k=1}^{t} k^m$ and $t = \frac{1}{2}n$ for even n or $\frac{1}{2}(n-1)$ for odd n.

(c) Particular series, all integers

$$\sum_{k=1}^{n} k = \frac{n^2}{2} + \frac{n}{2} \qquad\qquad \sum_{k=1}^{n} k^4 = \frac{n^5}{5} + \frac{n^4}{2} + \frac{n^3}{3} - \frac{n}{30}$$

$$\sum_{k=1}^{n} k^2 = \frac{n^3}{3} + \frac{n^2}{2} + \frac{n}{6} \qquad \sum_{k=1}^{n} k^5 = \frac{n^6}{6} + \frac{n^5}{2} + \frac{5n^4}{12} - \frac{n^2}{12}$$

$$\sum_{k=1}^{n} k^3 = \frac{n^4}{3} + \frac{n^3}{2} + \frac{n^2}{4} \qquad \sum_{k=1}^{n} k^6 = \frac{n^7}{7} + \frac{n^6}{2} + \frac{n^5}{2} - \frac{n^4}{6} + \frac{n}{42}$$

(d) Particular series, even integers

$$\sum_{k=1}^{n} (2k) = n^2 + n \qquad\qquad \sum_{k=1}^{n} (2k)^4 = \frac{(2n)^5}{10} + \frac{(2n)^4}{2} + \frac{2(2n)^3}{3} - \frac{8n}{15}$$

$$\sum_{k=1}^{n} (2k)^2 = \frac{4n^3}{3} + 2n^2 + \frac{2n}{3} \qquad \sum_{k=1}^{n} (2k)^5 = \frac{(2n)^6}{12} + \frac{(2n)^5}{2} + \frac{5(2n)^4}{6} - \frac{2(2n)^2}{15}$$

$$\sum_{k=1}^{n} (2k)^3 = 2n^4 + 4n^3 + 2n^2 \qquad \sum_{k=1}^{n} (2k)^6 = \frac{(2n)^7}{14} + \frac{(2n)^6}{2} + (2n)^5 - \frac{(2n)^4}{3} + \frac{16n}{21}$$

(e) Particular series, odd integers

$$\sum_{k=1}^{n} (2k-1) = n^2 \qquad\qquad \sum_{k=1}^{n} (2k-1)^4 = \frac{(2n)^5}{10} - \frac{(2n)^3}{3} + \frac{7(2n)}{30}$$

$$\sum_{k=1}^{n} (2k-1)^2 = \frac{4n^3}{3} - \frac{n}{3} \qquad \sum_{k=1}^{n} (2k-1)^5 = \frac{(2n)^6}{12} - \frac{5(2n)^4}{12} + \frac{7(2n)^2}{12}$$

$$\sum_{k=1}^{n} (2k-1)^3 = 2n^4 - n^2 \qquad \sum_{k=1}^{n} (2k-1)^6 = \frac{(2n)^7}{14} - \frac{(2n)^5}{7} + \frac{7(2n)^3}{6} - \frac{3(2n)}{42}$$

(f) General series, all integers $(r = \pm 1, \pm 2, \pm 3, \ldots)$

$$\sum_{k=1}^{n} (1 + rk)^m = (1+r)^m + (1+2r)^m + (1+3r)^m + \cdots + (1+nr)^m$$

$$= n + \binom{m}{1}r\sum_{k=1}^{n} k + \binom{m}{2}r^2\sum_{k=1}^{n} k^2 + \binom{m}{3}r^3\sum_{k=1}^{n} k^3 + \cdots + \binom{m}{m}r^m\sum_{k=1}^{n} k^m$$

$$= n + \binom{m}{1}r\mathscr{P}_{1,n} + \binom{m}{2}r^2\mathscr{P}_{2,n} + \binom{m}{3}r^3\mathscr{P}_{3,n} + \cdots + \binom{m}{m}r^m\mathscr{P}_{m,n}$$

where $\mathscr{P}_{1,n}, \mathscr{P}_{2,n}, \mathscr{P}_{3,n}, \mathscr{P}_{m,n}$ are the sums given in general form in (a) and in particlar forms in (c).

(a) Finite series of alternating powers of integers are exactly summable for $k, m, n = 1, 2, 3, \ldots$ as

$$\sum_{k=1}^{n} \beta k^m = 1^m - 2^m + 3^m - \cdots \alpha n^m = \frac{\bar{E}_m(n+1) - \alpha \bar{E}_m(0)}{2} = \bar{\mathscr{P}}_{m,n}$$

$$\alpha = (-1)^{n+1}$$
$$\beta = (-1)^{k+1}$$

where $\bar{E}_m(\ ) = $ Euler polynomial of order m (2.24) and $\bar{\mathscr{P}}_{m,n} = $ sum.

(b) Special series

$$\sum_{k=1}^{n} \beta (2k)^m = 2^m \bar{\mathscr{P}}_{m,n} \qquad \sum_{k=1}^{n} \beta (2k-1)^m = \bar{\mathscr{P}}^*_{m,n}$$

$$\theta = \begin{cases} 0 & \text{if } n \text{ is even} \\ 1 & \text{if } n \text{ is odd} \end{cases}$$

where $\bar{\mathscr{P}}^*_{m,n} = 2^m \bar{\mathscr{P}}_{m,n} - \binom{m}{1} 2^{m-1} \bar{\mathscr{P}}_{m-1,n} + \binom{m}{2} 2^{m-2} \bar{\mathscr{P}}_{m-2,n} - \cdots.$

(c) Particular series, all integers

$$\sum_{k=1}^{n} \beta k = \frac{\alpha n + \theta}{2} \qquad\qquad \sum_{k=1}^{n} \beta k^4 = \frac{\alpha n (n^3 + 2n^2 - 1)}{2}$$

$$\sum_{k=1}^{n} \beta k^2 = \frac{\alpha n (n+1)}{2} \qquad\qquad \sum_{k=1}^{n} \beta k^5 = \theta n^5 - \frac{(n+\theta)^5}{2} - \frac{5(n+\theta)^4}{4} - \frac{5(n+\theta)^2}{4}$$

$$\sum_{k=1}^{n} \beta k^3 = \frac{\alpha n^2 (2n+3) - \theta}{4} \qquad\qquad \sum_{k=1}^{n} \beta k^6 = \frac{\alpha n}{2} (n^5 + 3n^4 - 5n^2 + 3)$$

(d) Particular series, even integers

$$\sum_{k=1}^{n} \beta (2k) = \alpha n + \theta \qquad\qquad \sum_{k=1}^{n} \beta (2k)^4 = 8\alpha n (n^3 + 2n^2 - 1)$$

$$\sum_{k=1}^{n} \beta (2k)^2 = 2\alpha n (n+1) \qquad\qquad \sum_{k=1}^{n} \beta (2k)^5 = 32\theta n^5 - 16(n+\theta)^5$$
$$- 40(n+\theta)^4 - 40(n+\theta)^2$$

$$\sum_{k=1}^{n} \beta (2k)^3 = 2\alpha n^2 (2n+3) - 2\theta \qquad\qquad \sum_{k=1}^{n} \beta (2k)^6 = 32\alpha n (n^5 + 3n^4 - 5n^2 + 3)$$

(e) Particular series, odd integers

$$\sum_{k=1}^{n} \beta (2k-1) = \alpha n \qquad\qquad \sum_{k=1}^{n} \beta (2k-1)^3 = \alpha n (4n^2 - 3)$$

$$\sum_{k=1}^{n} \beta (2k-1)^2 = 2\alpha n^2 - \theta \qquad\qquad \sum_{k=1}^{n} \beta (2k-1)^4 = 4\alpha n^2 (2n^2 - 3) + 5\theta$$

(f) General series, all integers $(r = \pm 1, \pm 2, \pm 3, \ldots)$

$$\sum_{k=1}^{n} \beta (1 + rk)^m = (1+r)^m - (1+2r)^m + (1+3r)^m - \cdots \alpha (1+nr)^m$$

$$= \theta + \binom{m}{1} r \sum_{k=1}^{n} \beta k + \binom{m}{2} r^2 \sum_{k=1}^{n} \beta k^2 + \binom{m}{3} r^3 \sum_{k=1}^{n} \beta k^3 + \cdots + \binom{m}{m} r^m \sum_{k=1}^{n} \beta k^m$$

$$= \theta + \binom{m}{1} r \bar{\mathscr{P}}_{1,n} + \binom{m}{2} r^2 \bar{\mathscr{P}}_{2,m} + \binom{m}{3} r^3 \bar{\mathscr{P}}_{3,m} + \cdots + \binom{m}{m} r^m \bar{\mathscr{P}}_{m,n}$$

where $\bar{\mathscr{P}}_{1,n}, \bar{\mathscr{P}}_{2,n}, \bar{\mathscr{P}}_{3,n}, \bar{\mathscr{P}}_{m,n}$ are the sums given in general form in (a) and in particular form in (c).

(a) Finite series $(n = 1, 2, 3, 4, \ldots)$ is

$$\sum_{k=1}^{n} \frac{1}{k} = 1 + \frac{1}{2} + \frac{1}{3} + \cdots + \frac{1}{n} = \gamma + \psi(n) = \mathcal{Q}_{1,n}$$

where $\mathcal{Q}_{1,n} =$ sum (power 1, number of terms n) $\gamma = 0.577\,215\,664\,9 \cdots =$ Euler's constant (2.19), and

$$\psi(n) = \ln n + \frac{1}{2n} - \frac{1}{12n^2}\left(1 - \frac{1}{10n^2}\left(1 - \frac{10}{21n^2}\left(1 - \frac{21}{20n^2}\right)\right)\right) - \epsilon$$

is the digamma function (2.32), and for $n \geq 5$, $\epsilon \leq 1 \times 10^{-10}$.

(b) Example. To show the accuracy of the sum in (a) the first five terms (lower bound of the approximation) are computed first by direct summation as

$$\sum_{k=1}^{5} k = 1 + \tfrac{1}{2} + \tfrac{1}{3} + \tfrac{1}{4} + \tfrac{1}{5} = 2.283\,333\,333$$

Then using the sum formula,

$$\mathcal{Q}_{1,5} = \gamma + \ln 5 + \tfrac{1}{10} - \tfrac{1}{300}(1 - \tfrac{1}{250}(1 - \tfrac{10}{525}(1 - \tfrac{21}{500}))) = 2.283\,333\,333$$

(c) Special finite series

$$\sum_{k=1}^{n} \frac{1}{2k} = \frac{1}{2} + \frac{1}{4} + \frac{1}{6} + \cdots + \frac{1}{2n} = \frac{1}{2}[\gamma + \psi(n)] = \tfrac{1}{2}\mathcal{Q}_{1,n}$$

$$\sum_{k=1}^{n} \frac{1}{(2k-1)} = 1 + \frac{1}{3} + \frac{1}{5} + \cdots + \frac{1}{2n-1} = \frac{1}{2}[\gamma + \ln 4 + \psi(t)] = \tfrac{1}{2}\mathcal{Q}_{1,n}^{*}$$

where $\tfrac{1}{2}\mathcal{Q}_{1,n}$, $\tfrac{1}{2}\mathcal{Q}_{1,n}^{*}$ are the sums, $t = n - \frac{1}{2}$, and for $t > 5.5$, $\epsilon \leq 1 \times 10^{-10}$. The digamma function $\psi(t)$ is computed by the truncated series given in (a) for $\psi(n)$.

(d) Example. By direct summation,

$$\sum_{k=1}^{6} \frac{1}{2k-1} = 1 + \tfrac{1}{3} + \tfrac{1}{5} + \cdots + \tfrac{1}{11} = 1.878\,210\,678$$

and by the sum formula with $n = 6$, $t = 6 - \frac{1}{2} = 5.5$,

$$\tfrac{1}{2}\mathcal{Q}_{1,6}^{*} \cong \tfrac{1}{2}[\gamma + \ln 4 + \ln t] - \frac{1}{12t^2}\left(1 - \frac{1}{10t^2}\left(1 - \frac{10}{21t^2}\left(1 - \frac{21}{20t^2}\right)\right)\right) = 1.878\,210\,678$$

Examples (b) and (d) show the accuracy of the sum formulas for the lower bound of the range.

(e) Special values of $\mathcal{Q}_{1,n}$, $\mathcal{Q}_{1,n}^{*}$. $(\mathcal{Q}_{1,\infty} = \mathcal{Q}_{1,\infty}^{*} = \infty)$

$$\left.\begin{array}{l} \mathcal{Q}_{1,n} \cong \gamma + \ln n + \dfrac{1}{2n} - \dfrac{1}{12n^2} \\[3mm] \mathcal{Q}_{1,n}^{*} \cong \gamma + \ln 4t + \dfrac{1}{2t} - \dfrac{1}{12t^2} \end{array}\right\} \; n > 10^3 \qquad \left.\begin{array}{l} \mathcal{Q}_{1,n} \cong \gamma + \ln n + \dfrac{1}{2n} \\[3mm] \mathcal{Q}_{1,n}^{*} \cong \gamma + \ln 4t + \dfrac{1}{2t} \end{array}\right\} \; n > 10^4$$

(f) Superposition

$$\sum_{k=r}^{n} \frac{1}{k} = \frac{1}{r} + \frac{1}{r+1} + \frac{1}{r+2} + \cdots + \frac{1}{n-1} + \frac{1}{n} = \mathcal{Q}_{1,n} - \mathcal{Q}_{1,r-1}$$

(a) Finite series $(n = 1, 2, 3, \ldots)$ is

$$\sum_{k=1}^{n} \frac{\beta}{k} = 1 - \frac{1}{2} + \frac{1}{3} - \frac{1}{4} + \cdots + \frac{\alpha}{n} = \frac{1}{2}(\mathcal{Q}_{1,p}^* - \mathcal{Q}_{1,q}) = \bar{\mathcal{Q}}_{1,n}$$

where $\bar{\mathcal{Q}}_{1,n} =$ sum (power 1, number of terms n), $\beta = (1)^{k+1}$, $\alpha = (-1)^{n+1}$, $p =$ number of positive terms, $q =$ number of negative terms, $n = p + q$, and

$$\bar{\mathcal{Q}}_{1,n} = \tfrac{1}{2}[\psi(p - \tfrac{1}{2}) - \psi(q)] + \ln 2 - \epsilon$$

in which $\psi(p - \tfrac{1}{2})$, $\psi(q)$ are the respective digamma functions evaluated by the general formula given in $(5.13a)$, and for $n > 11$, $\epsilon \leq 1 \times 10^{-10}$.

(b) Special finite series

$$\sum_{k=1}^{n} \frac{\beta}{2k} = \frac{1}{2} - \frac{1}{4} + \frac{1}{6} + \cdots \frac{\alpha}{2n} = \tfrac{1}{2}\bar{\mathcal{Q}}_{1,n}$$

$$\sum_{k=1}^{n} \frac{\beta}{2k-1} = 1 - \frac{1}{3} + \frac{1}{5} - \cdots \frac{\alpha}{2n-1} = \tfrac{1}{2}\bar{\mathcal{Q}}_{1,n}^*$$

where $\bar{\mathcal{Q}}_{1,n}$ is the sum given in (a),

$$\tfrac{1}{2}\bar{\mathcal{Q}}_{1,n}^* = \frac{\pi}{4} - \ln \sqrt[4]{\frac{M}{N}} + \tfrac{1}{2}(M - N) - \tfrac{1}{3}(M^2 - N^2) + \tfrac{8}{15}(M^4 - N^4)$$

$$- \tfrac{32}{183}(M^6 - N^6) + \tfrac{256}{63}(M^8 - N^8) - \epsilon$$

in which $M = \dfrac{1}{4p - 3}$, $N = \dfrac{1}{4q - 1}$, and for $n \geq 11$, $\epsilon = 1 \times 10^{-10}$.

(c) Example. By direct summation,

$$\sum_{k=1}^{11} \frac{\beta}{2k-1} = 1 - \tfrac{1}{3} + \tfrac{1}{5} - \cdots \tfrac{1}{21} = 0.808\,078\,952$$

and by the sum formula with $n = 11$, $p = 6$, $q = 5$, $M = \tfrac{1}{21}$, $N = \tfrac{1}{19}$,

$$\tfrac{1}{2}\bar{\mathcal{Q}}_{1,n}^* = \frac{\pi}{4} - \ln \sqrt[4]{\frac{19}{21}} + \frac{1}{2}\left(\frac{1}{21} - \frac{1}{19}\right) - \frac{1}{3}\left(\frac{1}{21^2} - \frac{1}{19^2}\right) + \frac{8}{15}\left(\frac{1}{21^4} - \frac{1}{19^4}\right)$$

$$- \frac{32}{183}\left(\frac{1}{21^6} - \frac{1}{19^6}\right) + \frac{256}{63}\left(\frac{1}{21^8} - \frac{1}{19^8}\right) = 0.808\,078\,952$$

This example shows the accuracy of the sum formula.

(d) Special values of $\bar{\mathcal{Q}}_{1,n}, \mathcal{Q}_{1,n}$

$n > 10^3$	$\bar{\mathcal{Q}}_{1,n} \cong \ln 2 - \dfrac{1}{4n}$	$\tfrac{1}{2}\bar{\mathcal{Q}}_{1,n}^* \cong \dfrac{\pi}{4} - \ln \sqrt[4]{\dfrac{M}{N}} + \dfrac{M - N}{2}$
$n = \infty$	$\bar{\mathcal{Q}}_{1,\infty} = \ln 2$	$\tfrac{1}{2}\bar{\mathcal{Q}}_{1,\infty}^* = \dfrac{\pi}{4}$

(e) Superposition $(r = 1, 2, 3, \ldots)$

$$\sum_{k=r}^{n} \frac{\beta}{k} = \frac{1}{r} - \frac{1}{r+1} + \frac{1}{r+2} - \cdots \frac{\alpha}{r+n} = \bar{\mathcal{Q}}_{1,n} - \bar{\mathcal{Q}}_{1,r-1}$$

(a) Finite series $(m, n = 1, 2, 3, \ldots)$ is

$$\sum_{k=1}^{n} \frac{1}{k^m} = \frac{1}{1^m} + \frac{1}{2^m} + \frac{1}{3^m} + \cdots + \frac{1}{n^m} = Z(m) - \frac{\psi^{(m-1)}(n)}{\alpha(m-1)!} = \mathcal{D}_{m,n}$$

where $\mathcal{D}_{m,n} = $ sum (power m, number of terms n), $Z(m) = $ Riemann zeta function (2.26), and

$$\frac{\psi^{(m-1)}(n)}{\alpha(m-1)!} = \frac{1}{(m-1)n^{m-1}} - \frac{1}{2n^m}$$

$$+ \frac{m}{12n^{m-1}} \left(1 - \frac{(m+1)(m+2)}{10n^2} \left(1 - \frac{10(m+3)(m+4)}{21n^2} \left(1 - \frac{21(m+5)(m+6)}{20n^2} \right) \right) \right) - \epsilon$$

in which $\psi^{(m-1)}(n) = $ polygamma function of order $m - 1$ in n (2.34), and for $n \geq 10$, $\epsilon \leq 5 \times 10^{-9}$.

(b) Special finite series

$$\sum_{k=1}^{n} \frac{1}{(2k)^m} = \frac{1}{2^m} + \frac{1}{4^m} + \frac{1}{6^m} + \cdots + \frac{1}{(2n)^m} = \frac{1}{2^m} \mathcal{D}_{m,n}$$

$$\sum_{k=1}^{n} \frac{1}{(2k-1)^m} = \frac{1}{1^m} + \frac{1}{3^m} + \frac{1}{5^m} + \cdots + \frac{1}{(2n-1)^m} = \mathcal{D}_{m,2n} - \frac{1}{2^m} \mathcal{D}_{m,n} = \mathcal{D}_{m,n}^*$$

where $\mathcal{D}_{m,2n} = $ sum in (a) of $2n$ terms and $\mathcal{D}_{m,n} = $ sum in (a) of n terms.

(c) Special values of $\mathcal{D}_{m,n}$, $\mathcal{D}_{m,n}^*$

$\mathcal{D}_{m,n} = Z(m) - \dfrac{1}{(m-1)n^{m-1}}$	$\mathcal{D}_{m,n}^* = (1 - 2^{-m})\left[Z(m) - \dfrac{1}{(m-1)n^{m-1}} \right]$	$n^m > 10^8$
$\mathcal{D}_{m,\infty} = Z(m)$	$\mathcal{D}_{m,\infty}^* = (1 - 2^{-m})Z(m)$	$n = \infty$

Numerical values of $Z(m)$ for $m = 1, 2, 3, \ldots, 20$ are tabulated in (2.27c).

(d) Particular infinite series

$\sum_{k=1}^{\infty} \dfrac{1}{k^2} = \dfrac{\pi^2}{6}$	$\sum_{k=1}^{\infty} \dfrac{1}{(2k)^2} = \dfrac{\pi^2}{24}$	$\sum_{k=1}^{\infty} \dfrac{1}{(2k-1)^2} = \dfrac{\pi^2}{8}$
$\sum_{k=1}^{\infty} \dfrac{1}{k^4} = \dfrac{\pi^4}{90}$	$\sum_{k=1}^{\infty} \dfrac{1}{(2k)^4} = \dfrac{\pi^4}{1440}$	$\sum_{k=1}^{\infty} \dfrac{1}{(2k-1)^4} = \dfrac{\pi^4}{96}$
$\sum_{k=1}^{\infty} \dfrac{1}{k^6} = \dfrac{\pi^6}{945}$	$\sum_{k=1}^{\infty} \dfrac{1}{(2k)^6} = \dfrac{\pi^6}{60\,480}$	$\sum_{k=1}^{\infty} \dfrac{1}{(2k-1)^6} = \dfrac{\pi^6}{960}$
$\sum_{k=1}^{\infty} \dfrac{1}{k^8} = \dfrac{\pi^8}{9450}$	$\sum_{k=1}^{\infty} \dfrac{1}{(2k)^8} = \dfrac{\pi^8}{2\,419\,200}$	$\sum_{k=1}^{\infty} \dfrac{1}{(2k-1)^8} = \dfrac{\pi^8}{806\,400}$
$\sum_{k=1}^{\infty} \dfrac{1}{k^{10}} = \dfrac{\pi^{10}}{93\,555}$	$\sum_{k=1}^{\infty} \dfrac{1}{(2k)^{10}} = \dfrac{\pi^{10}}{95\,800\,320}$	$\sum_{k=1}^{\infty} \dfrac{1}{(2k-1)^{10}} = \dfrac{\pi^{10}}{2\,903\,040}$
$\sum_{k=1}^{\infty} \dfrac{1}{k^{2r}} = \dfrac{\alpha(2\pi)^{2r}}{(2r)!2} \bar{B}_{2r}$	$\sum_{k=1}^{\infty} \dfrac{1}{(2k)^{2r}} = \dfrac{\alpha\pi^{2r}}{(2r)!2} \bar{B}_{2r}$	$\sum_{k=1}^{\infty} \dfrac{1}{(2k-1)^{2r}} = \dfrac{\alpha(2^{2r}-1)\pi^{2r}}{(2r)!2} \bar{B}_{2r}$

where $r = 1, 2, 3, \ldots$, $\alpha = (-1)^{r+1}$, $\bar{B}_{2r} = $ Bernoulli number (2.22).

(a) Finite series $(m, n = 1, 2, 3, \ldots)$ is

$$\sum_{k=1}^{n} \frac{\beta}{k^m} = \frac{1}{1^m} - \frac{1}{2^m} + \frac{1}{3^m} - \cdots \frac{\alpha}{n^m} = \mathcal{Q}_{m,n} - 2^{m-1}\mathcal{Q}_{m,q} = \bar{\mathcal{Q}}_{m,n}$$

$$\alpha = (-1)^{n+1} \quad \beta = (-1)^{k+1}$$

where $\mathcal{Q}_{m,n}, \mathcal{Q}_{m,q}$ = sums given in (5.15a) with n = number of terms and q = number of negative terms.

(b) Special finite series

$$\sum_{k=1}^{n} \frac{\beta}{(2k)^m} = \frac{1}{2^m} - \frac{1}{4^m} + \frac{1}{6^m} - \cdots \frac{\alpha}{(2n)^m} = 2\,\bar{\mathcal{Q}}_{m,n}$$

$$\sum_{k=1}^{n} \frac{\beta}{(2k-1)^n} = \frac{1}{1^m} - \frac{1}{3^m} + \frac{1}{5^m} - \cdots \frac{\alpha}{(2n-1)^m} = \bar{\mathcal{Q}}^*_{m,n}$$

where $\bar{\mathcal{Q}}_{m,n}$ = sum in (a) and

$$\bar{\mathcal{Q}}^*_{m,n} = \bar{Z}(m) - \frac{1}{2(2n+1)^m}\left(1 + \frac{m}{2n+1}\left(1 - \frac{(m+1)(m+2)}{3(2n+1)^2}\left(1 - \frac{2(m+3)(m+4)}{3(2n+1)^2}\right)\right)\right) - \epsilon$$

in which $\bar{Z}(m)$ = complementary Riemann zeta function (2.27), and for $n \geq 10$, $\epsilon = 5 \times 10^{-9}$.

(c) Special values of $\bar{\mathcal{Q}}_{m,n}, \bar{\mathcal{Q}}^*_{m,n}$

$\bar{\mathcal{Q}}_{m,n} = (1 - 2^{1-m})Z(m) + \dfrac{1}{m-1}\left(\dfrac{1}{n^{m-1}} - \dfrac{2^{1-m}}{q^{m-1}}\right)$	$\bar{\mathcal{Q}}^*_{m,n} = \bar{Z}(m) - \dfrac{1}{2(2n+1)^m}$	$(2n+1)^{m+1} > 10^8$
$\bar{\mathcal{Q}}_{m,n} = (1 - 2^{1-m})Z(m)$	$\bar{\mathcal{Q}}^*_{m,n} = \bar{Z}(m)$	$n = \infty$

Numerical values of $Z(m)$ and $\bar{Z}(m)$ for $m = 1, 2, 3, \ldots, 20$ are tabulated in (2.27c).

(d) Particular infinite series

$\displaystyle\sum_{k=1}^{\infty} \frac{\beta}{k^2} = \frac{\pi^2}{12}$	$\displaystyle\sum_{k=1}^{\infty} \frac{\beta}{(2k)^2} = \frac{\pi^2}{48}$	$\displaystyle\sum_{k=1}^{\infty} \frac{\beta}{2k-1} = \frac{\pi}{4}$
$\displaystyle\sum_{k=1}^{\infty} \frac{\beta}{k^4} = \frac{7\pi^4}{720}$	$\displaystyle\sum_{k=1}^{\infty} \frac{\beta}{(2k)^4} = \frac{7\pi^4}{11\,520}$	$\displaystyle\sum_{k=1}^{\infty} \frac{\beta}{(2k-1)^3} = \frac{\pi^3}{32}$
$\displaystyle\sum_{k=1}^{\infty} \frac{\beta}{k^6} = \frac{31\pi^6}{30\,240}$	$\displaystyle\sum_{k=1}^{\infty} \frac{\beta}{(2k)^6} = \frac{31\pi^6}{1\,935\,360}$	$\displaystyle\sum_{k=1}^{\infty} \frac{\beta}{(2k-1)^5} = \frac{5\pi^5}{1536}$
$\displaystyle\sum_{k=1}^{\infty} \frac{\beta}{k^8} = \frac{127\pi^8}{1\,206\,600}$	$\displaystyle\sum_{k=1}^{\infty} \frac{\beta}{(2k)^8} = \frac{127\pi^8}{308\,889\,600}$	$\displaystyle\sum_{k=1}^{\infty} \frac{\beta}{(2k-1)^7} = \frac{61\pi^7}{30\,720}$
$\displaystyle\sum_{k=1}^{\infty} \frac{\beta}{k^{10}} = \frac{511\pi^{10}}{47\,900\,160}$	$\displaystyle\sum_{k=1}^{\infty} \frac{\beta}{(2k)^{10}} = \frac{511\pi^{10}}{49\,049\,764\,000}$	$\displaystyle\sum_{k=1}^{\infty} \frac{\beta}{(2k-1)^9} = \frac{277\pi^9}{8\,257\,536}$
$\displaystyle\sum_{k=1}^{\infty} \frac{\beta}{k^{2r}} = = \frac{\alpha(2^{2r}-2)\pi^{2r}}{(2r)!2}B_{2r}$	$\displaystyle\sum_{k=1}^{\infty} \frac{\beta}{(2k)^{2r}} = \frac{(\pi/2)^{2r}(2^{2r}-2)}{(2r)!2}\bar{B}_{2r}$	$\displaystyle\sum_{k=1}^{\infty} \frac{\beta}{(2k-1)^{2r-1}} = \frac{(\pi/2)^{2r-1}}{(2r)!2}\bar{E}_{2r}$

where $r = 1, 2, 3, \ldots, \bar{B}_{2r}$ = Bernoulli number (2.22), and $\bar{E}_{2r}$ = Euler number (2.24).

(a) Unit step series of factorial polynomials. When $m, n =$ positive intgers, $h = 1$, $x = m, m + 1$, $m + 2, \ldots, m + n$, then the factorial polynomial

$$X_1^{(m)} = x(x - 1)(x - 2) \cdots (x - m + 1) = m! \quad \text{and the series}$$

$$\sum_{x=m}^{m+n} X_1^{(m)} = \frac{m!}{0!} + \frac{(m + 1)!}{1!} + \frac{(m + 2)!}{2!} + \cdots + \frac{(m + n)!}{n!}$$

can be exactly summed for $n < \infty$ (5.01d), (5.03a) as

$$\sum_{x=m}^{m+n} X_1^{(m)} = \Delta^{-1} X_1^{(m)} \Big|_m^{m+n+1} = \frac{X_1^{(m+1)}}{m + 1} \Big|_m^{m+n+1} = \frac{(m + n + 1)_1^{(m+1)}}{m + 1}$$

since $m_1^{(m+1)} = m(m - 1)(m - 2) \cdots (m - m) = 0$.

(b) Example. With $m = 3$, $h = 1$, $n = 7$,

$$\sum_{x=3}^{10} X_1^{(3)} = 1 \cdot 2 \cdot 3 + 2 \cdot 3 \cdot 4 + 3 \cdot 4 \cdot 5 + \cdots + 8 \cdot 9 \cdot 10 = \frac{11 \cdot 10 \cdot 9 \cdot 8}{4} = 1980$$

(c) Unit step binomial coefficient series. Since the kth term in (a) above can be expressed as

$$\frac{(m + k)!}{k!} = m! \binom{m + k}{k} \qquad \text{or} \qquad \binom{m + k}{k} = \frac{(m + k)!}{k! m!}$$

the sum formula in (a) gives also the sum formula

$$\sum_{k=0}^{n} \binom{m + k}{k} = \binom{m}{0} + \binom{m + 1}{1} + \binom{m + 2}{2} + \cdots + \binom{m + n}{n} = \binom{m + n + 1}{m + 1}$$

(d) Multistep series of factorial polynomial. When $a, b, m, n =$ positive integers and $h = 2, 3, 4, \ldots$, then

$$X_h^{(m)} = x(x - h)(x - 2h) \cdots [x - (m - 1)h] \quad \text{and the series}$$

$$\sum_{k=a}^{a+nh} X_h^{(m)} = a^{(m)} + (a + h)^{(m)} + (a + 2h)^{(m)} + \cdots + (a + nh)^{(m)}$$

can be exactly summed for $n < \infty$ (5.01d), (5.03a) as

$$\sum_{x=a}^{a+nh} X_h^{(m)} = \Delta^{-1} X_h^{(m)} \Big|_a^{a+nh} = \frac{X_h^{(m+1)}}{(m + 1)h} \Big|_a^{b+h} = \frac{(b + h)_h^{(m+1)} - a_h^{(m+1)}}{(m + 1)h}$$

where $b = a + nh$.

(e) Example. With $m = 3$, $h = 2$, $a = 5$, $b = 15$

$$\sum_{x=5}^{15} X_2^{(3)} = 1 \cdot 3 \cdot 5 + 3 \cdot 5 \cdot 7 + 5 \cdot 7 \cdot 9 + \cdots + 11 \cdot 13 \cdot 15 = \frac{17_2^{(4)} - 5_2^{(4)}}{8} = 4560$$

(f) Multistep series of binomial coefficients. With $h = 2, 3, 4, \ldots$,

$$\sum_{k=0}^{n} \binom{m + kh}{kh} = \binom{m}{0} + \binom{m + h}{h} + \binom{m + 2h}{2h} + \cdots + \binom{m + nh}{nh} = \binom{m + n + h}{m + 1}$$

(a) Unit step series of inverse factorial polynomial. When in $(5.17a)$ $m = -r$, $h = 1$, and $x = 0, 1, 2, \ldots, n$, then the factorial polynomial

$$X_1^{(-r)} = \frac{1}{(x+1)(x+2)(x+3)\cdots(x+r)} = \frac{x!}{(x+r)!} \qquad \text{and the series}$$

$$\sum_{x=0}^{n} X_1^{(-r)} = \frac{0!}{r!} + \frac{1}{(r+1)!} + \frac{2!}{(r+2)!} + \cdots + \frac{n!}{(r+n)!}$$

can be exactly summed $(5.01d)$, $(5.03a)$ as

$$\sum_{x=0}^{n} X_1^{(-r)} = \Delta^{-1} X_1^{(-r)} \Big|_0^{n+1} = \frac{X_1^{(1-r)}}{1-r}\Big|_0^{n+1} = \frac{1}{r-1}\left[\frac{1}{(r-1)!} - \frac{(n+1)!}{(n+r)!}\right]$$

$$\sum_{x=0}^{\infty} X_1^{(-r)} = \lim_{n\to\infty}\left\{\frac{1}{r-1}\left[\frac{1}{(r-1)!} - \frac{1}{(r+n)_1^{(r-n)}}\right]\right\} = \frac{r}{r!(r-1)}$$

(b) Example. With $r = 3$, $n = 7$,

$$\sum_{x=0}^{7} X_1^{(-3)} = \frac{1}{1\cdot2\cdot3} + \frac{1}{2\cdot3\cdot4} + \frac{1}{3\cdot4\cdot5} + \cdots + \frac{1}{8\cdot9\cdot10} = \frac{1}{2}\left(\frac{1}{2!} - \frac{1}{10\cdot9}\right) = \frac{11}{45}$$

and for $n = \infty$, the sum is $\dfrac{3}{3!(2)} = \dfrac{1}{4}$.

(c) Unit step series of reciprocal binomial coefficients. Since the kth term of the series in (a) above can be expressed as

$$\frac{k!}{(r+k)!} = \frac{1}{r!\binom{r+k}{k}} \qquad \text{or} \qquad \frac{1}{\binom{r+k}{k}} = \frac{k!r!}{(r+k)!}$$

the sum formula in (a) gives also the sum formula

$$\sum_{k=0}^{n} \frac{1}{\binom{r+k}{k}} = \frac{r!}{r-1}\left[\frac{1}{(r-1)!} - \frac{(n+1)!}{(n+r)!}\right] \qquad\qquad \sum_{k=0}^{\infty}\frac{1}{\binom{r+k}{k}} = \frac{r}{r-1}$$

(d) Multistep series of reciprocal factorial polynomials. When a, b, n, r = positive integers and $h = 2, 3, 4, \ldots$, then

$$X_h^{(-r)} = \frac{1}{(x+h)(x+2h)(x+3h)\cdots(x+rh)} = \frac{1}{(x+rh)_h^{(r)}} \qquad \text{and the series}$$

$$\sum_{x=a}^{a+nh} X_h^{(-r)} = \frac{1}{(a+rh)_h^{(r)}} + \frac{1}{[a+(r+1)h]_h^{(r)}} + \frac{1}{[a+(r+2)h]_h^{(r)}} + \cdots + \frac{1}{[a+(r+n)h]_h^{(r)}}$$

can be exactly summed $(5.01d)$, $(5.03a)$ as

$$\sum_{k=a}^{b} X_h^{(-r)} = \Delta^{-1}X_h^{(-r)}\Big|_a^{b+h} = \frac{X_h^{(1-r)}}{(1-r)h}\Big|_a^{b+h} = \frac{1}{(r-1)h}\left\{\frac{1}{[a+(r-1)h]_h^{(r-1)}} - \frac{1}{[b+(r+1)h]_h^{(r-1)}}\right\}$$

$$\sum_{k=a}^{\infty} X_h^{(-r)} = \lim_{n\to\infty}\left\{\frac{1}{(r-1)h}\left[\frac{1}{[a+(r+1)h]_h^{(r-1)}} - \frac{1}{[b+rh]_h^{(r-1)}}\right]\right\} = \frac{1}{(r-1)h[a+(r+1)]_h^{(r-1)}}$$

(a) Basic series. When a, b = signed numbers, $m = 1, 2, 3, \ldots, p$, and $k = 1, 2, 3, \ldots, n$, then with $\lambda = h/b$, h = signed integer, fraction or decimal number,

$$\sum_{k=1}^{n} a(b + kh)^m = ab\left[n + \binom{m}{1}\lambda \mathscr{P}_{1,n} + \binom{m}{2}\lambda^2 \mathscr{P}_{2,n} + \cdots + \binom{m}{m}\lambda^m \mathscr{P}_{m,n}\right]$$

where $n < \infty$, and

$$\mathscr{P}_{1,n} = \sum_{k=1}^{n} k, \; \mathscr{P}_{2,m} = \sum_{k=1}^{n} k^2, \ldots, \mathscr{P}_{m,n} = \sum_{k=1}^{n} k^m$$

are the exact sums given in closed form in (5.11). Selected cases of sums of monotonic series of powers of decimal numbers are listed in (*b*) below.

(b) Particular cases

$$\sum_{k=1}^{n} a(b + kh) = \frac{abn}{2}[\lambda(n + 1) + 2]$$

$$\sum_{k=1}^{n} a(b + kh)^2 = \frac{ab^2 n}{6}[\lambda^2(n + 1)(2n + 1) + 6\lambda(n + 1) + 6]$$

$$\sum_{k=1}^{n} a(b + kh)^3 = \frac{ab^3 n}{4}[\lambda^3 n(n + 1)^2 + 2\lambda(n + 1)(2n + 1)(3n + 1) + 3\lambda(n + 1) + 4]$$

$$\sum_{k=1}^{n} a(b + 2kh) = abn[\lambda(n + 1) + 1]$$

$$\sum_{k=1}^{n} a(b + 2kh)^2 = \frac{ab^2 n}{3}[2\lambda^2(n + 1)(2n + 1) + 6\lambda(n + 1) + 3]$$

$$\sum_{k=1}^{n} a(b + 2kh)^3 = \frac{ab^3 n}{2}[4\lambda^3 n(n + 1)^3 + 4\lambda^2(n + 1)(2n + 1) + 3\lambda(n + 1) + 2]$$

$$\sum_{k=1}^{n} a[b + (2k - 1)h] = abn(\lambda n + 1)$$

$$\sum_{k=1}^{n} a[b + (2k - 1)h]^2 = \frac{ab^2 n}{3}[\lambda^2(4n^2 - 1) + 6\lambda n + 3]$$

$$\sum_{k=1}^{n} a[b + (2k - 1)h]^3 = ab^3 n[\lambda^3 n(2n^2 - 1) + \lambda^2(4n^2 - 1) + 3\lambda n + 1]$$

(c) General series. If a, b, k, m, n are the same as in (*a*) above, $r = 1, 2, 3, \ldots, s$, $h_1, h_2, h_3, \ldots, h_r$ are signed numbers, then with

$$A_k = a(b + kh_1 + k^2 h_2 + \cdots + k^r h_r)^m = ab^m(1 + c_1 k + c_2 k^2 + \cdots + c_{rm} k^{rm})$$

and for $n < \infty$,

$$\sum_{k=1}^{n} A_k = ab^m(n + c_1 \mathscr{P}_{1,n} + c_2 \mathscr{P}_{2,n} + \cdots + c_{rm} \mathscr{P}_{rm,n})$$

where $c_1, c_2, \ldots, c_{rm}$ are the coefficients of polynomial of rm degree and $\mathscr{P}_{1,n}, \mathscr{P}_{2,n}, \ldots, \mathscr{P}_{rm,n}$ are the exact sums in (5.11).

(a) Basic series. With notation of (5.19a) and with $\alpha = (-1)^{n+1}$, $\beta = (-1)^{k+1}$, the alternating series is

$$\sum_{k=1}^{n} \beta a(b + kh)^m = ab^m \left[\theta + \binom{m}{1} \lambda \bar{\mathscr{P}}_{1,n} + \binom{m}{2} \lambda^2 \bar{\mathscr{P}}_{2,n} + \cdots + \binom{m}{m} \lambda^m \bar{\mathscr{P}}_{m,n} \right] \qquad \theta = \begin{cases} 0 & n \text{ even} \\ 1 & n \text{ odd} \end{cases}$$

where

$$\bar{\mathscr{P}}_{1,n} = \sum_{k=1}^{n} \beta k, \ \bar{\mathscr{P}}_{2,n} = \sum_{k=1}^{n} \beta k^2, \ldots, \bar{\mathscr{P}}_{m,n} = \sum_{k=1}^{n} \beta k^m$$

are the exact sums given in closed form in (5.12). Selected cases of alternating series of powers of decimal numbers are listed in (b) below.

(b) Particular cases

$$\sum_{k=1}^{n} \beta a(b + kh) = \frac{abn}{2}[\lambda(\alpha + \theta) + 2]$$

$$\sum_{k=1}^{n} \beta a(b + kh)^2 = \frac{ab^2n}{2}[\lambda^2\alpha(n + 1) + 2\lambda(\alpha n + \theta) + 2]$$

$$\sum_{k=1}^{n} \beta a(b + kh)^3 = \frac{ab^3n}{4}\{\lambda^3[\alpha n(2n + 3) - \theta] + 6\lambda^2(n + 1) + 6\lambda(\alpha + \theta) + 4\}$$

$$\sum_{k=1}^{n} \beta a(b + 2kh) = abn(\lambda\alpha + 2\theta)$$

$$\sum_{k=1}^{n} \beta a(b + 2kh)^2 = ab^2n[2\lambda^2\alpha(n + 1) + 2\lambda\alpha + 3\theta]$$

$$\sum_{k=1}^{n} \beta a(b + 2kh)^3 = ab^3n[\lambda^3(4\alpha n^2 + 6\alpha n - \theta) + 6\lambda^2\alpha(n + 1) + 3\lambda(\alpha + \theta) + \theta]$$

$$\sum_{k=1}^{n} \beta a[b + (2k - 1)h] = abn(\lambda\alpha + 1)$$

$$\sum_{k=1}^{n} \beta a[b + (2k - 1)h]^2 = ab^2n[\lambda^2(\alpha n - \theta) + 2\lambda\alpha + 1]$$

$$\sum_{k=1}^{n} \beta a[b + (2k - 1)h]^3 = ab^3n[4\lambda^3(4n^3 - 3) + 3\lambda^2(2\alpha n - \theta) + 3\lambda\alpha + 1]$$

(c) General series. If a, b, k, m, n, r and $h_1, h_2, h_3, \ldots, h_r$ are the same as in (a) above, then with

$$B_k = \beta a(b + kh_1 + k^2h_2 + \cdots + k^rh_r)^m = \beta ab^m(1 + c_1k + c_2k^2 + \cdots + c_{rm}k^{rm})$$

and for $n < \infty$,

$$\sum_{k=1}^{n} B_k = ab^m(n\theta + c_1\bar{\mathscr{P}}_{1n} + c_2\bar{\mathscr{P}}_{2,n} + \cdots + c_{rm}\bar{\mathscr{P}}_{rm,n})$$

where $c_1, c_2, \ldots, c_{rm}$ are the coefficients of polynomial of rm degree and $\bar{\mathscr{P}}_{1,n}, \bar{\mathscr{P}}_{2,n}, \ldots, \bar{\mathscr{P}}_{rm,n}$ are the exact sums in (5.12).

(a) Monotonic series is defined as

$$\sum_{k=1}^{n} \frac{1}{a + bk} = \frac{1}{b} \sum_{k=1}^{n} \frac{1}{c + k} = \frac{1}{b}[\psi(n + c) - \psi(c)] = \mathcal{Q}_{1,n}(c)$$

where a, b = positive numbers, $c = a/b$, and $\psi(n + c)$, $\psi(c)$ = digamma functions (2.32) so that with $N = n + c$,

$$\mathcal{Q}_{1,n}(c) = \frac{1}{b}\left[\ln\frac{N}{c} - \frac{1}{2c} + \frac{1}{2N} + \sum_{j=1}^{r}\left(\frac{1}{c^{2j}} - \frac{1}{N^{2j}}\right)\frac{\bar{B}_{2j}}{2j}\right] + \epsilon_{2r+2}$$

in which $\bar{B}_{2j}$ = Bernoulli number (2.22) and

$$\left|\frac{n\bar{B}_{2r+2}}{N^{2r+2}}\right| < |\epsilon_{2r+2}| < \left|\frac{n\bar{B}_{2r+2}}{c^{2r+2}}\right|$$

For CA, if $c > 5$,

$$\sum_{k=1}^{n} \frac{1}{a + bk} = \frac{1}{b}\left[\ln\frac{N}{c} - \frac{1}{2c}\left(1 - \frac{1}{6c}\left(1 - \frac{1}{10c^2}\left(1 - \frac{10}{21c^2}\left(1 - \frac{21}{20c^2}\right)\right)\right)\right)\right.$$
$$\left. + \frac{1}{2N}\left(1 - \frac{1}{6N}\left(1 - \frac{1}{10N^2}\left(1 - \frac{10}{21N^2}\left(1 - \frac{21}{20N^2}\right)\right)\right)\right)\right] + \epsilon_{10}$$

For $c < 5$, $\psi(c)$ must be expressed by the small-argument formula of (2.33) and ϵ_{10} must be modified accordingly.

(b) Example. Given $a = 20$, $b = 1.25$, $n = 10$, $c = 16$, and $N = 26$,

$$\sum_{k=1}^{n} \frac{1}{a + bk} = \frac{1}{1.25}\left(\ln\frac{26}{16} - \phi_1 + \phi_2\right) + \epsilon_8 = 0.378\ 952\ 578\ 4$$

where $\phi_1 = \frac{1}{32}\left(1 - \frac{1}{96}\left(1 - \frac{1}{2560}\left(1 - \frac{10}{5376}\right)\right)\right) = 0.030\ 924\ 606\ 0$

$$\phi_2 = \frac{1}{52}\left(1 - \frac{1}{156}\left(1 - \frac{1}{6760}\left(1 - \frac{10}{14\ 196}\right)\right)\right) = 0.019\ 107\ 513\ 2$$

and $\left|\frac{10\bar{B}_8}{26^8} = +1.6(10)^{-12}\right| < |\epsilon_8| < \left|\frac{10\bar{B}_8}{16^8} = -7.8(10)^{-11}\right|$

(c) Alternating series is defined as

$$\sum_{k=1}^{n} \frac{\beta}{a + bk} = \frac{1}{b}\sum_{k=1}^{n}\frac{\beta}{c + k} = \frac{1}{b}\left(\frac{1}{c + 1} - \frac{1}{c + 2} - \cdots + \frac{\alpha}{c + n}\right) = \bar{\mathcal{Q}}_{1,n}(c)$$

where a, b, c are the same as in (a) above, $\alpha = (-1)^{n+1}$, $\beta = (-1)^{k+1}$, p = number of positive terms, q = number of negative terms, $n = p + q$, and

$$\bar{\mathcal{Q}}_{1,n}(c) = \frac{1}{2b}[\psi(p + c_1) - \psi(c_1) - \psi(q + c_2) + \psi(c_2)]$$

in which $c_1 = (a + b)/2b$, $c_2 = a/2b$, and $\psi(\ )$ = digamma function (2.32). For $n = \infty$,

$$\frac{1}{b}\sum_{k=1}^{\infty}\frac{\beta}{c + k} = \frac{1}{2b}\left[\psi\left(\frac{c}{2}\right) - \psi\left(\frac{c - 1}{2}\right)\right]$$

(a) Monotonic series is for $m = 1, 2, 3, \ldots$

$$\sum_{k=1}^{n} \frac{1}{(a + bk)^m} = \frac{1}{b^m} \sum_{k=1}^{n} \frac{1}{(c + k)^m} = \frac{1}{b^m} [\psi^{(m-1)}(c) - \psi^{(m-1)}(N)] = \mathcal{D}_{m,n}(c)$$

where a, b = positive real numbers, $c = a/b$, $N = c + n$, and $\psi^{(m-1)}(\ )$ is the polygamma function (2.34) so that

$$\psi^{(m-1)}(c) - \psi^{(m-1)}(N) = \sum_{k=1}^{\infty} \frac{1}{(c + k)^m} - \sum_{k=1}^{\infty} \frac{1}{(N + k)^m}$$

in which the polygamma functions can be evaluated by the formulas given (2.34), (2.35). An alternative and essentially the same procedure is the application of Euler–MacLaurin formulas (4.06),

$$\mathcal{D}_{m,n}(c) = \frac{1}{b^m} \left[\frac{1}{m - 1} \left(\frac{1}{c^{m-1}} - \frac{1}{N^{m-1}} \right) - \frac{1}{2} \left(\frac{1}{c^m} - \frac{1}{N^m} \right) + \sum_{j=1}^{r} \frac{\bar{B}_{2j}}{(2j)!} \omega_{2j-1} - \epsilon_{2r+2} \right]$$

where $\omega_{2j-1} = m(m + 1)(m + 2) \cdots (m + 2j - 2)\left(\dfrac{1}{c^{m+2j-1}} - \dfrac{1}{N^{m+2j-1}} \right)$, $\bar{B}_{2j}$ = Bernoulli number (2.22) and

$$\left| n \binom{m + 2r + 1}{2r + 2} \frac{\bar{B}_{2r+2}}{N^{m+2r+2}} \right| < |\epsilon_{2r+2}| < \left| n \binom{m + 2r + 1}{2r + 2} \frac{\bar{B}_{2r+2}}{c^{m+2r+2}} \right|$$

For $n = \infty$, $N = \infty$, and $\mathcal{D}_{m,\infty},(c) = \psi^{(m-1)}(c)$. For CA, with $n < \infty$,

$$\sum_{k=1}^{n} \frac{1}{(a + bk)^m} = \frac{1}{b^m} \left[\omega_0 + \frac{1}{12}\left(\omega_1 - \frac{1}{60}\left(\omega_3 + \frac{1}{42}\left(\omega_5 - \frac{\omega_7}{40} \right) \right) \right) + \epsilon_{10} \right]$$

(b) Alternating series in terms of a, b, m defined in (a) and with $\beta = (-1)^{km}$,

$$\sum_{k=1}^{n} \frac{\beta}{(a + bk)^m} = \frac{1}{b^m} \sum_{k=1}^{n} \frac{\beta}{(c + k)^m} = \bar{\mathcal{D}}_{m,n}(c)$$

can be resolved in a series of p positive terms and a series of g negative terms so that $n = p + q$ and

$$\bar{\mathcal{D}}_{m,n}(c) = \frac{1}{(2b)^m} \left[\sum_{k=1}^{p} \frac{1}{(c_1 + k)^m} - \sum_{k=1}^{q} \frac{1}{(c_2 + k)^m} \right]$$

$$= \frac{1}{(2b)^m} [\psi^{(m-1)}(c_1) - \psi^{(m-1)}(N_1) - \psi^{(m-1)}(c_2) + \psi^{(m-1)}(N_2)]$$

where $c_1 = (a + b)/2b$, $c_2 = a/2b$, $N_1 = p + c_1$, $N_2 = q + c_2$, and $\psi^{(m-1)}(\ )$ = polygamma function (2.34), which can be evaluated by the formulas given in (2.34), (2.35). For $n = \infty$, $N_1 = \infty$, $N_2 = \infty$,

$$\bar{\mathcal{D}}_{m,\infty}(c) = \frac{1}{(2b)} [\psi^{(m-1)}(c_1) - \psi^{(m-1)}(c_2)]$$

(a) Finite general series. When a, b are signed numbers,

$$\sum_{k=1}^{n} (a_1 + b_1 k)(a_2 + b_2 k) = A_0 + A_1 \mathscr{P}_{1,n} + A_2 \mathscr{P}_{2,n}$$

where $A_0 = a_1 a_2 n$, $A_1 = a_1 b_2 + a_2 b_2$, $A_2 = b_1, b_2$,

$$\sum_{k=1}^{n} (a_1 + b_1 k)(a_2 + b_2 k)(a_3 + b_3 k) = A_0 + A_1 \mathscr{P}_{1,n} + A_2 \mathscr{P}_{2,n} + A_3 \mathscr{P}_{3,n}$$

where $A_0 = a_1 a_2 a_3$, $A_1 = a_1 a_2 b_3 + a_2 a_3 b_1 + a_3 a_1 b_2$, $A_2 = b_1 b_2 a_3 + b_2 b_3 a_1 + b_3 b_1 a_2$, $A_3 = b_1 b_2 b_3$, and in general for $r = 1, 2, 3, \ldots$.

$$\sum_{k=1}^{n} (a_1 + b_1 k)(a_2 + b_2 k) \cdots (a_r + b_r k) = A_0 + A_1 \mathscr{P}_{1,n} + A_2 \mathscr{P}_{2,n} + \cdots + A_r \mathscr{P}_{r,n}$$

where $\mathscr{P}_{1,n}, \mathscr{P}_{2,n}, \ldots, \mathscr{P}_{r,n}$ are the exact sums of monotonic series of powers of integers (5.11) and $A_0, A_1, A_2, \ldots, A_r$ are the factors of the polynomial obtained as the product of the binomials.

(b) Example

$$\sum_{k=1}^{n} (2k - 1)(3k - 1) = \sum_{k=1}^{n} (1 - 5k + 6k^2) = n - 5\mathscr{P}_{1,n} + 6\mathscr{P}_{2,n}$$

where from (5.11d), $\mathscr{P}_{1,n} = n(n + 1)/2$ and $\mathscr{P}_{2,n} = n(n + 1)(n + 2)/6$.

(c) Two-term products

$$1 \cdot 2 + 2 \cdot 3 + 3 \cdot 4 + \cdots + n(n + 1) = n(n + 1)(2n + 4)/6$$

$$1 \cdot 3 + 2 \cdot 4 + 3 \cdot 5 + \cdots + n(n + 2) = n(n + 1)(2n + 7)/6$$

$$1 \cdot 1 + 2 \cdot 3 + 3 \cdot 5 + \cdots + n(2n - 1) = n(n + 1)(4n - 1)/6$$

$$1 \cdot 3 + 2 \cdot 5 + 3 \cdot 7 + \cdots + n(2n + 1) = n(n + 1)(4n + 5)/6$$

$$1 \cdot 2 + 3 \cdot 4 + 5 \cdot 6 + \cdots + 2n(2n - 1) = n(n + 1)(4n - 1)/3$$

$$2 \cdot 3 + 4 \cdot 5 + 6 \cdot 7 + \cdots + 2n(2n + 1) = n(n + 1)(4n + 5)/3$$

$$1 \cdot 3 + 3 \cdot 5 + 5 \cdot 7 + \cdots + (2n - 1)(2n + 1) = n(4n^2 + 6n - 1)/3$$

(d) Three-term products

$$1 \cdot 2 \cdot 3 + 2 \cdot 3 \cdot 4 + \cdots + n(n + 1)(n + 2) = n(n + 1)(n + 2)(n + 3)/4$$

$$1 \cdot 3 \cdot 5 + 2 \cdot 4 \cdot 6 + \cdots + n(n + 2)(n + 4) = n(n + 1)(n + 4)(n + 5)/4$$

$$1 \cdot 4 \cdot 7 + 2 \cdot 5 \cdot 8 + \cdots + n(n + 3)(n + 6) = n(n + 1)(n + 6)(n + 7)/4$$

(e) Power series

$$1 \cdot 2^2 + 2 \cdot 3^2 + 3 \cdot 4^2 + \cdots + n(n + 1)^2 = n(n + 1)(n + 2)(3n + 5)/12$$

$$1 \cdot 3^2 + 2 \cdot 5^2 + 3 \cdot 7^2 + \cdots + n(2n + 1)^2 = n(n + 1)(6n^2 + 14n + 7)/6$$

$$1 \cdot 2^3 + 2 \cdot 3^3 + 3 \cdot 4^3 + \cdots + n(n + 1) = n(n + 1)(12n^3 + 63n^2 + 107n + 58)/60$$

$$1 \cdot 3^3 + 2 \cdot 5^3 + 3 \cdot 7^3 + \cdots + n(2n + 1)^3 = n(n + 1)(48n^3 + 162n^2 + 158n + 37)/30$$

(a) Finite general series. With a, b defined in (5.23a) and $\alpha = (-1)^{n+1}$, $\beta = (-1)^{k+1}$, $\theta = 1$ if n is odd or $\theta = 0$ if n is even,

$$\sum_{k=1}^{n} \beta(a_1 + b_1 k)(a_2 + b_2 k) = B_0 + B_1 \bar{\mathscr{P}}_{1,n} + B_2 \bar{\mathscr{P}}_{2,n}$$

where $B_0 = a_1 a_2 \theta$ and $B_1 = A_1$, $B_2 = A_2$ in (5.23a),

$$\sum_{k=1}^{n} \beta(a_1 + b_1 k)(a_2 + b_2 k)(a_3 + b_3 k) = B_0 + B_1 \bar{\mathscr{P}}_{1,n} + B_2 \bar{\mathscr{P}}_{2,n} + B_3 \bar{\mathscr{P}}_{3n}$$

where $B_0 = a_1 a_2 a_3 \theta$ and $B_1 = A_1$, $B_2 = A_2$, $B_3 = A_3$ in (5.23a), and in general for $r = 1, 2, 3, \ldots$,

$$\sum_{k=1}^{n} \beta(a_1 + b_1 k)(a_2 + b_2 k) \cdots (a_r + b_r k) = B_0 + B_1 \bar{\mathscr{P}}_{1,n} + B_2 \bar{\mathscr{P}}_{2,n} + \cdots + B_r \bar{\mathscr{P}}_{r,n}$$

where $\bar{\mathscr{P}}_{1,n}, \bar{\mathscr{P}}_{2,n}, \ldots, \bar{\mathscr{P}}_{r,n}$ are the exact sums of alternating series of powers of integers (5.12) and $B_0, B_1, B_2, \ldots, B_r$ are the factors of the polynomial obtained as the product of binomials.

(b) Example

$$\sum_{k=1}^{n} \beta(2k - 1)(3k - 1) = \sum_{k=1}^{n} \beta(1 - 5k + 6k^2) = \theta - 5\bar{\mathscr{P}}_{1,n} + 6\bar{\mathscr{P}}_{2,n}$$

where from (5.12c), $\bar{\mathscr{P}}_{1,n} = (\alpha n + \theta)/2$, $\bar{\mathscr{P}}_{2,n} = \alpha n(n + 1)/2$.

(c) Two-term products

$$1 \cdot 2 - 2 \cdot 3 + 3 \cdot 4 - \cdots \alpha n(n + 1) \qquad = \tfrac{1}{2}[\alpha n(n + 2) + \theta]$$

$$1 \cdot 3 - 2 \cdot 4 + 3 \cdot 5 - \cdots \alpha n(n + 2) \qquad = \tfrac{1}{2}[\alpha n(n + 3) + 2\theta]$$

$$1 \cdot 1 - 2 \cdot 3 + 3 \cdot 5 - \cdots \alpha n(2n - 1) \qquad = \tfrac{1}{2}[\alpha n(2n + 1) - \theta]$$

$$1 \cdot 3 - 2 \cdot 5 + 3 \cdot 7 - \cdots \alpha n(2n + 1) \qquad = \tfrac{1}{2}[\alpha n(2n + 3) + \theta]$$

$$1 \cdot 2 - 3 \cdot 4 + 5 \cdot 6 - \cdots 2\alpha n(2n - 1) \qquad = [\alpha n(2n + 1) - \theta]$$

$$2 \cdot 3 - 4 \cdot 5 + 6 \cdot 7 - \cdots 2\alpha n(2n + 1) \qquad = [\alpha n(2n + 3) + \theta]$$

$$1 \cdot 3 - 3 \cdot 5 + 5 \cdot 7 - \cdots \alpha(2n - 1)(2n + 1) = 2\alpha n(n + 1) - \theta$$

(d) Three-term products

$$1 \cdot 2 \cdot 3 - 2 \cdot 3 \cdot 4 + \cdots \alpha n(n + 1)(n + 2) \qquad = \tfrac{1}{4}[\alpha n(n + 2)(2n + 5) + 3\theta]$$

$$1 \cdot 3 \cdot 5 - 2 \cdot 4 \cdot 6 + \cdots \alpha n(n + 2)(n + 4) \qquad = \tfrac{1}{4}[\alpha n(n + 4)(2n + 7) + 15\theta]$$

$$1 \cdot 4 \cdot 7 - 2 \cdot 5 \cdot 8 + \cdots \alpha n(n + 3)(n + 6) \qquad = \tfrac{1}{4}[\alpha n(n + 6)(2n + 9) + 35\theta]$$

(e) Power series

$$1 \cdot 2^2 - 2 \cdot 3^2 + 3 \cdot 4^2 - \cdots \alpha n(n + 1)^2 \qquad = \tfrac{1}{4}[\alpha n(n + 2)(2n + 3) + \theta]$$

$$1 \cdot 3^2 - 2 \cdot 5^2 + 3 \cdot 7^2 - \cdots \alpha n(2n + 1)^2 \qquad = \tfrac{1}{2}[\alpha n(4n^2 + 10n + 5) - \theta]$$

$$1 \cdot 2^3 - 2 \cdot 3^3 + 3 \cdot 4^3 - \cdots \alpha n(n + 1)^3 \qquad = \tfrac{1}{4}[\alpha n(2n^3 + 10n^2 + 15n + 6) - \theta]$$

$$1 \cdot 3^3 - 2 \cdot 5^3 + 3 \cdot 7^3 - \cdots \alpha n(2n + 1)^3 \qquad = \tfrac{1}{2}[\alpha n(8n^3 + 28n^2 + 24n - 1) - 5\theta]$$

(a) Monotonic series. When a_1, a_2, b_1, b_2 are positive numbers and $c_1 = a_1/b_1$, $c_2 = a_2/b_2$,

$$\sum_{k=1}^{n} \frac{1}{(a_1 + b_1 k)(a_2 + b_2 k)} = \frac{1}{b_1 b_2} \sum_{k=1}^{n} \frac{1}{(c_1 + k)(c_2 + k)} = \frac{\mathscr{D}_{1,n}(c_1) - \mathscr{D}_{1,n}(c)}{b_1 b_2 (c_2 - c_1)}$$

where by resolution in partial fractions,

$$\frac{1}{(c_1 + k)(c_2 + k)} = \frac{1}{c_2 - c_1} \left(\frac{1}{c_1 + k} - \frac{1}{c_2 + k} \right)$$

and according to (5.21a) with $N_1 = c_1 + n$, $N_2 = c_2 + n$,

$$\sum_{k=1}^{n} \frac{1}{c_1 + k} = \psi(N_1) - \psi(c_1) = \mathscr{D}_{1,n}(c_1) \qquad \sum_{k=1}^{n} \frac{1}{c_2 + k} = \psi(N_2) - \psi(c_2) = \mathscr{D}_{1,n}(c_2)$$

are the sums of general harmonic series. Similarly with $c_1 = a_1/b_1$, $c_2 = a_2/b_2$, $c_3 = a_3/b_3$,

$$\sum_{k=1}^{n} \frac{1}{(a_1 + b_1 k)(a_2 + b_2 k)(a_3 + b_3 k)} = \frac{(c_2 - c_3)\mathscr{D}_{1,n}(c_1) + (c_3 - c_1)\mathscr{D}_{1,n}(c_2) + (c_1 - c_2)\mathscr{D}_{1,n}(c_3)}{b_1, b_2, b_3 (c_1 - c_2)(c_2 - c_3)(c_3 - c_1)}$$

where $\mathscr{D}_{1,n}(c_1)$, $\mathscr{D}_{1,n}(c_2)$, $\mathscr{D}_{1,n}(c_3)$ have an analogical meaning as before.

(b) Special forms

$$\sum_{k=1}^{n} \frac{1}{(c + k)k} = \frac{1}{c}[\mathscr{D}_{1,n} - \mathscr{D}_{1,n}(c)] \qquad \sum_{k=1}^{n} \frac{1}{(c + k)k^2} = \frac{1}{c^2}[c\mathscr{D}_{2,n} - \mathscr{D}_{1,n}(c)]$$

$$\sum_{k=1}^{n} \frac{1}{(2k - 1)k} = \mathscr{D}_{1,n}^* - \mathscr{D}_{1,n} \qquad \sum_{k=1}^{n} \frac{1}{(2k - 1)k^2} = 2\mathscr{D}_{1,n}^* - 2\mathscr{D}_{1,n} - \mathscr{D}_{2,n}$$

$$\sum_{k=1}^{n} \frac{1}{(c + k)k^3} = \frac{1}{c^3}[c^2\mathscr{D}_{3,n} - c\mathscr{D}_{2,n} + \mathscr{D}_{1,n}(c)]$$

$$\sum_{k=1}^{n} \frac{1}{(2k - 1)k^3} = 4\mathscr{D}_{1,n}^* - 4\mathscr{D}_{1,n} - 2\mathscr{D}_{2,n} - \mathscr{D}_{3,n}$$

where $\mathscr{D}_{1,n}$, $\mathscr{D}_{1,n}^*$ are defined in (5.13), $\mathscr{D}_{m,n}$, $\mathscr{D}_{m,n}^*$ in (5.15) and $\mathscr{D}_{1,n}(c)$ is defined in (a) above.

(c) Alternating series. Although sum formulas of alternating series are available, their complexity makes their application impractical. It is more convenient to resolve a given alternating series in a series of p positive terms and another series of q negative terms ($n = p + q$), sum each series by the formula of (a), and subtract the sum of the series of negative terms from the series of positive terms as

$$\sum_{k=1}^{n} \frac{(-1)^{k+1}}{(a_1 + b_1 k)(a_2 + b_2 k)} = \frac{1}{4 b_1 b_2} \left[\sum_{k=1}^{p} \frac{1}{(d_1 + k)(d_2 + k)} - \sum_{k=1}^{q} \frac{1}{(e_1 + k)(e_2 + k)} \right]$$

where $d_1 = (c_1 - 1)/2$, $d_2 = (c_2 - 1)/2$, $e_1 = c_1/2$, $e_2 = c_2/2$, and each sum is expressed by the relations given in (a).

(a) Notation. p, q = positive integers n = number of terms $< \infty$

$\psi(\ )$ = digamma function (2.32) $\gamma = 0.577\,215\,667\cdots$ = Euler's constant (2.19)

$$\mathcal{D}_{1,n} = \sum_{k=1}^{n} \frac{1}{k} = \gamma + \psi(n) \qquad \mathcal{D}_{1,n}^{*} = \sum_{k=1}^{n} \frac{1}{2k-1} = \gamma + \psi(n - \tfrac{1}{2}) + 2\ln 2$$

(b) Two-fraction products

$$\frac{1}{1\cdot 2} + \frac{1}{2\cdot 3} + \frac{1}{3\cdot 4} + \cdots + \frac{1}{n(n+1)} = \frac{n}{n+1}$$

$$\frac{1}{1\cdot 3} + \frac{1}{3\cdot 5} + \frac{1}{5\cdot 7} + \cdots + \frac{1}{(2n-1)(2n+1)} = \frac{n}{2n+1}$$

$$\frac{1}{1\cdot 4} + \frac{1}{4\cdot 7} + \frac{1}{7\cdot 10} + \cdots + \frac{1}{(3n-2)(3n+1)} = \frac{n}{3n+1}$$

$$\frac{1}{1\cdot 3} + \frac{1}{12\cdot 4} + \frac{1}{3\cdot 5} + \cdots + \frac{1}{n(n+2)} = \frac{1}{2}\left(\frac{3}{2} - \frac{1}{n+1} - \frac{1}{n+2}\right)$$

$$\frac{1}{2\cdot 5} + \frac{1}{3\cdot 6} + \frac{1}{4\cdot 7} + \cdots + \frac{1}{(n+1)(n+4)} = \frac{1}{3}\left(\frac{13}{12} - \frac{1}{n+2} - \frac{1}{n+3} - \frac{1}{n-4}\right)$$

$$\frac{1}{(p+q)(p+2q)} + \frac{1}{(p+2q)(p+3q)} + \cdots + \frac{1}{(p+nq)[p+(n+1)q]} = \frac{nq}{(p+q)[p+(n+1)q]}$$

$$\frac{1}{2\cdot 3} + \frac{2}{2\cdot 4} + \frac{3}{4\cdot 5} + \cdots + \frac{n}{(n+1)(n+2)} = \mathcal{D}_{1,n} + \frac{1}{n+1} + \frac{2}{n+2} - 2$$

$$\frac{1}{2\cdot 5} + \frac{2}{3\cdot 6} + \frac{3}{4\cdot 7} + \cdots + \frac{n}{(n+1)(n+4)} = \mathcal{D}_{1,n} + \frac{1}{n+1} + \frac{4}{3}\left(\frac{1}{n+2} + \frac{1}{n+3} + \frac{1}{n+4}\right) - \frac{22}{9}$$

(c) Three-fraction products

$$\frac{1}{1\cdot 2\cdot 3} + \frac{1}{2\cdot 3\cdot 4} + \frac{1}{3\cdot 4\cdot 5} + \cdots + \frac{1}{n(n+1)(n+2)} = \frac{1}{2}\left(\frac{1}{2} - \frac{1}{n+1} + \frac{1}{n+2}\right)$$

$$\frac{1}{1\cdot 3\cdot 5} + \frac{1}{2\cdot 4\cdot 6} + \frac{1}{3\cdot 5\cdot 7} + \cdots + \frac{1}{n(n+2)(n+4)} = \frac{1}{8}\left(\frac{11}{12} - \frac{1}{n+1} - \frac{1}{n+2} + \frac{1}{n+3} + \frac{1}{n+4}\right)$$

$$\frac{1}{1\cdot 3\cdot 5} + \frac{1}{3\cdot 5\cdot 7} + \frac{1}{5\cdot 7\cdot 9} + \cdots + \frac{1}{(2n-1)(2n+1)(2n+3)} = \frac{1}{12}\left[1 - \frac{3}{(2n+1)(2n+3)}\right]$$

$$\frac{1}{1\cdot 4\cdot 7} + \frac{1}{4\cdot 7\cdot 10} + \frac{1}{7\cdot 10\cdot 13} + \cdots + \frac{1}{(3n-2)(3n+1)(3n+4)} = \frac{1}{24}\left[1 - \frac{4}{(3n+1)(3n+4)}\right]$$

$$\frac{1}{1\cdot 2\cdot 3} + \frac{1}{3\cdot 4\cdot 5} + \frac{1}{5\cdot 6\cdot 7} + \cdots + \frac{1}{(2n-1)(2n)(2n+1)} = \frac{1}{2}(\mathcal{D}_{1,n}^{*} - \mathcal{D}_{1,n}) - \frac{n}{2n-1}$$

$$\frac{1}{2\cdot 3\cdot 4} + \frac{2}{3\cdot 4\cdot 5} + \frac{3}{4\cdot 5\cdot 6} + \cdots + \frac{n}{(n+1)(n+2)(n+3)} = \frac{1}{4}\left(1 + \frac{2}{n+2} - \frac{6}{n+3}\right)$$

$$\frac{1}{2\cdot 4\cdot 6} + \frac{2}{3\cdot 5\cdot 7} + \frac{3}{4\cdot 6\cdot 8} + \cdots + \frac{n}{(n+1)(n+3)(n+5)} = \frac{1}{96}\left(17 + \frac{12}{n+2} + \frac{12}{n+3} - \frac{60}{n+3} - \frac{60}{n+4}\right)$$

$$\frac{1}{3\cdot 4\cdot 5} + \frac{2}{4\cdot 5\cdot 6} + \frac{3}{5\cdot 6\cdot 7} + \cdots + \frac{n}{(n+2)(n+3)(n+4)} = \frac{1}{4}\left(3 - \frac{8}{n+2} + \frac{2}{(n+1)(n+2)}\right)$$

(a) Monotonic series. The sum of the series in (5.25a) for $n \to \infty$ is

$$\sum_{k=1}^{\infty} \frac{1}{(a_1 + b_1 k)(a_2 + b_2 k)} = \lim_{n \to \infty} \sum_{k=1}^{n} \frac{1}{(a_1 + b_1 k)(a_2 + b_2 k)} = \frac{\psi(c_2) - \psi(c_1)}{b_1, b_2(c_2 - c_1)}$$

where $c_1 = a_1/b_1$, $c_2 = a_2/b_2$, and $\psi(\ ) = $ digamma function (2.32).

(b) General formulas in terms of $a, b, c =$ positive numbers, $m, p, q, r =$ positive integers, $\gamma = 0.577\ 215\ 664 \cdots$ (2.19),

$$\mathcal{Q}_{1,p} = \sum_{k=1}^{p} \frac{1}{k} \qquad \mathcal{Q}_{1,q} = \sum_{k=1}^{q} \frac{1}{k} \qquad \mathcal{Q}_{1,r} = \sum_{k=1}^{r} \frac{1}{k} \qquad Z(m) = \sum_{k=1}^{\infty} \frac{1}{k^m} = \text{Riemann zeta function}$$

(2.26) are

$$\sum_{k=1}^{\infty} \frac{1}{(p+k)k} = \frac{1}{p}\mathcal{Q}_{1,p} \qquad\qquad \sum_{k=1}^{\infty} \frac{1}{(c+k)k} = \frac{1}{c}\psi(c) + \gamma$$

$$\sum_{k=1}^{\infty} \frac{1}{(p+k)k^2} = \frac{1}{p^2}[pZ(2) - \mathcal{Q}_{1,p}] \qquad \sum_{k=1}^{\infty} \frac{1}{(c+k)k^2} = \frac{1}{c}[cZ(2) - \psi(c) - \gamma]$$

$$\sum_{k=1}^{\infty} \frac{1}{(p+k)(q+k)} = \frac{\psi(p) - \psi(q)}{p - q} \qquad \sum_{k=1}^{\infty} \frac{1}{(a+bk)[a+b(k+1)]} = \frac{1}{a+b}$$

$$\sum_{k=1}^{\infty} \frac{1}{(p+k)q+k)(r+k)} = \frac{(q-r)\mathcal{Q}_{1,p} + (r-p)\mathcal{Q}_{1,q} + (p-q)\mathcal{Q}_{1,r}}{(p-q)(q-r)(r-p)}$$

$$\sum_{k=1}^{\infty} \frac{1}{[a+b(k-1)](a+bk)[a+b(k+1)]} = \frac{1}{2ab(a+b)}$$

$$\sum_{k=1}^{\infty} \frac{1}{(p+k)k^m} = \frac{(-1)^m}{p^m}[pZ(2) - p^2 Z(3) + p^3 Z(4) - \cdots (-p)^{m-1}Z(m) - \mathcal{Q}_{1,p}]$$

$$\sum_{k=1}^{\infty} \frac{1}{(c+k)k^m} = \frac{(-1)^m}{c^m}[cZ(2) - c^2 Z(3) + c^3 Z(4) - \cdots (-c)^{m-1}Z(m) - \psi(c) - \gamma]$$

(c) Example

$$\sum_{k=1}^{\infty} \frac{1}{(3+k)k^3} = \frac{1}{4 \cdot 1^3} + \frac{1}{5 \cdot 2^3} + \frac{1}{6 \cdot 3^3} + \cdots$$

$$= \frac{(-1)^3}{3^3}[3Z(2) - 3^2 Z(3) - \psi(3) - \gamma] = 0.307\ 194\ 775$$

where from (2.26), $Z(2) = 1.644\ 934\ 067$, $Z(3) = 1.202\ 056\ 903$, and $\psi(3) + \gamma = 1 + \frac{1}{2} + \frac{1}{3} = 1.833\ 333\ 333$.

(d) Alternating series can be conveniently summed by the procedure described in (5.25c).

(e) Example. From (b) above, with $p = 1$, $c = \frac{1}{2}$ and by (2.32f)

$$\sum_{k=1}^{\infty} \frac{\beta}{k(k+1)} = \frac{1}{1 \cdot 2} - \frac{1}{2 \cdot 3} + \frac{1}{13 \cdot 4} - \cdots = \sum_{k=1}^{\infty} \frac{1}{k(k+1)} - 2\sum_{k=1}^{\infty} \frac{1}{2k(2k+1)}$$

$$\sum_{k=1}^{\infty} \frac{1}{k(k+1)} = 1 \qquad \text{and} \qquad 2\sum_{k=1}^{\infty} \frac{1}{2k(2k+1)} = \psi(\tfrac{1}{2}) + \gamma = 0.613\ 705\ 639$$

(a) Two-fraction products $(n = \infty)$

$$\frac{1}{1 \cdot 2} + \frac{1}{2 \cdot 3} + \frac{1}{3 \cdot 4} + \cdots = 1$$

$$\frac{1}{1 \cdot 2} + \frac{1}{2 \cdot 4} + \frac{1}{3 \cdot 6} + \cdots = \frac{\pi^2}{12}$$

$$\frac{1}{1 \cdot 3} + \frac{1}{2 \cdot 4} + \frac{1}{3 \cdot 5} + \cdots = \frac{3}{4}$$

$$\frac{1}{1 \cdot 3} + \frac{1}{2 \cdot 5} + \frac{1}{3 \cdot 7} + \cdots = 2 - \ln 2$$

$$\frac{1}{1 \cdot 4} + \frac{1}{2 \cdot 5} + \frac{1}{3 \cdot 6} + \cdots = \frac{11}{18}$$

$$\frac{1}{1 \cdot 4} + \frac{1}{2 \cdot 6} + \frac{1}{3 \cdot 8} + \cdots = \frac{1}{2}$$

$$\frac{1}{1 \cdot 5} + \frac{1}{2 \cdot 6} + \frac{1}{3 \cdot 7} + \cdots = \frac{25}{48}$$

$$\frac{1}{1 \cdot 5} + \frac{1}{2 \cdot 7} + \frac{1}{2 \cdot 9} + \cdots = \frac{8}{9} - \frac{2}{3} \ln 2$$

$$\frac{1}{2 \cdot 4} + \frac{1}{4 \cdot 6} + \frac{1}{6 \cdot 8} + \cdots = \frac{1}{4}$$

$$\frac{1}{2 \cdot 4} + \frac{1}{4 \cdot 8} + \frac{1}{6 \cdot 12} + \cdots = \frac{\pi^2}{48}$$

$$\frac{1}{3 \cdot 5} + \frac{1}{4 \cdot 6} + \frac{1}{5 \cdot 7} + \cdots = \frac{1}{24}$$

$$\frac{1}{3 \cdot 5} + \frac{1}{7 \cdot 9} + \frac{1}{11 \cdot 13} + \cdots = \frac{4 - \pi}{8}$$

$$\frac{1}{1 \cdot 2} + \frac{1}{2 \cdot 2^2} + \frac{1}{3 \cdot 2^3} + \cdots = \ln 2$$

$$\frac{1}{1 \cdot 3} + \frac{1}{2 \cdot 3^2} + \frac{1}{3 \cdot 3^3} + \cdots = \ln \frac{3}{2}$$

(b) Three-fraction products $(n = \infty)$

$$\frac{1}{1 \cdot 2 \cdot 3} + \frac{1}{2 \cdot 3 \cdot 4} + \cdots = \frac{1}{4}$$

$$\frac{1}{1 \cdot 2 \cdot 3} + \frac{1}{3 \cdot 4 \cdot 5} + \cdots = \ln 2 - \frac{1}{2}$$

$$\frac{1}{1 \cdot 2 \cdot 3} + \frac{1}{4 \cdot 5 \cdot 6} + \cdots = \frac{1}{4}\left(\frac{\pi}{\sqrt{3}} - \ln 3\right)$$

$$\frac{1}{1 \cdot 2 \cdot 3} + \frac{1}{5 \cdot 6 \cdot 7} + \cdots = \frac{1}{4} \ln 2$$

$$\frac{1}{1 \cdot 2 \cdot 3} + \frac{2}{3 \cdot 4 \cdot 5} + \cdots = \frac{1}{4}$$

$$\frac{1}{1 \cdot 2 \cdot 3} + \frac{3}{2 \cdot 3 \cdot 4} + \cdots = \frac{3}{4}$$

$$\frac{1}{1 \cdot 3 \cdot 5} + \frac{1}{2 \cdot 4 \cdot 6} + \cdots = \frac{11}{96}$$

$$\frac{1}{1 \cdot 3 \cdot 5} + \frac{1}{3 \cdot 5 \cdot 7} + \cdots = \frac{1}{12}$$

$$\frac{1}{1 \cdot 4 \cdot 7} + \frac{1}{2 \cdot 5 \cdot 8} + \cdots = \frac{43}{1080}$$

$$\frac{1}{1 \cdot 4 \cdot 7} + \frac{1}{4 \cdot 7 \cdot 10} + \cdots = \frac{1}{24}$$

$$\frac{1}{2 \cdot 3 \cdot 4} + \frac{1}{4 \cdot 5 \cdot 6} + \cdots = \frac{3}{4} - \ln 2$$

$$\frac{1}{2 \cdot 3 \cdot 4} + \frac{1}{6 \cdot 7 \cdot 8} + \cdots = \frac{\pi}{8} - \frac{1}{2} \ln 2$$

$$\frac{1}{2 \cdot 3 \cdot 4} + \frac{1}{3 \cdot 4 \cdot 5} + \cdots = \frac{1}{12}$$

$$\frac{1}{2 \cdot 4 \cdot 6} + \frac{1}{4 \cdot 6 \cdot 8} + \cdots = \frac{1}{32}$$

$$\frac{1}{3 \cdot 4 \cdot 5} + \frac{1}{4 \cdot 5 \cdot 6} + \cdots = \frac{1}{24}$$

$$\frac{1}{3 \cdot 5 \cdot 7} + \frac{1}{5 \cdot 7 \cdot 9} + \cdots = \frac{1}{60}$$

$$\frac{1}{2 \cdot 4 \cdot 6} + \frac{2}{3 \cdot 5 \cdot 7} + \cdots = \frac{7}{96}$$

$$\frac{1}{3 \cdot 4 \cdot 5} + \frac{2}{4 \cdot 5 \cdot 6} + \cdots = \frac{1}{6}$$

(c) Four-fraction products $(n = \infty)$

$$\frac{1}{1 \cdot 2 \cdot 3 \cdot 4} + \frac{1}{2 \cdot 3 \cdot 4 \cdot 5} + \cdots = \frac{1}{18}$$

$$\frac{1}{1 \cdot 2 \cdot 3 \cdot 4} + \frac{1}{5 \cdot 6 \cdot 7 \cdot 8} + \cdots = \frac{\ln 65 - \pi}{24}$$

$$\frac{1}{1 \cdot 3 \cdot 5 \cdot 7} + \frac{1}{3 \cdot 5 \cdot 7 \cdot 9} + \cdots = \frac{1}{45}$$

$$\frac{1}{1 \cdot 3 \cdot 5 \cdot 7} + \frac{1}{9 \cdot 11 \cdot 13 \cdot 15} + \cdots = \frac{(2 - \sqrt{2})\pi}{192}$$

(a) Notation

a = positive number k, r = positive integers n = number of terms

(b) Basic series ($n < \infty$, $a \neq 1$)

$\sum_{k=1}^{n} a^k = \dfrac{a(a^n - 1)}{a - 1}$	$\mathscr{A}_1$	$\sum_{k=1}^{n} \dfrac{1}{a^k} = \dfrac{a^n - 1}{(a - 1)a^n}$	$\mathscr{B}_1$
$\sum_{k=1}^{n} a^{2k} = \dfrac{a^2(a^{2n} - 1)}{a^2 - 1}$	$\mathscr{A}_2$	$\sum_{k=1}^{n} \dfrac{1}{a^{2k}} = \dfrac{a^{2n} - 1}{(a^2 - 1)a^{2n}}$	$\mathscr{B}_2$
$\sum_{k=1}^{n} a^{2k-1} = \dfrac{a(a^{2n} - 1)}{a^2 - 1}$	$\mathscr{A}_3$	$\sum_{k=1}^{n} \dfrac{1}{a^{2k-1}} = \dfrac{a^{2n} - 1}{(a^2 - 1)a^{2n-1}}$	$\mathscr{B}_3$
$\sum_{k=1}^{n} ka^k = \dfrac{a - (n + 1)a^{n+1} + na^{n+2}}{(a - 1)^2}$	$\mathscr{A}_4$	$\sum_{k=1}^{n} \dfrac{k}{a^k} = \dfrac{n - (n + 1)a + a^{n+1}}{(a - 1)^2 a^n}$	$\mathscr{B}_4$
$\sum_{k=1}^{n} a^{rk} = \dfrac{a^r(a^{rn} - 1)}{a^r - 1}$	$\mathscr{A}_5$	$\sum_{k=1}^{n} \dfrac{1}{a^{rk}} = \dfrac{a^{rn} - 1}{(a^{rn} - 1)a^{rn}}$	$\mathscr{B}_5$
$\sum_{k=1}^{n} ka^{rk} = \dfrac{a^r - (n + 1)a^{r(n+1)} + na^{r(n+2)}}{(a^r - 1)^2}$	$\mathscr{A}_6$	$\sum_{k=1}^{n} \dfrac{k}{a^{rk}} = \dfrac{n - (n + 1)a^r + a^{r(n+1)}}{(a^r - 1)^2 a^{rn}}$	$\mathscr{B}_6$
$\sum_{k=1}^{n} (2k - 1)a^k = 2\mathscr{A}_4 - \mathscr{A}_1$	$\mathscr{A}_7$	$\sum_{k=1}^{n} \dfrac{2k - 1}{a^k} = 2\mathscr{B}_4 - \mathscr{B}_1$	$\mathscr{B}_7$
$\sum_{k=1}^{n} (2k - 1)a^{rk} = 2\mathscr{A}_6 - \mathscr{A}_5$	$\mathscr{A}_8$	$\sum_{k=1}^{n} \dfrac{2k - 1}{a^{rk}} = 2\mathscr{B}_6 - \mathscr{B}_5$	$\mathscr{B}_8$

(c) **Derived sums.** If $f(k)$ and $f(k - 1)$ are functions of k and $k = 1, 2, 3, \ldots, n$, then

$$\sum_{k=1}^{n} [f(k) - f(k - 1)] = f(n) - f(0)$$

(d) Examples

$$\sum_{k=1}^{n} [k^2 a^k - (k - 1)^2 a^{k-1}] = n^2 a^n$$

or $\displaystyle\sum_{k=1}^{n} k^2 a^k - \sum_{k=1}^{n} k^2 a^{k-1} + \sum_{k=1}^{n} 2k a^{k-1} - \sum_{k=1}^{n} a^{k-1} = n^2 a^n$

from which $\left(1 - \dfrac{1}{a}\right) \displaystyle\sum_{k=1}^{n} k^2 a^k = -\dfrac{2}{a}\mathscr{A}_4 + \dfrac{1}{a}\mathscr{A}_1 + n^2 a^n$

and the final sum formula $\mathscr{A}_9$ is

$$\sum_{k=1}^{n} k^2 a^k = \frac{1}{a - 1}(n^2 a^{n+1} - 2\mathscr{A}_4 + \mathscr{A}_1)$$

where $\mathscr{A}_4$ and $\mathscr{A}_1$ are the basic sums given in (b) above. By a similar procedure,

$$\sum_{k=1}^{n} k^r a^k = \frac{1}{a - 1}\left[n^r a^{n+1} - \binom{r}{1}\mathscr{A}_{r+7} + \binom{r}{2}\mathscr{A}_{r+6} - \cdots (-1)^r \binom{r}{r - 1}\mathscr{A}_4 - (-1)^{r+1}\mathscr{A}_1\right]$$

where $\mathscr{A}_{r+7}, \mathscr{A}_{r+6}, \ldots$ are the sums of $k^{r-1}a^{k-1}, k^{r-2}a^{k-1}, \ldots$, respectively, and derived recurrently.

(a) Notation. a, k, m, n, r are defined in (5.29a), $\alpha = (-1)^{n+1}$, and $\beta = (-1)^{k+1}$.

(b) Basic series $(n < \infty)$

$\displaystyle\sum_{k=1}^{n} \beta a^k = \frac{a(1 + \alpha a^n)}{1 + a}$	$\mathscr{C}_1$	$\displaystyle\sum_{k=1}^{n} \frac{\beta}{a^k} = \frac{\alpha + a^n}{(1 + a)a^n}$	$\mathscr{D}_1$	
$\displaystyle\sum_{k=1}^{n} \beta a^{2k} = \frac{a^2(1 + \alpha a^{2n})}{1 + a^2}$	$\mathscr{C}_2$	$\displaystyle\sum_{k=1}^{n} \frac{\beta}{a^{2k}} = \frac{\alpha + a^{2n}}{(1 + a^2)a^{2n}}$	$\mathscr{D}_2$	
$\displaystyle\sum_{k=1}^{n} \beta a^{2k-1} = \frac{a(1 + \alpha a^{2n})}{1 + a^2}$	$\mathscr{C}_3$	$\displaystyle\sum_{k=1}^{n} \frac{\beta}{a^{2k-1}} = \frac{\alpha + a^{2n}}{(1 + a^2)a^{2n-1}}$	$\mathscr{D}_3$	
$\displaystyle\sum_{k=1}^{n} \beta k a^k = \frac{a + \alpha(n + 1)a^{n+1} + \alpha n a^{n+2}}{(1 + a)^2}$	$\mathscr{C}_4$	$\displaystyle\sum_{k=1}^{n} \frac{\beta k}{a^k} = \frac{\alpha n + \alpha(n + 1)a + a^{n-1}}{(1 + a)^2 a^n}$	$\mathscr{D}_4$	
$\displaystyle\sum_{k=1}^{n} \beta a^{rk} = \frac{a^r(1 + \alpha a^{rm})}{1 + a^r}$	$\mathscr{C}_5$	$\displaystyle\sum_{k=1}^{n} \frac{\beta}{a^{rk}} = \frac{\alpha + a^{rm}}{(1 + a^r)a^{rm}}$	$\mathscr{D}_5$	
$\displaystyle\sum_{k=1}^{n} \beta k a^{rk} = \frac{a^r + \alpha(n + 1)a^{r(n+1)} + \alpha n a^{r(n+2)}}{(1 + a^r)^2}$	$\mathscr{C}_6$	$\displaystyle\sum_{k=1}^{n} \frac{\beta k}{a^{rk}} = \frac{\alpha n + \alpha(n + 1)a^r + a^{r(n-1)}}{(1 + a^r)^2 a^{rn}}$	$\mathscr{D}_6$	
$\displaystyle\sum_{k=1}^{n} \beta(2k - 1)a^k = 2\mathscr{C}_4 - \mathscr{C}_1$	$\mathscr{C}_7$	$\displaystyle\sum_{k=1}^{n} \frac{\beta(2k - 1)}{a^k} = 2\mathscr{D}_4 - \mathscr{D}_1$	$\mathscr{D}_7$	
$\displaystyle\sum_{k=1}^{n} \beta(2k - 1)a^{rk} = 2\mathscr{C}_6 - \mathscr{C}_5$	$\mathscr{C}_8$	$\displaystyle\sum_{k=1}^{n} \frac{\beta(2k - 1)}{a^{rk}} = 2\mathscr{D}_6 - \mathscr{D}_6$	$\mathscr{D}_8$	

(c) Derived sums. If $g(k)$ and $g(k - 1)$ are alternating functions of k and $k = 1, 2, 3, \ldots, n$, then

$$\sum_{k=1}^{n} [g(k) - g(k - 1)] = g(n) - g(0)$$

(d) Examples

$$\sum_{k=1}^{n} [\beta k^2 a^k + \beta(k - 1)^2 a^{k-1}] = \alpha n^2 a^n$$

or $\displaystyle\sum_{k=1}^{n} \beta k^2 a^k + \sum_{k=1}^{n} \beta k^2 a^{k-1} - \sum_{k=1}^{n} 2\beta k a^{k-1} - \sum_{k=1}^{n} \beta a^{k-1} = \alpha n^2 a^n$

from which $\displaystyle\left(1 + \frac{1}{a}\right) \sum_{k=1}^{n} \beta k^2 a^k = \frac{2}{a}\mathscr{C}_4 - \frac{1}{a}\mathscr{C}_1 + \alpha n^2 a^n$

and the final sum formula $\mathscr{C}_9$ is

$$\sum_{k=1}^{n} \beta k^2 a^k = \frac{1}{1 + a}(\alpha n^2 a^{n+1} + 2\mathscr{C}_4 - \mathscr{C}_1)$$

By a similar procedure,

$$\sum_{k=1}^{n} \beta k^2 a^{rk} = \frac{1}{1 + a}\left[\alpha n^r a^{n+1} + \binom{r}{1}\mathscr{C}_{r+7} - \binom{r}{2}\mathscr{C}_{r+6} + \cdots (-1)^r \mathscr{C}_4 + (-1)^{r+1}\mathscr{C}_1\right]$$

where $\mathscr{C}_{r+7}, \mathscr{C}_{r+6}, \ldots$ are the sums of $\beta k^{r-1} a^{k-1}, \beta k^{r-2} a^{k-1}, \ldots$, respectively, and derived recurrently.

(a) **Notation.** a, k, n, α, β are defined in (5.29a) and (5.30a) and $a < \pi/2$.

(b) **Monotonic series of sine terms** $(n < \infty)$

$$\sum_{k=1}^{n} \sin ka = \frac{\sin\left[(n+1)a/2\right]\sin(na/2)}{\sin(a/2)}$$

$$\sum_{k=1}^{n} \sin^2 ka = \frac{n}{2} - \frac{\sin na \cos\left[(n+1)a\right]}{2\sin a}$$

$$\sum_{k=1}^{n} \sin 2ka = \frac{\sin\left[(n+1)a\right]\sin na}{\sin a}$$

$$\sum_{k=1}^{n} \sin^2 2ka = \frac{n}{2} - \frac{\sin 2na \cos\left[2(n+1)a\right]}{2\sin 2a}$$

$$\sum_{k=1}^{n} \sin(2k-1)a = \frac{\sin^2 na}{\sin a}$$

$$\sum_{k=1}^{n} \sin^2(2k-1)a = \frac{n}{2} - \frac{\sin 2na \cos 2na}{2\sin 2a}$$

$$\sum_{k=1}^{n} k\sin ka = \frac{\sin\left[(n+1)a\right]}{4\sin^2(a/2)} - \frac{(n+1)\cos\left[(2n+1)a/2\right]}{2\sin(a/2)}$$

$$\sum_{k=1}^{n} 2k\sin 2ka = \frac{\sin\left[2(n+1)a\right]}{2\sin^2 a} - \frac{(n+1)\cos\left[(2n+1)a\right]}{\sin a}$$

$$\sum_{k=1}^{n}(2k-1)\sin(2k-1)a = \frac{\cos a \sin(2na) - 2n\sin a \cos 2na}{\sin^2 a}$$

(c) **Alternating series of sine terms** $(n < \infty)$

$$\sum_{k=1}^{n} \beta\sin ka = \frac{\sin a + \alpha\sin na + \alpha\sin\left[(n+1)a\right]}{2(1+\cos a)}$$

$$\sum_{k=1}^{n} \beta\sin 2ka = \frac{\sin 2a + \alpha\sin 2na + \alpha\sin\left[2(n+1)a\right]}{2(1+\cos 2a)}$$

$$\sum_{k=1}^{n} \beta\sin(2k-1)a = \frac{\alpha\sin 2na}{2\cos a}$$

$$\sum_{k=1}^{n} \beta k\sin ka = -\frac{d}{da}\sum_{k=1}^{n}\beta\cos ka*$$

$$\sum_{k=1}^{n} 2\beta k\sin 2ka = -\frac{d}{da}\sum_{k=1}^{n}\beta\cos 2ka*$$

$$\sum_{k=1}^{n} \beta(2k-1)\sin(2k-1)a = -\frac{d}{da}\sum_{k=1}^{n}\beta\cos\left[(2k-1)a\right]*$$

$$\sum_{k=1}^{n} \beta\sin^2 ka = \frac{\alpha\cos\left[(2n+1)a\right]}{4\cos a} + \frac{2\theta-1}{4}$$

$$\sum_{k=1}^{n} \beta\sin^2 2ka = \frac{\alpha\cos\left[(4n+2)a\right]}{4\cos 2a} + \frac{2\theta-1}{4} \qquad \theta = \begin{cases} 0 & \text{if } n \text{ is even} \\ 1 & \text{if } n \text{ is odd} \end{cases}$$

$$\sum_{k=1}^{n} \beta\sin^2(2k-1)a = -\frac{1+\alpha\cos 4na}{4\cos 2a} + \frac{\theta}{2}$$

* Derivative of the respective sum formula in (5.32c).

(a) Notation. a, k, n, α, β are defined in (5.29a) and (5.30a) and $a < \pi/2$.

(b) Monotonic series of cosine terms $(n < \infty)$

$$\sum_{k=1}^{n} \cos ka = \frac{\cos \left[(n+1)a/2 \right] \sin (na/2)}{\sin (a/2)}$$

$$\sum_{k=1}^{n} \cos^2 ka = \frac{n}{2} + \frac{\sin na \cos \left[(n+1)a \right]}{2 \sin a}$$

$$\sum_{k=1}^{n} \cos 2ka = \frac{\sin \left[(n+1)a \right] \cos na}{\sin a}$$

$$\sum_{k=1}^{n} \cos^2 2ka = \frac{n}{2} + \frac{\sin 2na \cos \left[2(n+1)a \right]}{2 \sin 2a}$$

$$\sum_{k=1}^{n} \cos (2k-1)a = \frac{\sin 2na}{2 \sin a}$$

$$\sum_{k=1}^{n} \cos^2 (2k-1)a = \frac{n}{2} + \frac{\sin 2na \cos 2na}{2 \sin 2a}$$

$$\sum_{k=1}^{n} k \cos ka = \frac{\cos \left[(n+1)a \right] - 1}{4 \sin^2 (a/2)} + \frac{(n+1) \cos \left[(2n+1)a/2 \right]}{2 \sin (a/2)}$$

$$\sum_{k=1}^{n} 2k \cos 2ka = \frac{\cos \left[2(n+1)a \right] - 1}{2 \sin^2 a} + \frac{(n+1) \sin \left[(2n+1)a \right]}{\sin a}$$

$$\sum_{k=1}^{n} (2k-1) \cos (2k-1)a = \frac{n \sin (2na) \sin a - \sin^2 na \cos a}{\sin^2 a}$$

(c) Alternating series of cosine terms $(n < \infty)$

$$\sum_{k=1}^{n} \beta \cos ka = \frac{1 + \cos a + \alpha \cos na + \alpha \cos \left[(n+1)a \right]}{2(1 + \cos a)}$$

$$\sum_{k=1}^{n} \beta \cos 2ka = \frac{1 + \cos 2a + \alpha \cos 2na + \alpha \cos \left[2(n+1)a \right]}{2(1 + \cos 2a)}$$

$$\sum_{k=1}^{n} \beta \cos (2k-1)a = \frac{1 + \alpha \cos 2na}{2 \cos a}$$

$$\sum_{k=1}^{n} \beta k \cos ka = \frac{d}{da} \sum_{k=1}^{n} \beta \sin ka *$$

$$\sum_{k=1}^{n} 2\beta k \cos 2ka = \frac{d}{da} \sum_{k=1}^{n} \beta \sin 2ka *$$

$$\sum_{k=1}^{n} \beta (2k-1) \cos (2k-1)a = \frac{d}{da} \sum_{k=1}^{n} \beta \sin (2k-1)a *$$

$$\sum_{k=1}^{n} \beta \cos^2 ka = \frac{1}{4} - \frac{\alpha \cos (2n+1)a}{4 \cos a} + \frac{\theta}{2}$$

$$\sum_{k=1}^{n} \beta \cos^2 2ka = \frac{1}{4} - \frac{\alpha \cos \left[(4n+2)a \right]}{4 \cos 2a} + \frac{\theta}{2} \qquad\qquad \theta = \begin{cases} 0 & \text{if } n \text{ is even} \\ 1 & \text{if } n \text{ is odd} \end{cases}$$

$$\sum_{k=1}^{n} \beta \cos^2 (2k-1)a = \frac{1 + \alpha \cos 4na}{4 \cos 2a} + \frac{\theta}{2}$$

* Derivative of the respective sum formula in (5.31c).

(a) Notation used below is defined in (5.29a) and (5.30a) and b = positive number.

(b) Monotonic series of power terms $(a < 1, b > 1)$

$$\sum_{k=1}^{\infty} a^k = \frac{a}{1-a}$$

$$\sum_{k=1}^{\infty} a^{2k} = \frac{a^2}{1-a^2}$$

$$\sum_{k=1}^{\infty} a^{2k-1} = \frac{a}{1-a^2}$$

$$\sum_{k=1}^{\infty} ka^k = \frac{a}{(1-a)^2}$$

$$\sum_{k=1}^{\infty} (2k-1)a^k = \frac{(1+a)a}{(1-a)^2}$$

$$\sum_{k=1}^{\infty} \frac{1}{b^k} = \frac{1}{b-1}$$

$$\sum_{k=1}^{\infty} \frac{1}{b^{2k}} = \frac{1}{b^2-1}$$

$$\sum_{k=1}^{\infty} \frac{1}{b^{2k-1}} = \frac{b}{b^2-1}$$

$$\sum_{k=1}^{\infty} \frac{k}{b^k} = \frac{b}{(b-1)^2}$$

$$\sum_{k=1}^{\infty} \frac{2k-1}{b^k} = \frac{1+b}{(b-1)^2}$$

(c) Alternating series of power terms $(a < 1, b > 1)$

$$\sum_{k=1}^{\infty} \beta a^k = \frac{a}{1+a}$$

$$\sum_{k=1}^{\infty} \beta a^{2k} = \frac{a^2}{1+a^2}$$

$$\sum_{k=1}^{\infty} \beta a^{2k-1} = \frac{a}{1+a^2}$$

$$\sum_{k=1}^{\infty} \beta ka^k = \frac{a}{(1+a)^2}$$

$$\sum_{k=1}^{\infty} \beta(2k-1)a^k = \frac{a(1-a)}{(1+a)^2}$$

$$\sum_{k=1}^{\infty} \frac{\beta}{b^k} = \frac{1}{1+b}$$

$$\sum_{k=1}^{\infty} \frac{\beta}{b^{2k}} = \frac{1}{1+b^2}$$

$$\sum_{k=1}^{\infty} \frac{\beta}{b^{2k-1}} = \frac{b}{1+b^2}$$

$$\sum_{k=1}^{\infty} \frac{\beta k}{b^k} = \frac{b}{(1+b)^2}$$

$$\sum_{k=1}^{\infty} \frac{\beta(2k-1)}{b^k} = \frac{b-1}{(1+b)^2}$$

(d) Series of trigonometric terms

$\sum_{k=1}^{\infty} \dfrac{\sin ka}{k} = \dfrac{\pi - a}{2}$	$0 < a < 2\pi$	$\sum_{k=1}^{\infty} \dfrac{\sin(2k-1)a}{2k-1} = \dfrac{\pi}{4}$	$0 < a < \pi$
$\sum_{k=1}^{\infty} \dfrac{\cos ka}{k} = \ln\sqrt{2\sin a/2}$		$\sum_{k=1}^{\infty} \dfrac{\cos(2k-1)a}{2k-1} = -\ln\sqrt{\tan a/2}$	
$\sum_{k=1}^{\infty} a^k \sin kb = \dfrac{a\sin b}{1 - 2a\cos b + a^2}$	$a^2 < 1$	$\sum_{k=1}^{\infty} \dfrac{a^k \sin kb}{k} = \tan^{-1}\dfrac{a\sin b}{1 - a\cos b}$	$a^2 < 1$
$\sum_{k=1}^{\infty} a^k \cos kb = \dfrac{a\cos^2 b - a^2}{1 - 2a\cos b + a^2}$	$\|b\| < \infty$	$\sum_{k=1}^{\infty} \dfrac{a^k \cos kb}{k}$ $= -\ln\sqrt{1 - 2a\cos b + a^2}$	$0 < b < \pi$
$\sum_{k=1}^{\infty} \dfrac{\beta\sin ka}{k} = \dfrac{\pi}{4}$	$\|a\| < \pi$	$\sum_{k=1}^{\infty} \dfrac{\beta\sin ka}{2k-1} = \ln\sqrt{\tan\left(\dfrac{\pi}{4} + \dfrac{a}{2}\right)}$	$\|a\| < \dfrac{\pi}{a}$
$\sum_{k=1}^{\infty} \dfrac{\beta\cos ka}{k} = \ln\left(2\cos\dfrac{a}{2}\right)$		$\sum_{k=1}^{\infty} \dfrac{\beta\cos ka}{2k-1} = \dfrac{\pi}{4}$	$\dfrac{\pi}{2} < a < a$

6

ALGEBRAIC AND TRANSCENDENTAL EQUATIONS

(a) Definition. An equality of two algebraic functions $f(x)$ and $g(x)$ in one variable x,

$$f(x) = g(x)$$

is called the algebraic equation, which is valid for certain values of $x_1, x_2, \ldots, x_n$, designated as the roots.

(b) Canonical form. Any algebraic equation can be reduced by algebraic operations to the canonical form

$$\boxed{P(x) = a_0 + a_1 x + a_2 x^2 + \cdots + a_n x^n = 0}$$

which has the same roots as the equation (a) and the exponent n defines the degree of this equation.

(c) Example. $\sqrt{3 + 4x + 5x^2} = 1 + x^2$ becomes $3 + 4x + 5x^2 = 1 + 2x^2 + x^4$ and the canonical form is

$$P(x) = 2 + 4x + 3x^2 - x^4 = 0$$

(d) Roots of the algebraic equation may be real and separate, or real and repeated, or complex conjugate and separate, and/or complex conjugate and repeated. If $x_1, x_2, \ldots, x_n$ are the roots of $P(x) = 0$, then

$$P(x) = (x - x_1)(x - x_2) \cdots (x - x_n) = 0$$

(e) Superfluous roots. In transforming (a) to a canonical form (b), it may happen that (b) has one or more roots which do not satisfy (a). Two such cases may occur.

(f) Vanishing denominator. If $P(x)$, $Q(x)$ are polynomial and

$$\frac{P(x)}{Q(x)} = 0$$

then after multiplying by $Q(x)$, $P(x) = 0$. This canonical equation has all the roots of the original equation, and it may have in addition some superfluous ones.

(g) Example

$$\frac{P(x)}{Q(x)} = \frac{x^3 - 3x^2 + 3x - 1}{x^2 - 2x + 1} = 0 \qquad P(x) = x^3 - 3x^2 + 3x - 100$$

where the roots of $P(x)$ are $x_1 = 1$, $x_2 = 1$, $x_3 = 1$, but $P(x)/Q(x)$ has only one root, $x_1 = 1$.

(h) Irrational equation. If $f(x) = g(x)$ involves an irrational term, then again some superfluous roots may occur.

(i) Example. $\sqrt{x + 3} = x + 1$ becomes $x^2 + x - 2 = 0$, $x_1 = 1$, $x_2 = -2$, where the first root $x_1 = 1$ satisfies the first equation and x_2 is superfluous.

(j) Relations between roots and coefficients are given by the following equations:

$$\boxed{\begin{array}{llll}
\displaystyle\sum_{i=1}^{n} x_i = -\frac{a_{n-1}}{a_n} & \displaystyle\sum_{i,j=1}^{n} x_i x_j = \frac{a_{n-2}}{a_n} \quad (i < j) & \displaystyle\sum_{i,j,k=1}^{n} x_i x_j x_k = -\frac{a_{n-3}}{a_n} \quad (i < j < k) \\[4mm]
\displaystyle\sum_{i,j,k,l=1}^{n} x_i x_j x_k x_l = \frac{a_{n-4}}{a_n} \quad (i < j < k < l) & \cdots &
\end{array}}$$

(a) Quadratic equations in canonical form are

$ax^2 + bx + c = 0$	$x^2 + px + q = 0$
$x_{1,2} = \dfrac{-b \pm \sqrt{b^2 - 4ac}}{2a}$	$x_{1,2} = -\dfrac{p}{2} \pm \sqrt{\left(\dfrac{p}{2}\right)^2 - q}$

These roots depend upon the discriminant $D = b^2 - 4ac$ or $D = \left(\dfrac{p}{2}\right)^2 - q$.

If $D > 0$, the roots are real and unequal

If $D = 0$, the roots are real and equal

If $D < 0$, the roots are complex conjugate

(b) Examples. $(i = \sqrt{-1},\ i^2 = -1,\ i^3 = -i,\ i^4 = 1)$

$D > 0$	$D = 0$	$D < 0$
$x^2 - 2x - 8 = 0$	$x^2 - 8x + 16 = 0$	$x^2 - 2x + 8 = 0$
$x_{1,2} = 1 \pm \sqrt{1 + 8}$	$x_{1,2} = 4 \pm \sqrt{16 - 16}$	$x_{1,2} = 1 \pm \sqrt{1 - 8}$
$= 1 \pm 3$	$= 4$	$= 1 \pm \sqrt{7}\, i$

(c) Relationships of roots by $(6.01j)$ are

$$x_1 + x_2 = -\frac{b}{a} = -p \qquad x_1 \cdot x_2 = \frac{c}{a} = q$$

When the roots are real and one is much larger than the other one, the smaller one may be calculated with a higher degree of precision from the second relation above as

$$x_2 = \frac{c}{ax_1} = \frac{q}{x_1} \qquad (x_2 < x_1)$$

(d) Binomial equations $(h > 0)$

$x^n - h = 0:$ $\quad x_{1,2,3,\dots,n} = \sqrt[n]{h}\left(\cos\dfrac{2k\pi}{n} + i\sin\dfrac{2k\pi}{n}\right)$	$n = 2, 3, 4, \dots$
$x^n + h = 0:$ $\quad x_{1,2,3,\dots,n} = \sqrt[n]{h}\left(\cos\dfrac{(2k+1)\pi}{n} + i\sin\dfrac{(2k+1)\pi}{n}\right)$	$k = 0, 1, 2, \dots, n - 1$

(e) Example. The roots of $x^3 - 27 = 0$ are $x_1 = 3$ and $x_{2,3} = \frac{3}{2}(-1 \pm i\sqrt{3})$

(f) Trinomial equation $(n = 2, 3, \dots)$

$$ax^{2n} + bx^n + c = 0 \qquad\text{or}\qquad x^{2n} + px^n + q = 0$$

reduces by substitution $y = x^n$ to

$$ay^2 + by + c = 0 \qquad\qquad y^2 + py + q = 0$$

and yields $\quad x_{i,j}^n = \dfrac{-b \pm \sqrt{b^2 - 4ac}}{2a} \qquad\qquad x_{i,j}^n = -\dfrac{p}{2} \pm \sqrt{\left(\dfrac{p}{2}\right)^2 - q}$

the $2n$ roots of which are calculated by (d).

(a) General case. In canonical form,

$$ax^3 + bx^2 + cx + d = 0$$

reduces by substitution $x = y - b/3a$ to

$$y^3 + 3py + 2q = 0$$

where

$$p = \frac{c}{3a} - \frac{b^2}{9a^2} \qquad q = \frac{1}{2}\left(\frac{d}{a} - \frac{bc}{3a^2} + \frac{2b^3}{27a^3}\right)$$

and the roots of the original equation are

$$x_1 = y_1 - b/3a \qquad x_2 = y_2 - b/3a \qquad x_3 = y_3 - b/3a$$

in which y_1, y_2, y_3 are the roots of the reduced equation computed by the formulas given in (c), (d) below.

(b) Classification of roots. When a, b, c, d are real numbers and $D = p^3 + q^2$, then if,

$D > 0$, there is one real root and two conjugate complex roots

$D = 0$, there are three real roots of which at least two are equal

$D < 0$, there are three real unequal roots

(c) First root of the original equation is always real and occurs in one of the following forms.

Case	x_1		Conditions	
1	0		$d = 0$	
2	$-b/3a$		$q = 0$	
3	$\sqrt[3]{-q + \sqrt{D}} - \sqrt[3]{q + \sqrt{D}} - b/3a$		$p > 0$	
4	$-\sqrt[3]{2q} - b/3a$	$D > 0$	$p = 0$	
5	$-\sqrt[3]{q + \sqrt{D}} - \sqrt[3]{q - \sqrt{D}} - b/3a$		$p < 0$	$D = p^3 + q^2$
6	$-2\sqrt[3]{q} - b/3a$		$D = 0$	
7	$2\sqrt{\lvert p\rvert}\cos\left[\frac{1}{3}\cos^{-1}\frac{-q}{\sqrt{\lvert p\rvert^3}}\right] - \frac{b}{3a}$		$D < 0$	

(d) Second and third root of the original equation are the roots of the quadratic equation

$$x^2 + 2Bx + C = 0$$

where in terms of x_1 given in (c),

$$B = \frac{1}{2}\left(\frac{b}{a} + x_1\right) \qquad C = \left(\frac{b}{a} + x_1\right)x_1 + \frac{c}{a}$$

and $x_{2,3} = -B \pm \sqrt{B^2 - C}$

(e) Relations of roots by $(6.01j)$ are

$$x_1 + x_2 + x_3 = -\frac{b}{a}$$

$$\frac{1}{x_1} + \frac{1}{x_2} + \frac{1}{x_3} = -\frac{c}{d}$$

$$x_1 x_2 x_3 = -\frac{d}{a}$$

(f) Transcendental solutions of the original cubic equation in (*a*) are given below where $r = \pm\sqrt{|p|}$ and the sign of r and q must coincide.

Case	Roots	Conditions		
(1)	$x_1 = -2r\cosh\dfrac{\phi}{3} - \dfrac{b}{3a}$ $x_2 = r\cosh\dfrac{\phi}{3} + ir\sqrt{3}\sinh\dfrac{\phi}{3} - \dfrac{b}{3a}$ $x_3 = r\cosh\dfrac{\phi}{3} - ir\sqrt{3}\sinh\dfrac{\phi}{3} - \dfrac{b}{3a}$	$p < 0$	$D > 0$	$\phi = \cosh^{-1}\dfrac{q}{r^3}$
(2)	$x_1 = -2r\cos\dfrac{\phi}{3} - \dfrac{b}{3a}$ $x_2 = 2r\cos\left(\dfrac{\pi}{3} - \dfrac{\phi}{3}\right) - \dfrac{b}{3a}$ $x_3 = 2r\cos\left(\dfrac{\pi}{3} + \dfrac{\phi}{3}\right) - \dfrac{b}{3a}$	$p < 0$	$D \leq 0$	$\phi = \cos^{-1}\dfrac{q}{r^3}$
(3)	$x_1 = -2r\sinh\dfrac{\phi}{3} - \dfrac{b}{3a}$ $x_2 = r\sinh\dfrac{\phi}{3} + ir\sqrt{3}\cosh\dfrac{\phi}{3} - \dfrac{b}{3a}$ $x_3 = r\sinh\dfrac{\phi}{3} - ir\sqrt{3}\cosh\dfrac{\phi}{3} - \dfrac{b}{3a}$	$p > 0$	$D \gtreqless 0$	$\phi = \sinh^{-1}\dfrac{q}{r^3}$

(g) Example ($D > 0$). $x^3 + x^2 + x + 1 = 0$, where $a = 1$, $b = 1$, $c = 1$, $d = 1$, $p = \frac{2}{9}$, $q = \frac{20}{54}$, $D = 0.148\,148\,148 > 0$.

By (*c*-3), $x_1 = \sqrt[3]{-q + \sqrt{D}} - \sqrt[3]{q + \sqrt{D}} - b/3a = -1$

By (*d*), $x^2 + (1 - 1)x + 1 = 0$ $\qquad x_2 = i \qquad x_3 = -i$

(h) Example ($D = 0$). $x^3 - 5x^2 + 8x - 4 = 0$, where $a = 1$, $b = -5$, $c = 8$, $d = -4$, $p = -\frac{1}{9}$, $q = \frac{1}{27}$, $D = 0$.

By (*c*-6), $x_1 = -2\sqrt[3]{q} - b/3a = 2$

By (*d*), $x^2 - 3x + 2 = 0$ $\qquad x_2 = 2 \qquad x_3 = 1$

(i) Example ($D < 0$). $x^3 + 6x^2 + 11x + 6 = 0$, where $a = 1$, $b = 6$, $c = 11$, $d = 6$, $p = -\frac{1}{3}$, $q = 0$, $D = -\frac{1}{27}$, $r = \pm\sqrt{\frac{1}{3}}$.

By (*f*-2), $\phi = \cos^{-1}\dfrac{0}{\sqrt{\frac{1}{27}}} = \dfrac{\pi}{2}$ $\qquad$ and using $+$ for r,

$$x_1 = -2r\cos\dfrac{\phi}{3} - \dfrac{b}{3a} = -1 - 2 = -3$$

$$x_2 = 2r\cos\left(\dfrac{\pi}{3} - \dfrac{\phi}{3}\right) - \dfrac{b}{3a} = 1 - 2 = -1$$

$$x_3 = 2r\cos\left(\dfrac{\pi}{3} + \dfrac{\phi}{3}\right) - \dfrac{b}{3a} = 0 - 2 = -2$$

(a) General case in canonical form,

$$\boxed{a_4 x^4 + a_3 x^3 + a_2 x^2 + a_1 x + a_0 = 0}$$

reduces to

$$x^4 + b_3 x^3 + b_2 x + b_1 x + b_0 = 0$$

by dividing all coefficients by a_4.

(b) Equivalent form of the reduced equation in (a) can be taken as

$$(x^2 + Ax + B)^2 - (Cx + D)^2 = 0$$

or $\quad [x^2 + (A + C)x + B + D][x^2 + (A - C)x + B - D] = 0$

where $\quad 2A = b_3 \qquad\qquad\qquad 2AB - 2CD = b_1$

$\qquad\qquad A^2 + 2B - C^2 = b_2 \qquad\qquad B^2 - D^2 = b_0$

(c) Resolvent cubic equation in $y = 2B$, derived from the last four relations above by successive elimination of A, C, D, is

$$y^3 - b_2 y^2 + (b_1 b_3 - 4b_0) y + b_0(4b_2 - b_3^2) - b_1^2 = 0$$

The roots of the equation y_1, y_2, y_3 are abstracted by one of the procedures described in $(6.03c)$, $(6.03d)$, $(6.03f)$.

(d) Largest real root y_1 is then used for the calculation of A, B, C, D as

$$A = \tfrac{1}{2}b_3 \qquad\qquad B = \tfrac{1}{2}y_1 \qquad\qquad D = y_1^2 - b_0$$

$$C = \begin{cases} (\tfrac{1}{2}b_1 - AB)/D & \text{if } D \neq 0 \\ \sqrt{A^2 - b_2 + 2B} & \text{if } D = 0 \end{cases}$$

(e) Auxiliary quadratic equations from the product in (b), in terms of A, B, C, D calculated in (d) above,

$$x^2 + (A + C)x + B + D = 0 \qquad\qquad x^2 + (A - C)x + B - D = 0$$

yield the roots of the quartic equation in (a),

$$x_{1,2} = -\tfrac{1}{2}(A + C) \pm \sqrt{\tfrac{1}{4}(A + C)^2 - (B + D)} \qquad\qquad x_{3,4} = -\tfrac{1}{2}(A - C) \pm \sqrt{\tfrac{1}{4}(A - C)^2 - (B - D)}$$

(f) Relations of roots used for a numerical check are

$$y_1 = x_1 x_2 + x_3 x_4 \qquad\qquad y_2 = x_1 x_3 + x_2 x_4 \qquad\qquad y_3 = x_1 x_4 + x_2 x_3$$

(g) Classification of roots. If the coefficients of (a) are real, then the following classification is possible.

y roots	x roots
y_1, y_2, y_3 are real	four real roots four complex roots two real roots and two complex roots
y_1 is real, y_2 and y_3 are complex	four complex roots

(h) Example ($D < 0$). $x^4 + 10x^3 + 35x^2 + 50x + 24 = 0$, where $b_3 = 10$, $b_2 = 35$, $b_1 = 50$, $b_0 = 24$, reduces by (c) to

$$y^3 - 35y^2 + 404y - 1540 = 0$$

The roots of this resolvent equation are $y_1 = 14, y_2 = 11, y_3 = 10$. Taking $y_1 = 14$ into (d) as the largest real root,

$$A = 5 \qquad B = 7 \qquad C = -2 \qquad D = -5$$

and the auxiliary quadratic equations in (e) become

$$x^2 + 3x + 2 = 0 \qquad x_{1,2} = -1, -2 \qquad x^2 + 7x + 12 = 0 \qquad x_{3,4} = -3, -4$$

which must satisfy the relations in (f).

(i) Example ($D > 0$). $x^4 + 4x^3 + 7x^2 + 6x + 3 = 0$, where $b_3 = 4$, $b_2 = 7$, $b_1 = 6$, $b_0 = 3$, reduces by (c) to

$$y^3 - 7y^2 + 12y + 0 = 0$$

The roots of this resolvent equation are $y_1 = 4, y_2 = 3, y_3 = 0$. Taking $y_1 = 4$ into (d) as the largest real root,

$$A = 2 \qquad B = 2 \qquad C = 1 \qquad D = 1$$

and the auxiliary quadratic equations in (e) become

$$x^2 + x + 1 = 0 \qquad x_{1,2} = -\tfrac{1}{2}(1 \pm i\sqrt{3}) \qquad x^2 + 3x + 3 = 0 \qquad x_{3,4} = -\tfrac{1}{2}(3 \pm i\sqrt{3})$$

which must satisfy relations in (f).

(j) Symmetrical quartic equation

$$ax^4 + bx^3 + cx^2 + bx + a = 0$$

reduces after dividing by x^2 to

$$a\left(x^2 + \frac{1}{x^2}\right) + b\left(x + \frac{1}{x}\right) + c = 0$$

and with $\quad x + \dfrac{1}{x} = y \qquad x^2 + \dfrac{1}{x^2} = y^2 - 2$

becomes a quadratic equation in y,

$$ay^2 + by + c - 2a = 0$$

the roots of which when inserted in

$$x + \frac{1}{x} = y_1 \qquad x + \frac{1}{x} = y_2$$

yield the four desired roots of the original equation.

(k) Antisymmetrical quartic equation

$$ax^4 + bx^3 - bx - a = 0$$

reduces to $\quad a\left(x^2 - \dfrac{1}{x^2}\right) + b\left(x - \dfrac{1}{x}\right) = 0$

and can be solved by an analogical procedure as in (j).

(a) Case I

$$\boxed{x^2 + y^2 = a} \qquad \boxed{xy = b}$$

can be solved as $\quad x^2 + 2xy + y^2 = a + 2b \qquad\qquad x^2 - 2xy + y^2 = a - 2b$

$$x + y = \pm\sqrt{a + 2b} = \pm A \qquad x - y = \pm\sqrt{a - 2b} = \pm B$$

and

$$x_1 = \tfrac{1}{2}(A + B) \qquad x_2 = \tfrac{1}{2}(A - B) \qquad x_3 = -\tfrac{1}{2}(A - B) = -x_2 \qquad x_4 = -\tfrac{1}{2}(A + B) = -x_1$$

$$y_1 = \frac{2b}{A + B} \qquad y_2 = \frac{2b}{A - B} \qquad y_3 = -\frac{2b}{A - B} = -y_2 \qquad y_4 = -\frac{2b}{A + B} = y_1$$

(b) Case II

$$\boxed{\begin{aligned} a_1 x^2 + b_1 xy + a_1 y^2 + c_1 x + c_1 y + d_1 = 0 \\ a_2 x^2 + b_2 xy + a_2 y^2 + c_2 x + c_2 y + d_2 = 0 \end{aligned}}$$

are transformed by factoring to

$$a_1(x^2 + y^2) + b_1 xy + c_1(x + y) + d_1 = 0 \qquad a_2(x^2 + y^2) + b_2 xy + c_2(x + y) + d_2 = 0$$

and modified by $u = x + y$, $v = xy$ to

$$a_1 u^2 - (2a_1 - b_1)v + c_1 u + d_1 = 0 \qquad a_2 u^2 - (2a_2 - b_2)v + c_2 u + d_2 = 0$$

The elimination of v yields

$$\underbrace{[a_1(2a_2 - b_2) - a_2(2a_1 - b_1)]}_{A} u^2 + \underbrace{[c_1(2a_2 - b_2) - c_2(2a_1 - b_1)]}_{B} u$$

$$+ \underbrace{d_1(2a_2 - b_2) - d_2(2a_1 - b_1)}_{C} = 0$$

with roots $\quad u_1 = \dfrac{-B + \sqrt{B^2 - 4AC}}{2A} \qquad u_2 = \dfrac{-B - \sqrt{B^2 - 4AC}}{2A}$

$$v_1 = \frac{a_1 u_1^2 + c_1 u_1 + d_1}{2a_1 - b_1} \qquad v_2 = \frac{a_2 u_2^2 + c_2 u_2 + d_2}{2a_2 - b_2}$$

from which with $D_1 = \sqrt{u_1^2 - 4v_1}$ and $D_2 = \sqrt{u_2^2 - 4v_2}$,

$$x_1 = \tfrac{1}{2}(u_1 + D_1) \qquad x_2 = \tfrac{1}{2}(u_1 - D_1) \qquad x_3 = \tfrac{1}{2}(u_2 + D_2) \qquad x_4 = \tfrac{1}{2}(u_2 - D_2)$$

$$y_1 = \frac{2v_1}{u_1 + D_1} \qquad y_2 = \frac{2v_1}{u_1 - D_1} \qquad y_3 = \frac{2v_2}{u_2 + D_2} \qquad y_4 = \frac{2v_2}{u_2 - D_2}$$

(c) Case III

$$\boxed{a_2 x^2 + b_2 xy + c_2 y^2 + d_2 = 0} \qquad \boxed{a_1 x^2 + b_1 xy + c_1 y^2 + d_1 = 0}$$

can be transformed by elimination of d_1 and d_2 to

$$\underbrace{(a_1 d_2 - a_2 d_1)}_{A} x^2 + \underbrace{(b_1 d_2 - b_2 d_1)}_{B} xy + \underbrace{(c_1 d_2 - c_2 d_1)}_{C} y^2 = 0$$

and reduced by the substitution $x = uy$ to $Au^2 + Bu + C = 0$.

(a) Properties of binomials $(k = 1, 2, 3, \ldots)$

$$(x^{2k} + 1) : (x \pm 1) \quad = x^{2k-1} \mp x^{2k-2} + x^{2k-3} \mp \cdots 1$$

$$(x^{2k+1} + 1) : (x + 1) = x^{2k} \quad - x^{2k-1} + x^{2k-2} - \cdots 1$$

$$(x^{2k+1} - 1) : (x - 1) = x^{2k} \quad + x^{2k-1} + x^{2k-2} + \cdots + 1$$

$$(x^{2k} - 1) : (x^k + 1) = \quad x^k - 1 \qquad\qquad (x^{2k} - 1):(x^k - 1) = x^k + 1$$

(b) Symmetrical equation of third degree

$$ax^3 + bx^2 + bx + a = 0 \qquad \text{is transformed to} \qquad (x + 1)[ax^2 - (a - b)x + a] = 0$$

$$\text{and} \quad x_1 = -1 \qquad x_{2,3} = \frac{1}{2a}[(a - b) \pm \sqrt{(a - b)^2 - 4a^2}]$$

(c) Antisymmetrical equation of fourth degree

$$ax^4 + bx^3 - bx - a = 0 \qquad \text{is transformed to} \qquad (x^2 - 1)(ax^2 + bx + a) = 0$$

$$\text{and} \quad x_{1,2} = \pm 1 \qquad x_{3,4} = -\frac{b}{2a} \pm \sqrt{\left(\frac{b}{2a}\right)^2 - 1}$$

(d) Symmetrical equation of fifth degree

$$ax^5 + bx^4 + cx^3 + cx^2 + bx + a = 0$$

transforms to

$$a(x^5 + 1) + bx(x^3 + 1) + cx^2(x + 1) = 0$$

$$(x + 1)[ax^4 - \underbrace{(a - b)}_{\bar{B}}x^3 + \underbrace{(a + b + c)}_{\bar{C}}x^2 - \underbrace{(a - b)}_{\bar{B}}x + a] = 0$$

and with $y = x + 1/x$, the second part reduces to $ay^2 - \bar{B}y + \bar{C} - 2a = 0$, yields

$$y_{1,2} = \frac{\bar{B} \pm \sqrt{\bar{B}^2 - 4a(\bar{C} - 2a)}}{2a}$$

$$\text{and} \quad x_1 = -1 \qquad x_{2,3} = \frac{y_1}{2} \pm \sqrt{\frac{y_1^2}{4} - 1} \qquad x_{4,5} = \frac{y_2}{2} \pm \sqrt{\frac{y_2^2}{4} - 1}$$

(e) Antisymmetrical equation of fifth degree

$$ax^5 + bx^4 + cx^3 - cx^2 - bx - a = 0$$

transforms to

$$a(x^5 - 1) + bx(x^3 - 1) + cx^2(x - 1) = 0$$

$$(x - 1)[ax^4 + \underbrace{(a + b)}_{B}x^3 + \underbrace{(a + b + c)}_{C}x^2 + \underbrace{(a + b)}_{B}x + a] = 0$$

and similarly as in (d),

$$y_{1,2} = \frac{B \pm \sqrt{B^2 - 4a(C - 2a)}}{2a}$$

$$\text{and} \quad x_1 = 1 \qquad x_{2,3} = \frac{y_1}{2} \pm \sqrt{\frac{y_1^2}{4} - 1} \qquad x_{4,5} = \frac{y_2}{2} \pm \sqrt{\frac{y_2^2}{4} - 1}$$

(a) Numerical solution of any algebraic or transcendental equation $f(x) = 0$, should be preceded by the investigation of the graph of

$$y = f(x) \qquad \text{or} \qquad y = f_1(x), y = f_2(x)$$

the zero points or the points of intersection of which are the real roots $x_1, x_2, \ldots$ of the given equation as shown graphically below.

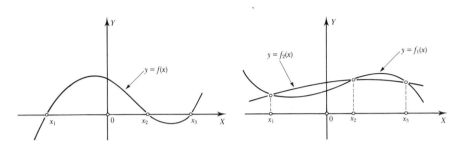

The plotting of the curve by computer and examination of the graph leads usually to good starting values for the computation of more accurate values.

(b) Intermediate-value theorem states that a continuous function $f(x)$ in some interval $[a, b]$, in which $f(a)$ and $f(b)$ are of opposite sign, has at least one or an odd number of real roots within this interval.

(c) Initial value of the approximation process may be then taken as the average of a and b,

$$m = \frac{a + b}{2} \cong x$$

(d) Example

$$f(x) = x^3 - 4x + 1 = 0 \qquad \text{in } a = 0, b = 1$$

has $\quad f(a) = 1 \qquad\qquad f(b) = -2 \qquad\qquad$ and $\quad x \cong \dfrac{0 + 1}{2} = 0.5$

The repetition of this process leads to a better approximation (6.08a).

(e) Methods used in the approximate calculations of roots of equations are based on the concept of interpolation and/or iteration. They are classified as general methods applicable to equations of all types and polynomial methods restricted to the solution of algebraic equations of higher degree (but applicable to all algebraic equations of $n > 1$).

(f) General methods introduced in the subsequent sections of this chapter are:

(1) Bisection method (6.08)
(2) Secant method (6.09)
(3) Tangent method (6.10)
(4) Iteration methods (6.11), (6.12)

(g) Polynomial methods of practical importance used in the search of real and complex roots of algebraic equations of higher degree and described in the final sections of this chapter are:

(1) Newton–Raphon's method (6.17)
(2) Bairstow's method (6.18)

Methods developed in the era of hand calculations such as Horner's method (Ref. 6.9), Graeffe's method (Ref. 6.9), and Bernoulli's method (Ref. 1.10) are of limited value in computer applications.

(a) Bisection method begins with the interval $[a_0, b_0]$ in which the function changes its sign so that

$$f(a_0) > 0 > f(b_0) \qquad \text{or} \qquad f(a_0) < 0 < f(b_0)$$

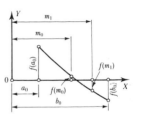

The midpoint of this interval,

$$m_0 = \tfrac{1}{2}(a_0 + b_0)$$

is the initial approximation of the root x. The next approximation in $f(a_0) > 0 > f(b_0)$ is

$$m_1 = \begin{cases} \tfrac{1}{2}(m_0 + b_0) & \text{if } f(m_0) > 0 \\ \tfrac{1}{2}(a_0 + m_0) & \text{if } f(m_0) < 0 \end{cases}$$

and in $f(a_0) < 0 < f(b_0)$ is

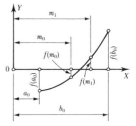

$$m_1 = \begin{cases} \tfrac{1}{2}(a_0 + m_0) & \text{if } f(m_0) > 0 \\ \tfrac{1}{2}(m_0 + b_0) & \text{if } f(m_0) < 0 \end{cases}$$

Repeated bisections reduce the interval of sign change until the value of the root x is obtained to a desired accuracy.

(b) Number of repetitions required to obtain a result of error ϵ in $[a_0, b_0]$ is

$$r \cong \frac{\ln |a_0 - b_0| - \ln \epsilon}{\ln 2}$$

(c) Minimum number of repetitions in (0, 1) for a prescribed error is given below.

ϵ	10^{-2}	10^{-3}	10^{-4}	10^{-5}	10^{-6}	10^{-7}	10^{-8}	10^{-9}	10^{-10}	10^{-11}	10^{-12}
r	7	10	14	17	20	24	27	30	34	37	40

(d) Numerical procedure uses only the coordinates of the endpoints of the interval for the calculation of the next approximation, and it always converges to the correct value. It requires usually 10 to 20 iterations, but this involves one function evaluation in one iteration step only. In searching for the interval of sign change, there is always the danger (if large steps are used) of overstepping some of the roots. The graph of $f(x)$ helps to eliminate this overstepping.

(e) Example. $f(a_0) < 0 < f(b_0)$.

$$f(x) = \tan x - x = 0 \qquad a_0 = 4.4 \qquad b_0 = 4.5 \qquad f(a_0) = -1.303\,368 \qquad f(b_0) = 0.137\,332$$

$$m_0 = \tfrac{1}{2}(4.400\,000 + 4.500\,000) = 4.450\,000 \qquad f(m_0) = -0.726\,731 < 0$$

$$m_1 = \tfrac{1}{2}(4.450\,000 + 4.500\,000) = 4.475\,000 \qquad f(m_1) = -0.341\,933 < 0$$

$$m_2 = \tfrac{1}{2}(4.475\,000 + 4.500\,000) = 4.487\,500 \qquad f(m_2) = -0.116\,078 < 0$$

$$\cdots \cdots \cdots \cdots \cdots \cdots \cdots \cdots \cdots \cdots \cdots$$

$$m_{10} = \tfrac{1}{2}(4.493\,263 + 4.493\,556) = 4.493\,409 \qquad f(m_{10}) = -0.000\,000\,4 < 0$$

and the correct root is $x = 4.493\,409$.

(a) Regula falsi method replaces the curve $y = f(x)$ in the interval of sign change $[a_0, b_0]$ by a secant connecting the endpoints of the curve in this interval. The point of intersection of this secant with the x axis is the initial approximation of the root of $f(x) = 0$. Analytically,

$$c_0 = \frac{a_0 f(b_0) - b_0 f(a_0)}{f(b_0) - f(a_0)}$$

(b) Next approximation depends upon the relations of $f(a_0), f(c_0), f(b_0)$ as shown graphically below.

Case I

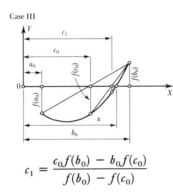

$$c_1 = \frac{c_0 f(a_0) - a_0 f(c_0)}{f(a_0) - f(c_0)}$$

Case II

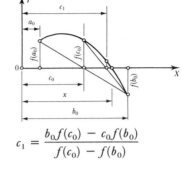

$$c_1 = \frac{b_0 f(c_0) - c_0 f(b_0)}{f(c_0) - f(b_0)}$$

Case III

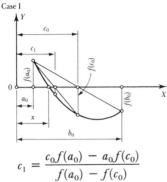

$$c_1 = \frac{c_0 f(b_0) - b_0 f(c_0)}{f(b_0) - f(c_0)}$$

Case IV

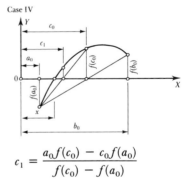

$$c_1 = \frac{a_0 f(c_0) - c_0 f(a_0)}{f(c_0) - f(a_0)}$$

(c) Repetition of this procedure leads to the desired root. This method converges usually (but not always) faster than the bisection method, but the error cannot be preassigned and the computer model is more involved.

(d) Examples

Case II: $f(a_0) > f(c_0) > 0 > f(b_0)$		Case III: $f(a_0) < f(c_0) < 0 < f(b_0)$	
$f(x) = x^3 - 4x^2 + 1 = 0 \quad a_0 = 0 \quad b_0 = 1$ $f(a_0) = 1.000\,000 \qquad f(b_0) = -2.000\,000$		$f(x) = \tan x - x = 0 \quad a_0 = 4.4 \quad b_0 = 4.5$ $f(a_0) = -1.303\,676 \qquad f(b_0) = 0.137\,332$	
$c_0 = 0.333\,333$	$f(c_0) = 0.592\,593$	$c_0 = 4.490\,470$	$f(c_0) = -0.058\,543$
$c_1 = 0.485\,714$	$f(c_1) = 0.170\,916$	$c_1 = 4.493\,318$	$f(c_1) = -0.001\,843$
. . . .	. . . .	. . . .	. . . .
$c_8 = 0.537\,401$	$f(c_8) = 0.000\,004$	$c_4 = 4.493\,409$	$f(c_4) = -0.000\,002$
Correct root $x = 0.537\,402$		Correct root $x = 4.493\,409$	

(a) Newton–Raphson's method substitutes the local tangent for the curve $y = f(x)$ in the segment of sign change $[a_0, b_0]$. The point of intersection of the tangent with the x axis is the initial approximation of the root of $f(x) = 0$. Analytically,

$$c_0 = a_0 - f(a_0)/f'(a) \qquad c_0 = b_0 - f(b_0)/f'(b_0)$$

where c_0 must fall within $[a_0, b_0]$ and $f'(a_0), f'(b_0)$ are the respective derivatives. If $f'(a_0) = 0$ or $f'(b_0) = 0$, the initial value of c_0 may be taken as $\frac{1}{2}(a_0 + b_0)$.

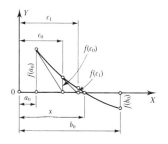 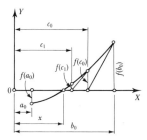

(b) Next approximations are

$$c_1 = c_0 - f(c_0)/f'(c_0) \qquad \text{if } f(c_0)f''(c_0) > 0$$

$$c_2 = c_1 - f(c_1)/f'(c_1) \qquad \text{if } f(c_1)f''(c_1) > 0$$

. . . .

$$x \cong c_r = c_{r-1} - f(c_{r-1})/f'(c_{r-1}) \qquad \text{if } f(c_{r-1})f''(c_{r-1}) > 0$$

where $c_1, c_2, \ldots, c_r$ are the first, second, $\ldots$, rth approximations of the root x. The method works if c_0 is close to x and $f'(x) \neq 0, f''(x) \neq 0$. The sufficient condition for the convergence is

$$\left| \frac{f(c_k)f''(c_k)}{[f'(c_k)]^2} \right| < 1$$

but it is not the final rate of convergence but the initial rate which controls the number of steps required.

(c) Example, $f(a_0) > 0 > f(b_0)$.

$$f(x) = x^3 - 4x^2 + 1 = 0, a_0 = 0, b_0 = 1, f(a) = 1, f'(x) = 3x^2 - 8x$$

and since $f'(a_0) = 0, c_0 = 0.5$

k	c_k	$f(c_k)$	$f'(c_k)$	$-f(c_k)/f'(c_k)$
0	0.500 000	0.125 000	−3.250 000	0.038 462
1	0.538 462	−0.003 643	−3.437 872	0.001 060
3	0.537 402	−0.000 002	−3.432 813	−0.000 000

(d) Example, $f(a_0) < 0 < f(b_0)$.

$$f(x) = \tan x - x = 0, a_0 = 4.4, b_0 = 4.5, f(b_0) = 0.137\,332, f'(x) = \sec^2 x - 1$$

and since $f'(b_0) = 21.504\,843, c_0 = 4.5$

k	c_k	$f(c_k)$	$f'(c_k)$	$-f(c_k)/f'(c_k)$
0	4.500 000	0.137 332	21.504 843	−0.006 386
1	4.493 614	0.004 132	20.229 717	−0.000 204
2	4.493 410	0.000 009	20.190 813	−0.000 001

(a) Direct iteration method is the simplest method of all, well-suited for computer applications. After locating the interval $[a_0, b_0]$ of sign change, the given function is modified from $f(x) = 0$ to

$$x = g(x)$$

By selecting the midpoint of $[a_0, b_0]$ as the starting value $c_0 = \frac{1}{2}(a_0 + b_0)$, the successive approximations become

$$c_1 = g(c_0), \, c_2 = g(c_1), \ldots, c_r = g(c_{r-1}) \cong x$$

To facilitate the convergence of this procedure, it is necessary that the absolute value of the first derivative of $g(x)$ be

$$|g'(x)| < 1$$

Since there are usually several forms of $g(x)$ available, the one of the least value of $|g'(x)|$ is the most suitable iteration model.

(b) Graphical representation of this process is shown by two graphs below.

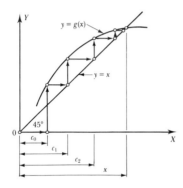
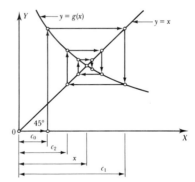

(c) Example, $f(a_0) > 0 > f(b_0)$.

$$f(x) = x^3 - 4x^2 + 1 = 0, \, a_0 = 0, \, b_0 = 1, \, c_0 = 0.5, \, g(x) = \tfrac{1}{2}\sqrt{x^3 + 1}.$$

$c_1 = \frac{1}{2}\sqrt{c_0^3 + 1} = 0.530\,330$	$c_5 = \frac{1}{2}\sqrt{c_x^3 + 1} = 0.537\,390$
$c_2 = \frac{1}{2}\sqrt{c_1^3 + 1} = 0.535\,993$	$c_6 = \frac{1}{2}\sqrt{c_5^3 + 1} = 0.537\,393$
$c_3 = \frac{1}{2}\sqrt{c_2^3 + 1} = 0.537\,118$	$c_7 = \frac{1}{2}\sqrt{c_6^3 + 1} = 0.537\,393$
$c_4 = \frac{1}{2}\sqrt{c_3^3 + 1} = 0.537\,344$	$c_8 = \frac{1}{2}\sqrt{c_1^3 + 1} = 0.537\,402$

(d) Example, $f(a_0) < 0 < f(b_0)$.

$$f(x) = \tan x - x = 0, \, a_0 = 4.4, \, b_0 = 4.5, \, c_0 = 4.45, \, g(x) = \tan^{-1}x + \pi.$$

$c_1 = \tan^{-1}c_0 + \pi = 4.491\,342$	$c_3 = \tan^{-1}c_3 + \pi = 4.493\,404$
$c_2 = \tan^{-1}c_1 + \pi = 4.493\,312$	$c_4 = \tan^{-1}c_4 + \pi = 4.493\,409$

(a) Roots of two equations,

$$f(x,y) = 0 \qquad \text{and} \qquad g(x,y)$$

can be computed by their iteration models,

$$x = \phi_1(x,y) \qquad \text{and} \qquad y = \phi_2(x,y)$$

by assuming a sufficiently close solution $x = a_0$ and $y = b_0$.

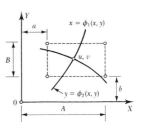

(b) Iteration procedure is then based on recurrent relations

$$a_k = \phi_1(a_{k-1}, b_{k-1}) \qquad\qquad b_k = \phi_2(a_{k-1}, b_{k-1})$$

(c) Convergence of this procedure is tested by the following conditions:

(1) In the neighborhood $R(a \le x \le A,\ b \le y \le B)$ there is one and only one pair of roots $x = u,\ y = v$.
(2) The functions ϕ_1, ϕ_2 are continuously differentiable in R.
(3) The initial approximations a_0, b_0 and all successive approximations belong to R.
(4) The following inequalities are valid in R,

$$\left| \frac{\partial \phi_1}{\partial x} + \frac{\partial \phi_2}{\partial x} \right| < 1 \qquad\qquad \left| \frac{\partial \phi_1}{\partial y} + \frac{\partial \phi_2}{\partial y} \right| < 1$$

(d) Example

$$f(x,y) = 3y^3 - 2x^3 - 10x^2 + 1 = 0 \qquad g(x,y) = y^3 - 3x^2 + 5y^2 - 5 = 0$$

and $a = 0.5,\ A = 0.7,\ b = 0.9,\ B = 1.1,\ a_0 = 0.6,\ b_0 = 1.0$. Taking

$$x = \phi_1 = \sqrt{\frac{3y^3 + 1}{2x + 10}} \qquad\qquad y = \phi_2 = \sqrt{\frac{3x^2 + 5}{y + 5}}$$

the partial derivatives in terms of $x = a_0, y = b_0$ are

$$\left| \frac{\partial \phi_1}{\partial x} + \frac{\partial \phi_2}{\partial x} \right| = \left| -\frac{1}{2x + 10} \sqrt{\frac{3y^3 + 1}{2x + 10}} + \frac{3x}{\sqrt{(3x^2 + 5)(y + 5)}} \right|_{a_0, b_0} = 0.31 < 1$$

$$\left| \frac{\partial \phi_1}{\partial y} + \frac{\partial \phi_2}{\partial y} \right| = \left| \frac{9y^2}{2\sqrt{(2x + 10)(3y^3 + 1)}} - \frac{1}{2(y + 5)} \sqrt{\frac{3x^2 + 5}{y + 5}} \right|_{a_0, b_0} = 0.50 < 1$$

The iteration process is convergent as shown below.

k	$\phi_1(a_k, b_k)$	$\phi_2(a_k, b_k)$	k	$\phi_1(a_k, b_k)$	$\phi_2(a_k, b_k)$
0	0.597 614	1.006 645	6	0.602 032	1.006 669
1	0.602 222	1.005 378	7	0.602 002	1.006 692
2	0.602 178	1.006 403	8	0.602 011	1.006 687
3	0.601 815	1.006 758	9	0.602 015	1.006 684
4	0.602 073	1.006 620	10	0.602 013	1.006 685
5	0.601 966	1.006 708	11	0.602 014	1.006 685

The roots $u = 0.602\,014,\ v = 1.006\,685$ must satisfy the initial equations

$$3v^3 - 2u^3 - 10u^2 + 1 = 5.32(10)^{-6} \cong 0 \qquad\qquad v^3 - 3u^2 + 5v^2 - 5 = 2.44(10)^{-7} \cong 0$$

(a) General properties of the polynomial equation

$$P(x) = a_0 + a_1 x + a_2 x^2 + \cdots + a_{n-1} x^{n-1} + a_n x^n = 0$$

are summarized in (6.01). Additional theorems related to the methods introduced in the subsequent sections are described below.

(b) Factor theorem.

If $x - c_i$ is a factor of $P(x)$, then c_i is the root of $P(x) = 0$. If p_m, q_m are real, $p_m^2 < q_m$, and $x^2 + 2p_m x + q_m$ is a factor of $P(x)$, then

$$c_j = -p_m + i\sqrt{q_m - p_m^2} \qquad\qquad c_k = -p_m - i\sqrt{q_m - p_m^2}$$

are the conjugate roots of $P(x)$.

(c) Factorization theorem.

A polynomial $P(x)$ with real coefficients can be uniquely factorized into a product of linear and quadratic factors with real coefficients. With $i = 1, 2, \ldots, r$ and $m = 1, 2, \ldots, s$,

$$P(x) = a_0 (x - c_1)^{\alpha_1} \cdots (x - c_r)^{\alpha_r} (x^2 + 2p_1 x + q_1)^{\beta_1} \cdots (x^2 + 2p_s x + q_s)^{\beta_s}$$

where α_r, β_s = positive integers and $\alpha_1 + \alpha_2 + \cdots + \alpha_r + 2(\beta_1 + \beta_2 + \cdots + \beta_s) = n$.

(d) Remainder theorem.

If $P(x) \neq 0$ is divided by $x - c$, where c is c_i, c_j or c_k defined in (c) so that

$$P(x) = (x - c)Q(x) + R$$

then for $x = c$ the remainder $R = P(c)$.

(e) Root theorem.

A polynomial equation of nth degree cannot have more than n roots, which can be real distinct, real repeated, complex conjugate distinct, complex conjugate repeated, or any combination of them. In particular:

(1) *All odd-degree equations* have at least one real root.
(2) *All symmetrical equations of odd degree* (6.06) have at least one real root $x_1 = +1$.
(3) *All antisymmetrical equations of odd degree* (6.06) have at least one real root $x_1 = -1$.
(4) *All antisymmetrical equations of even degree* (6.06) have at least one set of real antisymmetrical roots, $x_1 = +1$, $x_2 = -1$. All remaining roots must occur in antisymmetrical pairs and/or complex conjugate pairs.

(f) Deflation theorem.

Any odd-degree polynomial equation can be deflated to an even-degree polynomial equation

$$\frac{P(x)}{x - c_1} = Q(x) = 0$$

where the roots of $Q(x)$ in terms of p_m, q_m defined in (b) above occur in real pairs

$$c_{j,k} = -p_m \pm \sqrt{p_m^2 - q_m} \qquad \text{if } p_m^2 \geq q_m$$

or in complex conjugate pairs

$$c_{j,k} = -p_m \pm i\sqrt{q_m - p_m^2} \qquad \text{if } p_m^2 < q_m$$

or in their combinations.

(a) Synthetic division is an algebraic operation useful in calculating roots of polynomial equations and in evaluating polynomials and their derivatives at a particular point. Three forms of synthetic division are available: long form, short form, and matrix form. The first two forms are shown below in (*b*) and (*c*), and the third one is introduced in (6.15*c*), (6.15*e*), (6.15*g*).

(b) Long form without loss of generality can be demonstrated by the following particular case.

$$\frac{P(x)}{x-2} = (4x^3 + 2x^2 + 1) \div (x - 2) = 4x^2 + 10x + 20 + \frac{41}{x-2}$$

$$
\begin{array}{l}
-4x^3 + 8x^2 \\
\hline
\quad\quad 10x^2 + \ \ 0x \\
\quad\quad -10x^2 + 20x \\
\hline
\quad\quad\quad\quad 20x + \ \ 1 \\
\quad\quad\quad\quad -20x + 40 \\
\hline
\quad\quad\quad\quad\quad\quad 41
\end{array}
$$

and $P(x) = 4x^3 + 2x^2 + 1 = (4x^2 + 10x + 20)(x - 2) + 41$, where the remainder $R = P(2) = 41$ as required by the remainder theorem in (6.13*d*)

(c) Short form for the same case is

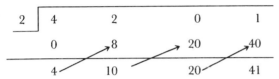

where the first row are the coefficients of $P(x)$, the second row is 0, $(4 + 0)(2) = 8$, $(2 + 8)(2) = 20$, $(0 + 20)(2) = 40$, and the last row is the sum in each column.

(d) Successive division of $P(x)$ by $x - a$ yields the factors of Taylor's series.

$$P(x) = r_0 + r_1(x - a) + r_2(x - a)^2 + r_3(x - a)^3 + \cdots + r_n(x - a)^n \qquad r_k = \frac{d^k[P(x)]}{k!\,dx^k}\bigg|_{x=a}$$

(e) Example. The calculation of r_k for $P(x) = 4x^3 + 2x^2 + 1 = 0$ at $x = 4$ is performed by short division (*c*).

$$
\begin{array}{c|cccc}
4 & 4 & 2 & 1 & 1 \\
  & 0 & 16 & 72 & 288 \\
\hline
  & 4 & 18 & 72 & \boxed{289} = r_0 \leftarrow P(4) \\
  & 0 & 16 & 136 & \\
4 & 4 & 34 & \boxed{208} & = r_1 \leftarrow \dfrac{P'(4)}{1} \\
  & 0 & 16 & & \\
4 & 4 & \boxed{50} & & = r_2 \leftarrow \dfrac{P''(4)}{2} \\
  & 0 & & & \\
4 & \boxed{4} & & & = r_3 \leftarrow \dfrac{P'''(4)}{6}
\end{array}
$$

and $P(x) = 289 + 208(x - 4) + 50(x - 4)^2 + 4(x - 4)^3$.

(a) Multiplication of polynomials. The product of two polynomials

$$a(x) = a_m x^m + a_{m-1} x^{m-1} + \cdots + a_2 x^2 + a_1 x + a_0$$

$$b(x) = b_n x^n + b_{n-1} x^{n-1} + \cdots + b_2 x^2 + b_1 x + b_0$$

where a_j $(j = m, m-1, \ldots, 2, 1, 0)$ and b_k $(k = n, n-1, \ldots, 2, 1, 0)$ are signed numbers, can be evaluated by the method of detached coefficients shown in (b). The first term of the product

$$[a(x)][b(x)] = c(x)$$

is $a_m b_n x^{m+n}$ and the last one is $a_0 b_0$.

(b) Example

$$c(x) = (x^3 + 4x^2 + 3x - 1)(4x^2 - 5x + 6)$$

	4	16	12	$-$	4		
		$-$ 5	$-$ 20	$-$ 15	5		
			6	24	18	$-$ 6	

$$c(x) = 4x^5 + 11x^4 - 2x^3 + 5x^2 + 23x - 6$$

(c) Division of polynomial by binomial

$$\frac{a_m x^m + a_{m-1} x^{m-1} + a_{m-2} x^{m-2} + \cdots + a_2 x^2 + a_1 x + a_0}{b_1 x + b_0}$$

$$= r_m x^{m-1} + r_{m-1} x^{m-2} + r_{m-2} x^{m-3} + \cdots + r_2 x + r_1 + R$$

where $r_m, r_{m-1}, r_{m-2}, \ldots, r_2, r_1, r_0$ are the coefficients of the matrix given below and

$$R = \frac{b_1 r_0}{b_1 x + b_0}$$

is the remainder. The coefficient matrix equation is

$$
\begin{bmatrix} r_m \\ r_{m-1} \\ r_{m-2} \\ \vdots \\ r_2 \\ r_1 \\ r_0 \end{bmatrix}
= \frac{1}{b_1}
\begin{bmatrix}
a_m \\
a_{m-1} & -b_0 \\
a_{m-2} & 0 & -b_0 \\
\vdots & \vdots & \vdots & \vdots & \vdots & & \cdot \\
a_2 & 0 & 0 & 0 & 0 & 0 & -b_0 \\
a_1 & 0 & 0 & 0 & 0 & 0 & 0 & -b_0 \\
a_0 & 0 & 0 & 0 & 0 & 0 & 0 & 0 & -b_0
\end{bmatrix}
\begin{bmatrix} 1 \\ r_m \\ r_{m-1} \\ \vdots \\ r_3 \\ r_2 \\ r_1 \end{bmatrix}
$$

where $r_m = a_m/b_1, \ldots, r_k = (a_k - b_0 r_{k-1})/b_1, \ldots, r_0 = (a_0 - b_0 r_1)/b_1$.

(d) Example

$$\frac{4x^5 + 8x^4 + 16x^2 + 20x + 30}{2x + 6} = r_5 x^4 + r_4 x^3 + r_3 x^2 + r_2 x + r_1 + \frac{2r_0}{2x + 6}$$

where $r_5 = \tfrac{1}{2}(4) = 2$, $r_4 = \tfrac{1}{2}[8 - 6(2)] = -2, \ldots, r_0 = \tfrac{1}{2}[30 - 6(40)] = 105$.

(e) Division of polynomial by trinomial

$$\frac{a_m x^m + a_{m-1}x^{m-1} + a_{m-2}x^{m-2} + \cdots + a_2 x^2 + a_1 x + a_0}{b_2 x^2 + b_1 x + b_0}$$

$$= r_m x^{m-2} + r_{m-1}x^{m-3} + r_{m-2}x^{m-4} + \cdots + r_3 x + r_2 + \frac{b_2(r_1 x + r_0)}{b_2 x^2 + b_1 x + b_0}$$

where $r_m, r_{m-1}, r_{m-2}, \ldots, r_2, r_1, r_0$ are the coefficients of the matrix given below and the fraction is the remainder. The coefficient matrix is

$$\begin{bmatrix} r_m \\ r_{m-1} \\ r_{m-2} \\ \vdots \\ r_2 \\ r_1 \\ r_0 \end{bmatrix} = \frac{1}{b_2} \begin{bmatrix} a_m & & & & & & \\ a_{m-1} & -b_1 & & & & & \\ a_{m-2} & -b_0 & -b_1 & & & & \\ \vdots & \vdots & \vdots & \vdots & \vdots & \ddots & \\ a_2 & 0 & 0 & 0 & 0 & -b_0 & -b_1 \\ a_1 & 0 & 0 & 0 & 0 & 0 & -b_0 & -b_1 \\ a_0 & 0 & 0 & 0 & 0 & 0 & 0 & -b_0 & -b_1 \end{bmatrix} \begin{bmatrix} 1 \\ r_m \\ r_{m-1} \\ \vdots \\ r_3 \\ r_2 \\ r_1 \end{bmatrix}$$

where $r_m = a_m/b_2$, $r_{m-1} = (a_{m-1} - b_1 r_m)/b_2$, $r_{m-2} = (a_{m-2} - b_0 r_m - b_1 r_{m-1})/b_2, \ldots, r_k = (a_k - b_0 r_{k+2} - b_1 r_{k+1})/b_2, \ldots, r_1 = (a_1 - b_0 r_3 - b_1 r_2)/b_2$, $r_0 = (a_0 - b_0 r_2 - b_1 r_1)/b_2$.

(f) Example

$$\frac{4x^5 + 8x^4 + 16x^2 + 20x + 30}{4x^2 + 2x + 6} = r_5 x^3 + r_4 x^2 + r_3 x + r_2 + \frac{4(r_1 x + r_0)}{4x^2 + 2x + 6}$$

where from

$$\begin{bmatrix} r_5 \\ r_4 \\ r_3 \\ r_2 \\ r_1 \\ r_0 \end{bmatrix} = \frac{1}{4} \begin{bmatrix} 4 & & & & \\ 8 & -2 & & & \\ 0 & -6 & -2 & & \\ 16 & 0 & -6 & -2 & \\ 20 & 0 & 0 & -6 & -2 \\ 30 & 0 & 0 & 0 & -6 & -2 \end{bmatrix} \begin{bmatrix} 1 \\ r_5 \\ r_4 \\ r_3 \\ r_2 \\ r_1 \end{bmatrix}$$

$$r_5 = \tfrac{1}{4}(4) = 1 \qquad r_3 = \tfrac{1}{4}[0 - 6(1) - 2(\tfrac{3}{2})] = -\tfrac{9}{2} \qquad r_1 = \tfrac{1}{4}[20 - 6(-\tfrac{9}{2}) - 2(4)] = \tfrac{39}{4}$$

$$r_4 = \tfrac{1}{4}[8 - 2(1)] = \tfrac{3}{2} \qquad r_2 = \tfrac{1}{4}[16 - 6(\tfrac{3}{2}) - 2(-\tfrac{9}{2})] = 4 \qquad r_0 = \tfrac{1}{4}[30 - 6(4) - 2(\tfrac{39}{4})] = \tfrac{11}{8}$$

and the remainder $R = (39x + 5.5)/(4x^2 + 2x + 6)$.

(a) Roots (zeros) of polynomial equation are classified in (6.13e). Their existence and relations to the algebraic curve $y = P(x)$ are illustrated by the following examples.

(b) Example of real distinct roots. The roots of

$$P(x) = x^3 - 6x^2 + 11x - 6 = (x - 1)(x - 2)(x - 3) = 0$$

are $x_1 = 1$, $x_2 = 2$, $x_3 = 3$, shown in the adjacent graph.

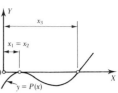

(c) Example of real repeated roots. The roots of

$$P(x) = x^3 - 4x^2 + 5x - 2 = (x - 1)^2(x - 2) = 0$$

are $x_{1,2} = 1$, $x_3 = 2$, shown in the adjacent graph, where the repeated roots x_1 and x_2 coincide.

(d) Example of real and complex conjugate roots. The roots of

$$P(x) = x^3 + 3x^2 + 7x - 5 = (x - 1)(x^2 - 2x + 5) = 0$$

are $x_1 = 1$, $x_2 = 1 + 2i$, $x_3 = 1 - 2i$. The real root x_1 is shown in the adjacent graph, whereas the complex conjugate roots $x_{2,3} = 1 \pm 2i$ escape the graphical representation in the real plane.

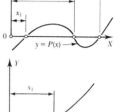

(e) Processing of polynomials. The approximate location of real roots is found by graphs as shown above and/or by the search of sign change in $P(x)$ according to (6.07b). For this purpose any of the methods introduced in (6.08) to (6.11) can be used. Since the number of all roots n is given by the degree of $P(x)$ and the number of real roots n_r is determined by the process just mentioned, the number of complex roots is

$$n_c = n - n_r$$

Once all real roots are known, $P(x)$ is deflated by successive division to a new polynomial equation of n_c degree, which is either quadratic (6.02), quartic (6.04), or of higher even degree and must be solved for the complex roots by one of the special methods introduced in (6.17), (6.18).

6.17 NEWTON–RAPHSON'S METHOD

(a) General procedure of finding the real roots of $P(x) = 0$ in (6.13a) is based on Newton–Raphson's method (6.10) combined with the short synthetic division as illustrated by the following particular case.

(1) *Initial estimate* of the first root x_1 of $x^3 - 14x^2 + 9x + 180 = 0$ is assumed to be

$$x_{0,1}^2 - 14x_{0,1} + 9 = 0 \qquad x_{0,1} = 7 + \sqrt{49 - 9} = 13.325$$

(2) *First approximation* in terms of $x_{0,1}$ by (6.14e),

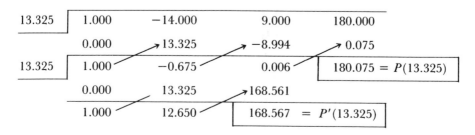

is according to (6.10), $x_{1,1} = 13.325 - \dfrac{180.075}{168.567} = 12.257$.

(3) *Second approximation* in terms of $x_{1,1}$ by (6.14e)

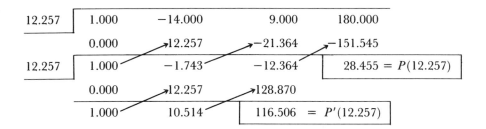

is similarly as in (2), $x_{2,1} = 12.257 - \dfrac{28.455}{116.506} = 12.013$.

(4) *Last approximation* in terms of $x_{2,1}$ by (6.14e),

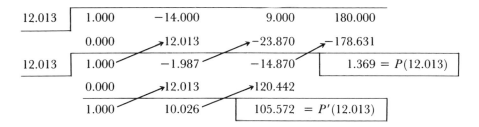

is similarly as in (3), $x_{3,1} = 12.013 - \dfrac{1.369}{105.572} = 12.000 = x_1$.

(5) *Deflated polynomial* $(x^3 - 14x^2 + 9x + 180) \div (x - 12)$ is

12.000	1.000	−14.000	9.000	180.000
	0.000	12.000	−24.000	−180.000
	$1.000x^2$	$-2.000x$	−15.000 =	0.000

the roots of which are the remaining roots of the polynomial equation

$$x_2 = 1 + \sqrt{1 + 15} = 5 \qquad\qquad x_3 = 1 - \sqrt{1 + 15} = -3$$

(b) Multiple roots. If x_i is one of the roots of $P(x)$ in (6.13a), $P'(x_j) = 0$, but $P''_j(x_j) \neq 0$, then x_j is a double root; if $P'(x_j) = 0$, $P''(x_j) = 0$, but $P'''_j(x_j) \neq 0$, then x_j is a triple root; and so on.

(c) Example

$$P(x) = x^4 - 4x^3 + 6x^2 - 4x + 1 = 0 \qquad \boxed{P(1) = 0}$$

$$P'(x) = 4x^3 - 12x^2 + 12x - 4 = 0 \qquad \boxed{P'(1) = 0} \qquad P'''(x) = 24x - 24 = 0 \qquad \boxed{P'''(1) = 0}$$

$$P''(x) = 12x^2 - 24x + 12 = 0 \qquad \boxed{P''(1) = 0} \qquad P^{iv}(x) = 24 \qquad \boxed{P^{iv}(1) = 24}$$

and $x_1 = x_2 = x_3 = x_4 = 1$ is a quadruple root of $P(x)$.

(a) Division of an even-degree polynomial

$$P_m(x) = x^m + a_1 x^{m-1} + a_2 x^{m-2} + \cdots + a_{m-1}x + a_m$$

by a quadratic factor $Q(x) = x^2 + px + q$ yields

$$P_m(x) = Q(x)P_{m-2}(x) + b_{m-1}x + b_m$$

where $P_{m-2}(x) = x^{m-2} + b_1 x^{m-1} + b_2 x^{m-2} + \cdots + b_{m-3}x + b_{m-2}$

and b_{m-1}, b_m are algebraic functions of p, q. The goal is now to find the values of p, q such that $b_{m-1} = 0, b_m = 0$.

(b) Approximations of p, q which satisfy this condition are given by the following recurrent relations,

$p_1 = p_0 + \Delta p_1$	$q_1 = q_0 + \Delta q_1$	$\Delta p_k = \dfrac{\nabla_{k,p}}{\Delta_k} = \dfrac{b_{k,m-1}c_{k,m-2} - b_{k,m}c_{k,m-3}}{c_{k,m-2}^2 - d_{k,m-1}c_{k,m-3}}$
$p_2 = p_1 + \Delta p_2$	$q_2 = q_1 + \Delta q_2$	
$\cdots\cdots\cdots\cdots$	$\cdots\cdots\cdots\cdots$	$\Delta q_k = \dfrac{\nabla_{k,q}}{\Delta_k} = \dfrac{b_{k,m}c_{k,m-2} - b_{k,m-1}d_{k,m-1}}{c_{k,m-2} - d_{k,m-1}c_{k,m-3}}$
$p_r = p_{r-1} + \Delta p_r$	$q_r = q_{r-1} + \Delta q_r$	

in which p_0, q_0 are the initial estimates of p, q and $k = 1, 2, \ldots, r$.

(c) Coefficients in these formulas are again related by recurrent relations (with the subscript k omitted but implied) as

$b_0 = a_0 = 1$	$c_0 = b_0 = 1$
$b_1 = a_1 - pb_0$	$c_1 = b_1 - pc_0$
$b_2 = a_2 - pb_1 - qb_0$	$c_2 = b_2 - pc_1 - qc_0$
$b_3 = a_3 - pb_2 - qb_1$	$c_3 = b_3 - pc_2 - qc_1$
$\cdots\cdots\cdots\cdots$	$\cdots\cdots\cdots\cdots$
$b_m = a_m - pb_{m-1} - qb_{m-2}$	$c_{m-1} = b_{m-1} - pc_{m-2} - qc_{m-3}$

$$d_{k,m-1} = c_{k,m-1} - b_{k,m-1}$$

or tabulated in the short synthetic division shown below.

		1	a_1	a_2	$\cdots$	a_{m-3}	a_{m-2}	a_{m-1}	a_m
$-p$		0	$-p$	$-pb_1$	$\cdots$	$-pb_{m-4}$	$-pb_{m-3}$	$-pb_{m-2}$	$-pb_{m-1}$
$-q$		0	0	$-q$	$\cdots$	$-qb_{m-5}$	$-qb_{m-4}$	$-qb_{m-3}$	$-qb_{m-2}$
Σ		1	b_1	b_2	$\cdots$	b_{m-3}	b_{m-2}	b_{m-1}	b_m
$-p$		0	$-p$	$-pc_1$	$\cdots$	$-pc_{m-4}$	$-pc_{m-3}$	$-pc_{m-2}$	
$-q$		0	0	$-q$	$\cdots$	$-qc_{m-5}$	$-qc_{m-4}$	$-qc_{m-3}$	
Σ		1	c_1	c_2	$\cdots$	c_{m-3}	c_{m-2}	c_{m-1}	$c_{m-1} - b_{m-1}$

(d) Coefficients b_j $(j = 1, 2, \ldots, m)$ in (c), required for the calculation of $\Delta p, \Delta q$ and for the deflated polynomial equation $P_{m-2}(x) = 0$, are expressible in closed form as

$$
\begin{bmatrix} b_1 \\ b_2 \\ b_3 \\ \cdots \\ b_{m-1} \\ b_m \end{bmatrix}
=
\underbrace{\begin{bmatrix}
F_1 & 1 & & & & & \\
F_2 & F_1 & 1 & & & & \\
F_3 & F_2 & F_1 & 1 & & & \\
\cdots & \cdots & \cdots & \cdots & \cdots & & \\
F_{m-1} & F_{m-2} & F_{m-3} & F_{m-4} & \cdots & 1 & \\
F_m & F_{m-1} & F_{m-2} & F_{m-3} & \cdots & F_1 & 1
\end{bmatrix}}_{\mathbf{F}}
\begin{bmatrix} 1 \\ a_1 \\ a_2 \\ \cdots \\ a_{m-1} \\ a_m \end{bmatrix}
$$
$$\underbrace{}_{\mathbf{b}} \qquad\qquad\qquad\qquad\qquad\qquad\qquad\qquad \underbrace{}_{\mathbf{a}}$$

where a_j $(j = 1, 2, \ldots, m)$ are the coefficients of $P_m(x)$ and in terms of $M = q/p^2$,

$F_1 = -p$	$F_6 = p^6(1 - M(5 - M^2(6 - M^3)))$
$F_2 = p^2(1 - M)$	$F_7 = -p^7(1 - M(6 - M^2(10 - 4M^3)))$
$F_3 = -p^3(1 - 2M)$	$F_8 = p^8(1 - M(7 - M^2(15 - M^3(10 - M^4))))$
$F_4 = p^4(1 - M(3 - M^2))$	$F_9 = -p^9(1 - M(8 - M^2(21 - M^3(20 - 5M^4))))$
$F_5 = -p^5(1 - M(4 - 3M^2))$	$F_{10} = p^{10}(1 - M(9 - M^2(28 - M^3(35 - M^4(15 - M^5)))))$

or recurrently

$$F_j = pF_{j-1} - qF_{j-2}$$

(e) Coefficient c_j $(j = 1, 2, \ldots, m - 1)$ in (c) required for the calculation of $\Delta p, \Delta q$ are again expressible in closed form as

$$\mathbf{c} = \mathbf{Fb}$$

where $\mathbf{c}$ is a column matrix of $1, c_1, c_2, \ldots, c_{m-1}$ and $\mathbf{F}$ and $\mathbf{b}$ are the matrices given in (d) above.

(f) Factors p, q are then calculated to a desired accuracy by means of (b) and inserted in the quadratic equation $Q(x) = 0$, whose roots are the first two roots of $P_m(x) = 0$. The next quadratic equation is obtained by the repetition of the same process from the deflated equation $P_{m-2}(x) = 0$, and so on.

(g) General procedure of finding the quadratic roots of $P_m(x) = 0$ in (a) is shown in the particular case below.

(1) *Initial values of p, q in*

$$P_4(x) = x^4 - 8x^3 + 39x^2 - 62x + 50 = 0 \qquad (m = 4)$$

are taken as $p_0 = -\frac{62}{39} \to -1.5$, $q_0 = \frac{50}{39} \to 1.5$.

(2) *First approximations of p, q are computed in terms of a_j, $b_{1,j}$, $c_{1,j}$, and $d_{1,j}$ tabulated below.*

	1.000	−8.000	39.000	−62.000	50.000
$-p_0 = +1.5$		1.500	−9.750	41.625	−15.938
$-q_0 = -1.5$			−1.500	9.750	−41.625
	1.000	−6.500 $b_{1,2}$	27.750 $b_{1,3}$	−10.625 $b_{1,4}$	−7.563
$-p_0 = +1.5$		1.500	−7.500	28.125	
$-q_0 = -1.5$			−1.500	7.500	
	1.000 $c_{1,1}$	−5.000 $c_{1,2}$	18.750 $c_{1,3}$	25.000 $d_{1,3}$	35.625

$$p_1 = p_0 + \Delta p_1 = -1.947 \qquad q_1 = q_0 + \Delta q_1 = +1.947$$

(3) *Second approximations of p, q are computed in terms of a_j, $b_{2,j}$, $c_{2,j}$, and $d_{2,j}$ tabulated below.*

	1.000	−8.000	39.000	−62.000	50.000
$-p_1 = +1.947$		1.947	−11.785	49.196	−1.984
$-q_1 = -1.947$			−1.947	11.785	−49.197
	1.000	−6.053 $b_{2,2}$	25.268 $b_{2,3}$	−1.019 $b_{2,4}$	−1.183
$-p_1 = +1.947$		1.974	−7.994	29.842	
$-q_1 = -1.947$			−1.947	7.994	
	1.000 $c_{2,1}$	−4.106 $c_{2,2}$	15.327 $c_{2,3}$	36.817 $d_{2,3}$	37.836

$$p_2 = p_1 + \Delta p_2 = -1.999 \qquad q_2 = q_1 + \Delta q_2 = +1.999$$

(4) *Quadratic factor in (a) with $p_2 \cong p = -2$, $q_2 = q \cong 2$ becomes*

$$Q_{1,2}(x) = x^2 - 2x + 2 = 0$$

and the first two roots of $P_4(x) = 0$ are

$$x_1 = -1 + \sqrt{1-2} = -1 + i \qquad x_2 = -1 - \sqrt{1-2} = -1 - i$$

(5) *Deflated equation,*

$$P_2(x) = x^2 - 6x + 25 = 0$$

where $p = -8 + 2 = -6 = b_1$, $q = 39 + 2(-6) - 2(1) = 25 = b_2$, and the remaining roots of $P_4(x) = 0$ are

$$x_3 = 3 + \sqrt{9-25} = 3 + 4i \qquad x_4 = 3 - \sqrt{9-25} = 3 - 4i$$

(h) Convergence and improvement of convergence is discussed by Hildebrand (Ref. 1.10, pp. 472–477) and Hamming (Ref. 1.09, pp. 108–111).

7
MATRIX
EQUATIONS

(a) Algebraic form. A system of n simultaneous linear equations given as

$$a_{11}x_1 + a_{12}x_2 + \cdots + a_{1n}x_n = b_1$$
$$a_{21}x_1 + a_{22}x_2 + \cdots + a_{2n}x_n = b_2$$
$$\cdots\cdots\cdots\cdots\cdots\cdots\cdots\cdots\cdots$$
$$a_{n1}x_1 + a_{n2}x_2 + \cdots + a_{nn}x_n = b_n$$

where $a_{11}, a_{12}, \ldots, a_{nn}$ and $b_1, b_2, \ldots, b_n$ are constants and $x_1, x_2, \ldots, x_n$ are the unknowns, is said to be consistent if there is a solution for the unknowns; otherwise, inconsistent. The system has a unique nonzero solution if the determinant

$$D = \begin{vmatrix} a_{11} & a_{12} & \cdots & a_{1n} \\ a_{12} & a_{22} & \cdots & a_{2n} \\ \cdots & \cdots & \cdots & \cdots \\ a_{n1} & a_{n2} & \cdots & a_{nn} \end{vmatrix} \neq 0$$

and at least one of the b terms is different from zero.

(b) Matrix form of the same system is

$$\underbrace{\begin{bmatrix} a_{11} & a_{12} & \cdots & a_{1n} \\ a_{21} & a_{22} & \cdots & a_{2n} \\ \cdots & \cdots & \cdots & \cdots \\ a_{n1} & a_{n2} & \cdots & a_{nn} \end{bmatrix}}_{\mathbf{A}} \underbrace{\begin{bmatrix} x_1 \\ x_2 \\ \cdots \\ x_n \end{bmatrix}}_{\mathbf{X}} = \underbrace{\begin{bmatrix} b_1 \\ b_2 \\ \cdots \\ b_n \end{bmatrix}}_{\mathbf{B}}$$

where $\mathbf{A}$ is a square matrix $[n \times n]$, $\mathbf{X}$ is a column matrix $[n \times 1]$, and $\mathbf{B}$ is a column matrix $[n \times 1]$.

(c) Classification of solutions. The values of x_j which satisfy the given system are the solutions (roots) of the system, which fall in the following categories:

(1) *Unique solution.* If $D \neq 0$, $\mathbf{B} \neq 0$, the system has a unique solution in which some, but not all, x_j may be zero.

(2) *Trivial solution.* If $D \neq 0$, $\mathbf{B} = 0$, the system has only one solution, $x_1 = x_2 = \cdots = x_n = 0$.

(3) *Infinitely many solutions.* If $D = 0$, $\mathbf{B} = 0$, the system is called *homogeneous* and it has infinitely many solutions, one of which is the trivial solution.

(4) *No solution.* If $D = 0$, $\mathbf{B} \neq 0$, the system has no solution.

(d) Methods of solution introduced in the subsequent sections of this chapter fall into two categories:

(1) *Direct methods* producing an exact solution by using a finite number of arithmetic operations

(2) *Iterative methods* producing an approximate solution of desired accuracy by yielding a sequence of solutions, which converges to the exact solution as the number of iterations tends to infinity

(a) Diagonal system. If $\mathbf{A}$ in $(7.01b)$ is a diagonal matrix so that

$$\begin{bmatrix} a_{11} & & & \\ & a_{22} & & \\ & & \cdot & \\ & & & a_{nn} \end{bmatrix}\begin{bmatrix} x_1 \\ x_2 \\ \cdots \\ x_n \end{bmatrix} = \begin{bmatrix} b_1 \\ b_2 \\ \cdots \\ b_n \end{bmatrix}$$

then $x_1 = b_1/a_{11}, x_2 = b_2/a_{22}, \ldots, x_n = b_n/a_{nn}$.

(b) Lower triangular system. If $\mathbf{A}$ in $(7.01b)$ is a lower triangular matrix so that

$$\begin{bmatrix} l_{11} & & & \\ l_{21} & l_{22} & & \\ \cdots & \cdots & & \\ l_{n1} & l_{n2} & \cdots & l_{nn} \end{bmatrix}\begin{bmatrix} x_1 \\ x_2 \\ \cdots \\ x_n \end{bmatrix} = \begin{bmatrix} b_1 \\ b_2 \\ \cdots \\ b_n \end{bmatrix}$$

then $x_1 = b_1/l_{11}, x_2 = (b_2 - l_{21}x_1)/l_{22}, \ldots, x_n = \left(b_n - \displaystyle\sum_{j=1}^{n-1} l_{nj}x_j\right)\Big/l_n$.

(c) Upper triangular system. If $\mathbf{A}$ in $(7.01b)$ is an upper triangular matrix so that

$$\begin{bmatrix} t_{11} & t_{12} & \cdots & t_{1n} \\ & t_{22} & \cdots & t_{2n} \\ & & \cdots & \\ & & & t_{nn} \end{bmatrix}\begin{bmatrix} x_1 \\ x_2 \\ \cdots \\ x_n \end{bmatrix} = \begin{bmatrix} b_1 \\ b_2 \\ \cdots \\ b_n \end{bmatrix}$$

then $x_n = b_n/t_{nn}, x_{n-1} = (b_{n-1} - t_{n-1,n}x_n)/t_{n-1,n-1}, \ldots, x_1 = \left(b_1 - \displaystyle\sum_{j=2}^{n} t_{1j}x_j\right)\Big/t_{11}$.

7.03 CRAMER'S RULE

(a) Determinant solution of a given system $(7.01a)$ is of the form

$$x_1 = \frac{D_1}{D}, x_2 = \frac{D_2}{D}, \ldots, x_n = \frac{D_n}{D}$$

where D is the determinant of $\mathbf{A}$ in $(7.01b)$ and $D_1, D_2, \ldots, D_n$ are the augmented determinants defined below.

(b) Augmented determinants obtained from D by removing the respective column in $\mathbf{A}$ and replacing it by the column $\mathbf{B}$ as

$$D_1 = \begin{vmatrix} b_1 & a_{12} & \cdots & a_{1n} \\ b_2 & a_{22} & \cdots & a_{2n} \\ \cdots & \cdots & & \cdots \\ b_n & a_{n2} & \cdots & a_{nn} \end{vmatrix}, D_2 = \begin{vmatrix} a_{11} & b_1 & \cdots & a_{1n} \\ a_{21} & b_2 & \cdots & a_{2n} \\ \cdots & \cdots & & \cdots \\ a_{n1} & b_{n1} & \cdots & a_{nn} \end{vmatrix}, \ldots, D_n = \begin{vmatrix} a_{11} & a_{12} & \cdots & b_1 \\ a_{21} & a_{22} & \cdots & b_2 \\ \cdots & \cdots & & \cdots \\ a_{n1} & a_{n2} & \cdots & b_n \end{vmatrix}$$

Although simple in appearance, this method is too laborious for practical use in cases of $n > 4$. The number of required operations is $n(3 + n^2 + n^3)/3$. The application of this method is shown in $(7.04b)$.

(a) Solution by inversion is the most direct and symbolically the most elegant method. For the set defined in (7.01b) the matrix solution is

$$\mathbf{X} = \mathbf{A}^{-1}\mathbf{B}$$

where

$$\mathbf{A}^{-1} = \frac{1}{D} \begin{bmatrix} A_{11} & A_{21} & \cdots & A_{n1} \\ A_{12} & A_{22} & \cdots & A_{n2} \\ \cdots\cdots\cdots\cdots\cdots \\ A_{1n} & A_{2n} & \cdots & A_{nn} \end{bmatrix}$$

is the inverse of $\mathbf{A}$ discussed in Ref. 1.20. The solution by inversion requires n^3 arithmetic operations and is numerically more involved than the determinant solution (7.03) or the other methods introduced in the subsequent sections. If, however, there are several $\mathbf{B}$ matrices involved, the inverse $\mathbf{A}^{-1}$ may be applied for each particular $\mathbf{B}$ with a minimum amount of numerical work as shown below.

(b) Example. If three conditions are involved in (7.01b), such as

$$\underbrace{\begin{bmatrix} 4 & -2 & 1 \\ 3 & 6 & -4 \\ 2 & 1 & 8 \end{bmatrix}}_{\mathbf{A}} \underbrace{\begin{bmatrix} x_{11} \\ x_{21} \\ x_{31} \end{bmatrix}}_{\mathbf{X}_1} = \underbrace{\begin{bmatrix} 12 \\ -25 \\ 32 \end{bmatrix}}_{\mathbf{B}_1} \qquad \underbrace{\begin{bmatrix} 4 & -2 & 1 \\ 3 & 6 & -4 \\ 2 & 1 & 8 \end{bmatrix}}_{\mathbf{A}} \underbrace{\begin{bmatrix} x_{12} \\ x_{22} \\ x_{32} \end{bmatrix}}_{\mathbf{X}_2} = \underbrace{\begin{bmatrix} 4 \\ 10 \\ 22 \end{bmatrix}}_{\mathbf{B}_2} \qquad \underbrace{\begin{bmatrix} 4 & -2 & 1 \\ 3 & 6 & -4 \\ 2 & 1 & 8 \end{bmatrix}}_{\mathbf{A}} \underbrace{\begin{bmatrix} x_{13} \\ x_{23} \\ x_{33} \end{bmatrix}}_{\mathbf{X}_3} = \underbrace{\begin{bmatrix} 20 \\ -30 \\ 40 \end{bmatrix}}_{\mathbf{B}_3}$$

then

$$\mathbf{X}_1 = \mathbf{A}^{-1}\mathbf{B}_1 \qquad\qquad \mathbf{X}_2 = \mathbf{A}^{-1}\mathbf{B}_2 \qquad\qquad \mathbf{X}_3 = \mathbf{A}^{-1}\mathbf{B}_3$$

where $\mathbf{A}^{-1}$ is calculated only once by the procedure described in (a) and

$$\begin{bmatrix} x_{11} & x_{12} & x_{13} \\ x_{21} & x_{22} & x_{23} \\ x_{31} & x_{32} & x_{33} \end{bmatrix} = \frac{1}{263} \begin{bmatrix} 52 & 17 & 2 \\ -32 & 30 & 19 \\ -9 & -8 & 30 \end{bmatrix} \begin{bmatrix} 12 & 4 & 20 \\ -25 & 10 & -30 \\ 32 & 22 & 40 \end{bmatrix}$$

from which the particular solutions are

$$\begin{bmatrix} x_{11} \\ x_{21} \\ x_{31} \end{bmatrix} = \begin{bmatrix} 1.000 \\ -2.000 \\ 4.000 \end{bmatrix} \qquad\qquad \begin{bmatrix} x_{12} \\ x_{22} \\ x_{32} \end{bmatrix} = \begin{bmatrix} 1.604 \\ 2.243 \\ 2.068 \end{bmatrix} \qquad\qquad \begin{bmatrix} x_{13} \\ x_{23} \\ x_{33} \end{bmatrix} = \begin{bmatrix} 12.319 \\ -2.965 \\ 4.790 \end{bmatrix}$$

By Cramer's method (7.03) with $\mathbf{B}_1$ and $D = 263$,

$$D_1 = \begin{vmatrix} 12 & -2 & 1 \\ -25 & 6 & -4 \\ 32 & 1 & 8 \end{vmatrix} = 263 \qquad D_2 = \begin{vmatrix} 4 & 12 & 1 \\ 3 & -25 & -4 \\ 2 & 32 & 8 \end{vmatrix} = -526 \qquad D_3 = \begin{vmatrix} 4 & -2 & 12 \\ 3 & 6 & -25 \\ 2 & 1 & 32 \end{vmatrix} = 1052$$

is

$$x_1 = \frac{263}{263} = 1 \qquad\qquad x_2 = \frac{-526}{263} = -2 \qquad\qquad x_3 = \frac{1052}{263} = 4$$

The same procedure must be repeated for $\mathbf{B}_2$ and $\mathbf{B}_3$.

(a) Banded system defined in general form as

$$
\begin{bmatrix}
1 & r_{12} & & & \\
r_{21} & 1 & r_{23} & & \\
& r_{23} & 1 & r_{34} & \\
& & \cdot & \cdot & \cdot \\
& & & r_{n,n-1} & 1
\end{bmatrix}
\begin{bmatrix}
x_1 \\ x_2 \\ x_3 \\ \cdots \\ x_n
\end{bmatrix}
=
\begin{bmatrix}
m_1 \\ m_2 \\ m_3 \\ \cdots \\ m_n
\end{bmatrix}
$$

where $\quad r_{12} = a_{12}/a_{11}, r_{21} = a_{21}/a_{22}, \dots, r_{n,n-1} = a_{n,n-1}/a_{nn} \quad$ and $\quad m_1 = b_1/a_{11}, m_2 = b_2/a_{22}, \dots, m_m = b_n/a_{nn}$ can be inverted by geometric series (Ref. 7.01). Several particular solutions are given in closed form below in terms of

$$
D = 1 - r_{12}r_{21} \qquad D_{13} = 1 - r_{12}r_{21} - r_{23}r_{32} \qquad D_{34} = 1 - r_{34}r_{43} \qquad D_{14} = 1 - \frac{r_{23}r_{32}}{D_{12}D_{34}}
$$

(b) Second-order system

If $0 < r_{ij} < 1$,

$$
\begin{bmatrix} x_1 \\ x_2 \end{bmatrix} =
\begin{bmatrix}
\dfrac{1}{D_{12}} & -\dfrac{r_{12}}{D_{12}} \\
-\dfrac{r_{21}}{D_{12}} & \dfrac{1}{D_{12}}
\end{bmatrix}
\begin{bmatrix} m_1 \\ m_2 \end{bmatrix}
$$

If $r_{ij} = \frac{1}{2}$,

$$
\begin{bmatrix} x_1 \\ x_2 \end{bmatrix} =
\begin{bmatrix}
\dfrac{4}{3} & -\dfrac{2}{3} \\
-\dfrac{2}{3} & \dfrac{4}{3}
\end{bmatrix}
\begin{bmatrix} m_1 \\ m_2 \end{bmatrix}
$$

(c) Third-order system

If $0 < r_{ij} < 1$,

$$
\begin{bmatrix} x_1 \\ x_2 \\ x_3 \end{bmatrix} =
\begin{bmatrix}
1 + \dfrac{r_{12}r_{21}}{D_{13}} & -\dfrac{r_{12}}{D_{13}} & \dfrac{r_{12}r_{23}}{D_{13}} \\
-\dfrac{r_{21}}{D_{13}} & \dfrac{1}{D_{13}} & -\dfrac{r_{23}}{D_{13}} \\
\dfrac{r_{32}r_{21}}{D_{13}} & -\dfrac{r_{32}}{D_{13}} & 1 + \dfrac{r_{32}r_{23}}{D_{13}}
\end{bmatrix}
\begin{bmatrix} m_1 \\ m_2 \\ m_3 \end{bmatrix}
$$

If $r_{ij} = \frac{1}{2}$

$$
\begin{bmatrix} x_1 \\ x_2 \\ x_3 \end{bmatrix} =
\begin{bmatrix}
\dfrac{3}{2} & -1 & \dfrac{1}{2} \\
-1 & 2 & -1 \\
\dfrac{1}{2} & -1 & \dfrac{3}{2}
\end{bmatrix}
\begin{bmatrix} m_1 \\ m_2 \\ m_3 \end{bmatrix}
$$

(d) Fourth-order system.

If $0 < r_{ij} < 1$,

$$
\begin{bmatrix} x_1 \\ x_2 \\ x_3 \\ x_4 \end{bmatrix} =
\begin{bmatrix}
1 + \dfrac{r_{12}r_{21}}{D_{12}D_{14}} & -\dfrac{r_{12}}{D_{12}D_{14}} & \dfrac{r_{12}}{r_{32}}\left(\dfrac{1}{D_{14}} - 1\right) & -\dfrac{r_{12}r_{34}}{r_{32}}\left(\dfrac{1}{D_{14}} - 1\right) \\
-\dfrac{r_{21}}{D_{12}D_{14}} & \dfrac{1}{D_{12}D_{14}} & -\dfrac{1}{r_{32}}\left(\dfrac{1}{D_{14}} - 1\right) & \dfrac{r_{34}}{r_{32}}\left(\dfrac{1}{D_{14}} - 1\right) \\
\dfrac{r_{21}}{r_{23}}\left(\dfrac{1}{D_{14}} - 1\right) & -\dfrac{1}{r_{23}}\left(\dfrac{1}{D_{14}} - 1\right) & \dfrac{1}{D_{14}D_{34}} & -\dfrac{r_{34}}{D_{14}D_{34}} \\
-\dfrac{r_{21}r_{43}}{r_{23}}\left(\dfrac{1}{D_{14}} - 1\right) & \dfrac{r_{43}}{r_{23}}\left(\dfrac{1}{D_{14}} - 1\right) & -\dfrac{r_{43}}{D_{14}D_{34}} & 1 + \dfrac{r_{43}r_{34}}{D_{14}D_{34}}
\end{bmatrix}
\begin{bmatrix} m_1 \\ m_2 \\ m_3 \\ m_4 \end{bmatrix}
$$

(a) Forward procedure consists of eliminating one unknown at a time in the initial system (7.01a), starting with x_1 and closing with the last equation in x_n. This is essentially the transformation of the initial set to the special case (7.02c).

(b) Backward procedure is starting with x_n and closing with the first equation in x_1, thus transforming the initial set to the special case (7.02b)

(c) Number of required arithmetic operations in Gauss' and Cramer's procedure is, respectively,

$$N_G = n(n^2 + 3n - 1)/6 \cong n^3/6 \qquad N_C = n(n^3 + n^2 + 3)/3 \cong n^4/3$$

which shows the advantage of the Gauss method

(d) Particular case illustrates the forward procedure.

$$\begin{bmatrix} a_{110} & a_{120} & a_{130} \\ a_{210} & a_{220} & a_{230} \\ a_{310} & a_{320} & a_{330} \end{bmatrix} \begin{bmatrix} x_1 \\ x_2 \\ x_3 \end{bmatrix} = \begin{bmatrix} b_{10} \\ b_{20} \\ b_{30} \end{bmatrix} \begin{matrix} \to E_{10} \\ \to E_{20} \\ \to E_{30} \end{matrix}$$

$$\begin{bmatrix} a_{221} & a_{231} \\ a_{321} & a_{331} \end{bmatrix} \begin{bmatrix} x_2 \\ x_3 \end{bmatrix} = \begin{bmatrix} b_{21} \\ b_{31} \end{bmatrix} \begin{matrix} \to E_{21} = E_{10}/a_{110} - E_{20}/a_{210} \\ \to E_{31} = E_{10}/a_{110} - E_{30}/a_{310} \end{matrix}$$

$$a_{332}x_3 = b_{32} \quad \to E_{32} = E_{21}/a_{221} - E_{31}/a_{331}$$

and yields

$$\begin{bmatrix} a_{110} & a_{120} & a_{130} \\ 0 & a_{221} & a_{231} \\ 0 & 0 & a_{332} \end{bmatrix} \begin{bmatrix} x_1 \\ x_2 \\ x_3 \end{bmatrix} = \begin{bmatrix} b_{10} \\ b_{21} \\ b_{32} \end{bmatrix} \qquad \begin{matrix} x_3 = b_{32}/a_{332} \\ x_2 = (b_{21} - a_{231}x_3)/a_{221} \\ x_1 = (b_{10} - a_{120}x_2 - a_{130}x_3)/a_{110} \end{matrix}$$

(e) Example introduced in (7.04b) is solved by the procedure described in (d) above.

k	a_{i1k}	a_{i2k}	a_{i3k}	b_{ik}	E_{ik}	F_{ik}
0	4 3 2	−2 6 1	1 −4 8	12 −25 32	E_{10} E_{20} E_{30}	15 −20 43
1		−2.5000 −1.0000	1.5833 −3.7500	11.333 −13.000	$E_{21} = E_{10}/4 - E_{20}/3$ $E_{31} = E_{10}/4 - E_{30}/2$	$10.416 = F_{21}$ $-17.750 = F_{31}$
2			−4.3833	−17.533	$E_{32} = -E_{21}/2.5 + E_{31}$	$-21.916 = F_{32}$

$$x_3 = \frac{-17.533}{-4.3833} = 4 \qquad x_2 = \frac{11.333 - 1.583\,3(4)}{-2.5} = -2 \qquad x_1 = \frac{12 + 2(-2) - 4}{4} = 1$$

Numerical check requires $F_{10}/4 - F_{20}/3 = F_{21}$, $F_{10}/4 - F_{30}/2 = F_{31}$, $-F_{21}/2.5 + F_{31}/1 = F_{32}$, where F_{ik} is the sum of elements in the respective row.

(a) Solution by successive transformation is essentially a matrix elimination procedure based on the idea of finding transformation matrices that by multiplication transform matrix $\mathbf{A}$ in $(7.01b)$ into its inverse $\mathbf{A}^{-1}$ $(7.04a)$. The number of arithmetic operations is again n^3 as in $(7.04a)$.

(b) Particular case illustrates this procedure.

$$\mathbf{T}_1\mathbf{A}_{ij0} = \begin{bmatrix} 1/a_{110} & 0 & 0 \\ -a_{210}/a_{110} & 1 & 0 \\ -a_{310}/a_{110} & 0 & 1 \end{bmatrix} \begin{bmatrix} a_{110} & a_{120} & a_{130} \\ a_{210} & a_{220} & a_{230} \\ a_{310} & a_{320} & a_{330} \end{bmatrix} = \begin{bmatrix} 1 & a_{121} & a_{131} \\ 0 & a_{221} & a_{231} \\ 0 & a_{321} & a_{331} \end{bmatrix} = \mathbf{A}_{ij1}$$

$$\mathbf{T}_2\mathbf{A}_{ij1} = \begin{bmatrix} 1 & -a_{121}/a_{221} & 0 \\ 0 & 1/a_{221} & 0 \\ 0 & -a_{321}/a_{221} & 1 \end{bmatrix} \begin{bmatrix} 1 & a_{121} & a_{131} \\ 0 & a_{221} & a_{231} \\ 0 & a_{321} & a_{331} \end{bmatrix} = \begin{bmatrix} 1 & 0 & a_{132} \\ 0 & 1 & a_{232} \\ 0 & 0 & a_{332} \end{bmatrix} = \mathbf{A}_{ij2}$$

$$\mathbf{T}_3\mathbf{A}_{ij2} = \begin{bmatrix} 1 & 0 & -a_{132}/a_{332} \\ 0 & 1 & -a_{232}/a_{332} \\ 0 & 0 & 1/a_{332} \end{bmatrix} \begin{bmatrix} 1 & 0 & a_{132} \\ 0 & 1 & a_{232} \\ 0 & 0 & a_{332} \end{bmatrix} = \begin{bmatrix} 1 & 0 & 0 \\ 0 & 1 & 0 \\ 0 & 0 & 1 \end{bmatrix} = \mathbf{I}$$

and with $\mathbf{A} = \mathbf{A}_{ij0}$, $\mathbf{T}_3\mathbf{T}_2\mathbf{T}_1\mathbf{A} = \mathbf{I}$ or $\mathbf{A}^{-1} = \mathbf{T}_1\mathbf{T}_2\mathbf{T}_3$.

(c) Example introduced in $(7.04b)$ is solved by the prodcedure described in (b) above. In the table below, a_{ijk} = element of $\mathbf{A}$, c_{ijk} = element of $\mathbf{A}^{-1}$, and E_{ik}, F_{ik} have the same meaning as in $(7.06e)$.

k	a_{i1k}	a_{i2k}	a_{i3k}	c_{i1k}	c_{i2k}	c_{13k}	E_{ik}	F_{1k}
0	4	-2	1	1	0	0	E_{10}	4
	3	6	-4	0	1	0	E_{20}	6
	2	1	8	0	0	1	E_{30}	12
1	1.0000	-0.5000	0.2500	0.2500	0	0	$E_{11} = E_{10}/4$	1
	0	7.5000	-4.7500	-0.7500	1.0000	0	$E_{21} = E_{20} - 3E_{10}/4$	3
	0	2.0000	7.5000	-0.5000	0	1.0000	$E_{31} = E_{30} - 2E_{10}/4$	10
2	1.0000	0	-0.0667	0.2000	0.0667	0	$E_{12} = E_{11} - \dfrac{-0.5}{7.5}E_{21}$	1.2
	0	1.0000	-0.6333	-0.1000	0.1333	0	$E_{22} = E_{21}/7.5$	0.4
	0	0	8.7667	-0.3000	-0.2667	1.0000	$E_{32} = E_{31} - \dfrac{2}{7.5}E_{21}$	9.2
3	1.0000	0	0	0.1977	0.0646	0.0076	$E_{13} = E_{12} + \dfrac{0.0667}{8.7667}E_{23}$	1.27
	0	1.0000	0	-0.1217	0.1141	0.0722	$E_{23} = E_{22} + \dfrac{0.6333}{8.7667}E_{23}$	1.06
	0	0	1.0000	-0.0342	-0.0304	0.1141	$E_{33} = E_{23}/8.7667$	1.05

$$\underbrace{\qquad\qquad\qquad}_{\mathbf{I}} \qquad \underbrace{\qquad\qquad\qquad\qquad}_{\mathbf{A}^{-1}}$$

and $\mathbf{A}^{-1}$ above must be identical to $\mathbf{A}^{-1}$ in $(7.04b)$. The numerical check used in $(7.06e)$ is also applicable in this case.

(a) Decomposition process, developed by Cholesky in algebraic form, by Banachiewicz in matrix form, and applied by Crout to machine calculations, assumes that the initial system in $(7.01b)$ can be reduced to the upper triangular system $(7.02c)$,

$$
\begin{bmatrix}
1 & t_{12} & t_{13} & \cdots & t_{1n} \\
 & 1 & t_{23} & \cdots & t_{2n} \\
 & & 1 & \cdots & t_{3n} \\
 & & & \cdots & \cdots \\
 & & & & 1
\end{bmatrix}
\underbrace{\begin{bmatrix} x_1 \\ x_2 \\ x_3 \\ \cdots \\ x_n \end{bmatrix}}_{}
=
\underbrace{\begin{bmatrix} c_1 \\ c_2 \\ c_3 \\ \cdots \\ c_n \end{bmatrix}}_{}
$$

$$\underbrace{}_{\mathbf{T}} \quad \mathbf{X} \quad \mathbf{C}$$

and that there exists a lower triangular matrix **L** so that

$$
\begin{bmatrix}
l_{11} \\
l_{21} & l_{22} \\
l_{31} & l_{32} & l_{33} \\
\\
l_{n1} & l_{n2} & l_{n3} & \cdots & l_{nn}
\end{bmatrix}
\begin{bmatrix}
1 & t_{12} & t_{13} & \cdots & t_{1n} \\
 & 1 & t_{23} & \cdots & t_{2n} \\
 & & 1 & \cdots & t_{3n} \\
 & & & \cdots & \cdots \\
 & & & & 1
\end{bmatrix}
\begin{bmatrix} x_1 \\ x_2 \\ x_3 \\ \cdots \\ x_n \end{bmatrix}
=
\begin{bmatrix} b_1 \\ b_2 \\ b_3 \\ \cdots \\ b_n \end{bmatrix}
$$

$$\mathbf{L} \qquad\qquad \mathbf{T} \qquad\qquad \mathbf{X} \qquad \mathbf{B}$$

where **B** is the initial column matrix in $(7.01b)$.

(b) Relations of these matrices

$$\mathbf{LT = A} \qquad \mathbf{LC = B}$$

can be expressed in compact form as

$$
\begin{bmatrix}
a_{11} & a_{12} & a_{13} & \cdots & a_{1n} & b_1 \\
a_{21} & a_{22} & a_{23} & \cdots & a_{2n} & b_2 \\
 & & & \cdots & & \\
a_{n1} & a_{n2} & a_{n3} & \cdots & a_{nn} & b_n
\end{bmatrix}
=
\begin{bmatrix}
l_{11} \\
l_{21} & l_{22} \\
\cdots & \cdots \\
l_{n1} & l_{n2} & l_{n3} & \cdots & l_{nn}
\end{bmatrix}
\begin{bmatrix}
1 & t_{12} & t_{13} & \cdots & t_{1n} & c_1 \\
 & 1 & t_{23} & \cdots & t_{2n} & c_2 \\
 & & & \cdots & & \\
 & & & & 1 & c_n
\end{bmatrix}
$$

$$\underbrace{}_{\mathbf{A}} \quad \underbrace{}_{\mathbf{B}} \qquad \underbrace{}_{\mathbf{L}} \qquad\qquad \underbrace{}_{\mathbf{T}} \quad \underbrace{}_{\mathbf{C}}$$

from which $l_{11} = a_{11} \qquad t_{12} = a_{12}/a_{11} \qquad t_{13} = a_{13}/a_{11} \qquad \cdots \qquad t_n = a_{1n}/a_{11}$

$$l_{21} = a_{21}$$

$$l_{31} = a_{31}$$

$$t_{ij} = (a_{ij} - l_{i1}t_{1j} - l_{i2}t_{2j} - \cdots - l_{i,i-1}t_{i-1,j})/l_{ii}$$

$$\cdots\cdots$$

$$l_{ij} = a_{ij} - l_{i1}t_{1j} - l_{i2}t_{2j} - \cdots - l_{i,j-1}t_{j-1,j}$$

$$l_{n1} = a_{n1}$$

and $\qquad c_1 = b_1/l_{11} \qquad c_2 = (b_2 - l_{21}c_1)/l_{22} \qquad c_3 = (b_3 - l_{31}c_1 - l_{32}c_2)/l_{33}$

$$\cdots \qquad c_n = (b_n - l_{n1}c_1 - l_{n2}c_2 - l_{n3}c_3 - \cdots - l_{n,n-1}c_{n-1})/l_{nn}$$

(c) Final matrix equation is then the matrix equation given in (a) above, the elements t_{ij} and c_i of which are now known from (b) and the unknowns of which can be calculated by back substitution as in $(7.02c)$.

(a) Unsymmetrical system used in (7.03c) is solved below as an illustration of the Cholesky method (7.08). First, from the given system

$$\begin{bmatrix} 4 & -2 & 1 \\ 3 & 6 & -4 \\ 2 & 1 & 8 \end{bmatrix} \begin{bmatrix} x_1 \\ x_2 \\ x_3 \end{bmatrix} = \begin{bmatrix} 12 \\ -25 \\ 32 \end{bmatrix}$$

the following identity is constructed according to (7.08b):

$$\mathbf{L} + \mathbf{T} - \mathbf{I} = \begin{bmatrix} l_{11} & t_{12} & t_{13} \\ l_{21} & l_{22} & t_{23} \\ l_{31} & l_{32} & l_{33} \end{bmatrix} = \begin{bmatrix} 4.000\ 0 & -0.500\ 0 & 0.250\ 00 \\ 3.000\ 0 & 7.500\ 0 & -0.633\ 33 \\ 2.000\ 0 & 2.000\ 0 & 8.766\ 70 \end{bmatrix}$$

(b) Coefficients of this identity,

$l_{11} = 4, l_{21} = 3, l_{31} = 2$ are the elements of the first column in **A**

$t_{12} = -\frac{2}{4}, t_{13} = \frac{1}{4}$ are the second and third elements of the first row in **A** divided by a_{11},

and the remaining coefficients are calculated from the following schemes:

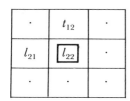

$$l_{22} = a_{22} - l_{21}t_{12}$$
$$= 6 - (3)(0.5) = 7.500$$

$$t_{23} = (a_{23} - l_{21}t_{13})/l_{22}$$
$$= [-4 - (3)(0.25)]/(7.5) = -0.633\ 33$$

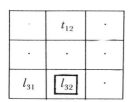

$$l_{32} = a_{32} - l_{31}t_{12}$$
$$= 1 - (2)(-0.5) = 2.000$$

$$l_{33} = a_{33} - l_{31}t_{13} - l_{32}t_{23}$$
$$= 8 - (2)(0.25) - (2)(-0.633\ 33) = 8.7667$$

The c coefficients with $b_1 = 12, b_2 = -25, b_3 = 32$ are by (7.08b)

$$c_1 = b_1/l_{11} = 3.000\ 0 \qquad c_2 = (b_2 - l_{21}c_1)/l_{22} = -4.533\ 33$$
$$c_3 = (b_3 - l_{31}c_1 - l_{32}c_2)/l_{33} = 4.000\ 0$$

(c) Final matrix equation in terms of t_{ij} and c_j obtained in (b) is

$$\begin{bmatrix} 1 & -0.500\ 00 & 0.250\ 00 \\ & 1 & -0.633\ 33 \\ & & 1 \end{bmatrix} \begin{bmatrix} x_1 \\ x_2 \\ x_3 \end{bmatrix} = \begin{bmatrix} 3.000\ 00 \\ -4.533\ 33 \\ 4.000\ 00 \end{bmatrix}$$

and by backsubstitution $x_3 = 4, x_2 = -2, x_1 = 1$.

(d) Number of significant figures lost in these computations is approximately $0.3n$ to $0.5n$. For the system of order $n = 3$, with required three-digit accuracy, all calculations must be carried out with at least five significant figures as shown above.

(a) Three-band system introduced in (7.05a) reduces by (7.08) to

$$
\begin{bmatrix}
1 & t_{12} & & & \\
 & 1 & t_{23} & & \\
 & & 1 & t_{34} & \\
 & & & \cdots & \\
 & & & & 1
\end{bmatrix}
\begin{bmatrix}
x_1 \\ x_2 \\ x_3 \\ \cdots \\ x_n
\end{bmatrix}
=
\begin{bmatrix}
d_1 \\ d_2 \\ d_3 \\ \cdots \\ d_n
\end{bmatrix}
$$

where

$$t_{12} = r_{12}, t_{23} = r_{23}/(1 - r_{21}t_{12}), t_{34} = r_{34}/(1 - r_{32}t_{23}), \ldots,$$

and

$$d_1 = m_1, d_2 = -r_{21}m_1 + m_2, d_3 = -r_{32}d_2 + m_3, \ldots, d_n = -r_{n,n-1}d_{n-1} + m_n.$$

The reduced equation shown above is again the special case solved in (7.02c).

(b) Five-band system formed by the governing relation

$$\boxed{r_{k,k-2}x_{k-2} + r_{k,k-1}x_{k-1} + x_k + r_{k,k+1}x_{k+1} + r_{k,k+2}x_{k+2} = m_k}$$

reduces by (7.08) to

$$
\begin{bmatrix}
1 & t_{12} & t_{13} & & \\
 & 1 & t_{23} & t_{24} & \\
 & & 1 & t_{34} & t_{35} \\
 & & \cdot & \vdots & \vdots & \vdots \\
 & & & & 1
\end{bmatrix}
\begin{bmatrix}
x_1 \\ x_2 \\ x_3 \\ \cdots \\ x_n
\end{bmatrix}
=
\begin{bmatrix}
d_1 \\ d_2 \\ d_3 \\ \cdots \\ d_n
\end{bmatrix}
$$

where

$$t_{12} = r_{12} \qquad l_{21} = r_{21} \qquad l_{22} = 1 - r_{21}r_{12} \qquad t_{13} = r_{13} \qquad l_{31} = r_{31}$$

$$l_{42} = r_{32} - r_{31}r_{12} \qquad t_{23} = (r_{23} - r_{21}r_{13})/l_{22} \qquad t_{24} = r_{24}/l_{22}$$

$$l_{33} = 1 - r_{31}r_{13} - l_{32}t_{23}$$

$$l_{43} = r_{43} - r_{42}t_{23} \qquad t_{34} = (r_{34} - l_{32}t_{23})/l_{33} \qquad t_{35} = r_{35}/l_{33}$$

$$l_{44} = 1 - r_{42}t_{24} - l_{43}t_{34}$$

$$\cdots\cdots\cdots\cdots\cdots\cdots\cdots\cdots\cdots\cdots\cdots\cdots\cdots\cdots\cdots$$

$$l_{k+1,k} = r_{k+1,k} - r_{k+1,k-1}t_{k-1,k} \quad t_{k,k+1} = (r_{k,k+1} - l_{k,k-1}t_{k-1,k})/l_{kk} \quad t_{k,k+2} = r_{k,k+2}/l_{kk}$$

$$l_{kk} = 1 - r_{k,k-2}t_{k-2,k} - l_{k,k-1}t_{k-1,k}$$

$$d_1 = m_1, d_2 = (m_2 - r_{21}d_1)/l_{22}, d_3 = (m_3 - r_{31}d_1 - l_{32}d_2)/l_{33}, \ldots,$$

$$d_n = (m_n - r_{n,n-2}d_{n-2} - l_{n,n-1}d_{n-1})/l_{nn}$$

As in (a), the reduced equation is the special case solved in (7.02c).

Matrix Equations

(a) Decomposition process developed by Banachiewicz for the solution of symmetrical systems assumes that the initial matrices $\mathbf{A}$ and $\mathbf{B}$ can be resolved as

$$\mathbf{A} = \mathbf{U}^T\mathbf{U} \qquad \mathbf{B} = \mathbf{U}^T\mathbf{C}$$

and the reduced matrix equation becomes

$$\mathbf{UX} = \mathbf{C}$$

where $\mathbf{A}, \mathbf{B}$ are defined in $(7.01b)$ and

$$\mathbf{U} = \begin{bmatrix} u_{11} & u_{12} & u_{13} & \cdots & u_{1n} \\ & u_{22} & u_{23} & \cdots & u_{2n} \\ & & u_{33} & \cdots & u_{3n} \\ & & & \cdots & \cdots \\ & & & & u_{nn} \end{bmatrix} \qquad \mathbf{C} = \begin{bmatrix} c_1 \\ c_2 \\ c_3 \\ \cdots \\ c_n \end{bmatrix}$$

(b) Coefficients of of $\mathbf{U}$ and $\mathbf{C}$ are

$$u_{11} = \sqrt{a_{11}} \qquad u_{22} = \sqrt{a_{22} - u_{12}^2} \qquad u_{33} = \sqrt{a_{33} - u_{13}^2 - u_{23}^2}$$

$$u_{12} = a_{12}/u_{11} \qquad u_{23} = (a_{23} - u_{12}u_{13})/u_{22} \qquad u_{34} = (a_{34} - u_{13}u_{14} - u_{23}u_{24})/u_{33}$$

$$u_{13} = a_{13}/u_{11} \qquad u_{24} = (a_{24} - u_{12}u_{14})/u_{22} \qquad u_{35} = (a_{35} - u_{13}u_{15} - u_{23}u_{25})/u_{33}$$

$$u_{1n} = a_{1n}/u_{nn} \qquad u_{2n} = (a_{2n} - u_{12}u_{1n})/u_{22} \qquad u_{3n} = (a_{3n} - u_{13}u_{1n} - u_{23}u_{2n})/u_{nn}$$

$$u_{ii} = \sqrt{a_{ii} - u_{1i}^2 - u_{2i}^2 - \cdots - u_{i-1,i}^2} \qquad i > 1$$

$$u_{ij} = (a_{ij} - u_{1i}u_{1j} - u_{2i}u_{2j} - \cdots - u_{i-1,i}u_{i-1,j})/u_{ii} \qquad i < j, i > 1$$

$$c_1 = b_1/u_{11} \qquad c_2 = (b_2 - c_1u_{12})/u_{22} \qquad c_3 = (b_3 - c_2u_{23} - c_1u_{13})/u_{33}$$

$$c_i = (b_i - c_{i-1}u_{i-1,i} - \cdots - c_2u_{2i} - c_1u_{1i})/u_{ii}$$

(c) Final matrix equation is (a) in terms of calculated coefficients in (b) above. The solution of this system follows the special case $(7.02c)$.

(d) Example. The reduction of the initial symmetrical system

$$\begin{bmatrix} 4 & 3 & 2 & 1 \\ 3 & 4 & 3 & 2 \\ 2 & 3 & 4 & 3 \\ 1 & 2 & 3 & 4 \end{bmatrix} \begin{bmatrix} x_1 \\ x_2 \\ x_3 \\ x_4 \end{bmatrix} = \begin{bmatrix} 20 \\ 28 \\ 32 \\ 30 \end{bmatrix}$$

by (a) in terms of (b) is

$$\begin{bmatrix} 2.000\,00 & 1.500\,00 & 1.000\,00 & 0.500\,00 \\ & 1.322\,88 & 1.133\,89 & 0.944\,91 \\ & & 1.309\,31 & 1.091\,09 \\ & & & 1.290\,99 \end{bmatrix} \begin{bmatrix} x_1 \\ x_2 \\ x_3 \\ x_4 \end{bmatrix} = \begin{bmatrix} 10.000\,00 \\ 9.827\,04 \\ 8.292\,32 \\ 5.163\,99 \end{bmatrix}$$

and $x_4 = 4.000\,0$, $x_3 = 3.000\,0$, $x_2 = 2.000\,0$, $x_1 = 1.000\,0$.

(a) Matrix equation (7.10b) can be expressed in partition form with $n = p + q$ as

$$\left[\begin{array}{c|c} \mathbf{A}_{pp} & \mathbf{A}_{pq} \\ \hline \mathbf{A}_{qp} & \mathbf{A}_{qq} \end{array}\right]\left[\begin{array}{c} \mathbf{X}_p \\ \mathbf{X}_q \end{array}\right] = \left[\begin{array}{c} \mathbf{B}_p \\ \mathbf{B}_q \end{array}\right]$$

where $\quad \mathbf{A}_{pp}$ = matrix of a_{ij}, $[p \times p]$, $i = 1, 2, \ldots, p, j = 1, 2, \ldots, p$

$\quad\quad \mathbf{A}_{pq}$ = matrix of a_{ik}, $[p \times q]$, $i = 1, 2, \ldots, p, k = p + 1, p + 2, \ldots, n$

$\quad\quad \mathbf{A}_{qp}$ = matrix of a_{kj}, $[q \times p]$, $k = p + 1, p + 2, \ldots, n, j = 1, 2, \ldots, p$

$\quad\quad \mathbf{A}_{qq}$ = matrix of a_{kl}, $[q \times q]$, $k = p + 1, p + 2, \ldots, n, l = p + 1, p + 2, \ldots, n$

$\quad\quad \mathbf{X}_p = [x_1 \quad x_2 \quad \cdots \quad x_p]^T \quad\quad\quad \mathbf{X}_q = [x_{p+1} \quad x_{p+2} \quad \cdots \quad x_n]^T$

$\quad\quad \mathbf{B}_p = [b_1 \quad b_2 \quad \cdots \quad b_p]^T \quad\quad\quad \mathbf{B}_q = [b_{p+1} \quad b_{p+2} \quad \cdots \quad b_n]^T$

(b) Submatrix form of (a) yields the final solution,

$$\left[\begin{array}{c} \mathbf{X}_p \\ \hline \mathbf{X}_q \end{array}\right] = \left[\begin{array}{c|c} \mathbf{M} & -\mathbf{M}\mathbf{A}_{pq}\mathbf{A}_{qq}^{-1} \\ \hline -\mathbf{N}\mathbf{A}_{qp}\mathbf{A}_{pp}^{-1} & \mathbf{N} \end{array}\right]\left[\begin{array}{c} \mathbf{B}_p \\ \mathbf{B}_q \end{array}\right]$$

where $\mathbf{A}_{pp}^{-1}, \mathbf{A}_{qq}^{-1}$ are the inverses of $\mathbf{A}_{pp}, \mathbf{A}_{qq}$, respectively, and

$$\mathbf{M} = [\mathbf{A}_{pp} \quad -\mathbf{A}_{pq}\mathbf{A}_{qq}^{-1}\mathbf{A}_{qp}]^{-1} \quad\quad \mathbf{N} = [\mathbf{A}_{qq} \quad -\mathbf{A}_{qp}\mathbf{A}_{pp}^{-1}\mathbf{A}_{pq}]^{-1}$$

(c) Example. The system in $(7.11d)$ is solved below by (b). With

$$\mathbf{A}_{pp} = \mathbf{A}_{qq} = \begin{bmatrix} 4 & 3 \\ 3 & 4 \end{bmatrix} \quad\quad \mathbf{A}_{qp}\mathbf{A}_{pp}^{-1} = \begin{bmatrix} 2 & 3 \\ 1 & 2 \end{bmatrix}\begin{bmatrix} \frac{4}{7} & -\frac{3}{7} \\ -\frac{3}{7} & \frac{4}{7} \end{bmatrix} = \begin{bmatrix} -\frac{1}{7} & \frac{6}{7} \\ -\frac{2}{7} & \frac{5}{7} \end{bmatrix}$$

$$\mathbf{A}_{pp}^{-1} = \mathbf{A}_{qq}^{-1} = \begin{bmatrix} \frac{4}{7} & -\frac{3}{7} \\ -\frac{3}{7} & \frac{4}{7} \end{bmatrix} \quad\quad \mathbf{A}_{pq}\mathbf{A}_{qq}^{-1} = \begin{bmatrix} 2 & 1 \\ 3 & 2 \end{bmatrix}\begin{bmatrix} \frac{4}{7} & -\frac{3}{7} \\ -\frac{3}{7} & \frac{4}{7} \end{bmatrix} = \begin{bmatrix} \frac{5}{7} & -\frac{2}{7} \\ \frac{6}{7} & -\frac{1}{7} \end{bmatrix}$$

$$\mathbf{A}_{qp}\mathbf{A}_{pp}^{-1}\mathbf{A}_{pq} = \begin{bmatrix} -\frac{1}{7} & \frac{6}{7} \\ -\frac{2}{7} & \frac{5}{7} \end{bmatrix}\begin{bmatrix} 2 & 1 \\ 3 & 2 \end{bmatrix} = \begin{bmatrix} \frac{16}{7} & \frac{11}{7} \\ \frac{11}{7} & \frac{8}{7} \end{bmatrix} \quad\quad \mathbf{A}_{pp} - \mathbf{A}_{pq}\mathbf{A}_{qq}^{-1}\mathbf{A}_{qp} = \begin{bmatrix} \frac{20}{7} & \frac{10}{7} \\ \frac{10}{7} & \frac{12}{7} \end{bmatrix}$$

$$\mathbf{A}_{pq}\mathbf{A}_{qq}^{-1}\mathbf{A}_{qp} = \begin{bmatrix} \frac{5}{7} & -\frac{2}{7} \\ \frac{6}{7} & -\frac{1}{7} \end{bmatrix}\begin{bmatrix} 2 & 3 \\ 1 & 2 \end{bmatrix} = \begin{bmatrix} \frac{8}{7} & \frac{11}{7} \\ \frac{11}{7} & \frac{16}{7} \end{bmatrix} \quad\quad \mathbf{A}_{qq} - \mathbf{A}_{qp}\mathbf{A}_{pp}^{-1}\mathbf{A}_{pq} = \begin{bmatrix} \frac{12}{7} & \frac{10}{7} \\ \frac{10}{7} & \frac{20}{7} \end{bmatrix}$$

$$\mathbf{M} = \begin{bmatrix} \frac{20}{7} & \frac{10}{7} \\ \frac{10}{7} & \frac{12}{7} \end{bmatrix}^{-1} = \begin{bmatrix} 0.6 & -0.5 \\ -0.5 & 1.0 \end{bmatrix} \quad\quad \mathbf{N} = \begin{bmatrix} \frac{12}{7} & \frac{10}{7} \\ \frac{10}{7} & \frac{20}{7} \end{bmatrix}^{-1} = \begin{bmatrix} 1.0 & -0.5 \\ -0.5 & 0.6 \end{bmatrix}$$

$$-\mathbf{M}\mathbf{A}_{pq}\mathbf{A}_{qq}^{-1} = -\begin{bmatrix} 0.6 & -0.5 \\ -0.5 & 1.0 \end{bmatrix}\begin{bmatrix} \frac{5}{7} & -\frac{2}{7} \\ \frac{6}{7} & -\frac{1}{7} \end{bmatrix} = \begin{bmatrix} 0 & 0.1 \\ -0.5 & 0 \end{bmatrix}$$

$$-\mathbf{N}\mathbf{A}_{qp}\mathbf{A}_{pp}^{-1} = -\begin{bmatrix} 1.0 & -0.5 \\ -0.5 & 0.6 \end{bmatrix}\begin{bmatrix} -\frac{1}{7} & \frac{6}{7} \\ -\frac{2}{7} & \frac{5}{7} \end{bmatrix} = \begin{bmatrix} 0 & -0.5 \\ 0.1 & 0 \end{bmatrix}$$

the final solution is

$$\begin{bmatrix} x_1 \\ x_2 \\ x_3 \\ x_4 \end{bmatrix} = \left[\begin{array}{cc|cc} 0.6 & -0.5 & 0 & 0.1 \\ -0.5 & 1.0 & -0.5 & 0 \\ \hline 0 & -0.5 & 1.0 & -0.5 \\ 0.1 & 0 & -0.5 & 0.6 \end{array}\right]\begin{bmatrix} 20 \\ 28 \\ 32 \\ 30 \end{bmatrix} = \begin{bmatrix} 1 \\ 2 \\ 3 \\ 4 \end{bmatrix}$$

(a) Iteration system

$$
\begin{bmatrix}
1 & r_{12} & r_{13} & \cdots & r_{1n} \\
r_{21} & 1 & r_{23} & \cdots & r_{2n} \\
r_{31} & r_{32} & 1 & \cdots & r_{3n} \\
 & & & \cdots & \\
r_{n1} & r_{n2} & r_{n3} & \cdots & 1
\end{bmatrix}
\begin{bmatrix}
x_1 \\ x_2 \\ x_3 \\ \cdots \\ x_n
\end{bmatrix}
=
\begin{bmatrix}
m_1 \\ m_2 \\ m_3 \\ \cdots \\ m_n
\end{bmatrix}
$$

obtained from (7.01*b*) by dividing each equation by a_{ii} so that

$$r_{12} = a_{12}/a_{11}, r_{13} = a_{13}/a_{11}, r_{14} = a_{14}/a_{11}, \ldots, r_{1n} = a_{1n}/a_{11}, m_1 = b_1/a_{11}$$

$$r_{21} = a_{21}/a_{22}, r_{23} = a_{23}/a_{22}, r_{24} = a_{24}/a_{22}, \ldots, r_{2n} = a_{2n}/a_{22}, m_2 = b_2/a_{22}$$

$$r_{31} = a_{31}/a_{33}, r_{32} = a_{32}/a_{33}, r_{34} = a_{34}/a_{33}, \ldots, r_{3n} = a_{3n}/a_{33}, m_3 = b_3/a_{33}$$

$$\cdots\cdots\cdots\cdots\cdots\cdots\cdots\cdots\cdots\cdots\cdots\cdots\cdots\cdots\cdots\cdots\cdots$$

$$r_{n1} = a_{n1}/a_{nn}, r_{n2} = a_{n2}/a_{nn}, r_{n3} = a_{n3}/a_{nn}, \ldots, r_{n-1,n} = a_{n-1,n}/a_{nn}, m_n = b_n/a_n n$$

is called a strong diagonal system if the sum of off-diagonal elements r_{ij} $(i \neq j)$ in each row is less than 1. If this condition is satisfied, the system can be solved by the Gauss–Seidel iteration method described below.

(b) Starting values of this iteration process may be taken as

$$x_1^{(0)} = m_1, x_2^{(0)} = m_2, x_3^{(0)} = m_3, \ldots, x_n^{(0)} = m_n$$

or any other set of values used as estimates of the final solution.

(c) Successive approximations starting with this trial solution become

$$x_1^{(1)} = x_1^{(0)} - \sum_{j=1}^{n} r_{1j}x_j^{(0)}, x_2^{(1)} = x_2^{(0)} - \sum_{j-1}^{n} r_{2j}x_j^{(0)}, \ldots, x_n^{(1)} = x_n^{(0)} - \sum_{j=1}^{n} r_{nj}x_j^{(0)}$$

$$x_1^{(2)} = x_1^{(1)} - \sum_{j=1}^{n} r_{1j}x_j^{(1)}, x_2^{(2)} = x_2^{(1)} - \sum_{j=1}^{n} r_{2j}x_j^{(1)}, \ldots, x_n^{(2)} = x_n^{(1)} - \sum_{j=1}^{n} r_{nj}x_j^{(1)}$$

$$x_1^{(3)} = x_1^{(2)} - \sum_{j=1}^{n} r_{1j}x_j^{(2)}, x_2^{(3)} = x_2^{(2)} - \sum_{j=1}^{n} r_{2j}x_j^{(2)}, \ldots, x_n^{(3)} = x_n^{(2)} - \sum_{j=1}^{n} r_{nj}x_j^{(2)}$$

$$\cdots\cdots\cdots\cdots\cdots\cdots\cdots\cdots\cdots\cdots\cdots\cdots\cdots\cdots\cdots\cdots\cdots$$

and the final solution is

$$x_1 = \sum_{k=1}^{s} x_1^{(k)}, x_2 = \sum_{k=1}^{s} x_2^{(k)}, \ldots, x_n = \sum_{k=1}^{s} x_n^{(k)}$$

where s is the number of required iterations. Under certain conditions this process is rapidly convergent, but is may also converge very slowly.

(d) Matrix formulation of this process is

$$\mathbf{X} = [\mathbf{I} + \mathbf{R} + \mathbf{R}^2 + \mathbf{R}^3 + \cdots]\mathbf{M} = [\mathbf{I} + \mathbf{R}[\mathbf{I} + \mathbf{R}[\mathbf{I} + \mathbf{R}[\mathbf{I} + \cdots]]]]\mathbf{M}$$

$$\text{where} \quad \mathbf{R} =
\begin{bmatrix}
0 & -r_{12} & -r_{13} & \cdots & -r_{1n} \\
-r_{21} & 0 & -r_{23} & \cdots & -r_{2n} \\
 & & & & \\
-r_{n1} & r_{n2} & -r_{n3} & \cdots & 0
\end{bmatrix}
\qquad
\mathbf{M} =
\begin{bmatrix}
m_1 \\ m_2 \\ \cdots \\ m_n
\end{bmatrix}
$$

and the sum of the matrix power series is the inverse of the matrix $\mathbf{A}$ in (7.01*b*).

(a) Tabular form of the iteration method (7.13) in cases of three-band and five-band strong diagonal matrix equations is called the carry-over method, illustrated by a numerical example below (Ref. 7.09).

(b) Example. For

$$\begin{bmatrix} 1 & 0.17 & 0 & 0 \\ 0.30 & 1 & 0.20 & 0 \\ 0 & 0.29 & 1 & 0.21 \\ 0 & 0 & 0.50 & 1 \end{bmatrix} \begin{bmatrix} x_1 \\ x_2 \\ x_3 \\ x_4 \end{bmatrix} = \begin{bmatrix} -14.0 \\ -30.0 \\ -21.4 \\ -50.0 \end{bmatrix}$$

the carry-over table is arranged in the following form.

	x_1	x_2	x_3	x_4
Carry-over factors	→ −0.30	← → −0.17 −0.29	← → −0.20 −0.50	← −0.21
Starting values	−14.000	−30.000	−21.400	−50.000
First iteration	+5.100		+8.70 +10.500	
Second iteration		+2.670 +0.440		+1.100
Third iteration	−0.529		−0.902 −0.230	
Fourth iteration		+0.159 +0.226		+0.566
Fifth iteration	−0.068		−0.112 −0.118	
Sixth iteration		+0.020 +0.046		+0.115
Final solution	−9.494	−26.430	−3.562	−48.219

The error of the tabulated values can be checked by the system as

$$x_1 = -14.000 - (0.17)(-26.430) = -9.5069$$

$$x_2 = -30.000 - (0.30)(-9.494) - (0.20)(-3.562) = -26.4394$$

$$x_3 = -21.400 - (0.29)(-26.430) - (0.21)(-48.219) = -3.6093$$

$$x_4 = -50.000 - (0.30)(-3.562) = -48.219$$

For better accuracy additional iterations may be included.

8
EIGENVALUE
EQUATIONS

(a) Matrix equation of the form,

$$
\underbrace{\begin{bmatrix} a_{11} - \lambda & a_{12} & \cdots & a_{1n} \\ a_{21} & a_{22} - \lambda & \cdots & a_{2n} \\ & & \cdots & \\ a_{n1} & a_{n2} & \cdots & a_{nn} - \lambda \end{bmatrix}}_{\mathbf{A} - \mathbf{I}\lambda} \underbrace{\begin{bmatrix} x_1 \\ x_2 \\ \cdots \\ x_n \end{bmatrix}}_{\mathbf{X}} = \underbrace{\begin{bmatrix} 0 \\ 0 \\ \cdots \\ 0 \end{bmatrix}}_{\mathbf{0}}
$$

where a_{ij} = given constants, λ, x_i = unknowns, and $i, j = 1, 2, 3, \ldots, n$, is called the eigenvalue matrix equation.

(b) Nontrivial solution of this homogeneous matrix equation exists, if and only if the determinant

$$|\mathbf{A} - \mathbf{I}\lambda| = 0$$

The expansion of this determinant is a polynomial of nth degree in λ which is known as the *characteristic equation* (eigenvalue equation) of **A**. The roots $\lambda_1, \lambda_2, \ldots, \lambda_n$ of this equation are called the *eigenvalues*.

(c) Eigenvalues of a hermitian matrix (normal matrix) are real, of an antihermitian matrix (antinormal matrix) are zero or pure imaginary, and of a unitary matrix (orthogonal matrix) are all equal to 1.

(d) Eigenvector. Corresponding to each eigenvalue λ_k there is a set of values of X_{ik} which form the eigenvector (column matrix)

$$\mathbf{X}_k = \{X_{1k}, X_{2k}, \ldots, X_{nk}\} \qquad (k = 1, 2, 3, \ldots, n)$$

Since the eigenvectors are the solution of a homogeneous system of equations, only the relative values of their components are determined, giving their direction but not their magnitude, so that

$$L_k = \sqrt{X_{1k}^2 + X_{2k}^2 + \cdots + X_{nk}^2}$$

is the *apparent vector length*.

(e) Numerical methods for the solution of eigenvalue problems fall into three categories: (1) *Direct methods*, (2) *transformation methods*, (3) *iterative methods*. The most important methods in each category are described in this chapter.

(f) Example. From the given matrix equation (a),

$$
\begin{bmatrix} 2 - \lambda & -1 & 0 \\ -1 & 2 - \lambda & -1 \\ 0 & -1 & 2 - \lambda \end{bmatrix} \begin{bmatrix} X_1 \\ X_2 \\ X_3 \end{bmatrix} = \begin{bmatrix} 0 \\ 0 \\ 0 \end{bmatrix}
$$

the eigenvalue determinant equation (b) is $\lambda^3 - 6\lambda^2 + 10\lambda - 4 = 0$, whose roots are $\lambda_1 = 2 + \sqrt{2} = 3.414\,214$, $\lambda_2 = 2$, $\lambda_3 = 2 - \sqrt{2} = 0.585\,786$.

For λ_1 with $X_{11} = 1$,

$$
\begin{bmatrix} -\sqrt{2} & -1 & 0 \\ -1 & -\sqrt{2} & -1 \\ 0 & -1 & -\sqrt{2} \end{bmatrix} \begin{bmatrix} 1 \\ X_{21} \\ X_{31} \end{bmatrix} = \begin{bmatrix} 0 \\ 0 \\ 0 \end{bmatrix}
$$

$$\mathbf{X}_1 = \{1, \sqrt{2}, 1\} \qquad L_1 = 2$$

For λ_2 with $X_{12} = 1$,

$$
\begin{bmatrix} 0 & -1 & 0 \\ -1 & 0 & -1 \\ 0 & -1 & -1 \end{bmatrix} \begin{bmatrix} 1 \\ X_{22} \\ X_{32} \end{bmatrix} = \begin{bmatrix} 0 \\ 0 \\ 0 \end{bmatrix}
$$

$$\mathbf{X}_2 = \{1, 0, -1\} \qquad L_2 = \sqrt{2}$$

and similarly for λ_3, with $X_{13} = 1$, $\mathbf{X}_3 = \{1, \sqrt{2}, 1\}$, $L_3 = 2$.

(a) First orthogonality condition. If X_i, X_j are two eigenvectors of order $[n \times 1]$ corresponding to two eigenvalues λ_i, λ_j, respectively, and L_i, L_j are their apparent lengths defined in (8.01d), then

$$X_i^T X_j = \begin{cases} 0 & \text{if } i \neq j \\ L_i^2 & \text{if } i = j \end{cases}$$

is called the first orthogonality condition.

(b) Second orthogonality condition. If A is a symmetrical matrix (normal matrix) of order $[n \times n]$, X_i, X_j, L_i, L_j and λ_i, λ_j are the same as in (a), then

$$X_i^T A X_j = \begin{cases} 0 & \text{if } i \neq j \\ L_i^2 \lambda_i & \text{if } i = j \end{cases}$$

is called the second orthogonality condition.

(c) First normalization condition. If the unit eigenvectors are defined in terms of (a) as

$$u_i = \frac{1}{L_i} X_i \qquad u_j = \frac{1}{L_j} X_j$$

then

$$u_i^T u_j = \begin{cases} 0 & \text{if } i \neq j \\ 1 & \text{if } i = j \end{cases}$$

is called the first normalization condition.

(d) Second normalization condition. If u_i, u_j are the same as in (c) above and A is the matrix defined in (b), then

$$u_i^T A u_j = \begin{cases} 0 & \text{if } i \neq j \\ \lambda_i & \text{if } i = j \end{cases}$$

is called the second normalization condition.

(e) Eigenvector spectra. The matrix containing all eigenvectors of A and the matrix containing all unit eigenvectors of A are called the natural eigenvector spectrum and the normalized eigenvector spectrum, respectively; they are

$$X = \begin{bmatrix} X_{11} & X_{12} & \cdots & X_{1n} \\ X_{21} & X_{22} & \cdots & X_{2n} \\ & & & \\ X_{n1} & X_{n2} & \cdots & X_{nn} \end{bmatrix} \qquad u = \begin{bmatrix} u_{11} & u_{12} & \cdots & u_{1n} \\ u_{12} & u_{22} & \cdots & u_{2n} \\ & & & \\ u_{n1} & u_{n2} & \cdots & u_{nn} \end{bmatrix}$$

(f) Example. For the matrix A in (8.01f).

$$X^T X = \begin{bmatrix} 4 & & \\ & 2 & \\ & & 4 \end{bmatrix} \qquad X^T A X = \begin{bmatrix} 4(2 + \sqrt{2}) & & \\ & 4 & \\ & & 4(2 - \sqrt{2}) \end{bmatrix}$$

$$u^T u = \begin{bmatrix} 1 & & \\ & 1 & \\ & & 1 \end{bmatrix} \qquad u^T A u = \begin{bmatrix} 2 + \sqrt{2} & & \\ & 2 & \\ & & 2 - \sqrt{2} \end{bmatrix}$$

The result of all congruent transformations defined in (a) to (d) is a diagonal matrix.

(a) Correspondence. The square matrix **B** is correspondent to another square matrix **A** of the same order if there is a nonsingular square matrix **C** such that

$$\mathbf{B} = \mathbf{CA}$$

(b) Contracorrespondence. The square matrix **B** is contracorrespondent to another square matrix **A** of the same order if there is a nonsingular matrix **C** such that

$$\mathbf{B} = \mathbf{C}^T\mathbf{A}$$

(c) Congruence. The square matrix **B** is congruent to another square matrix **A** of the same order if there is a nonsingular matrix **Q** such that

$$\mathbf{B} = \mathbf{Q}^T\mathbf{AQ}$$

This transformation is called the congruent transformation, which may take special form under special conditions.

(1) If **A** is a symmetrical matrix, then **B** is also a symmetrical matrix.
(2) If **A** is a unit matrix **I** and **B** is also a unit matrix **I**, then

$$\mathbf{Q}^T = \mathbf{Q}^{-1}$$

and **Q** is said to be an orthogonal matrix.
(3) If **B** is a diagonal matrix, such that

$$\text{Diag } \mathbf{B} = \mathbf{Q}^T\mathbf{AQ}$$

then the columns of **Q** are said to be orthogonal.
(4) If **B** is a diagonal matrix, the square roots of whose coefficients are real numbers, such that

$$\text{Diag } \mathbf{B} = \text{Diag } \mathbf{L} \text{ Diag } \mathbf{L}$$

and

$$\underbrace{\text{Diag } [1/L]\mathbf{Q}^T}_{\mathbf{E}^T}\mathbf{AQ}\underbrace{\text{Diag } [1/L]}_{\mathbf{E}} = \mathbf{I}$$

the matrix **E** is said to be normalized with respect to **A** and

$$\mathbf{E}^{-1} = \mathbf{E}^T\mathbf{A}$$

(d) Contracongruence. The square matrix **B** is contracongruent to another square matrix **A** if there is a nonsingular matrix **Q** such that

$$\mathbf{B} = \mathbf{QAQ}^T$$

(e) Similarity. The square matrix **B** is similar to another square matrix **A** if there is a nonsingular matrix **S** such that

$$\mathbf{B} = \mathbf{S}^{-1}\mathbf{AS} \qquad \text{or} \qquad \mathbf{A} = \mathbf{SBS}^{-1}$$

where **A** and **B** have the same eigenvalues and their respective unit vectors $\mathbf{u}_i$, $\mathbf{v}_i$ are related as

$$\mathbf{u}_i = \mathbf{Sv}_i \qquad \mathbf{v}_i = \mathbf{S}^{-1}\mathbf{u}_i$$

(f) Inverse matrices $\mathbf{A}^{-1}, \mathbf{B}^{-1}$ have the same unit eigenvectors $\mathbf{u}_i, \mathbf{v}_i$, but their respective eigenvalues are $1/\lambda_i$.

(g) Diagonal transformation. If **S** is the matrix of unit eigenvectors $\mathbf{S} = \mathbf{u}$, then the similarity transformation in (e) reduces **A** to a diagonal matrix

$$\mathbf{S}^{-1}\mathbf{AS} = \text{Diag } [\lambda_i]$$

(a) Gerschgorin's theorem. The largest eigenvalue in modulus of the square matrix $\mathbf{A}$ cannot exceed the largest sum of the moduli of the coefficients of any row or any column. Analytically, if

$$R_{\max} = \sum_{j=1}^{n} |a_{ij}| \qquad\qquad C_{\max} = \sum_{i=1}^{n} |a_{ij}|$$

then

$$|\lambda| \le R_{\max} \quad \text{for } R_{\max} \ge C_{\max} \qquad |\lambda_i| \le C_{\max} \quad \text{for } C_{\max} \ge R_{\max}$$

(b) Brauer's theorem. If P_k is the sum of all the moduli of the coefficients of one row or column excluding a_{kk}, then every eigenvalue of $\mathbf{A}$ lies inside on the boundary of at least one of the circles

$$|\lambda - a_{kk}| = P_k$$

The eigenvalues of $\mathbf{A}$ are the diagonal coefficients if $\mathbf{A}$ is a diagonal matrix $\mathbf{D}$ of a lower triangular matrix $\mathbf{L}$ or an upper triangular matrix $\mathbf{U}$.

(c) Cayley–Hamilton's theorem states that a matrix $\mathbf{A}$ satisfies its own characteristic equation, so that if

$$p(\lambda) = \lambda^n + p_1\lambda^{n-1} + p_2\lambda^{n-2} + \cdots + p_{n-1}\lambda^1 + p_n = 0$$

then

$$p(\mathbf{A}) = \mathbf{A}^n + p_1\mathbf{A}^{n-1} + p_2\mathbf{A}^{n-2} + \cdots + p_{n-1}\mathbf{A} + p_n\mathbf{I} = 0$$

where $p_1, p_2, \ldots, p_n$ are the unknown coefficients. With $\mathbf{Y}_0 = [1 \quad 0 \quad 0 \quad \cdots \quad 0]^T$,

$$\mathbf{Y}_1 = \mathbf{AY}_0, \mathbf{Y}_2 = \mathbf{AY}_1, \ldots, \mathbf{Y}_n = \mathbf{AY}_{n-1}$$

the polynomial matrix equation becomes

$$p_1\mathbf{Y}_{n-1} + p_2\mathbf{Y}_{n-2} + \cdots + p_{n-1}\mathbf{Y}_1 + p_n\mathbf{Y}_0 = -\mathbf{Y}_n$$

and yields the unknown coefficients p_i.

(d) Example. Using the matrix $\mathbf{A}$ of $(8.01f)$,

$$\mathbf{Y}_0 = \begin{bmatrix} 1 \\ 0 \\ 0 \end{bmatrix} \qquad \mathbf{Y}_1 = \mathbf{AY}_0 = \begin{bmatrix} 2 \\ -1 \\ 0 \end{bmatrix} \qquad \mathbf{Y}_2 = \mathbf{AY}_1 = \begin{bmatrix} 5 \\ -4 \\ 1 \end{bmatrix} \qquad \mathbf{Y}_3 = \mathbf{AY}_2 = \begin{bmatrix} 14 \\ -14 \\ 6 \end{bmatrix}$$

from which

$$\begin{bmatrix} 5 & 2 & 1 \\ -4 & -1 & 0 \\ 1 & 0 & 0 \end{bmatrix} \begin{bmatrix} p_1 \\ p_2 \\ p_3 \end{bmatrix} = \begin{bmatrix} -14 \\ 14 \\ -6 \end{bmatrix}$$

The solution of this set yields the coefficients $p_1 = -6, p_2 = 10, p_3 = -4$, and

$$p(\lambda) = \lambda^3 - 6\lambda^2 + 10\lambda - 4 = 0$$

The roots of this cubic equation, calculated in $(8.01f)$, $\lambda_1 = 2 + \sqrt{2}, \lambda_2 = 2, \lambda_3 = 2 - \sqrt{2}$, must satisfy the conditions stated in (a) and (b) above.

(a) Iteration procedure called the power method is the most widely used process for the estimation of the largest eigenvalue of $\mathbf{A}$. Assuming that $\mathbf{A}$ is real, and

$$|\lambda_1| \geq |\lambda_2| \geq |\lambda_3| \geq \cdots \geq |\lambda_n|$$

the iteration sequence of eigenvectors is taken as

$$\mathbf{X}_1^{(1)} = \mathbf{A}\mathbf{X}_1^{(0)}, \mathbf{X}_1^{(2)} = \mathbf{A}\mathbf{X}_1^{(1)}, \ldots, \mathbf{X}_1^{(k)} = \mathbf{A}\mathbf{X}_1^{(k-1)}$$

where

$$\mathbf{X}_1^{(0)} = \{X_{1,1}^{(0)}, X_{1,2}^{(0)}, \ldots, X_{1,n}^{(0)}\}$$

is an arbitrary initial vector, at least one component of which is nonzero (usually, the initial vector is taken with all components equal to unity). Using the iteration sequence defined above, the process stops when the unit vectors

$$\mathbf{u}_1^{(k)} \cong \mathbf{u}_1^{(k-1)}$$

If this condition is satisfied, the largest eigenvalue is

$$\boxed{\lambda_1 \cong X_{1,1}^{(k)}/X_{1,1}^{(k-1)} \cong X_{1,2}^{(k)}/X_{1,2}^{(k+1)} \cong \cdots \cong X_{1,n}^{(k)}/X_{1,n}^{(k+1)}}$$

and the corresponding eigenvector is

$$\mathbf{X}_1 \cong \mathbf{X}_1^{(k)}$$

This process is well suitable for matrices of higher order.

(b) Example. Using the matrix $\mathbf{A}$ in $(8.01f)$ and selecting the arbitrary initial vector $\mathbf{X}_1^{(0)}$, the iteration process becomes

$$\mathbf{A}\mathbf{X}_1^{(0)} = \mathbf{X}_1^{(1)} \rightarrow \begin{bmatrix} 2 & -1 & 0 \\ -1 & 2 & -1 \\ 0 & -1 & 2 \end{bmatrix} \begin{bmatrix} 1 \\ -1 \\ 1 \end{bmatrix} = \begin{bmatrix} 3 \\ -4 \\ 3 \end{bmatrix} \qquad \mathbf{u}_1^{(1)} = \begin{bmatrix} 0.514\,496 \\ -0.685\,994 \\ 0.514\,496 \end{bmatrix}$$

$$\mathbf{A}\mathbf{X}_1^{(1)} = \mathbf{X}_1^{(2)} \rightarrow \begin{bmatrix} 2 & -1 & 0 \\ -1 & 2 & -1 \\ 0 & -1 & 2 \end{bmatrix} \begin{bmatrix} 3 \\ -4 \\ 3 \end{bmatrix} = \begin{bmatrix} 10 \\ -14 \\ 10 \end{bmatrix} \qquad \mathbf{u}_1^{(2)} = \begin{bmatrix} 0.502\,519 \\ -0.703\,527 \\ 0.502\,519 \end{bmatrix}$$

$$\mathbf{A}\mathbf{X}_1^{(2)} = \mathbf{X}_1^{(3)} \rightarrow \begin{bmatrix} 2 & -1 & 0 \\ -1 & 2 & -1 \\ 0 & -1 & 2 \end{bmatrix} \begin{bmatrix} 34 \\ -14 \\ 10 \end{bmatrix} = \begin{bmatrix} 34 \\ -48 \\ 34 \end{bmatrix} \qquad \mathbf{u}_1^{(3)} = \begin{bmatrix} 0.500\,433 \\ -0.706\,494 \\ 0.500\,433 \end{bmatrix}$$

The last two iteration cycles yield

$X_{1,j}^{(7)}$	$u_{1,j}^{(7)}$	$X_{1,j}^{(7)}/X_{1,j}^{(6)}$	$X_{1,j}^{(8)}$	$u_{1,j}^{(8)}$	$X_{1,j}^{(8)}/X_{1,j}^{(7)}$
15 760	0.500 000	3.414 211	53 808	0.500 000	3.414 213
−22 288	0.707 107	3.414 215	−76 096	0.707 107	3.414 214
15 760	0.500 000	3.414 215	53 808	0.500 000	3.414 213

From this table,

$$\lambda_1 \cong X_{1,1}^{(8)}/X_{1,1}^{(7)} \cong X_{1,2}^{(8)}/X_{1,2}^{(7)} \cong X_{1,3}^{(8)}/X_{1,3}^{(7)} \cong 3.414\,214$$

$$\mathbf{u}_1 \cong \mathbf{u}_1^{(8)} = \{0.500\,000, -0.707\,107, 0.500\,000\}$$

which is the same result as in $(8.01f)$.

(a) Iteration procedure called the inverse power method is usually more powerful than the power method and yields the smallest eigenvalue of $\mathbf{A}$. Assuming $\mathbf{A}$ is real, and

$$|\lambda_n| \le |\lambda_{n-1}| \le \cdots \le |\lambda_2| \le |\lambda_1|$$

the iteration sequence of eigenvectors is taken as

$$\mathbf{Y}_n^{(1)} = \mathbf{A}^{-1}\mathbf{Y}_n^{(0)}, \ \mathbf{Y}_n^{(2)} = \mathbf{A}^{-1}\mathbf{Y}_n^{(1)}, \dots, \mathbf{Y}_n^{(k)} = \mathbf{A}^{-1}\mathbf{Y}_n^{(k-1)}$$

where

$$\mathbf{Y}_n^{(0)} = \{Y_{n,1}^{(0)}, Y_{n,2}^{(0)}, \dots, Y_{n,n}^{(0)}\}$$

is again an arbitrary initial vector, at least one component of which is nonzero (usually, the initial vector is taken with all components equal to unity), and $\mathbf{A}^{-1}$ is the inverse of $\mathbf{A}$ in (8.01*a*). Using the iteration sequence defined above, the process stops when the unit vectors

$$\mathbf{v}_n^{(k)} \cong \mathbf{v}_n^{(k-1)}$$

If this condition is satisfied, the smallest eigenvalue is

$$\boxed{\lambda_n \cong Y_{n,1}^{(k-1)}/Y_{n,1}^{(k)} \cong Y_{n,2}^{(k-1)}/Y_{n,2}^{(k)} \cong \cdots \cong Y_{n,n}^{(k-1)}/Y_{n,n}^{(k)}}$$

and the corresponding eigenvector is

$$\mathbf{Y}_n \cong \mathbf{Y}_n^{(k)}$$

This process is also suitable for matrices of higher order.

(b) Example. Using the inverse of the matrix $\mathbf{A}$ in (8.01*f*) and selecting the arbitrary initial vector $\mathbf{Y}_n^{(0)}$, the iteration sequence becomes

$$\mathbf{A}^{-1}\mathbf{Y}^{(0)} = \mathbf{Y}_3^{(1)} \rightarrow \frac{1}{4}\begin{bmatrix} 3 & 2 & 1 \\ 2 & 4 & 2 \\ 1 & 2 & 3 \end{bmatrix}\begin{bmatrix} 1 \\ 1 \\ 1 \end{bmatrix} = \frac{1}{4}\begin{bmatrix} 8 \\ 8 \\ 6 \end{bmatrix} \qquad \mathbf{v}_n^{(1)} = \begin{bmatrix} 0.514\,496 \\ 0.685\,994 \\ 0.514\,496 \end{bmatrix}$$

$$\mathbf{A}^{-1}\mathbf{Y}_3^{(1)} = \mathbf{Y}_3^{(2)} \rightarrow \frac{1}{16}\begin{bmatrix} 3 & 2 & 1 \\ 2 & 4 & 2 \\ 1 & 2 & 3 \end{bmatrix}\begin{bmatrix} 6 \\ 8 \\ 6 \end{bmatrix} = \frac{1}{16}\begin{bmatrix} 40 \\ 56 \\ 40 \end{bmatrix} \qquad \mathbf{v}_n^{(2)} = \begin{bmatrix} 0.502\,519 \\ 0.703\,526 \\ 0.502\,519 \end{bmatrix}$$

$$\mathbf{A}^{-1}\mathbf{Y}_3^{(2)} = \mathbf{Y}_3^{(3)} \rightarrow \frac{1}{64}\begin{bmatrix} 3 & 2 & 1 \\ 2 & 4 & 2 \\ 1 & 2 & 3 \end{bmatrix}\begin{bmatrix} 40 \\ 56 \\ 40 \end{bmatrix} = \frac{1}{64}\begin{bmatrix} 272 \\ 384 \\ 272 \end{bmatrix} \qquad \mathbf{v}_n^{(3)} = \begin{bmatrix} 0.500\,430 \\ 0.706\,494 \\ 0.500\,430 \end{bmatrix}$$

The last two iteration cycles yield,

$Y_{n,j}^{(7)}$	$v_{n,j}^{(7)}$	$Y_{n,j}^{(6)}/Y_{n,j}^{(7)}$	$Y_{n,j}^{(8)}$	$v_{n,j}^{(8)}$	$Y_{n,j}^{(7)}/Y_{n,j}^{(8)}$
61.573 77	0.500 000	0.585 787	105.113.0	0.500 000	0.585 786
87.078 44	0.707 107	0.585 786	148.652 2	0.707 107	0.585 786
61.573 77	0.500 000	0.585 787	105.113 0	0.500 000	0.585 786

From this table,

$$\lambda_3 \cong Y_{1,3}^{(7)}/Y_{1,3}^{(8)} \cong Y_{2,3}^{(7)}/Y_{2,3}^{(8)} \cong Y_{3,3}^{(7)}/Y_{3,3}^{(8)} \cong 0.585\,786$$

$$\mathbf{v}_3 \cong \mathbf{v}_3^{(8)} \cong \{0.500\,000, 0.707\,107, 0.500\,000\}$$

which is the same result as in (8.01*f*).

(a) Transformation of the eigenvalue determinant equation in (8.01a) to the reduced determinant equation proposed by Krylov,

$$|\mathbf{K}(\lambda)| = \begin{vmatrix} k_{11} - \lambda & k_{12} & \cdots & k_{1n} \\ k_{21} - \lambda^2 & k_{22} & \cdots & k_{2n} \\ \multicolumn{4}{c}{\dotfill} \\ k_{n1} - \lambda^n & k_{n2} & \cdots & k_{nn} \end{vmatrix} = 0$$

leads to the direct formulation of the eigenvalue polynomial equation in λ,

$$\boxed{(k_{11} - \lambda)C_{11} + (k_{21} - \lambda^2)C_{21} + \cdots + (k_{n1} - \lambda^n)C_{n1} = 0}$$

where $C_{11}, C_{21}, \ldots, C_{n1}$ are the cofactors of $|\mathbf{K}(\lambda)|$ and $k_{11}, k_{12}, \ldots, k_{nn}$ are the coefficients given below.

(b) k coefficients are calculated by the following recurrent formulas:

$$[k_{11} \quad k_{12} \quad \cdots \quad k_{1n}] = [a_{11} \quad a_{12} \quad \cdots \quad a_{1n}]$$

$$[k_{21} \quad k_{22} \quad \cdots \quad k_{2n}] = [k_{11} \quad k_{12} \quad \cdots \quad k_{1n}] \begin{bmatrix} a_{11} & a_{12} & \cdots & a_{1n} \\ a_{21} & a_{22} & \cdots & a_{2n} \\ \multicolumn{4}{c}{\dotfill} \\ a_{n1} & a_{n2} & \cdots & a_{nn} \end{bmatrix}$$

$$[k_{n1} \quad k_{n2} \quad \cdots \quad k_{nn}] = [k_{n-1,1} \quad k_{n-1,2} \quad \cdots \quad k_{n-1,n}] \begin{bmatrix} a_{11} & a_{12} & \cdots & a_{1n} \\ a_{21} & a_{22} & \cdots & a_{2n} \\ \multicolumn{4}{c}{\dotfill} \\ a_{n1} & a_{n2} & \cdots & a_{nn} \end{bmatrix}$$

(c) Example. Using the matrix equation in (8.01f), k coefficients are

$$[k_{11} \quad k_{12} \quad k_{13}] = [2 \quad -1 \quad 0]$$

$$[k_{21} \quad k_{22} \quad k_{33}] = [2 \quad -1 \quad 0] \begin{bmatrix} 2 & -1 & 0 \\ -1 & 2 & -1 \\ 0 & -1 & 2 \end{bmatrix} = [5 \quad -4 \quad 1]$$

$$[k_{31} \quad k_{32} \quad k_{33}] = [5 \quad -4 \quad 1] \begin{bmatrix} 2 & -1 & 0 \\ -1 & 2 & -1 \\ 0 & -1 & 2 \end{bmatrix} = [14 \quad -14 \quad 6]$$

and $\quad |\mathbf{K}(\lambda)| = \begin{vmatrix} 2 - \lambda & -1 & 0 \\ 5 - \lambda^2 & -4 & 1 \\ 14 - \lambda^3 & -14 & 6 \end{vmatrix} = 0$

in which

$$C_{11} = (-1)^2 \begin{bmatrix} -4 & 1 \\ -14 & 6 \end{bmatrix} = 10 \qquad C_{21} = (-1)^3 \begin{bmatrix} -1 & 0 \\ -14 & 6 \end{bmatrix} = 6 \qquad C_{31} = (-1)^4 \begin{bmatrix} -10 \\ -41 \end{bmatrix} = -1$$

The eigenvalue equation in λ is then

$$(2 - \lambda)(-10) + (5 - \lambda^2)(6) + (14 - \lambda^3)(-1) = 0$$

and reduces to

$$\lambda^3 - 6\lambda^2 + 10\lambda - 4 = 0$$

which is the same result as in (8.01f).

(a) Transformation of the eigenvalue determinant equation in (8.01a) to the reduced determinant equation proposed by Danielvsky,

$$|D - I\lambda| = \begin{vmatrix} d_{11} - \lambda & d_{12} & d_{13} & \cdots & d_{1n} \\ 1 & -\lambda & 0 & \cdots & 0 \\ 0 & 1 & -\lambda & \cdots & 0 \\ \cdots & \cdots & \cdots & \cdots & \cdots \\ 0 & 0 & 1 & \cdots & -\lambda \end{vmatrix} = 0$$

leads to the direct formulation of the eigenvalue polynomial equation

$$\lambda^n - d_{11}\lambda^{n-1} - d_{12}\lambda^{n-2} - \cdots - d_{1n} = 0$$

where $d_1, d_2, \ldots, d_n$ are the coefficients of the first row.

(b) Sequence of matrix operations required for this transformation is illustrated below.

(1) *First reduction*

$$B = M_4 A M_4^{-1} = \begin{bmatrix} b_{11} & b_{12} & b_{13} & b_{14} \\ b_{21} & b_{22} & b_{23} & b_{24} \\ b_{31} & b_{32} & b_{33} & b_{34} \\ 0 & 0 & 1 & 0 \end{bmatrix}$$

where **A** is given in (8.01a) and

$$M_4 = \begin{bmatrix} 0 & 0 & 0 & 0 \\ 0 & 1 & 0 & 0 \\ a_{41} & a_{42} & a_{43} & a_{44} \\ 0 & 0 & 0 & 1 \end{bmatrix} \qquad M_4^{-1} = \begin{bmatrix} 1 & 0 & 0 & 0 \\ 0 & 1 & 0 & 0 \\ -a_{41}/a_{43} & -a_{42}/a_{43} & 1/a_{43} & -a_{44}/a_{43} \\ 0 & 0 & 0 & 1 \end{bmatrix}$$

(2) *Second reduction*

$$C = M_3 B M_3 = \begin{bmatrix} c_{11} & c_{12} & c_{13} & c_{14} \\ c_{21} & c_{22} & c_{23} & c_{24} \\ 0 & 1 & 0 & 0 \\ 0 & 0 & 1 & 0 \end{bmatrix}$$

where **B** is given in (1) above and

$$M_3 = \begin{bmatrix} 1 & 0 & 0 & 0 \\ b_{31} & b_{32} & b_{33} & b_{34} \\ 0 & 0 & 1 & 0 \\ 0 & 0 & 0 & 1 \end{bmatrix} \qquad M_3^{-1} = \begin{bmatrix} 1 & 0 & 0 & 0 \\ -b_{31}/b_{32} & 1/b_{32} & -b_{33}/b_{32} & -b_{34}/b_{32} \\ 0 & 0 & 1 & 0 \\ 0 & 0 & 0 & 1 \end{bmatrix}$$

(3) *Last reduction*

$$D = M_2 C M_2^{-1} = \begin{bmatrix} d_{11} & d_{12} & d_{13} & d_{14} \\ 1 & 0 & 0 & 0 \\ 0 & 1 & 0 & 0 \\ 0 & 0 & 1 & 0 \end{bmatrix}$$

where **C** is given in (2) above and

$$M_2 = \begin{bmatrix} c_{21} & c_{22} & c_{23} & c_{24} \\ 0 & 1 & 0 & 0 \\ 0 & 0 & 1 & 0 \\ 0 & 0 & 0 & 1 \end{bmatrix} \qquad M_2^{-1} = \begin{bmatrix} 1/c_{21} & -c_{22}/c_{21} & -c_{23}/c_{21} & -c_{24}/c_{21} \\ 0 & 1 & 0 & 0 \\ 0 & 0 & 1 & 0 \\ 0 & 0 & 0 & 1 \end{bmatrix}$$

(a) Direction cosines. Since the discovery of the existence of principal axes by Euler, the direction cosines of these axes have been calculated by the simultaneous solution of three homogeneous linear equations,

$$
\begin{bmatrix}
T_{xx} - T_p & T_{xy} & T_{xz} \\
T_{yx} & T_{yy} - T_p & T_{yz} \\
T_{zx} & T_{zy} & T_{zz} - T_p
\end{bmatrix}
\begin{bmatrix}
\alpha_p \\
\beta_p \\
\gamma_p
\end{bmatrix}
=
\begin{bmatrix}
0 \\
0 \\
0
\end{bmatrix}
$$

and one quadratic equation,

$$\alpha_p^2 + \beta_p^2 + \gamma_p^2 = 1$$

where $\alpha_p, \beta_p, \gamma_p$ are the direction cosines of the principal axes, T_p is the eigenvalue, and the remaining terms are the components of the tensor $\mathbf{T}$. As $p = 1, 2, 3$, three sets of solutions are required, each yielding six roots, from which the three applicable ones must be selected by means of the orthogonality condition. In actual engineering problems, where small and large values T_{ij} are involved, the traditional procedure becomes quite tedious for hand calculations and is not very efficient for machine computations (Ref. 8.02).

(b) Geometric method. Physically, the matrix equation in (a) represents a bundle of planes in $\alpha_p, \beta_p, \gamma_p$ coordinates, whose line of intersection is the principal axis. The direction normals of these planes are in the cartesian vector form,

$$\mathbf{N}_{1p} = (T_{xx} - T_p)\mathbf{i} + T_{xy}\mathbf{j} + T_{xz}\mathbf{k}$$

$$\mathbf{N}_{2p} = T_{yx}\mathbf{i} + (T_{yy} - T_p)\mathbf{j} + T_{yz}\mathbf{k}$$

$$\mathbf{N}_{3p} = T_{zx}\mathbf{i} + T_{zy}\mathbf{j} + (T_{zz} - T_p)\mathbf{k}$$

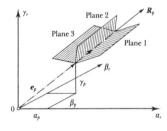

and the vector of the principal axis $\mathbf{R}_p$ must be normal to $\mathbf{N}_{1p}, \mathbf{N}_{2p}, \mathbf{N}_{3p}$ so that

$$\mathbf{R}_p = \mathbf{N}_{1p} \times \mathbf{N}_{2p} = \mathbf{N}_{2p} \times \mathbf{N}_{3p} = \mathbf{N}_{3p} \times \mathbf{N}_{1p}$$

Thus the vector product of any two rows of $[\mathbf{T} - \lambda \mathbf{T}_p]$ selected in cyclic order gives the vector of the respective principal axis.

(c) Unit vector of this axis is, then,

$$\mathbf{e}_p = \mathbf{R}_p / R_p = \alpha_p \mathbf{i} + \beta_p \mathbf{j} + \gamma_p \mathbf{k}$$

where R_p is the vector length which as a scalar is always positive and $\alpha_p, \beta_p, \gamma_p$ are the desired direction cosines. Their closed algebraic form is then

$$\alpha_p = \cos(R_p, X) = \frac{a_p}{d_p} \qquad \beta_p = \cos(R_p, Y) = -\frac{b_p}{d_p} \qquad \gamma_p = \cos(R_p, Z) = \frac{c_p}{d_p}$$

where in terms of $\mathbf{N}_{1p} \times \mathbf{N}_{2p}$,

$$
a_p = \begin{vmatrix} T_{xy} & T_{xz} \\ T_{yy} - T_p & T_{yz} \end{vmatrix}
\qquad
b_p = \begin{vmatrix} T_{xx} - T_p & T_{xz} \\ T_{yx} & T_{yz} \end{vmatrix}
\qquad
c_p = \begin{vmatrix} T_{xx} - T_p & T_{xy} \\ T_{yx} & T_{yy} - T_p \end{vmatrix}
$$

$$d_p = +\sqrt{a_p^2 + b_p^2 + c_p^2}$$

This geometric solution bypasses the solution of the simultaneous equations, prevents the large number of round-off errors, and yields the magnitudes of direction cosines and their correct signs automatically by simple arithmetic.

(a) Statement of problem. The principal stresses and direction cosines of principal axes of the stress field given by the normal stresses in ksi as

$$s_{xx} = 10 \text{ ksi} \qquad s_{yy} = -5 \text{ ksi} \qquad s_{zz} = -5 \text{ ksi}$$

and by the tangential stresses

$$t_{xy} = 2 \text{ ksi} \qquad t_{yz} = 2 \text{ ksi} \qquad t_{zx} = 0$$

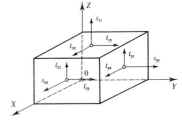

shown in the adjacent figure are calculated by the geometric method (8.09*b*).

(b) Principal stresses are the roots of the eigenvalue determinant (8.09*a*),

$$\begin{vmatrix} 10 - s_p & 2 & 0 \\ 2 & -5 - s_p & 2 \\ 0 & 2 & -5 - s_p \end{vmatrix} = 0 \qquad (p = 1, 2, 3)$$

which reduces to

$$s_p^3 - 83 s_p - 230 = 0$$

and yields $s_1 = 10.266\,59$ ksi, $s_2 = -7.120\,23$ ksi, and $s_3 = -3.146\,36$ ksi.

(c) Direction normals and direction cosines defined in (8.09*b*), (8.09*c*) are:

(1) *In terms of* s_1, $a_1 = 4$

$\quad \mathbf{N}_{11} = -0.266\,59\mathbf{i} + 2\mathbf{j}$ $\quad b_1 = 5.331\,80 \times 10^{-1}$ $\quad \alpha_1 = 9.910\,80 \times 10^{-1}$

$\quad \mathbf{N}_{21} = 2\mathbf{i} - 15.266\,59\mathbf{j} + 2\mathbf{k}$ $\quad c_1 = 6.992\,02 \times 10^{-2}$ $\quad \beta_1 = 1.321\,07 \times 10^{-1}$

$\quad \mathbf{N}_{31} = 2\mathbf{j} - 15.266\,59\mathbf{k}$ $\quad d_1 = 4.035\,98$ $\quad \gamma_1 = 1.732\,42 \times 10^{-2}$

(2) *In terms of* s_2, $a_2 = 4$

$\quad \mathbf{N}_{12} = 17.120\,23\mathbf{i} + 2\mathbf{j}$ $\quad b_2 = 3.424\,05 \times 10$ $\quad \alpha_2 = 8.467\,39 \times 10^{-2}$

$\quad \mathbf{N}_{22} = 2\mathbf{i} + 2.120\,23\mathbf{j} + 2\mathbf{k}$ $\quad c_2 = 3.229\,88 \times 10$ $\quad \beta_2 = -7.248\,18 \times 10^{-1}$

$\quad \mathbf{N}_{32} = 2\mathbf{j} + 2.120\,23\mathbf{k}$ $\quad d_2 = 4.724\,01 \times 10$ $\quad \gamma_2 = 6.837\,17 \times 10^{-1}$

(3) *In terms of* s_3, $a_3 = 4$

$\quad \mathbf{N}_{13} = 13.146\,36\mathbf{i} + 2\mathbf{j}$ $\quad b_3 = 2.629\,27 \times 10$ $\quad \alpha_3 = 1.028\,66 \times 10^{-1}$

$\quad \mathbf{N}_{23} = 2\mathbf{i} - 1.853\,65\mathbf{j} + 2\mathbf{k}$ $\quad c_3 = 2.836\,87 \times 10$ $\quad \beta_3 = -6.761\,56 \times 10^{-1}$

$\quad \mathbf{N}_{33} = 2\mathbf{j} - 1.853\,65\mathbf{k}$ $\quad d_3 = 3.888\,56 \times 10$ $\quad \gamma_3 = -7.295\,54 \times 10^{-1}$

(d) Numerical check can be performed by

$$\mathbf{e}_1 \cdot \mathbf{e}_2 = \alpha_1 \alpha_2 + \beta_1 \beta_2 + \gamma_1 \gamma_2 = 0 \qquad (\epsilon = 1.06 \times 10^{-5})$$

$$\mathbf{e}_2 \cdot \mathbf{e}_3 = \alpha_2 \alpha_3 + \beta_2 \beta_3 + \gamma_2 \gamma_3 = 0 \qquad (\epsilon = 8.37 \times 10^{-6})$$

$$\mathbf{e} \cdot \mathbf{e}_1 = \alpha_3 \alpha_1 + \beta_3 \beta_1 + \gamma_3 \gamma_1 = 0 \qquad (\epsilon = 1.54 \times 10^{-5})$$

and also by any of the equations in (8.09*a*), such as

$$(s_{xx} - s_1)\alpha_1 + t_{xy}\beta_1 + t_{xz}\gamma = 0 \qquad (\epsilon = 1.98 \times 10^{-6})$$

$$t_{yx}\alpha_2 + (s_{yy} - s_2)\beta_2 + t_{yz}\gamma_2 = 0 \qquad (\epsilon = 6.43 \times 10^{-7})$$

$$t_{zx}\alpha_3 + t_{zy}\beta_3 + (s_{zz} - s_3)\gamma_3 = 0 \qquad (\epsilon = 1.47 \times 10^{-6})$$

(a) Difference matrix equation defined as

$$
\begin{bmatrix}
a - \lambda & -1 \\
-1 & a - \lambda & -1 \\
 & -1 & a - \lambda & -1 \\
 & & \cdot & \cdot & \cdot \\
 & & & \cdot & \cdot & \cdot \\
 & & & & \cdot & \cdot & \cdot \\
 & & & & & -1 & a - \lambda & -1 \\
 & & & & & & -1 & a - \lambda
\end{bmatrix}
\begin{bmatrix}
x_1 \\ x_2 \\ x_3 \\ \cdot \\ \cdot \\ \cdot \\ x_{n-1} \\ x_n
\end{bmatrix}
=
\begin{bmatrix}
0 \\ 0 \\ 0 \\ \cdot \\ \cdot \\ \cdot \\ 0 \\ 0
\end{bmatrix}
$$

yields the closed-form eigenvalues

$$\lambda_k = a - 2 \cos \phi_k \qquad (k = 1, 2, \ldots, n)$$

where

$$\phi_k = \frac{k\pi}{n + 1}$$

(b) Eigenvector corresponding to λ_k is

$$\mathbf{X}_k = \{\sin \phi_k, \sin 2\phi_k, \sin 3\phi_k, \ldots, \sin n\phi_k\}$$

and its components are independent of a.

(c) Example. For the matrix equation

$$
\begin{bmatrix}
4 - \lambda & -1 \\
-1 & 4 - \lambda & -1 \\
 & -1 & 4 - \lambda & -1 \\
 & & -1 & 4 - \lambda & -1 \\
 & & & -1 & 4 - \lambda
\end{bmatrix}
\begin{bmatrix}
x_1 \\ x_2 \\ x_3 \\ x_4 \\ x_5
\end{bmatrix}
=
\begin{bmatrix}
0 \\ 0 \\ 0 \\ 0 \\ 0
\end{bmatrix}
$$

the eigenvalues in terms of $k = 1, 2, \ldots, 5$, $n = 5$,

$$\phi_1 = 30° \qquad \phi_2 = 60° \qquad \phi_3 = 90° \qquad \phi_4 = 120° \qquad \phi_5 = 150°$$

are

$$\lambda_1 = 4 - \sqrt{3} \qquad \lambda_2 = 3 \qquad \lambda_3 = 4 \qquad \lambda_4 = 5 \qquad \lambda_5 = 4 + \sqrt{3}$$

and the eigenvectors in terms of the same ϕ_k are

$$
\begin{bmatrix}
\frac{1}{2} & \frac{1}{2}\sqrt{3} & 1 & \frac{1}{2}\sqrt{3} & \frac{1}{2} \\
\frac{1}{2}\sqrt{3} & \frac{1}{2}\sqrt{3} & 0 & -\frac{1}{2}\sqrt{3} & -\frac{1}{2}\sqrt{3} \\
1 & 0 & -1 & 0 & 1 \\
\frac{1}{2}\sqrt{3} & \frac{1}{2}\sqrt{3} & 0 & \frac{1}{2}\sqrt{3} & -\frac{1}{2}\sqrt{3} \\
\frac{1}{2} & \frac{1}{2}\sqrt{3} & 1 & -\frac{1}{2}\sqrt{3} & \frac{1}{2}
\end{bmatrix}
$$

which shows the independence of eigenvectors of $a = 4$.

9
SERIES OF
FUNCTIONS

(a) Sequence of functions in x is a set of n functions $f_1(x), f_2(x), \ldots, f_n(x)$ arranged in a prescribed order and formed according to a definite rule. The sequence is finite if $n < \infty$ or infinite if $n = \infty$.

(b) Finite series of functions is the sum of finite sequence defined in $a \leq x \leq b$ as

$$\sum_{k=1}^{n} f_k(x) = f_1(x) + f_2(x) + \cdots + f_n(x) = S_n(x)$$

where $f_k(x)$ is the kth function, $n < \infty$, and $S_n(x)$ is the sum.

(c) Infinite series of functions is the sum of infinite sequence defined in $a \leq x \leq b$ as

$$\sum_{k=1}^{\infty} f_k(x) = f_1(x) + f_2(x) - \cdots = S(x)$$

where $n = \infty$ and the sum $S(x)$ may or may not exist (see below).

(d) Region of convergence of the series in (c) above is the set of values of the argument x for which the series converges.. That is, if

$$S_n(x) = \sum_{k=1}^{n} f_k(x) \qquad \text{and} \qquad \lim_{n \to \infty} S(x) = S(x)$$

exists for $a \leq x \leq b$, then the series is said to be convergent at all points of this closed interval. In some cases, the region of convergence is an open interval $a < x < b$, which indicates that at the endpoints $x = a$, $x = b$ the series is divergent.

(e) Uniformly convergent series. If for any preassigned positive number ϵ, however small, there exists an integer N such that

$$|S(x) - S_n(x)| < \epsilon$$

for all $n > N$, then the series is uniformly convergent, and if the sum of absolute values of the same functions for any x in the region of convergence is convergent, the series is also *absolutely convergent.*

(f) Continuity. If the functions of the uniformly convergent series are continuous in the region of convergence of the series, then their sum is also a continuous function in the same region.

(g) Differentiability and integrability. A uniformly convergent series of functions can be differentiated and/or integrated term by term in the region of its convergence, and the sum of its derivatives and/or the sum of its integrals is equal to the derivative and/or is equal to the integral of its sum, respectively, so that

$$\frac{d}{dx}\left[\sum_{k=1}^{\infty} f_k(x)\right] = \sum_{k=1}^{\infty}\left[\frac{df_k(x)}{dx}\right] \qquad \int_0^x \left[\sum_{k=1}^{\infty} f_k(x)\right] = \sum_{k=1}^{\infty}\left[\int_0^x f_k(x)\, dx\right]$$

in $a \leq x \leq b$.

(h) Remainder. The difference between $S(x)$ and $S_n(x)$ of a uniformly convergent series is the remainder R_{n+1},

$$R_{n+1} = S(x) - S_n(x) = f_{n+1}(x) + f_{n+2}(x) + \cdots$$

as shown in (9.03).

(a) Power series in the real (or complex) variable x is in the standard form

$$\sum_{k=0}^{n} a_k x^k = a_0 + a_1 x + a_2 x^2 + \cdots + a_n x^n$$

where $a_0, a_1, a_2, \ldots, a_n$ are real or complex numbers independent of x and $n < \infty$ for a finite series or $n = \infty$ for an infinite series. Two *nested forms* are shown in (1.07), and the application of nesting is presented below.

(b) Example

$$\sum_{k=0}^{n} \frac{x^k}{k!} = 1 + \frac{x}{1!} + \frac{x^2}{2!} + \frac{x^3}{3!} + \cdots + \frac{x^n}{n!}$$

$$= 1 + x\left(\frac{1}{1!} + x\left(\frac{1}{2!} + x\left(\frac{1}{3!} + \cdots + x\right)\right)\right) = \overset{n}{\underset{k=0}{\Lambda}}\left[\frac{1}{k!} + x\right]$$

$$= 1 + \frac{x}{1}\left(1 + \frac{x}{2}\left(1 + \frac{x}{3}\left(1 + \cdots + \frac{x}{n}\right)\right)\right) = \overset{n}{\underset{k=1}{\Lambda}}\left[1 + \frac{x}{k}\right]$$

(c) Convergence and divergence. The power series in (a) is convergent if

$$\lim_{n\to\infty}\left|\frac{a_{n+1}x^{n+1}}{a_n x^n}\right| = \lim_{n\to\infty}\left|\frac{a_{n+1}}{a_n}\right| |x| = \frac{1}{r\,|x|} < 1$$

and is divergent if

$$\lim_{n\to\infty}\left|\frac{a_{n+1}x^{n+1}}{a_n x^n}\right| = \lim_{n\to\infty}\left|\frac{a_{n+1}}{a_n}\right| |x| = \frac{1}{r\,|x|} > 1$$

where r is a number (including 0).

(d) Interval of convergence of the power series in (a) is

$$-r < x < r$$

where $r = \lim_{n\to\infty}\left|\frac{a_n}{a_{n+1}}\right|$

and r is said to be its *radius of convergence*. The series is convergent in this interval and diverges outside this interval. It may or may not converge at the endpoints of this interval.

(e) Examples. For the series in (b),

$$r = \lim_{n\to\infty}\left|\frac{a_n}{a_{n+1}}\right| = \lim_{n\to\infty}\left|\frac{1/n!}{1/(n+1)!}\right| = |\infty| \qquad (-\infty < x < \infty)$$

(f) Uniform and absolute convergence. The power series which converges in the interval $a < x < b$ converges absolutely and uniformly for every value of x within this interval.

(g) Differentiability and integrability. A power series can be differentiated and integrated term by term in any closed interval if and only if this interval lies entirely within the interval of uniform convergence of the power series.

(a) Standard Taylor's series. If the function $y = f(x)$ is single-valued and continuous, has all derivatives up to $d^n y/dx^n = f^n(x)$ in $a \le x \le b$ and the derivative $f^{n+1}(x)$ in $a < x < b$, then it can be represented in this interval by the power series in x as

$$f(x) = f(a) + \frac{x-a}{1!}f'(a) + \frac{(x-a)^2}{2!}f''(a) + \cdots + \frac{(x-a)^n}{n!}f^{(n)}(a) + R_{n+1}$$

where $R_{n+1} \le \begin{cases} \dfrac{f^{(n+1)}(\eta)}{(n+1)!}(x-a)^{n+1} & \text{(Lagrange's form)} \\ \dfrac{f^{(n+1)}(\eta)}{n!}(x-\eta)^n(x-a) & \text{(Cauchy's form)} \end{cases}$

are the estimated remainders and η lies in (a, x).

(b) Nested Taylor's series for $f(x)$ is

$$f(x) = f(a) + (x-a)\bigwedge_{k=1}^{n}\left[f^{(k)}(a) + \frac{x-a}{k+1}\right] + R_{n+1}$$

$$= f(a) + (x-a)\left[f'(a) + \frac{x-a}{2}\left[f''(a) + \frac{x-a}{3}\left[f'''(a) + \cdots + \frac{x-a}{n}f^{(n)}(a)\right]\right]\right] + R_{n+1}$$

where R_{n+1} is the same as in (a).

(c) Standard MacLaurin's series is a special form of Taylor's series in (a) for $a = 0$, written as

$$f(x) = f(0) + \frac{1}{1!}f'(0) + \frac{x^2}{2!}f''(0) + \cdots + \frac{x^n}{n!}f^{(n)}(0) + R_{n+1}$$

where $R_{n+1} = \begin{cases} \dfrac{f^{(n+1)}(\eta)}{(n+1)!}x^{n+1} & \text{(Lagrange's form)} \\ \dfrac{f^{(n+1)}(\eta)}{n!}(x-\eta)^n x & \text{(Cauchy's form)} \end{cases}$

are the estimated remainders and η lies in $(0, x)$.

(d) Nested MacLaurin's series for $f(x)$ is

$$f(x) = f(0) + x\bigwedge_{k=1}^{n}\left[f^{(k)}(0) + \frac{x}{k+1}\right] + R_{n+1}$$

$$= f(0) + x\left[f'(0) + \frac{x}{2}\left[f''(0) + \frac{x}{3}\left[f'''(0) + \cdots + \frac{x}{n}f^{(n)}(0)\right]\right]\right] + R_{n+1}$$

where R_{n+1} is the same as in (c).

(e) Modified Taylor's series. If $f(a)$ is known and it is required to find the value of $f(a+h)$, where h is a small increment of a, then the series in (a) becomes

$$f(a+h) = f(a) + \frac{h}{1!}f'(a) + \frac{h^2}{2!}f''(a) + \frac{h^3}{3!}f'''(a) + \cdots + \frac{h^n}{n!}f^{(n)}(a) + R_{n+1}$$

where $R_{n+1} = \dfrac{h^{n+1}}{(n+1)!}f^{(n+1)}(a+\theta h)$ $(0 < \theta < 1)$

The nested form of this series follows the pattern of (b).

(f) Gregory–Newton's series is the diffrence form of the Taylor's series, written as

$$f(x) = f(a) + \frac{(x-a)\Delta f(a)}{1!h} + \frac{(x-a)^2\Delta^2 f(a)}{2!h^2} + \frac{(x-a)^3\Delta^3 f(a)}{3!h^3}$$

$$+ \cdots + \frac{(x-a)^n\Delta^n f(a)}{n!h^n} + R_{n+1}$$

where for $k = 1, 2, \ldots, n$, $\Delta^k f(a)$ is the kth forward difference of $f(a)$ and $h = \Delta x$ is the difference interval. The remainder

$$R_{n+1} \cong \frac{(x-a)^{n+1}\Delta^{n+1} f(\eta)}{(n+1)!h^{n+1}}$$

where η is the same as in (c). The difference form can also be extended to the MacLaurin's series with $a = 0$. Both forms are particularly useful in cases where $f^{(k)}(x)$ is a very involved expression.

9.04 CONVERGENCE AND APPLICATIONS

(a) Convergence. The necessary and sufficient condition for the series in (9.03) to converge at x and to have a sum $S(x) = f(x)$ is

$$\lim_{n \to \infty} R_{n+1} = 0$$

If the series converges at x but

$$\lim_{n \to \infty} R_{n+1} \neq 0$$

then $S(x) \neq f(x)$ and the series does not represent completely and sufficiently $f(x)$ in the given interval.

(b) Applications. The expansion of a function in a power series is used for two practical purposes:

(1) Development of a substitute function, which approximates the given function in a prescribed interval

(2) Numerical evaluation of the given function for a certain value of its argument

(c) Substitute function has in many instances several advantages over the given function:

(1) Power series is a polynomial in x of the nth degree and as such is an algebraic (elementary) function, representing a transcendental or a special function.

(2) Operations with these polynomials are governed by the laws of ordinary algebra, which makes them particularly suitable to differentiation and integration.

(d) Evaluation of a function by the power series for a certain argument can be accomplished to a desired degree of accuracy by using the appropriate number of terms (by using a truncated power series).

(a) Definitions $(u = bx, v = bx \ln a)$.

$$y_1 = e^u \qquad y_2 = e^{-u} \qquad y_3 = a^u = e^v \qquad y_4 = a^{-u} = e^{-v}$$

where $e = 2.718\,281\,828\cdots(2.18)$, a, b = constants $\neq 0$, $\ln a$ = natural logarithm of a, x = real or complex variable.

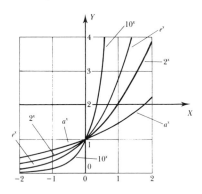 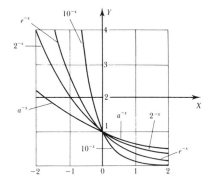

(b) Derivatives and integrals $(n = 1, 2, \ldots)$.

$$\frac{d^n y_1}{dx^n} = b^n e^u \qquad \int_0^x\!\!\int\cdots\int_0^x y_1(dx)^n = \frac{1}{b^n}\left[e^u - \sum_{k=0}^{n-1}\frac{u^k}{k!}\right]$$

$$\frac{d^n y_2}{dx^n} = (-b)^n e^{-u} \qquad \int_0^x\!\!\int\cdots\int_0^x y_2(dx)^n = \frac{1}{(-b)^n}\left[e^{-u} - \sum_{k=0}^{n-1}\frac{(-u)^k}{k!}\right]$$

$$\frac{d^n y_3}{dx^n} = v^n a^u x^{-n} \qquad \int_0^x\!\!\int\cdots\int_0^x y_3(dx)^n = \frac{x^n}{v^n}\left[a^u - \sum_{k=0}^{n-1}\frac{v^k}{k!}\right]$$

$$\frac{d^n y_4}{dx^n} = (-v)^n a^{-u} x^{-n} \qquad \int_0^x\!\!\int\cdots\int_0^x y_4(dx)^n = \frac{x^n}{(-v)^n}\left[a^{-1} - \sum_{k=0}^{n-1}\frac{(-v)^k}{k!}\right]$$

(c) Standard series

$$e^{bx} = \sum_{k=0}^{\infty}\frac{u^k}{k!} = 1 + \frac{u}{1!} + \frac{u^2}{2!} + \cdots + \frac{u^n}{n!} + R_{n+1} \qquad (0 < u < \infty)$$

$$e^{-bx} = \sum_{k=0}^{\infty}\frac{(-u)^k}{k!} = 1 - \frac{u}{1!} + \frac{u^2}{2!} - \cdots (-1)^n\frac{u^n}{n!} + R^*_{n+1} \qquad (-\infty < u < 0)$$

where $\quad \dfrac{u^{n+1}}{(n+1)!} \le R_{n+1} \le \dfrac{u^{n+1}}{(n+1)!}S_n \qquad \dfrac{(-u)^{n+1}}{(n+1)!} \le R^*_{n+1} \le \dfrac{(-u)^{n+1}}{(n+1)!}S^*_n$

and S_n, S^*_n is the sum of n terms of the series preceding R_{n+1}, R^*_{n+1}, respectively,

$$a^{bx} = \sum_{k=0}^{\infty}\frac{v^k}{k!} = 1 + \frac{v}{1!} + \frac{v^2}{2!} + \cdots + \frac{v^n}{n!} + R_{n+1} \qquad (0 \le v < \infty)$$

$$a^{-bx} = \sum_{k=0}^{\infty}\frac{(-v)^k}{k!} = 1 - \frac{v}{1!} + \frac{v^2}{2!} - \cdots (-1)^n\frac{v^n}{n!} + R^*_{n+1} \qquad (-\infty < v \le 0)$$

where R_{n+1}, R^*_{n+1} have an analogical meaning.

(d) Nested series $(u = bx, v = bx \ln a)$

$$e^{bx} = \bigwedge_{k=1}^{\infty}\left[1 + \frac{u}{k}\right] = \left(+ u\left(1 + \frac{u}{2}\left(1 + \frac{u}{3}\left(1 + \cdots + \frac{u}{n}\right)\right)\right)\right) + R_{n+1}$$

$$e^{-bx} = \bigwedge_{k=1}^{\infty}\left[1 - \frac{u}{k}\right] = \left(1 - u\left(1 - \frac{u}{2}\left(1 - \frac{u}{3}\left(1 - \cdots - \frac{u}{n}\right)\right)\right)\right) + R_{n+1}^{*}$$

$$a^{bx} = \bigwedge_{k=1}^{\infty}\left[1 + \frac{v}{k}\right] = \left(1 + v\left(1 + \frac{v}{2}\left(1 + \frac{v}{3}\left(1 + \cdots + \frac{v}{n}\right)\right)\right)\right) + R_{n+1}$$

$$a^{-bx} = \bigwedge_{k=1}^{\infty}\left[1 - \frac{v}{k}\right] = \left(1 - v\left(1 - \frac{v}{2}\left(1 - \frac{v}{3}\left(1 - \cdots - \frac{v}{n}\right)\right)\right)\right) + R_{n+1}^{*}$$

where R_{n+1}, R_{n+1}^{*} are the respective remainder estimated in (c).

(e) Small argument. The numerical evaluation of the series in (c), (d) is theoretically possible for all values of bx; computationally, the respective series becomes inconvenient for $u > 2$ and $v > 2$. The range of practicality of these series is given in (f) by the number of terms required for $10D$ accuracy and the respective remainders.

(f) Range of practicality

Range		$0 \le u \le 0.2$	$0.2 < u \le 0.5$	$0.5 < u \le 1$	$1 < u \le 2$	$2 < u \le 3$
e^u	n	7	9	12	17	22
	R_{n+1}	8×10^{-11}	4×10^{-10}	4×10^{-10}	3×10^{-10}	5×10^{-10}
e^{-u}	n	6	9	11	15	18
	R_{n+1}	2×10^{-12}	1×10^{-11}	7×10^{-10}	4×10^{-10}	8×10^{-9}

The same ranges apply for a^{bx}, a^{-bx}, where u is replaced by $v = bx \ln a$ in the table above.

(g) Large argument. For $2 < bx \le 10$ or $2 < v \le 10$, the values e^{bx}, a^{bx} are calculated as

$$e^{bx} = e^{M+c} = e^{M}e^{c} \qquad a^{bx} = e^{v} = e^{N+d} = e^{N}e^{d}$$

where M, N = integers and c, d = fractions $(c < 1, d < 1)$. In these formulas

$$e^{M} = \underbrace{e \cdot e \cdot e \cdots e}_{M \text{ times}} \qquad e^{N} = \underbrace{e \cdot e \cdot e \cdots e}_{N \text{ times}}$$

and e^{c}, e^{d} are computed by (c), (d). The same applies in cases of negative arguments.

(h) Example

$$e^{5.6789} = e^{5}e^{0.6789} = 292.627\ 362\ 7$$

where $e^{5} = 148.413\ 159\ 1$, and by the series of 12 terms $e^{0.6789} = 1.971\ 707\ 660$.

(i) Very large argument. For $bx > 10, v > 10$ the logarithmic resolution of the argument must be used.

$$e^{bx} = e^{(M+c)\ln 10} = 10^{M}e^{d}$$

where $\ln 10 = 2.302\ 585\ 093$, M is an integer, and c, d = fractions $(c < 1, d < 1)$ so that $M + c = bx/(\ln 10)$, $d = c \ln 10$.

(a) Definitions

$$y_1 = \ln x \qquad\qquad y_3 = \ln(1 + x) \qquad\qquad y_5 = \log x$$

$$y_2 = \ln(a + x) \qquad\quad y_4 = \ln(1 - x) \qquad\quad y_6 = \log(a + x)$$

when $\ln(\ )$ = natural logarithm, $\log(\ )$ = decadic logarithm, a = constant $\neq 0$, and x = real or complex variable.

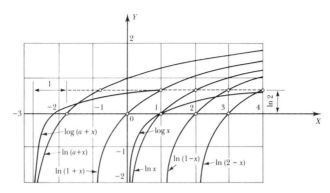

(b) Derivatives and integrals ($n = 1, 2, 3, \ldots$)

$$\frac{d^n y_1}{dx^n} = \frac{(-1)^{n-1}(n-1)!}{x^n} \qquad\qquad \int_0^x\!\!\cdots\!\int_0^x y_1\,(dx)^n = \frac{x^n}{n!}\left(\frac{n\ln x}{n!} - \sum_{k=1}^n \frac{1}{k!}\right)$$

$$\frac{d^n y_2}{dx^n} = \frac{(-1)^{n-1}(n-1)!}{(a+x)^n} \qquad \int_0^x\!\!\cdots\!\int_0^x y_2\,(dx)^n = \frac{x^n}{n!}\left[\frac{n\ln(a+x)}{n!} - \sum_{k=1}^n \frac{1}{k!}\right]$$

$$\frac{d^n y_5}{dx^n} = \frac{(-1)^{n-1}(n-1)!\log e}{x^n} \qquad \int_0^x\!\!\cdots\!\int_0^x y_5\,(dx)^n = \frac{x^n}{n!\ln 10}\left(\frac{n\ln x}{n!} - \sum_{k=1}^n \frac{1}{k!}\right)$$

$$\frac{d^n y_6}{dx^n} = \frac{(-1)^{n-1}(n-1)!\log e}{(a+x)^n} \quad \int_0^x\!\!\cdots\!\int_0^x y_6\,(dx)^n = \frac{x^n}{n!\ln 10}\left[\frac{n\ln(a+x)}{n!} - \sum_{k=1}^n \frac{1}{k!}\right]$$

(c) Standard series

$$\ln(1 + x) = \sum_{k=1}^{\infty} \frac{\beta x^k}{k} = x - \frac{x^3}{3} + \frac{x^5}{5} - \cdots \alpha\frac{x^n}{n} - \alpha R_{n+1} \qquad\qquad (-1 < x < 1)$$

$$\ln(a + x) = \ln a + 2\sum_{k=1}^{\infty} \frac{w^{2k-1}}{2k-1} = \ln a + 2\left(w + \frac{w^3}{3} + \cdots + \frac{w^{2n-1}}{2n-1}\right) + R^*_{n+1} \ (-a < x < \infty)$$

where $\quad \alpha = (-1)^{n+1} \qquad \beta = (-1)^{k+1} \qquad w = \dfrac{x}{2a+x}$

and $\quad R_{n+1} \leq \dfrac{x^n}{n} \qquad R^*_{n+1} = \leq \dfrac{w^{2n-1}}{2n-1}$

$$\log(1 + x) = \frac{\ln(1+x)}{\ln 10} \quad (-1 < x \leq 1) \qquad \log(a+x) = \frac{\ln(a+x)}{\ln 10} \quad (-a < x < \infty)$$

where the series for $\ln(1 + x)$ and $\ln(a + x)$ are given above and $\ln 10 = 2.302\,585\,093$.

(d) Nested series $\left(w = \dfrac{x}{2a + x}\right)$

$$\ln (1 + x)x \bigwedge_{k=1}^{\infty} \left[\frac{1}{k} - x\right] = x\left(1 - x\left(\frac{1}{2} - x\left(\frac{1}{3} - x\left(\frac{1}{4} - \cdots - \frac{x}{n}\right)\right)\right)\right) + R_{n+1}$$

$$\ln (a + x) = \ln a + 2w \bigwedge_{k=1}^{\infty} \left[\frac{1}{2k - 1} + w^2\right]$$

$$= \ln a + 2w\left(1 + w^2\left(\frac{1}{3} + w^2\left(\frac{1}{5} + w^2\left(\frac{1}{7} + \cdots + \frac{w^2}{2n - 1}\right)\right)\right)\right) + R_{n+1}^{*}$$

$$\log (1 + x) = \frac{x}{\ln 10} \bigwedge_{k=1}^{\infty} \left[\frac{1}{k} - x\right]$$

$$\log (a + x) = \frac{\ln a}{\ln 10} + \frac{2w}{\ln 10} \bigwedge_{k=1}^{\infty} \left[\frac{1}{2k - 1} + w^2\right]$$

where R_{n+1}, R_{n+1}^{*} are the respective remainders estimated in (c).

(e) Very small argument. The first series in (a) converges very slowly, and its practical application is limited to the region $-0.25 \le x \le 0.25$. The table (f) gives the number of terms required for $10D$ accuracy and the respective remainders.

(f) Range of practicality of $\ln (1 + x)$ series

Range		$0 < |x| \le 0.05$	$0.05 < |x| \le 0.15$	$0.15 < |x| < 0.25$	$0.25 < |x| \le 0.50$
$\ln (1 + x)$	n	6	9	13	26
	R_{n+1}	1×10^{-10}	5×10^{-10}	2×10^{-10}	2×10^{-10}

(g) Small argument. For $0.25 < |x| \le 1$ the argument $1 + x$ is modified as

$$\ln\left[2\left(\frac{1 + x}{2}\right)\right] = \ln 2 + \ln (1 - c)$$

where $\ln 2 = 0.693\ 147\ 180\ 6$ is given in (2.20) and $c = (1 - x)/2$.

(h) Example

$$\ln 1.789 = \ln 2 + \ln \left(1 - \frac{1 - 0.789}{2}\right)$$

$$= \ln 2 + \ln (1 - 0.1055) - 0.581\ 656\ 804\ 5$$

where $c = 0.1055$,

$$\ln (1 - c) = -c\left(1 + c\left(\frac{1}{2} + c\left(\frac{1}{3} + c\left(\frac{1}{4} + \cdots + \frac{c}{9}\right)\right)\right)\right) = -0.111\ 490\ 376\ 0$$

(i) Large argument. The second series in (d) can be used very efficiently for the evaluation of $\ln (a + x)$ if $a + x > 1$ by the following reduction.

$$\ln (a + x) = \ln\left[2^M\left(\frac{a + x}{2^M}\right)\right] = M \ln 2 + \ln (N + c)$$

where M, N are integers, c is a fracion ($c < 1$), and $\ln 2 = 0.693\ 147\ 180\ 6$.

(a) Definitions $(u = ax)$

$$y_1 = \sin ax = \sin u \qquad\qquad y_2 = \cos ax = \cos u$$

where $a = $ constant $\neq 0$ and $x = $ real or complex variable.

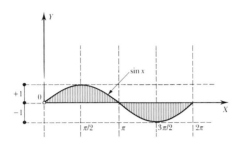

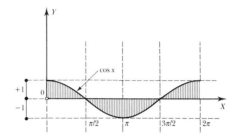

(b) Standard and nested series $(-\infty < u < \infty)$

$$\sin ax = \sum_{k=1}^{\infty} \frac{\beta u^{2k-1}}{(2k-1)!} = u\left(1 - \frac{u^2}{2\cdot 3}\left(1 - \frac{u^2}{4\cdot 5}\left(1 - \cdots - \frac{u^2}{(2n-2)(2n-1)}\right)\right)\right) + \alpha R_{2n+1}$$

$$\cos ax = \sum_{k=1}^{\infty} \frac{\beta u^{2k-1}}{(2k-2)!} = \left(1 - \frac{u^2}{1\cdot 2}\left(1 - \frac{u^2}{3\cdot 4}\left(1 - \cdots - \frac{u^2}{(2n-3)(2n-2)}\right)\right)\right) - \alpha R_{2n}^*$$

where $\quad \alpha = (-1)^{n+1} \qquad \beta = (-1)^{k+1} \qquad$ and $\qquad R_{2n+1} \leq \dfrac{u^{2n+1}}{(2n+1)!} \qquad R_{2n}^* \leq \dfrac{u^{2n}}{(2n)!}$

(c) Small argument.
Numerical evaluation of the series in (c) is theoretically possible for all values of u; computationally, their practical applications are restricted to $-\pi/2 \leq u \leq \pi/2$ as shown in (d).

(d) Range of practicality of sine and cosine series

Range		$0 \leq \|u\| \leq 0.3$	$0.3 < \|u\| \leq 0.6$	$0.6 < \|u\| \leq 0.9$	$0.9 < \|u\| < 1.2$	$1.2 < \|u\| \leq \dfrac{\pi}{2}$
$\sin u$	n	4	5	6	7	8
	R_{2n+1}	5×10^{-11}	1×10^{-10}	4×10^{-11}	1×10^{-11}	1×10^{-10}
$\cos u$	n	4	5	6	7	8
	R_{2n}^*	6×10^{-11}	4×10^{-11}	5×10^{-10}	1×10^{-10}	7×10^{-11}

(e) Large argument.
For $|u| > \pi/2$, the reduction of the argument to the small argument discussed in (c), (d) is obtained from the following relations, in which $k = 0, 1, 2, \ldots$:

$$\sin u = \sin (k\pi + v) = (-1)^k \sin v \qquad\qquad \cos u = \cos (k\pi + v) = (-1)^k \cos v$$

(f) Example.
If $u = 35.678$, then $35.678/\pi = 11.356\,660\,12$ and $k = 11$, and $v = 0.356\,660\,12\pi$. The range in (d) requires $n = 7$, and

$$\sin 35.678 = (-1)^{11} \sin 1.120\,480\,813$$

$$= -v\left(1 - \frac{v^2}{2\cdot 3}\left(1 - \frac{v^2}{4\cdot 5}\left(1 - \frac{v^2}{6\cdot 7}\left(1 - \frac{v^2}{8\cdot 9}\left(1 - \frac{v^2}{10\cdot 11}\left(1 - \frac{v^2}{12\cdot 13}\right)\right)\right)\right)\right)\right) + R_{15}$$

$$= -[0.900\,309\,819\,9 + (<4 \times 10^{-12})]$$

(g) Equivalents used in the evaluation of higher derivatives and multiple integrals of sin ax and cos ax in (h) to (j) below are

n = order of derivative or multiplicity of integral $A = (-1)^{n/2}$ $B = (-1)^{(n+1)/2}$

$k = 1, 2, 3, \ldots$ $S_k = (ka)^n \sin kax$ $C_k = (ka)^n \cos kax$

(h) Derivatives of $\sin^k ax$

	$d^n(\sin^k ax)/dx^n$	
$\sin^k ax$	n even	n odd
$\sin ax$	AS_1	$-BC_1$
$\sin^2 ax$	$-\dfrac{A}{2}C_2$	$-\dfrac{B}{2}S_2$
$\sin^3 ax$	$-\dfrac{A}{4}(S_3 - 3S_1)$	$-\dfrac{B}{4}(C_3 - 3C_1)$
$\sin^4 ax$	$\dfrac{A}{8}(C_4 - 4C_2)$	$\dfrac{B}{8}(S_4 - 4S_2)$
$\sin^5 ax$	$\dfrac{A}{16}(S_5 - 5S_3 + 10S_1)$	$\dfrac{B}{16}(C_5 - 5C_3 + 10C_1)$

(i) Derivatives of $\cos^k ax$

	$d^n(\cos^k ax)/dx^n$	
$\cos^k ax$	n even	n odd
$\cos ax$	AC_1	BS_1
$\cos^2 ax$	$\dfrac{A}{2}C_2$	$\dfrac{B}{2}S_2$
$\cos^3 ax$	$\dfrac{A}{4}(C_3 + 3C_1)$	$\dfrac{B}{4}(S_3 + 3S_1)$
$\cos^4 ax$	$\dfrac{A}{8}(C_4 + 4C_2)$	$\dfrac{B}{8}(S_4 + 4S_2)$
$\cos^5 ax$	$\dfrac{A}{16}(C_5 + 5C_3 + 10C_1)$	$\dfrac{B}{16}(S_5 + 5S_3 + 10S_1)$

(j) Integrals of sin ax **and** cos ax

	$\displaystyle\int_0^x\!\!\int_0^x \cdots \int f(x)(dx)^n \quad (n > 1)$	
$f(x)$	n even	n odd
$\sin ax$	$\dfrac{A}{a^n}\left[\sin ax + \displaystyle\sum_{k=1}^{n/2}(-1)\dfrac{(ax)^{2k-1}}{(2k-1)!}\right]$	$\dfrac{B}{a^n}\left[\cos ax - \displaystyle\sum_{k=0}^{(n-1)/2}(-1)^k\dfrac{(ax)^{2k}}{(2k)!}\right]$
$\cos ax$	$\dfrac{A}{a^n}\left[\cos ax - \displaystyle\sum_{k=0}^{(n-2)/2}(-1)^k\dfrac{(ax)^{2k}}{(2k)!}\right]$	$-\dfrac{B}{a^n}\left[\sin ax + \displaystyle\sum_{k=1}^{(n-1)/2}(-1)^k\dfrac{(ax)^{2k-1}}{(2k-1)!}\right]$

(a) Definitions ($u = ax$)

$$y_1 = \sinh ah = \sin u = \tfrac{1}{2}(e^u - e^{-u})$$

$$y_2 = \cosh ax = \cosh u = \tfrac{1}{2}(e^u + e^{-u})$$

where a, x are the same as in (9.07a).

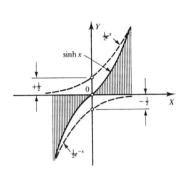

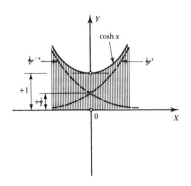

(b) Standard and nested series ($-\infty < u < \infty$)

$$\sinh ax = \sum_{k=1}^{\infty} \frac{u^{2k-1}}{(2k-1)!} = u\left(1 + \frac{u^2}{2 \cdot 3}\left(1 + \frac{u^2}{4 \cdot 5}\left(1 + \cdots + \frac{u^2}{(2n-2)(2n-1)}\right)\right)\right) + R_{2n+1}$$

$$\cosh ax = \sum_{k=1}^{\infty} \frac{u^{2k-2}}{(2k-2)!} = \left(1 + \frac{u^2}{1 \cdot 2}\left(1 + \frac{u^2}{3 \cdot 4}\left(1 + \cdots + \frac{u^2}{(2n-3)(2n-2)}\right)\right)\right) + R_{2n}^*$$

where $\quad R_{2n+1} \leq \dfrac{u^{2n-1}}{3(2n-1)!}$ $\qquad R_{2n}^* \leq \dfrac{2u^{2n-2}}{3(2n-2)!}$

(c) Small argument.
Numerical evaluation of the series in (b) is theoretically possible for all values of u; computationally, their practical applications are restricted to $-3 \leq u \leq 3$ as shown in (d) below.

(d) Range of practicality of sinh and cosh series

Range		$0 < |u| \leq 0.2$	$0.2 < |u| \leq 0.5$	$0.5 < |u| \leq 1.0$	$1.0 < |u| \leq 2.0$	$2.0 < |u| \leq 3.0$
sinh u	n	3	5	6	8	10
	R_{2n+1}	1×10^{-12}	1×10^{-11}	1×10^{-10}	3×10^{-10}	2×10^{-10}
cosh u	n	4	5	6	9	11
	R_{2n}^*	6×10^{-11}	2×10^{-10}	2×10^{-10}	4×10^{-11}	3×10^{-11}

(e) Large argument.
For $3 < |u| < 20$,

$$\sinh ax = \tfrac{1}{2}(e^{M+c} - e^{-M-c}) \qquad \cosh ax = \tfrac{1}{2}(e^{M+c} + e^{-M-c})$$

where M = integer, c = fraction ($c < 1$), and e^{M+c}, e^{-M-c} are evaluated by (9.05i).

(f) Very large argument.
For $|u| \geq 20$, the contribution of e^{-ax} becomes insignificant, and $\sinh ax \cong \cosh ax \cong \tfrac{1}{2}e^u$ and the evaluation is done again by (9.05i).

(g) Equivalents used in the evaluation of higher derivatives and multiple integrals of sinh ax and cosh ax in (h, i, j) below are:

n = order of derivative or multiplicity of integral

$k = 1, 2, 3, \ldots$ $\bar{S}_k = (ka)^n \sinh kah$ $\bar{C}_k = (ka)^n \cosh kax$

(h) Derivatives of $\sinh^k ax$

	$d(\sinh^k ax)/dx^n$	
$\sinh^k ax$	n even	n odd
$\sinh ax$	$\bar{S}_1$	$\bar{C}_1$
$\sinh^2 ax$	$\dfrac{1}{2}\bar{C}_2$	$\dfrac{1}{2}\bar{S}_2$
$\sinh^3 ax$	$\dfrac{1}{4}(\bar{S}_3 - 3\bar{S}_1)$	$\dfrac{1}{4}(\bar{C}_3 - 3\bar{C}_1)$
$\sinh^4 ax$	$\dfrac{1}{8}(\bar{C}_4 - 4\bar{C}_2)$	$\dfrac{1}{8}(\bar{S}_4 - 4\bar{S}_2)$
$\sinh^5 ax$	$\dfrac{1}{16}(\bar{S}_5 - 5\bar{S}_3 + 10\bar{S}_1)$	$\dfrac{1}{16}(\bar{C}_5 - 5\bar{C}_3 + 10\bar{C}_1)$

(i) Derivative of $\cosh^k ax$

	$d(\cosh^k ax)/dx^n$	
$\cosh^k ax$	n even	n odd
$\cosh ax$	$\bar{C}_1$	$\bar{S}_1$
$\cosh^2 ax$	$\dfrac{1}{2}\bar{C}_2$	$\dfrac{1}{2}\bar{S}_2$
$\cosh^3 ax$	$\dfrac{1}{4}(\bar{C}_3 + 3\bar{C}_1)$	$\dfrac{1}{4}(\bar{S}_3 + 3\bar{S}_1)$
$\cosh^4 ax$	$\dfrac{1}{8}(\bar{C}_4 + 4\bar{C}_2)$	$\dfrac{1}{8}(\bar{S}_4 + 4\bar{S}_2)$
$\cosh^5 ax$	$\dfrac{1}{16}(\bar{C}_5 + 5\bar{C}_3 + 10\bar{C}_1)$	$\dfrac{1}{16}(\bar{S}_5 + 5\bar{S}_3 + 10\bar{S}_1)$

(j) Integrals of sinh ax **and** cosh ax

	$\displaystyle\int_0^x\int_0^x \cdots \int f(x)(dx)^n$	
$f(x)$	n even	n odd
$\sinh ax$	$a^{-n}\left[\sinh ax - \displaystyle\sum_{k=1}^{n/2} \dfrac{(ax)^{2k-1}}{(2k-1)!}\right]$	$a^{-n}\left[\cosh ax - \displaystyle\sum_{k=0}^{(n-1)/2} \dfrac{(ax)^{2k}}{(2k)!}\right]$
$\cosh ax$	$a^{-n}\left[\cosh ax - \displaystyle\sum_{k=1}^{n/2} \dfrac{(ax)^{2k}}{(2k)!}\right]$	$a^{-n}\left[\sinh ax - \displaystyle\sum_{k=0}^{(n-1)/2} \dfrac{(ax)^{2k-1}}{(2k-1)!}\right]$

(a) Notation $(u = bx)$.

B_k = auxiliary Bernoulli number (2.22)

(b) Definition

$$y = \tan bx = \tan u$$

where b = constant and x = real or complex variable.

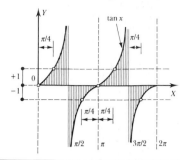

(c) Standard and nested series $(-\tfrac{1}{2}\pi < u < \tfrac{1}{2}\pi)$

$$\tan bx = \sum_{k=1}^{\infty} a_k u^{2k-1} = u + a_2 u^3 + a_3 u^5 + \cdots + a_n u^{2n-1} + R_{n+1}$$

$$= \bigwedge_{k=2}^{\infty} [1 + A_k u^2] = u(1 + A_2 u^2(1 + A_3 u^2(1 + \cdots + A_n u^2))) + R_{n+1}$$

$$a_k = \frac{4^k(4^k - 1)}{(2k)!} B_k$$

$$A_k = \frac{a_k}{a_{k-1}}$$

where $R_{n+1} < a_n u^{2n-1}$ and the numerical values a_k and A_k are tabulated below.

k	a_k	A_k	k	a_k	A_k
1	1.000 000 000 0	1.000 000 000 0	6	0.008 863 235 5	0.405 278 591 7
2	0.333 333 333 3	0.333 333 333 3	7	0.003 592 128 0	0.405 284 052 3
3	0.133 333 333 3	0.400 000 000 0	8	0.001 455 834 3	0.405 284 639 1
4	0.053 968 253 9	0.404 761 904 4	9	0.000 590 027 4	0.405 284 722 3
5	0.021 869 488 5	0.405 228 758 2	10	0.000 239 129 1	0.405 284 737 6

(d) Small argument. The numerical evaluation of the series is theoretically possible for all values of u in its interval of convergence; computationally, the series becomes inconvenient for $|u| > 0.5$ as shown below.

Range		$0 \le |u| \le 0.1$	$0.1 < |u| \le 0.2$	$0.2 < |u| \le 0.3$	$0.3 < |u| \le 0.4$	$0.4 < |u| \le \pi/4$
$\tan u$	n	4	5	6	8	11
	R_{n+1}	2×10^{-11}	1×10^{-10}	5×10^{-10}	1×10^{-10}	4×10^{-10}

(e) Intermediate argument. For $|u| \le 1.5$, the continued fraction

$$\tan u = \cfrac{u}{1 - \cfrac{u^2}{3 - \cfrac{u^2}{5 - \cdots}}}$$

offers a very efficient computational model in which N is the last integer used $(N = 1, 2, 5, \ldots)$.

Range		$0.5 < |u| \le 0.7$	$0.7 < |u| \le 0.9$	$0.9 < |u| \le 1.1$	$1.1 < |u| \le 1.3$	$1.3 < |u| \le 1.5$
$\tan u$	N	11	13	13	13	15
	R_{N+2}	4×10^{-11}	5×10^{-10}	3×10^{-10}	2×10^{-9}	3×10^{-8}

(f) Large argument. For $|u| > 1.5$, the reduction of the argument to a small argument is obtained from the following relations.

$$\tan\left[(4k + 1)\frac{\pi}{2} \pm u\right] = \tan\left[(4k + 3)\frac{\pi}{2} \pm u\right] = \frac{\mp 1}{\tan u} \qquad \tan\left[(4k + 2)\frac{\pi}{2} \pm u\right] = \tan\left[(4k + 4)\frac{\pi}{2} \pm u\right] = \pm\tan u$$

(a) Notation $(u = bx)$

$$\alpha = (-1)^{n+1} \qquad\qquad \beta = (-1)^{k+1}$$

a_k, A_k, B_k are defined in (9.09)

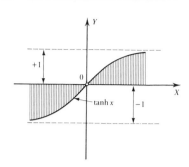

(b) Definition

$$\boxed{y = \tanh bx = \tanh u}$$

where $b = $ constant $\neq 0$ and $x = $ real or complex variable.

(c) Standard and nested series $(-\tfrac{1}{2}\pi < u < \tfrac{1}{2}\pi)$

$$\tanh bx = \sum_{k=1}^{\infty} \beta a_k u^{2k-1} = u - a_2 u^3 + a_3 u^5 - \cdots \alpha a_n u^{2n-1} - \alpha R_{n+1}$$

$$= u \bigwedge_{k=1}^{\infty} [1 - A_k u^2] = u(1 - A_2 u^2(1 - A_3 u^2(1 - \cdots - A_n u^2))) - \alpha R_{n+1}$$

where $R_{n+1} < a_n u^{2n-1}$ and the numerical values a_k and A_k are tabulated in (9.09c).

(d) Small argument. The numerical evaluation of the series is theoretically possible for all values of u in the interval of convergence; computationally, the series becomes inconvenient for $|u| > 0.5$ as shown below.

Range		$0 < |u| \leq 0.1$	$0.1 < |u| \leq 0.2$	$0.2 < |u| \leq 0.3$	$0.3 < |u| \leq 0.4$	$0.4 < |u| \leq 0.5$
$\tanh u$	n	4	5	6	8	9
	R_{n+1}	2×10^{-11}	1×10^{-10}	1×10^{-10}	1×10^{-10}	4×10^{-10}

(e) Intermediate argument. For $|u| \leq 1.5$, the continued fraction

$$\tanh u = \cfrac{u}{1 + \cfrac{u^2}{3 + \cfrac{u^2}{5 + \cdots}}}$$

offers a very efficient computational model in which N is the last integer used $(N = 1, 3, 5, \ldots)$,

Range		$0.5 < |u| \leq 0.7$	$0.7 < |u| \leq 0.9$	$0.9 < |u| \leq 1.1$	$1.1 < |u| \leq 1.3$	$1.3 < |u| \leq 1.5$
$\tanh u$	N	11	13	13	13	15
	R_{N+2}	4×10^{-11}	5×10^{-10}	3×10^{-10}	2×10^{-9}	3×10^{-8}

(f) Large argument. In case of $|u| > 1.5$, the function can be expressed as

$$\tanh u = \frac{e^u - e^{-u}}{e^u + e^{-u}} = \frac{1 - e^{-2u}}{1 + e^{-2u}}$$

where e^{-2u} is computed by the procedure introduced in (9.05). If $|u| > 10$, $\tanh u \cong 1$.

(a) Notation $(u = bx)$

$$u = 1 - v \text{ if } v > 0 \qquad u = 1 + v \text{ if } v < 0$$

$$a_k = \frac{(2k)!}{4^k(2k+1)(k!)^2} \qquad A_k = \frac{a_k}{a_{k-1}}$$

$$c_k = \frac{(2k)!}{8^k(2k+1)(k!)^2} \qquad C_k = \frac{c_k}{c_{k-1}}$$

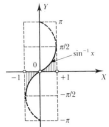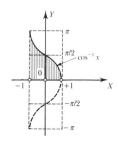

The numerical values a_k, A_k, c_k, C_k are tabulated in (h).

(b) Definitions

$$y_1 = \sin^{-1} bx = \sin^{-1} u = \arcsin u = \tfrac{1}{2}\pi - \arccos u$$

$$y_2 = \cos^{-1} bx = \cos^{-1} u = \arccos u = \tfrac{1}{2}\pi - \arcsin u$$

where $b = \text{constant} \neq 0$ and $x \neq$ real or complex variable.

(c) Standard and nested series $(-1 < u < 1)$

$$\arcsin u = \sum_{k=0}^{\infty} a_k u^{2k+1} = u + a_1 u^3 + a_2 u^5 + \cdots + a_n u^{2n+1} + R_{n+1}$$

$$= u \bigwedge_{k=1}^{\infty} [1 + A_k u^2] = u(1 + A_1 u^2(1 + A_2 u^2(1 + \cdots + A_n u^2)) + R_{n+1}$$

where $\quad R_{n+1} \leq \dfrac{2}{2n+3}\dbinom{2n+2}{n+1}\left(\dfrac{u}{2}\right)^{2n+3}$ $\qquad$ and $\qquad$ $\arccos u = \tfrac{1}{2}\pi - \arcsin u.$

(d) Range of practicality is restricted to $|u| < 0.5$.

Range		$0 \leq |u| \leq 0.1$	$0.1 < |u| \leq 0.2$	$0.2 < |u| \leq 0.3$	$0.3 < |u| \leq 0.4$	$0.4 < |u| \leq 0.5$
arcsin u	n	4	5	8	9	10
	R_{n+1}	3×10^{-11}	4×10^{-10}	2×10^{-10}	2×10^{-10}	4×10^{-10}

(e) Auxiliary standard and nested series $(-1 < v < 1)$

$$\arcsin v = \frac{\pi}{2} - \sqrt{2u}\left(1 + \sum_{k=1}^{\infty} c_k u^n\right) = \frac{\pi}{2} - \sqrt{2u}(1 + c_1 u + c_2 u^2 + \cdots + c_n u^n + R_{n+1})$$

$$= \frac{\pi}{2} - \sqrt{2u} \bigwedge_{k=1}^{\infty} [1 + c_k u] = \frac{\pi}{2} - \sqrt{2u}[1 + C_1 u(1 + C_2 u(1 + \cdots + C_n u))) + R_{n+1}]$$

where $\quad R_{n+1} \leq \dfrac{2}{2n+1}\dbinom{2n-1}{n}u^n$ $\qquad$ and $\qquad$ $\arccos v = \tfrac{1}{2}\pi - \arcsin v.$

(f) Range of practicality of this series is restricted to $-1 < v < -0.5$ and $0.5 < v < 1$.

Range		$0 \leq |u| \leq 0.1$	$0.1 < |u| \leq 0.2$	$0.2 < |u| \leq 0.3$	$0.3 < |u| \leq 0.4$	$0.4 < |u| \leq 0.5$
arcsin $(1 \pm u)$	n	5	7	8	9	9
	R_{n+1}	1×10^{-10}	7×10^{-11}	2×10^{-10}	7×10^{-10}	8×10^{-10}

(g) Negative argument

$$\arcsin(-bx) = \arcsin(-u) = -\arcsin u = -\tfrac{1}{2}\pi + \arccos u$$

$$\arccos(-bx) = \arccos(-u) = \pi - \arccos u = \tfrac{1}{2}\pi + \arcsin u$$

(h) Table of constants

k	a_k	A_k	c_k	C_k
1	0.166 666 666 7	0.166 666 666 7	0.083 333 333 3	0.083 333 333 3
2	0.075 000 000 0	0.450 000 000 0	0.018 750 000 0	0.225 000 000 0
3	0.044 642 857 1	0.595 238 095 2	0.005 580 357 1	0.297 619 047 6
4	0.030 381 944 4	0.680 555 555 6	0.001 898 871 5	0.340 277 777 8
5	0.022 372 159 9	0.736 363 636 3	0.000 699 129 9	0.368 181 818 2
6	0.017 352 764 4	0.775 641 025 6	0.000 271 136 9	0.387 820 512 8
7	0.013 964 843 7	0.804 761 904 8	0.000 109 100 3	0.402 380 952 4
8	8.011 551 800 9	0.827 205 882 4	0.000 045 124 2	0.413 602 941 2
9	0.009 761 609 5	0.845 029 239 8	0.000 019 065 6	0.422 514 619 9
10	0.008 390 335 8	0.859 523 809 5	0.000 008 193 6	0.429 761 904 8

(i) Example

$$\operatorname{arcos} 0.9 = \frac{\pi}{2} - \arcsin 0.9 \qquad \text{with } u = 1 - 0.9 = 0.1 \text{ by } (e)$$

$$= \sqrt{0.2}\,(1 + 0.1C_1(1 + 0.1C_2(1 + 0.1C_3(1 + 0.1C_4(1 + 0.1C_5))))) + R_6$$

$$= 0.451\,026\,811\,8 \qquad \text{and} \qquad R_6 < 1 \times 10^{-10}$$

(j) Continued fraction

$$\arcsin u = \cfrac{u\sqrt{1 - u^2}}{1 - \cfrac{1 \cdot 2u^2}{3 - \cfrac{1 \cdot 2u^2}{5 - \cfrac{3 \cdot 4u^2}{7 - \cfrac{3 \cdot 4u^2}{9 - \cdots}}}}}$$

offers a very simple and efficient model for small values of u in which N is the last integer used ($N = 1, 3, 5, \ldots$) as shown below.

Range		$0 \le \lvert u \rvert \le 0.1$	$0.1 < \lvert u \rvert \le 0.2$	$0.2 < \lvert u \rvert \le 0.3$	$0.3 < \lvert u \rvert \le 0.4$	$0.4 < \lvert u \rvert \le 0.05$
arcsin u	N	11	11	13	13	15
	R_{N+2}	1×10^{-11}	1×10^{-11}	1×10^{-11}	2×10^{-10}	5×10^{-10}

The same fraction may be used for the evaluation of arccos u by means of the relation in (*b*).

(a) Notation $(u = bx)$

$$\alpha = (-1)^{n+1} \qquad\qquad \beta = (-1)^{k+1}$$

$$A_k = \frac{(2k - 1)^2}{2k(2k + 1)} \qquad\qquad C_k = \frac{(k - 1)(2k - 1)}{2k^2}$$

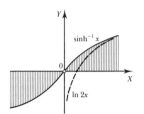

(b) Definitions

$$y_1 = \sinh^{-1} bx = \sinh^{-1} u = \operatorname{arcsinh} u = \ln\left(u + \sqrt{u^2 + 1}\right)$$

$$y_2 = \cosh^{-1} bx = \cosh^{-1} u = \operatorname{arccosh} u = \pm\ln\left(u + \sqrt{u^2 - 1}\right)$$

where $b = $ constant $\neq 0$ and $x = $ real or complex variable.

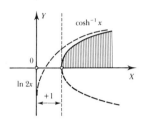

(c) Standard and nested series $(-1 < u < 1)$

$$\operatorname{arcsinh} u = u \bigwedge_{k=2}^{\infty} [1 - A_k u^2] = u(1 - A_2 u^2(1 - A_3 u^2(1 - \cdots - A_n u^2))) - \alpha R_{n+1}$$

where $R_{n+1} \leq \dfrac{1}{4^{n+1}(2n + 3)}\dbinom{2n + 2}{n + 1}u^{2n+1}.$

(d) Range of practicality of this series is restricted to $|u| \leq 0.4$.

Range		$0 \leq |u| \leq 0.1$	$0.1 < |u| \leq 0.2$	$0.2 < |u| \leq 0.3$	$0.3 < |u| \leq 0.4$
arcsinh u	n	4	5	8	9
	R_{n+1}	3×10^{-11}	4×10^{-10}	2×10^{-10}	2×10^{-10}

(e) Logarithmic transform of arcsinh u for $0.4 < |u| < 3$, with $v_1 = u + \sqrt{u^2 + 1} - 2$ and $w_1 = v_1/(4 + v_1)$, is

$$\operatorname{arcsinh} u = \ln\left(u + \sqrt{u^2 + 1}\right) = \ln\left(2 + v_1\right)$$

$$= \ln 2 + 2w_1\left(1 + \tfrac{1}{3}w_1^2\left(1 + \tfrac{3}{5}w_1^2\left(1 + \tfrac{5}{7}w_1^2\left(1 + \cdots + \frac{2n - 3}{2n - 1}w_1^2\right)\right)\right)\right) + R_{n+1}$$

where $R_{n+1} \leq \dfrac{2}{2n + 1}w_1^{2n+1}.$

(f) Logarithmic transform of arccosh u for $1 < |u| < 3$, with $v_2 = u + \sqrt{u^2 - 1} - 2$ and $w_2 = v_2/(4 + v_2)$, is

$$\operatorname{arccosh} u = \ln\left(u + \sqrt{u^2 - 1}\right) = \ln\left(2 + v_2\right)$$

$$= \ln 2 + 2w_2\left(1 + \tfrac{1}{3}w_2^2\left(1 + \tfrac{3}{5}w_2^2\left(1 + \tfrac{5}{7}w_2^2\left(1 + \cdots + \frac{2n - 3}{2n - 1}w_2^2\right)\right)\right)\right) + R_{n+1}$$

where $R_{n+1} \leq \dfrac{2}{2n + 1}w_2^{2n+1}.$

(g) Example. With $u = 2$, $v_2 = \sqrt{3}$, and $w_2 = 0.302\ 169\ 479\ 3$,

$$\operatorname{arccosh} 2 = \ln 2 + w_2(1 + \tfrac{1}{3}w_2^2(1 + \tfrac{3}{5}w_2^2(1 + \tfrac{5}{7}w_2^2(1 + \cdots + \tfrac{13}{15}w_2^2)))) + R_9$$

$$= 1.316\ 957\ 897 + 0$$

(h) Range of practicality of logarithm transforms in $(e,)$ (f) is shown below.

Range		$0.4 < \|u\| < 0.6$	$0.6 < \|u\| < 0.8$	$0.8 < \|u\| < 1.0$	$1.0 < \|u\| < 2.0$	$2.0 < \|u\| < 3.0$
arcsinh u	n	4	4	5	10	15
	R_{n+1}	1×10^{-10}	1×10^{-10}	1×10^{-10}	4×10^{-10}	2×10^{-10}
arccosh u	n	$\cdots$	$\cdots$	$\cdots$	8	13
	R^*_{n+1}	$\cdots$	$\cdots$	$\cdots$	1×10^{-11}	7×10^{-10}

(i) Large arguments. For $-\infty < u < -3, \ 3 < u < \infty$,

$$\operatorname{arcsinh} u = \ln 2u + \tfrac{1}{4}u^{-2} \bigwedge_{k=2}^{\infty} [1 - C_k u^{-2}]$$

$$= \ln 2u + \tfrac{1}{4}u^{-2}(1 - C_2 u^{-2}(1 - C_3 u^{-2}(1 - \cdots - C_n u^{-2}))) - \alpha R_{n+1}$$

$$\operatorname{arccosh} u = \ln 2u - \tfrac{1}{4}u^{-2} \bigwedge_{k=2}^{\infty} [1 + C_k u^{-2}]$$

$$= \ln 2u - \tfrac{1}{4}u^{-2}(1 + C_2 u^{-2}(1 + C_3 u^{-2}(1 + \cdots + C_n u^{-2}))) + R_{n+1}$$

where $\quad R_{n+1} \cong \dfrac{(2n + 1)! u^{2n+1}}{4^{n+1}[(2n + 1)!]^2}.$

(j) Table of coefficients

k	2	3	4	5	6	7	8	9	10	11
C_k	$\dfrac{1 \cdot 3}{2 \cdot 4}$	$\dfrac{2 \cdot 5}{3 \cdot 6}$	$\dfrac{3 \cdot 7}{4 \cdot 8}$	$\dfrac{4 \cdot 9}{5 \cdot 10}$	$\dfrac{5 \cdot 11}{6 \cdot 12}$	$\dfrac{6 \cdot 13}{7 \cdot 14}$	$\dfrac{7 \cdot 15}{8 \cdot 16}$	$\dfrac{8 \cdot 17}{9 \cdot 18}$	$\dfrac{9 \cdot 19}{10 \cdot 20}$	$\dfrac{10 \cdot 21}{11 \cdot 22}$

Both series converge very rapidly, and only a few terms are required for $10D$ accuracy.

(k) Example. With $n = 4$ and $\ln 20$ calculated by $(9.06i)$,

$$\operatorname{arcsinh} 10 = \ln 20 + \frac{10^{-2}}{4}\left(1 - \tfrac{3}{8}(10)^{-2}(1 - \tfrac{10}{18}(10)^{-2}(1 - \tfrac{21}{32}(10)^{-2}))\right) + R_5$$

$$= 2.998\,222\,950\,3 \quad\text{and}\quad R_5 = 0$$

For $10\,000 < |u| < \infty$, $\operatorname{arcsinh} |u| \cong \operatorname{arccosh} |u| \cong \ln |2u|$.

(l) Continued fraction

$$\operatorname{arcsinh} u = \cfrac{u\sqrt{1 + u^2}}{1 + \cfrac{1 \cdot 2u^2}{3 + \cfrac{1 \cdot 2u^2}{5 + \cfrac{3 \cdot 4u^2}{7 + \cfrac{3 \cdot 4u^2}{9 + \cdots}}}}}$$

offers again a very simple and efficient model for small values of u $[0 < |u| \leq 0.5]$, where N is the last integer used.

(m) Example. With $N = 1, 3, 5, 7, 9, 11, 13$, $\operatorname{arcsinh} 0.5 = 0.481\,211\,824\,3 + 8 \times 10^{-10}$

(a) Notation ($u = bx$)

$$s_1 = \frac{u}{\sqrt{1 + u^2}} \qquad s_2 = \frac{u}{\sqrt{1 - u^2}}$$

$$\alpha = (-1)^{n+1} \qquad \beta = (-1)^{k+1}$$

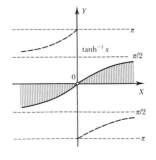

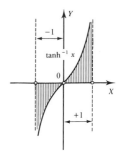

(b) Definitions

$$y_1 = \tan^{-1} bx = \tan^{-1} u = \arctan u$$

$$y_2 = \tanh^{-1} bx = \tanh^{-1} u = \operatorname{arctanh} u$$

where b = constant $\neq 0$ and x = real or complex variable.

(c) Standard and nested series ($-1 < u < 1$).

$$\arctan u = \sum_{k=1}^{\infty} \frac{\beta u^{2k-1}}{2k-1} = u - \frac{u^3}{3} + \frac{u^5}{5} - \cdots \alpha \frac{u^{2n-1}}{2n-1} - \alpha R_{n+1}$$

$$= u \bigwedge_{k=2}^{\infty} \left[1 - \frac{2k-3}{2k-1} u^{-2} \right] = u \left(1 - \frac{u^2}{3} \left(1 - \frac{3}{5} u^2 \left(1 - \cdots - \frac{2n-3}{2n-1} u^2 \right) \right) \right) - \alpha R_{n+1}$$

$$\operatorname{arctanh} u = \sum_{k=1}^{\infty} \frac{u^{2k-1}}{2k-1} = u + \frac{u^3}{3} + \frac{u^5}{5} + \cdots + \frac{u^{2n-1}}{2n-1} + R_{n+1}$$

$$= u \bigwedge_{k=2}^{\infty} \left[1 + \frac{2k-3}{2k-1} u^2 \right] = u \left(1 + \frac{u^2}{3} \left(1 + \frac{3}{5} u^2 \left(1 + \cdots + \frac{2n-3}{2n-1} u^2 \right) \right) \right) + R_{n+1}$$

where $R_{n+1} \cong \dfrac{u^{2n+1}}{2n+1}$.

(d) Small argument. The range of practicality of these series is $-0.5 \leq u \leq 0.5$.

Range		$0 < \|u\| \leq 0.1$	$0.1 < \|u\| \leq 0.2$	$0.2 < \|u\| \leq 0.3$	$0.3 < \|u\| \leq 0.4$	$0.4 < \|u\| \leq 0.5$
arctan u	n	4	6	8	10	13
or arctanh u	R_{n+1}	1×10^{-10}	6×10^{-11}	8×10^{-11}	2×10^{-10}	2×10^{-10}

For the remaining values in $0.5 < |u| < 1.0$, the transformations

$$\arctan u = \arcsin \frac{u}{\sqrt{1 + u^2}} = \arcsin s_1$$

$$= \frac{\pi}{2} - \sqrt{2 t_1} \bigwedge_{k=2}^{n} \left[1 + \frac{(2k-3)^2 t_1}{4(k-1)(2k-1)} \right] - R_{n+1}$$

$$\operatorname{arctanh} u = \operatorname{arcsinh} \frac{u}{\sqrt{1 - u^2}} = \operatorname{arcsinh} s_2$$

$$= \ln 2 + 2 t_2 \bigwedge_{k=2}^{n} \left[1 + \frac{2k-3}{2k-1} t_2^2 \right] + R_{n+1}^*$$

with $t_1 = 1 - s_1$ and $t_2 = \dfrac{s_1 + \sqrt{1 + s_1^2} - 2}{s_1 + \sqrt{1 + s_1^2} + 2}$

offer a rapid evaluation.

(e) Example. With $u = 0.8$, $s_1 = \dfrac{0.8}{\sqrt{1.64}}$, $t_1 = 0.375\,304\,952\,4$, and $n = 9$.

$$\arctan 0.8 = \frac{\pi}{2} - \sqrt{2t_1}\left(1 + \frac{1}{3\cdot 4}t_1\left(1 + \frac{3^2}{5\cdot 8}t_1\left(1 + \cdots + \frac{15^2}{17\cdot 32}t_1\right)\right)\right) + R_{10}$$

$$= 0.674\,740\,945\,1 \quad\text{and}\quad R_{10} < 2.6 \times 10^{-9}$$

(f) Large argument. For $|u| > 1$ only arctangent can be considered, and in $1 < |u| < 3$ the transformation introduced in (*d*) yields good results. For $|u| \ge 3$,

$$\boxed{\begin{aligned}
\arctan u &= \pm\frac{\pi}{2} - \sum_{k=1}^{\infty} \frac{\beta}{(2k-1)u^{2k-1}} \\
&= \pm\frac{\pi}{2} - \frac{1}{u} + \frac{1}{3u^3} - \frac{1}{5u^5} + \cdots \frac{\alpha}{(2k-1)u^{2n-1}} - \alpha R_{n+1} \\
&= \pm\frac{\pi}{2} - \frac{1}{u}\overset{\infty}{\underset{k=2}{\Lambda}}\left[1 - \frac{2k-3}{(2k-1)u^2}\right] \\
&= \pm\frac{\pi}{2} - \frac{1}{u}\left(1 - \frac{1}{3u^2}\left(1 - \frac{3}{5u^2}\left(1 - \cdots - \frac{2n-3}{(2n-1)u_2}\right)\right)\right) - \alpha R_{n+1}
\end{aligned}}$$

where the sign of $\pi/2$ is $+$ for $u \ge 3$ and $-$ for $u \le 3$. The remainder is estimated as

$$R_{n+1} \cong \frac{2n-1}{(2n+1)^{2n+1}}$$

and the rapid convergence of the series is shown below.

Range		$3 \le |u| \le 5$	$5 < |u| \le 10$	$10 < |u| \le 20$	$20 < |u| \le 50$
arctan u	n	8	6	4	3
	R_{n+1}	7×10^{-9}	7×10^{-10}	8×10^{-11}	1×10^{-9}

For $|u| > 50$, $\arctan u \cong \dfrac{\pi}{2} - \dfrac{1}{u}\left(1 - \dfrac{1}{3u^2}\right)$.

(g) Continued fractions

$$\arctan u = \cfrac{u}{1 + \cfrac{u^2}{3 + \cfrac{(2u)^2}{5 + \cfrac{(3u)^2}{7 + \cdots}}}}
\qquad\qquad
\operatorname{arctanh} u = \cfrac{u}{1 - \cfrac{u^2}{3 - \cfrac{(2u)^2}{5 - \cfrac{(3u)^2}{7 + \cdots}}}}$$

are limited in their practical application to $0 < |u| \le 0.5$ as shown below where N is the last integer used.

Range		$0.0 < |u| \le 0.1$	$0.1 < |u| \le 0.2$	$0.2 < |u| \le 0.3$	$0.3 < |u| \le 0.4$	$0.4 < |u| \le 0.5$
arctan u	N	7	9	11	13	15
	R_{N+2}	5×10^{-12}	3×10^{-11}	4×10^{-11}	5×10^{-11}	7×10^{-11}
arctanh u	N	9	11	13	19	19
	R_{N+2}	0	5×10^{-13}	2×10^{-12}	3×10^{-13}	3×10^{-12}

(a) Summation theorem. Two power series in x can be added or subtracted term by term for each value of x common to their interval of convergence.

$$\sum_{k=0}^{\infty} a_k x^k \pm \sum_{k=0}^{\infty} b_k x^k = \sum_{k=0}^{\infty} (a_k \pm b_k) x^k$$

Their sum is a new power series in x which converges at least in the common interval of convergence of the two series.

(b) Multiplication theorem. Two power series in x can be multiplied term by term for each value of x common to their interval of convergence.

$$\left(\sum_{k=0}^{\infty} a_k x^k\right)\left(\sum_{k=0}^{\infty} b_k x^k\right) = \sum_{k=0}^{\infty} c_k x^k$$

Their product is a new power series in x which converges at least in the common interval of convergence of the two series. The coefficients c_k of the new series are defined by the following matrix equation.

$$
\begin{bmatrix} c_0 \\ c_1 \\ c_2 \\ c_3 \\ c_4 \\ c_5 \\ \cdots \end{bmatrix}
=
\begin{bmatrix}
a_0 & & & & & \\
a_1 & a_0 & & & & \\
a_2 & a_1 & a_0 & & & \\
a_3 & a_2 & a_1 & a_0 & & \\
a_4 & a_3 & a_2 & a_1 & a_0 & \\
a_5 & a_4 & a_3 & a_2 & a_1 & a_0 \\
\cdots & & & & &
\end{bmatrix}
\begin{bmatrix} b_0 \\ b_1 \\ b_2 \\ b_3 \\ b_4 \\ b_5 \\ \cdots \end{bmatrix}
$$

(c) Division theorem. The quotient of two power series in x in their common interval of convergence,

$$\frac{\displaystyle\sum_{k=0}^{\infty} a_k x^k}{\displaystyle\sum_{k=0}^{\infty} b_k x^k} = \sum_{k=0}^{\infty} d_k x^k$$

is a new power series in x, and its interval of convergence (if any) cannot be determined from the interval of convergence of the two series. The coefficients d_k of the new series are defined by the following matrix equation.

$$
\begin{bmatrix} d_0 \\ d_1 \\ d_2 \\ d_3 \\ d_4 \\ d_5 \\ \cdots \end{bmatrix}
=
\frac{1}{b_0}
\begin{bmatrix}
a_0 & & & & & \\
a_1 & -b_1 & & & & \\
a_2 & -b_2 & -b_1 & & & \\
a_3 & -b_3 & -b_2 & -b_1 & & \\
a_4 & -b_4 & -b_3 & -b_2 & -b_1 & \\
a_5 & -b_5 & -b_4 & -b_3 & -b_2 & -b_1 \\
\cdots & & & & &
\end{bmatrix}
\begin{bmatrix} 1 \\ d_0 \\ d_1 \\ d_2 \\ d_3 \\ d_4 \\ \cdots \end{bmatrix}
$$

where each d_k is evaluated by using the preceding $d_{k-1}, d_{k-2}, \ldots, d_1, d_0$.

(a) Notation

$$a, b = \text{constants} > 0 \qquad M_{1,k} = \frac{(\sqrt{2})^k \cos k\phi}{k!} \qquad N_{1,k} = \frac{(-\sqrt{2})^k \cos k\phi}{k!}$$

$$c = \sqrt{a^2 + b^2} \qquad M_{2,k} = \frac{(\sqrt{2})^k \sin k\phi}{k!} \qquad N_{2,k} = -\frac{(-\sqrt{2})^k \sin k\phi}{k!}$$

$$\omega = \arctan \frac{b}{a} \qquad M_{3,k} = \frac{(\sqrt{2})^{k+1} \sin (k+1)\phi}{k!} \qquad N_{3,k} = -\frac{(-\sqrt{2})^{k+1} \cos (k+1)\phi}{k!}$$

$$\phi = \tfrac{1}{4}\pi \qquad M_{4,k} = \frac{(\sqrt{2})^{k+1} \cos (k+1)\phi}{k!} \qquad N_{4,k} = -\frac{(-\sqrt{2})^{k+1} \sin (k+1)\phi}{k!}$$

(b) Winkler's functions of the first kind $(-\infty < x < \infty)$

$$e^{ax} \cos bx = \sum_{k=0}^{\infty} \frac{(cx)^k \cos k\omega}{k!} \qquad e^{ax} (\cos bx + \sin bx) = \sum_{k=0}^{\infty} \frac{\sqrt{2}(cx)^k \sin (\phi + k\omega)}{k!}$$

$$e^{ax} \sin bx = \sum_{k=0}^{\infty} \frac{(cx)^k \sin k\omega}{k!} \qquad e^{ax} (\cos bx - \sin bx) = \sum_{k=0}^{\infty} \frac{\sqrt{2}(cx)^k \cos (\phi + k\omega)}{k!}$$

$$e^{ax} \cos ax = \sum_{k=0}^{\infty} M_{1,k}(ax)^k \qquad e^{ax} (\cos ax + \sin ax) = \sum_{k=0}^{\infty} M_{3,k}(ax)^k$$

$$e^{ax} \sin ax = \sum_{k=0}^{\infty} M_{2,k}(ax)^k \qquad e^{ax} (\cos ax - \sin ax) = \sum_{k=0}^{\infty} M_{4,k}(ax)^k$$

(c) Winkler's functions of the second $(\infty < x < \infty)$

$$e^{-ax} \cos bx = \sum_{k=0}^{\infty} \frac{(-cx)^k \cos k\omega}{k!} \qquad e^{-ax} (\cos bx + \sin bx) = \sum_{k=0}^{\infty} \frac{\sqrt{2}(-cx)^k \cos (\phi + k\omega)}{k!}$$

$$e^{-ax} \sin bx = \sum_{k=0}^{\infty} \frac{(-cx)^k \sin k\omega}{k!} \qquad e^{-ax} (\cos bx - \sin bx) = \sum_{k=0}^{\infty} \frac{\sqrt{2}(-cx)^k \sin (\phi + k\omega)}{k!}$$

$$e^{-ax} \cos ax = \sum_{k=0}^{\infty} N_{1,k}(ax)^k \qquad e^{-ax} (\cos ax + \sin ax) = \sum_{k=0}^{\infty} N_{3,k}(ax)^k$$

$$e^{-ax} \sin ax = \sum_{k=0}^{\infty} N_{2,k}(ax)^k \qquad e^{-ax} (\cos ax - \sin ax) = \sum_{k=0}^{\infty} N_{4,k}(ax)^k$$

(d) Factors

k	$M_{1,k}$	$M_{2,k}$	$N_{1,k}$	$N_{2,k}$
0	1	0	1	0
1	1/1!	1/1!	−1/1!	1/1!
2	0	2/2!	0	−2/2!
3	−2/3!	2/3!	2/3!	2/3!
4	−4/4!	0	−4/4!	0
5	−4/5!	−4/5!	4/5!	−4/5!
6	0	−8/6!	0	8/6!
7	8/7!	−8/7!	−8/7!	−8/7!
8	16/8!	0	16/8!	0
9	16/9!	16/9!	−16/9!	16/9!
10	0	32/10!	0	−32/10!
11	−32/11!	32/11!	32/11!	32/11!
12	−64/12!	0	−64/12!	0
13	−64/13!	−64/13!	64/13!	−64/13!
14	0	−128/14!	0	128/14!
15	128/15!	−128/15!	−128/15!	−128/15!

(a) Exponential theorem. If S is a power series in x, such as

$$S = a_0 + a_1 x + a_2 x^2 + a_3 x^3 + \cdots = \sum_{k=0}^{\infty} a_k x^k$$

then

$$\boxed{e^S = e^{a_0} \sum_{k=1}^{\infty} c_k x^k}$$

where $e = 2.718\ 281\ 828 \cdots$ (2.18), $c_0 = 1$, $c_1 = a_1$, and for $k > 1$,

$$c_k = \frac{1}{k!} \sum_{r=1}^{k=1} \binom{k-1}{r} (k-r)! c_r$$

or in matrix form,

$$
\begin{bmatrix} c_1 \\ c_2 \\ c_3 \\ c_4 \\ c_5 \\ \cdots \end{bmatrix}
=
\begin{bmatrix}
a_1 \\
a_2 & \dfrac{1!}{2!} a_1 \\
a_3 & \dfrac{2!}{3!} 2a_2 & \dfrac{1!}{3!} a_1 \\
a_4 & \dfrac{3!}{4!} 3a_3 & \dfrac{2!}{4!} 3a_2 & \dfrac{1!}{4!} a_1 \\
a_5 & \dfrac{4!}{5!} 4a_4 & \dfrac{3!}{5!} 6a_3 & \dfrac{2!}{5!} 4a_2 & \dfrac{1!}{5!} a_1 \\
\cdots
\end{bmatrix}
\begin{bmatrix} 1 \\ c_1 \\ c_2 \\ c_3 \\ c_4 \\ \cdots \end{bmatrix}
$$

The coefficients c_k of the new series, which is convergent in the same interval as S, are evaluated by using the preceding $c_{k-1}, c_{k-2}, \ldots, c_2, c_1$.

(b) Logarithmic theorem. In terms of S defined in (a) above,

$$\boxed{\ln (S - a_0) = \sum_{k=1}^{\infty} c_k x^k}$$

where $c_1 = a_1$ and for $k > 0$,

$$c_k = a_k - \frac{1}{k} \sum_{r=1}^{k-1} r a_{k-r} c_k$$

or in matrix form,

$$
\begin{bmatrix} c_1 \\ c_2 \\ c_3 \\ c_4 \\ c_5 \\ \cdots \end{bmatrix}
=
\begin{bmatrix}
a_1 \\
a_2 & -a_1/2 \\
a_3 & -a_2/3 & -2a_1/3 \\
a_4 & -a_3/4 & -2a_2/4 & -3a_1/4 \\
a_5 & -a_4/5 & -2a_3/5 & -3a_2/5 & -4a_1/5 \\
\cdots
\end{bmatrix}
\begin{bmatrix} 1 \\ c_1 \\ c_2 \\ c_3 \\ c_4 \\ \cdots \end{bmatrix}
$$

The coefficients c_k of the new series, which is convergent in the same interval as S, are calculated by the same procedure as in (a).

(a) Notation $a = \text{constant} > 0$ $B_k = $ auxiliary Bernoulli number (A.12)

$$A_k = \sum_{r=1}^{k} \frac{A_{k-r}\cos r\pi}{2k(2k-2r)!(2r-1)!} \qquad M_k = \frac{4^k(4^k-1)}{(2k)(2k):}B_k \qquad T_r = \left[\frac{d^r \tan ax}{dx^r}\right]_{ax=0}$$

$$D_k = \sum_{r=1}^{k} \frac{D_{k-r}\sin \frac{1}{2}r\pi}{k(k-r)!(r-1)!} \qquad N_k = \frac{4^k}{(2k)(2k)!}B_k$$

$$C_k = \sum_{r=1}^{k} \frac{C_{k-r}T_r}{k(k-r)!(r-1)!} \qquad P_k = \frac{4^k(4^k-2)}{(2k)(2k)!}B_k$$

(b) Exponential functions $(k = 0, 1, 2, \ldots, u = ax)$

$$e^{\cos u} = e\left(1 - \frac{u^2}{2!} + \frac{4u^4}{4!} - \frac{31u^6}{6!} + \cdots + A_k u^{2k} + \cdots\right) \qquad (-\infty < u < \infty)$$

$$e^{\sin u} = 1 + u + \frac{u^2}{2!} - \frac{3u^4}{4!} - \frac{8u^5}{5!} - \cdots + D_k u^k + \cdots \qquad (-\infty < u < \infty)$$

$$e^{\tan u} = 1 + u + \frac{u^2}{2!} + \frac{3u^3}{3!} + \frac{9u^4}{4!} + \cdots + C_k u^k + \cdots \qquad (-\tfrac{1}{2}\pi < u < \tfrac{1}{2}\pi)$$

(c) Coefficients A_k, B_k, C_k

k	A_k		D_k		C_k	
0	1.000 000 000	(+00)	1.000 000 000	(+00)	1.000 000 000	(+00)
1	−5.000 000 000	(−01)	1.000 000 000	(+00)	1.000 000 000	(+00)
2	1.666 666 667	(−01)	5.000 000 000	(−01)	5.000 000 000	(−01)
3	−4.305 555 556	(−02)	0.000 000 000	(+00)	5.000 000 000	(−01)
4	9.399 801 587	(−03)	−1.250 000 000	(−01)	3.750 000 000	(−01)
5	−1.806 657 848	(−03)	−6.666 666 667	(−02)	3.333 333 333	(−01)
6	3.138 799 536	(−04)	−4.166 666 667	(−03)	2.458 333 333	(−01)
7	−5.016 685 851	(05)	1.111 111 111	(−03)	1.902 777 778	(−01)
8	7.470 220 791	(−06)	5.381 944 444	(−03)	1.373 263 889	(−01)
9	−1.046 251 198	(−06)	1.763 668 430	(−04)	1.132 082 231	(−01)

(d) Logarithmic functions $(k = 1, 2, 3, \ldots, u = ax)$

$$\ln(\cos u) = -\left(\frac{u^2}{2} + \frac{u^4}{12} + \frac{u^6}{45} + \frac{17u^8}{2520} + \cdots + M_k u^{2k} + \cdots\right) \qquad (-\tfrac{1}{2}\pi < u < \tfrac{1}{2}\pi)$$

$$\ln(\sin u) = \ln u - \left(\frac{u^2}{6} + \frac{u^4}{180} + \frac{u^6}{2835} + \cdots + N_k u^{2k} + \cdots\right) \qquad (-\infty < u < \infty)$$

$$\ln(\tan u) = \ln u + \left(\frac{u^2}{3} + \frac{7u^4}{90} + \frac{62u^6}{2835} + \cdots + P_k u^{2k} + \cdots\right) \qquad (-\tfrac{1}{2}\pi < u < \tfrac{1}{2}\pi)$$

(e) Coefficients M_k, N_k, P_k

k	M_k		N_k		P_k	
1	5.000 000 000	(−01)	1.666 666 667	(−01)	3.333 333 333	(−01)
2	8.333 333 333	(−02)	5.555 555 556	(−03)	7.777 777 778	(−02)
3	2.222 222 222	(−02)	3.527 336 861	(−04)	2.186 948 845	(−02)
4	6.746 031 746	(−03)	2.645 502 645	(−05)	6.719 576 721	(−03)
5	2.186 948 854	(−03)	2.137 779 916	(−06)	2.184 811 074	(−03)
6	7.386 029 608	(−04)	1.803 670 234	(−07)	7.384 225 938	(−04)
7	2.565 805 743	(−04)	1.566 139 132	(−08)	2.565 649 126	(−04)
8	9.098 964 916	(−05)	1.388 413 049	(−09)	9.098 826 078	(−05)
9	3.277 930 227	(−05)	1.250 435 918	(−10)	3.277 917 723	(−05)
10	1.179 645 571	(−05)	1.140 257 564	(−11)	1.195 644 431	(−05)

(a) Power theorem. The mth power of the power series S in x is

$$\left(\sum_{k=0}^{\infty} a_k x^k\right)^m = a_0^m(1 + b_1 x + b_2 x^2 + \cdots)^m = a_0^m \sum_{k=0}^{\infty} c_k x^k$$

where m = signed number, $b_k = a_k/a_0$, $c_0 = 1$, and for $k > 0$,

$$c_k = \sum_{r=1}^{k}\left[\frac{r}{k}(m+1) - 1\right]b_r c_{k-r} = \sum_{r=1}^{k} w_{kr} b_r c_{k-r}$$

or in matrix form,

$$
\begin{bmatrix} c_1 \\ c_2 \\ c_3 \\ c_4 \\ c_5 \\ \cdots \end{bmatrix}
=
\begin{bmatrix}
w_{11}b_1 \\
w_{22}b_2 & w_{21}b_1 \\
w_{33}b_3 & w_{32}b_2 & w_{31}b_1 \\
w_{44}b_4 & w_{43}b_3 & w_{42}b_2 & w_{41}b_1 \\
w_{55}b_5 & w_{54}b_4 & w_{53}b_3 & w_{52}b_2 & w_{51}b_1 \\
\cdots & \cdots & \cdots & \cdots & \cdots & \cdots
\end{bmatrix}
\begin{bmatrix} c_0 \\ c_1 \\ c_2 \\ c_3 \\ c_4 \\ \cdots \end{bmatrix}
$$

The coefficients c_k of the new series, which is convergent in the same interval as S, are evaluated by using the preceding $c_{k-1}, c_{k-2}, \ldots, c_2, c_1, c_0$.

(b) Factors w_{kr} $(p, q, r = 1, 2, 3, \ldots)$

m	w_{kr}	m	w_{kr}	m	w_{kr}	m	w_{kr}
2	$(3r-k)/k$	-1	-1	$1/2$	$(3r-2k)/2k$	$-1/2$	$(r-2k)/2k$
3	$(3r-k)/k$	-2	$-(r+k)/k$	$1/3$	$(4r-3k)/3k$	$-1/3$	$(2r-3k)/3k$
4	$(5r-k)/k$	-3	$-(2r+k)/k$	$1/4$	$(5r-4k)/4k$	$-1/4$	$(3r-4k)/4k$
p	$[r(p+1)-k]/k$	$-p$	$-[r(p-1)+k]/k$	p/q	$[r(p+q)-kq]/kq$	$-p/q$	$[r(q+p)-kq]/kq$

(c) Example

$$\sqrt{\sin x} = \left(\frac{x}{1!} - \frac{x^3}{3!} + \frac{x^5}{5!} - \cdots\right)^{1/2} = \sqrt{x}\sum_{k=0}^{\infty} c_k x^k$$

where $b_1 = 0, b_2 = -1/3!, b_3 = 0, b_4 = 1/5!, b_5 = 0, \ldots$

$$w_{11} = \tfrac{1}{2} \quad w_{22} = \tfrac{1}{2} \quad w_{21} = -\tfrac{1}{4} \quad w_{33} = \tfrac{1}{2} \quad w_{32} = 0 \quad w_{31} = -\tfrac{1}{2}$$

$$w_{44} = \tfrac{1}{2} \quad w_{43} = \tfrac{1}{8} \quad w_{42} = -\tfrac{2}{8} \quad w_{41} = -\tfrac{5}{8}$$

$$w_{55} = \tfrac{1}{2} \quad w_{54} = \tfrac{2}{10} \quad w_{53} = -\tfrac{1}{10} \quad w_{52} = -\tfrac{4}{10} \quad w_{51} = -\tfrac{7}{10}$$

and the coefficients c_k are given by (a) as

$$c_0 = 1 \qquad c_1 = \left(\tfrac{1}{2}\right)(0)(1) = 0 \qquad c_2 = \left(\tfrac{1}{2}\right)\left(-\tfrac{1}{3!}\right)(1) + \left(-\tfrac{1}{4}\right)(0)(0) = -\frac{1}{12}$$

$$c_3 = \left(\tfrac{1}{2}\right)(0)(1) + (0)\left(-\tfrac{1}{3!}\right)(0) + \left(-\tfrac{3}{6}\right)(0)\left(-\tfrac{1}{12}\right) = 0$$

$$c_4 = \left(\tfrac{1}{2}\right)\left(\tfrac{1}{5!}\right)(1) + \left(\tfrac{1}{8}\right)(0)(0) + \left(-\tfrac{1}{4}\right)\left(-\tfrac{1}{3!}\right) - \frac{1}{12} + \left(-\tfrac{5}{8}\right)(0)(0) = \frac{1}{1440}$$

$$c_5 = \left(\tfrac{1}{2}\right)(0)(1) + \left(\tfrac{2}{10}\right)\left(\tfrac{1}{5!}\right)(0) + \left(-\tfrac{1}{10}\right)(0)\left(-\tfrac{1}{12}\right) + \left(-\tfrac{4}{10}\right)\left(-\tfrac{1}{3!}\right)(0) + \left(-\tfrac{7}{10}\right)(0)(0) = 0$$

from which, for $x = 0.5$,

$$\sqrt{\sin 0.5} = \sqrt{0.5}(1 - \tfrac{1}{12}0.5^2 + \tfrac{1}{1440}0.5^4) + R_6 = 0.692\,406\,080 - 4.7 \times 10^{-7}$$

10
SPECIAL
FUNCTIONS

(a) Origin. J. Liouville proved in 1833 that the following eight integrals cannot be expressed in terms of elementary functions and can only be evaluated by integrating their power series expansions. They are:

$\text{Si}(x)$ = circular sine integral	$\text{Ci}(x)$ = circular cosine integral
$\bar{\text{Si}}(x)$ = hyperbolic sine integral	$\bar{\text{Ci}}(x)$ = hyperbolic cosine integral
$\text{Ei}(x)$ = alternating exponential integral	$\text{Li}(x)$ = logarithmic integral
$\bar{\text{Ei}}(x)$ = monotonic exponential integral	$\text{erf}(x)$ = error function

Their analytical forms are given below in (b), their graphs are shown in (10.02), and their numerical values are tabulated in $(A.13)$ to $(A.15)$.

(b) Analytical forms

$$\text{Si}(x) = \int_0^x \frac{\sin t}{t}\,dt = \sum_{k=1}^\infty \frac{\beta x^{2k-1}}{(2k-1)(2k-1)!} \qquad (-\infty < x < \infty)$$

$$\text{Ci}(x) = \int_\infty^x \frac{\cos t}{t}\,dt = \gamma + \ln x - \sum_{k=1}^\infty \frac{\beta x^{2k}}{2k(2k)!} \qquad (0 < x < \infty)$$

$$\bar{\text{Si}}(x) = \int_0^x \frac{\sinh t}{t}\,dt = \sum_{k=1}^\infty \frac{x^{2k-1}}{(2k-1)(2k-1)!} \qquad (-\infty < x < \infty)$$

$$\bar{\text{Ci}}(x) = \int_\infty^x \frac{\cosh t}{t}\,dt = \gamma + \ln x + \sum_{k=1}^\infty \frac{x^{2k}}{2k(2k)!} \qquad (0 < x < \infty)$$

$$\text{Ei}(x) = \int_\infty^x \frac{e^{-t}}{t}\,dt = \gamma + \ln x - \sum_{k=1}^\infty \frac{\beta x^k}{k(k!)} \qquad (0 < x < \infty)$$

$$\bar{\text{Ei}}(x) = \int_{-\infty}^x \frac{e^t}{t}\,dt = \gamma + \ln x + \sum_{k=1}^\infty \frac{x^k}{k(k!)} \qquad (-\infty < x < \infty)$$

$$\text{Li}(x) = \int_0^x \frac{1}{\ln t}\,dt = \gamma + \ln[\ln x] + \sum_{k=1}^\infty \frac{(\ln x)^k}{k(k!)} \qquad (0 < x < \infty)$$

$$\text{erf}(x) = \frac{2}{\sqrt\pi}\int_0^x e^{-t^2}\,dt = \frac{2}{\sqrt\pi}\sum_{k=1}^\infty \frac{\beta x^{2k-1}}{(k-1)!(2k-1)} \qquad (-\infty < x < \infty)$$

where $\gamma = 0.577\,216\,665$ (2.19) and $\beta = (-1)^{k+1}$. Also

$$\bar{\text{Ei}}(x) = \bar{\text{Ci}}(x) + \bar{\text{Si}}(x) \qquad \text{Ei}(x) = \bar{\text{Ei}}(-x) \qquad \text{Li}(x) = \bar{\text{Ei}}(\ln x) \qquad \text{erfc}(x) = 1 - \text{erf}(x)$$

(c) Related functions are the exponential integrals of order r, defined in $0 < x < \infty$ as

$$E_1(x) = \int_x^\infty \frac{e^{-t}}{t}\,dt = -\gamma - \ln x + \sum_{k=1}^\infty \frac{\beta x^k}{k(k!)} = -\text{Ei}(-x)$$

and in general,

$$E_{r+1}(x) = \int_1^\infty \frac{e^{-xt}}{t^{r+1}}\,dt = \frac{1}{r}[e^{-x} - xE_r(x)]$$
$$= \frac{e^{-x}}{r}\left(1 - \frac{x}{r-1}\left(1 - \frac{x}{r-2}\left(1 - \cdots - \frac{x}{1}\left(1 - xe^x E_1(x)\right)\right)\right)\right)$$

where $r = 1, 2, 3, \ldots$, and $E_0(x) = e^{-x}/x$.

(a) Special values

$\text{Si}(0) = 0$	$\text{Ci}(0) = -\infty$	$\bar{\text{Si}}(0) = 0$	$\bar{\text{Ci}}(0) = -\infty$
$\text{Si}(\infty) = \frac{1}{2}\pi$	$\text{Ci}(\infty) = 0$	$\bar{\text{Si}}(\infty) = \infty$	$\bar{\text{Ci}}(\infty) = \infty$
$\text{Si}(-x) = -\text{Si}(x)$	$\text{Ci}(-x) = \text{Ci}(x)$	$\bar{\text{Si}}(-x) = -\bar{\text{Si}}(x)$	$\bar{\text{Ci}}(-x) = \bar{\text{Ci}}(x)$
$\bar{\text{Ei}}(0) = -\infty$	$E_r(0) = 1/(r-1)$	$\text{Li}(1) = -\infty$	$\text{erf}(0) = 0$
$\bar{\text{Ei}}(\infty) = +\infty$	$E_r(\infty) = 0$	$\text{Li}(\infty) = \infty$	$\text{erf}(\infty) = \frac{1}{2}\sqrt{\pi}$
$\bar{\text{Ei}}(-x) = -\text{Ei}(x)$			$\text{erf}(-x) = -\text{erf}(x)$

(b) Graph of $\text{Si}(x)$

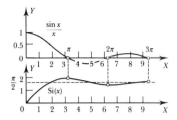

(c) Graph of $\text{Ci}(x)$

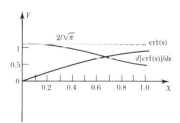

(d) Graphs of $\text{Ei}(x)$ and $\bar{\text{Ei}}(x)$

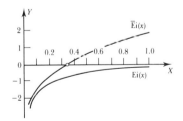

(e) Graphs of $\text{erf}(x)$ and $d[\text{erf}(x)]/dx$

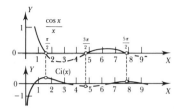

(f) Extreme values of $\text{Si}(x)$ **and** $\text{Ci}(x)$. The sine integral has maxima at $x = (2r-1)\pi$ and minima at $x = 2r\pi$. The cosine integral has maxima at $x = \frac{1}{2}(4r-3)\pi$ and minima at $x = \frac{1}{2}(4r-1)\pi$, for $r = 1, 2, 3, \ldots$, as shown in (b), (c) above. Analytically,

$$\text{Si}(M) = \tfrac{1}{2}\pi + \frac{\alpha}{M}\left(1 - \frac{1 \cdot 2}{M^2}\left(1 - \frac{3 \cdot 4}{M^2}\left(1 - \cdots - \frac{(2m-1)2m}{M^2}\right)\right)\right)$$

$$\text{Ci}(N) = -\frac{\alpha}{N}\left(1 - \frac{1 \cdot 2}{N^2}\left(1 - \frac{3 \cdot 4}{N^2}\left(1 - \cdots - \frac{(2m-1)2m}{N^2}\right)\right)\right)$$

where $M = m\pi$, $N = (m - \frac{1}{2})\pi$, $m = 7, 8, 9, \ldots$, and $\alpha = (-1)^{m+1}$. For lower m the numerical values are listed below.

m	1	2	3	4	5	6
$\text{Si}(M)$	1.851 937	1.418 152	1.674 762	1.492 161	1.633 965	1.518 034
$\text{Ci}(N)$	0.472 000	−0.198 410	0.123 770	−0.089 564	0.070 065	−0.057 501

(a) Small arguments. The range of practicality is restricted to $0 < x \leq 1$ for

$$\text{Si}(x) = x\left(1 - \frac{x^2}{2 \cdot 9}\left(1 - \frac{3x^2}{4 \cdot 25}\left(1 - \frac{5x^2}{6 \cdot 49}\left(1 - \cdots - \frac{(2n-3)x^2}{(2n-2)(2n-1)^2}\right)\right)\right)\right) + R_{n+1}$$

$$\text{Ci}(x) = L_1 - \left(\frac{x^2}{4}\left(1 - \frac{2x^2}{3 \cdot 16}\left(1 - \frac{4x^2}{5 \cdot 36}\left(1 - \cdots - \frac{(2n-2)x^2}{(2n-1)(2n+2)^2}\right)\right)\right)\right) + R_{n+1}$$

$$\text{Ei}(x) = L_1 - x\left(1 - \frac{x}{4}\left(1 - \frac{2x}{9}\left(1 - \cdots - \frac{(n-1)x}{n^2}\right)\right)\right) + R_{n+1}$$

$$\bar{\text{Ei}}(x) = L_1 + x\left(1 + \frac{x}{4}\left(1 + \frac{2x}{9}\left(1 + \cdots + \frac{(n-1)x}{n^2}\right)\right)\right) + R_{n+1}$$

$$\text{Li}(x) = L_2 + u\left(1 + \frac{u}{4}\left(1 + \frac{2u}{9}\left(1 + \cdots + \frac{(n-1)u}{n^2}\right)\right)\right) + R_{n+1}$$

$$\text{erf}(x) = \frac{2x}{\sqrt{\pi}}\left(1 - \frac{x^2}{1 \cdot 3}\left(1 - \frac{3x^2}{2 \cdot 5}\left(1 - \frac{5x^2}{3 \cdot 7}\left(1 - \cdots - \frac{(2n-3)x^2}{(n-1)(2n-1)}\right)\right)\right)\right) + R_{n+1}$$

where $L_1 = \gamma + \ln x$, $u = \ln x$, $L_2 = \gamma + \ln u$, and $\gamma = 0.577\,216\,665$ (2.19). The range of practicality of these series is shown below.

Range		$0 < x \leq 0.1$	$0.1 < x \leq 0.2$	$0.2 < x \leq 0.3$	$0.3 < x \leq 0.4$	$0.4 < x \leq 0.5$	$0.5 < x \leq 1.0$
Si(x)	n	3	3	3	4	4	5
	R_{n+1}	0	3×10^{-10}	6×10^{-9}	0	6×10^{-10}	4×10^{-10}
Ci(x)	n	2	3	3	3	4	5
	R_{n+1}	2×10^{-1}	0	2×10^{-10}	2×10^{-9}	0	2×10^{-9}
$\bar{\text{Ei}}(x)$	n	6	6	6	7	9	10
	R_{n+1}	2×10^{-10}	3×10^{-10}	6×10^{-9}	2×10^{-9}	6×10^{-10}	2×10^{-9}
erf(x)	n	3	4	5	6	8	11
	R_{n+1}	3×10^{-9}	3×10^{-9}	2×10^{-9}	8×10^{-10}	2×10^{-11}	1×10^{-9}

(b) Example

$$\text{Si}(0.5(\,= 0.5\left(1 - \frac{(0.5)^2}{18}\left(1 - \frac{3(0.5)^2}{100}\left(1 - \frac{5(0.5)^2}{294}\right)\right)\right) + R_5$$

$$= 0.493\,107\,417 + 6(10)^{-10}$$

(c) Large arguments. For $1 < x < 30$, the following approximate formulas developed by C. Hastings (Ref. 9.03) yield good results.

$$\text{Si}(x) = \frac{\pi}{2} - A(x)\cos x - B(x)\sin x \qquad\qquad \text{Ci}(x) = A(x)\sin x - B(x)\cos x$$

$$\text{Ei}(x) = -a(x)\frac{e^{-x}}{x} \qquad\qquad \bar{\text{Ei}}(x) = b(x)\frac{e^x}{x} \qquad\qquad \text{erf}(x) = c(x)$$

where $A(x)$, $B(x)$, $a(x)$, $b(x)$, $c(x)$ are the rational polynomials listed in (*d*) and $\alpha = (-1)^{n+1}$.

(d) Rational polynomials used in (*c*) are

$$A(x) = \frac{(A_1 + x(A_2 + x(A_3 + x(A_4 + x))))}{x(A_5 + x(A_6 + x(A_7 + x(A_8 + x))))} + |\epsilon = 5 \times 10^{-7}|$$

$$B(x) = \frac{(B_1 + x(B_2 + x(B_3 + x(B_4 + x))))}{x^2(B_5 + x(B_6 + x(B_7 + x(B_8 + x))))} + |\epsilon = 3 \times 10^{-7}|$$

$$a(x) = \frac{(a_1 - x(a_2 - x(a_3 - x(a_4 - x))))}{a_5 - x(a_6 - x(a_7 - x(a_8 - x))))} + |\epsilon = 2 \times 10^{-8}|$$

$$b(x) = \frac{(b_1 + x(b_2 + x(b_3 + x(b_4 + x))))}{(b_5 + x(b_6 + x(b_7 + x(b_8 + x))))} + |\epsilon = 2 \times 10^{-8}|$$

$$c(x) = 1 - u(c_1 - u(c_2 - u(c_3 - u(c_4 - c_5 u)))) e^{-x^2} + |\epsilon = 1.5 \times 10^{-7}|$$

in which A_k, B_k, a_k, b_k, c_k are the numerical coefficients tabulated below and

$$u = \frac{1}{1 + 0.327\ 591\ 111 x}$$

(e) Coefficients of rational polynomials

k	A_k	B_k	$a_k = b_k$	c_k
1	38.102 495	21.821 899	0.267 773 733 4	0.254 829 592
2	335.677 320	352.018 498	8.634 760 892 5	0.284 496 736
3	265.187 033	302.757 865	18.059 016 973 0	1.421 413 741
4	38.027 264	42.242 855	8.573 328 740 1	1.453 152 027
5	157.105 423	449.690 326	3.958 496 922 8	1.061 405 429
6	570.236 280	1114.978 885	21.099 653 082 7	
7	322.624 911	482.485 984	25.632 956 148 0	
8	40.021 433	48.196 927	9.573 322 345 4	

(f) Very large arguments. For $x > 30$, the following asymptotic series yield fast and accurate results.

$$\text{Si}(x) = \frac{\pi}{2} - \bar{A}(x)\cos x - \bar{B}(x)\sin x \qquad \text{Ci}(x) = \bar{A}(x)\sin x - \bar{B}(x)\cos x$$

$$\text{Ei}(x) = -\bar{a}(x)\frac{e^{-x}}{x} \qquad \overline{\text{Ei}}(x) = \bar{b}(x)\frac{e^x}{x} \qquad \text{erf}(x) = \bar{c}(x)$$

where with $\alpha = (-1)^{n+1}$,

$$\bar{A}(x) = \frac{1}{x}\left(1 - \frac{1\cdot 2}{x^2}\left(1 - \frac{3\cdot 4}{x^2}\left(1 - \frac{5\cdot 6}{x^2}\left(1 - \cdots - \frac{(2n-1)2n}{x^2}\right)\right)\right)\right) - \alpha\epsilon$$

$$\bar{B}(x) = \frac{1}{x^2}\left(1 - \frac{2\cdot 3}{x^2}\left(1 - \frac{4\cdot 5}{x^2}\left(1 - \frac{6\cdot 7}{x^2}\left(1 - \cdots - \frac{2n(2n+1)}{x^2}\right)\right)\right)\right) - \alpha\epsilon$$

$$\bar{a}(x) = \left(1 - \frac{1}{x}\left(1 - \frac{2}{x}\left(1 - \frac{3}{x}\left(1 - \cdots - \frac{n}{x}\right)\right)\right)\right) - \alpha\epsilon$$

$$\bar{b}(x) = \left(1 + \frac{1}{x}\left(1 + \frac{2}{x}\left(1 + \frac{3}{x}\left(1 + \cdots + \frac{n}{x}\right)\right)\right)\right) + \alpha\epsilon$$

$$\bar{c}(x) = 1 - \frac{e^{-x^2}}{x}\left[\left(1 - \frac{1}{2x^2}\left(1 - \frac{3}{2x^2}\left(1 - \frac{5}{2x^2}\left(1 - \cdots - \frac{2n-1}{2x^2}\right)\right)\right)\right) - \alpha\epsilon\right]$$

and for $n = 5$, $\epsilon \leq 5 \times 10^{-7}$.

(a) Basic forms. Related to the error functions are Fresnel integrals defined in $-\infty < x < \infty$ with $\phi = \pi/2$ as

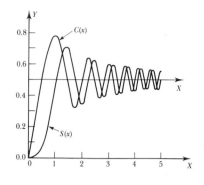

$$S(x) = \int_0^x \sin{(\phi t^2)}\, dt$$

$$= \sum_{k=0}^{\infty} \frac{(-1)^k \phi^{2k+1} x^{4k+3}}{(2k+1)!(4k+3)} = -S(-x)$$

$$C(x) = \int_0^x \cos{(\phi t^2)}\, dt$$

$$= \sum_{k=0}^{\infty} \frac{(-1)^k \phi^{2k} x^{4k+1}}{(2k)!(4k+1)} = -C(-x)$$

Their graphs are shown in the adjacent figure, and their special values are

$$S(-\infty) = C(-\infty) = -\tfrac{1}{2} \qquad S(0) = C(0) = 0 \qquad S(\infty) = C(\infty) = \tfrac{1}{2}$$

(b) Related integrals are

$$S_1(x) = -S_1(-x) = \frac{1}{\sqrt{\phi}} \int_0^x \sin{t^2}\, dt = \frac{1}{\sqrt{\phi}} \sum_{k=0}^{\infty} \frac{(-1)^k x^{4k+3}}{(2k+1)!(4k+3)}$$

$$C_1(x) = -C_1(-x) = \frac{1}{\sqrt{\phi}} \int_0^x \cos{t^2}\, dt = \frac{1}{\sqrt{\phi}} \sum_{k=0}^{\infty} \frac{(-1)^k x^{4k+1}}{(2k)!(4k+1)}$$

$$S_2(x) = -S_2(-x) = \frac{1}{2\sqrt{\phi}} \int_0^x \frac{\sin t}{\sqrt{t}}\, dt = \sqrt{\frac{x}{\phi}} \sum_{k=0}^{\infty} \frac{(-1)^k x^{2k+1}}{(2k+1)!(4k+3)}$$

$$C_2(x) = -C_2(-x) = \frac{1}{2\sqrt{\phi}} \int_0^x \frac{\cos t}{\sqrt{t}}\, dt = \sqrt{\frac{x}{\phi}} \sum_{k=0}^{\infty} \frac{(-1)^k x^{2k}}{(2k)!(4k+1)}$$

and

$$\int_0^{\infty} \frac{\sin t}{\sqrt{t}}\, dt = \int_0^{\infty} \frac{\cos t}{\sqrt{t}}\, dt = \sqrt{\frac{\pi}{2}}$$

(c) Evaluation by series listed above in (a), (b) converges very rapidly for $-1 \le x \le 1$ with $n = 5$ and $R_6 \le 6 \times 10^{10}$. For larger arguments,

$$S(x) = \tfrac{1}{2} - A \sin{(\phi x^2)} - B \cos{(\phi x^2)} \qquad C(x) = \tfrac{1}{2} - A \cos{(\phi x^2)} + B \sin{(\phi x^2)}$$

$$S_1(x) = \tfrac{1}{2} - \bar{A} \sin{(x^2)} - \bar{B} \cos{(x^2)} \qquad C_1(x) = \tfrac{1}{2} - \bar{A} \cos{(x^2)} + \bar{B} \sin{(x^2)}$$

where with $M = (x\sqrt{\pi})^4$ and $N = (x\sqrt{2})^4$,

$$A = \frac{1}{\pi^2 x^3}\left(1 - \frac{3 \cdot 5}{M}\left(1 - \frac{7 \cdot 9}{M}\left(1 - \cdots - \frac{((4n-1)(4n+1))}{M}\right)\right)\right)$$

$$B = \frac{1}{\pi x}\left(1 - \frac{1 \cdot 3}{M}\left(1 - \frac{5 \cdot 7}{M}\left(1 - \cdots - \frac{(4n-3)(4n-1)}{M}\right)\right)\right)$$

$$\bar{A} = \frac{1}{2x^3\sqrt{2\pi}}\left(1 - \frac{3 \cdot 5}{N}\left(1 - \frac{7 \cdot 9}{N}\left(1 - \cdots - \frac{(4n-1)(4n+1)}{N}\right)\right)\right)$$

$$\bar{B} = \frac{1}{x\sqrt{2\pi}}\left(1 - \frac{1 \cdot 3}{N}\left(1 - \frac{5 \cdot 7}{N}\left(1 - \cdots - \frac{(4n-3)(4n-1)}{N}\right)\right)\right)$$

(if $n = 5$, $|\epsilon| \le 5 \times 10^{-8}$)

(a) General forms. Integrals of the form

$$\int R(t, \sqrt{a_0 + a_1 t + a_2 t^2 + a_3 t^3}) \, dt \qquad\qquad \int R(t, \sqrt{a_0 + a_1 t + a_2 t^2 + a_3 t^3 + a_4 t^4}) \, dt$$

where $R(\ )$ is a rational function in t and the fourth-order polynomial in t has no repeating roots, are called the *elliptic* and *hyperbolic integrals*, which in general cannot be expressed in terms of elementary functions and can only be evaluated by integrating their power series. The name elliptic integral was introduced by A. M. Legendre in 1794 in solving the circumference of an ellipse.

(b) Incomplete elliptic integrals obtained by suitable transformations are:

(1) *Integral of the first kind*

$$F(k, x) = \int_0^x \frac{dt}{\sqrt{(1 - t^2)(1 - k^2 t^2)}}$$

$$F(k, \phi) = \int_0^\phi \frac{d\theta}{\sqrt{1 - k^2 \sin^2 \theta}}$$

(2) *Integral of the second kind*

$$E(k, x) = \int_0^x \sqrt{\frac{1 - k^2 t^2}{1 - t^2}} \, dt$$

$$E(k, \phi) = \int_0^\phi \sqrt{1 - k^2 \sin^2 \theta} \, d\theta$$

(3) *Integral of the third kind*

$$\Pi(k, a, x) = \int_0^x \frac{dt}{(1 + at^2)\sqrt{(1 - t^2)(1 - k^2 t^2)}}$$

$$\Pi(k, a, \phi) = \int_0^\phi \frac{d\theta}{(1 + a \sin^2 \theta)\sqrt{1 - k^2 \sin^2 \theta}}$$

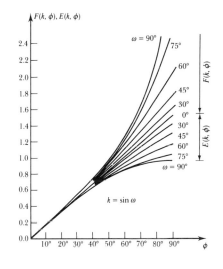

where $k = \sin \omega =$ given modulus $(0 \leq k \leq 1)$, $x =$ independent variable $(-1 \leq x \leq 1)$, $\phi = \sin^{-1} x =$ independent variable $(-\frac{1}{2}\pi \leq \phi \leq \frac{1}{2}\phi)$, and $a =$ given real number.

(c) Complete elliptic integrals obtained from (b) for $x = 1$ or $\phi = \frac{1}{2}\pi$ are

$$F(k, 1) = F(k, \tfrac{1}{2}\pi) = K \qquad E(k, 1) = E(k, \tfrac{1}{2}\pi) = E \qquad \Pi(k, a, 1) = \Pi(k, a, \tfrac{1}{2}\pi) = \Pi$$

and their complementary counterparts are

$$F(k', 1) = F(k', \tfrac{1}{2}\pi) = K' \qquad R(k', 1) = E(k', \tfrac{1}{2}\pi) = E' \qquad \Pi(k', a, 1) = \Pi(k', a, \tfrac{1}{2}\pi) = \Pi'$$

where $k' = \sqrt{1 - k^2} =$ complementary modulus $(0 \leq k' \leq 1)$.

(d) Reflection and reduction formulas

$$F(k, -\phi) = -F(k, \phi) \qquad\qquad F(k, r\pi \pm \phi) = 2rK \pm F(k, \phi)$$

$$E(k, -\phi) = -E(k, \phi) \qquad\qquad E(k, r\pi \pm \phi) = 2rE \pm E(k, \phi)$$

and $\quad KK' + \frac{1}{2}\pi = KE' + EK'$

where $r = 1, 2, \ldots$ and K, K', E, E' are the complete elliptic integrals defined in (c).

10.06 ELLIPTIC INTEGRALS, NUMERICAL EVALUATION

(a) Notation $k = \sin \omega = $ given modulus $(10.05b)$

$K, E = $ complete elliptic integrals $(10.05c)$

$F(k, \phi), E(k, \phi) = $ incomplete elliptic integrals $(10.05b)$

$$A = \frac{1 - \sqrt{1 - k^2}}{1 + \sqrt{1 - k^2}} \qquad k_1 = \frac{2}{1 + \sqrt{1 - k^2}} \qquad k_2 = \frac{2}{1 + \sqrt{1 - k_1^2}} \qquad k_3 = \frac{2}{1 + \sqrt{1 - k_2^2}} \cdots$$

(b) Special values

ϕ ⟍ k	$k = 0$		$k = 1$	
$\phi = 0$	$F(0, 0) = 0$	$E(0, 0) = 0$	$F(1, 0) = 0$	$E(1, 0) = 0$
$-\frac{1}{2}\pi < \phi < \frac{1}{2}\pi$	$F(0, \phi) = \phi$	$E(0, \phi) = \phi$	$F(1, \phi) = \ln [\tan (\frac{1}{4}\pi + \frac{1}{2}\phi)]$	$E(1, \phi) = \sin \phi$
$\phi = \frac{1}{2}\pi$	$K = \frac{1}{2}\pi$	$E = \frac{1}{2}\pi$	$K = \infty$	$E = 1$

(c) Evaluation of complete integral of the first kind
For $0° < \omega < 60°$

$$K = \frac{\pi}{2}(1 + A)\left(1 + \frac{A^2}{4}\left(1 + \frac{9A^2}{16}\left(1 + \frac{25A^2}{36}\left(1 + \frac{49A^2}{64}\left(1 + \frac{81A^2}{100}\right)\right)\right)\right)\right) + (\epsilon \le 5 \times 10^{-8})$$

For $60° \le \omega < 90°$
$$K = \frac{\pi}{2}(1 + k_1)(1 + k_2)(1 + k_3)(1 + k_4) + (\epsilon \le 5 \times 10^{-8})$$

(d) Examples. If $k = 0.7$, $\omega = \sin^{-1} 0.7 = 0.775\,397\,497 = 44.427°$, $\cos \omega = 0.714\,142\,843$, and $A = 0.166\,763\,907$, then by the first formula in (c), $K = 1.845\,693\,998 + (\epsilon = 0)$. If $k = \sqrt{0.95}$, $k' = \sqrt{1 - k^2} = 0.223\,606\,798$, $\omega = 77.079°$, and

$$1 + k_1 = 1 + \frac{2}{1 + k'} = 1.634\,512\,005 \qquad\qquad k_1' = \sqrt{1 - k_1^2} = 0.772\,913\,007$$

$$1 + k_2 = 1 + \frac{2}{1 + k_1'} = 1.128\,086\,935 \qquad\qquad k_2' = \sqrt{1 - k_2^2} = 0.991\,762\,944$$

$$1 + k_3 = 1 + \frac{2}{1 + k_2'} = 1.004\,135\,560 \qquad\qquad k_3' = \sqrt{1 - k_3^2} = 0.999\,991\,449$$

$$1 + k_4 = 1 + \frac{2}{1 + k_3'} = 1.000\,004\,276 \qquad\qquad k_4' = \sqrt{1 - k_4^2} = 1.000\,000\,000$$

then by the second formula in (c), $K = 2.908\,337\,248 + (\epsilon = 0)$.

(e) Evaluation of complete integral of the second kind
For $0° < \omega < 60°$

$$E = \frac{\pi}{2(1 + A)}\left(1 + \frac{A^2}{4}\left(1 + \frac{A^2}{16}\left(1 + \frac{9A^2}{36}\left(1 + \frac{25A^2}{64}\left(1 + \frac{49A^2}{100}\right)\right)\right)\right)\right) + (\epsilon \le 5 \times 10^{-8})$$

For $60° \le \omega < 90°$
$$E = K(1 - \tfrac{1}{2}k^2(1 + \tfrac{1}{2}k_1(1 + \tfrac{1}{2}k_2(1 + \tfrac{1}{2}k_3(1 + \tfrac{1}{2}k_4))))) + (\epsilon \le 5 \times 10^{-8})$$

where k, k_1, k_2, k_3, k_4, and K are the same as in (c).

(f) Examples. If $k = 0.7$ and ω, $\cos\omega$, and A are the same as in (d), then by the first formula in (e),

$E = 1.355\,661\,136 + (\epsilon = 1 \times 10^{-9})$

If $k = \sqrt{0.95}$ and k_1, k_2, k_3, k_4 are the same as in (d), then by the second formula in (e),

$E = 1.060\,473\,727 + (\epsilon = 5 \times 10^{-10})$

(g) Evaluation of incomplete integral of the first kind $(m = k^2)$

For $0° < \omega < 20°$

$$F(k, \phi) = \frac{2\phi}{\pi} K - \frac{m \sin 2\phi}{8} \left(1 + \frac{9ma_1}{16} \left(1 + \frac{25ma_2}{36} \left(1 + \frac{49ma_3}{64} \right) \right) \right) + (\epsilon \le 5 \times 10^{-7})$$

where K is computed by (c) and

$$a_1 = 1 + \frac{2}{3} \sin^2 \phi \qquad a_2 = 1 + \frac{8}{15} \frac{\sin^4 \phi}{a_1} \qquad a_3 = 1 + \frac{16}{35} \frac{\sin^6 \phi}{a_2}$$

For $20° \le \omega < 90°$

$$F(k, \phi) = \sqrt{\frac{k_1 k_2 k_3 k_4}{4}} \ln \left[\tan \left(\frac{\pi}{4} + \frac{\phi_4}{2} \right) \right] + (\epsilon = 5 \times 10^{-7})$$

where with given k and ϕ,

$$k_1 = \frac{2\sqrt{k}}{1 + k}, k_2 = \frac{2\sqrt{k_1}}{1 + k_1}, \dots, k_4 = \frac{2\sqrt{k_3}}{1 + k_3} \cong 1$$

and $\quad \phi_1 = \frac{1}{2}[\phi + \sin^{-1}(k \sin \phi)], \dots, \phi_4 = \frac{1}{2}[\phi_3 + \sin^{-1}(k_3 \sin \phi_3)]$

(h) Evaluation of incomplete integral of the second kind $(m = k^2)$

For $0° < \omega < 20°$

$$E(k, \phi) = \frac{2\phi}{\pi} E + \frac{m \sin 2\phi}{8} \left(1 + \frac{3ma_1}{16} \left(1 + \frac{5ma_2}{36} \left(1 + \frac{7ma_3}{64} \right) \right) \right) + (\epsilon \le 5 \times 10^{-7})$$

where E is computed by (e) and a_1, a_2, a_3 are the same as in (g) above.

For $20° \le \omega < 90°$

$$
\begin{aligned}
E(k, \phi) &= F(k, \phi)(1 - \tfrac{1}{2}m(1 + \tfrac{1}{2}k_1(1 + \tfrac{1}{2}k_2(1 + \tfrac{1}{2}k_3(1 + \tfrac{1}{2}k_4))))) \\
&+ \tfrac{1}{2}k\sqrt{k_1} \, (\sin \phi_1 + \tfrac{1}{2}\sqrt{k_2} \, (\sin \phi_2 + \tfrac{1}{2}\sqrt{k_3} \, (\sin \phi_3 + \tfrac{1}{2}\sqrt{k_4} \sin \phi_4))) \\
&\hspace{8cm} + (\epsilon \le 5 \times 10^{-7})
\end{aligned}
$$

where with given k and ϕ the remaining symbols are the same as in (g) above.

(i) Examples. With $\omega = 10°$, $\phi = 60°$, $m = \sin^2 10° = 0.030\,153\,690$, $a_1 = 1.5$, $a_2 = 1.2$, $a_2 = 1.160\,714\,286$, $K = 1.582\,842\,804$, and $E = 1.558\,887\,197$,

$F(\sin 10°, 60°) = 1.055\,228\,536 - 0.003\,343\,417 = 1.051\,879\,119 + (\epsilon = 5 \times 10^{-10})$

$E(\sin 10°, 60°) = 1.039\,258\,131 + 0.003\,292\,055 = 1.042\,550\,186 + (\epsilon = 3 \times 10^{-7})$

where K, E are taken from (A.20). With $\omega = 30°$, $\phi = 60°$,

	$r = 1$	$r = 2$	$r = 3$	$r = 4$
k_r	0.942 809 042	0.999 566 630	0.999 999 977	1.000 000 000
ϕ_r	42.829 453 1	41.345 660 8	41.334 738 2	41.334 737 57

$F(\sin 30°, 60°) = 1.089\,550\,670 + (\epsilon = 0) \qquad E(\sin 30°, 60°) = 1.007\,555\,567 + (\epsilon = 0)$

(a) Origin and definitions. General integrals of the Bessel's differential equation

$$x^2 \frac{d^2y}{dx^2} + x \frac{dy}{dx} + (x^2 - r^2)x = 0$$

are
$$y = \begin{cases} C_1 J_r(x) + C_2 J_{-r}(x) & r \neq 0, 1, 2, \ldots \\ C_1 J_r(x) + C_2 Y_r(x) & r = \text{positive real number} \end{cases}$$

where $J_r(x)$ = Bessel function of the first kind of order r
$Y_r(x)$ = Bessel function of the second kind of order r
C_1, C_2 = constants of integration

(b) Bessel functions of the first kind are defined with $u = \frac{1}{2}x$ as

$$J_r(x) = u^r \sum_{k=0}^{\infty} \frac{(-1)^k u^{2k}}{k! \Gamma(r + 1 + k)} \qquad J_{-r}(x) = u^{-r} \sum_{k=0}^{\infty} \frac{(-1)^k u^{2k}}{k! \Gamma(k + 1 - r)}$$

$$J_{-r}(x) = (-1)^r J_r(x) \qquad \text{for } r = 0, 1, 2, \ldots, \text{only}$$

where r = positive integer or fraction and $\Gamma(\ ) $ = gamma function (2.28).

(c) Bessel functions of the second kind are defined in general as

$$Y_r(x) = \begin{cases} \dfrac{J_r(k) \cos rk - J_{-r}(x)}{\sin rx} & r \neq 0, 1, 2, \ldots \\ \lim_{p \to r} \dfrac{J_p(x) \cos px - J_{-p}(x)}{\sin px} & r = 0, 1, 2, \ldots \end{cases}$$

For $r = 0, 1, 2, \ldots$, the limit must be evaluated by L'Hopital's rule and yields with $u = \frac{1}{2}x$,

$$Y_r(x) = \frac{1}{\pi} \left\{ 2(\gamma + \ln u) J_r(x) - u^{-r} \sum_{k=0}^{r-1} \frac{(r - k - 1)!}{k!} u^{2k} \right.$$
$$\left. + u^r \sum_{k=0}^{\infty} \frac{(-1)^k [\psi(k) + \psi(r + k) + 2\gamma]}{k! \, (r + k)!} u^{2k} \right\}$$

$$Y_{-r}(x) = (-1)^r Y_r(x) \qquad \text{for } r = 0, 1, 2, \ldots, \text{only}$$

where by (2.32d),

$$\gamma + \psi(k) = 1 + \tfrac{1}{2} + \tfrac{1}{3} \qquad \gamma + \psi(r + k) = 1 + \tfrac{1}{2} + \tfrac{1}{3} + \cdots + \frac{1}{r + k}$$

$$\psi(0) = -\gamma \qquad \gamma = 0.577\,215\,665 \ (2.19)$$

The Bessel function Y_r is also called the Neumann function $N_r(x)$.

(d) Recurrent relations for all $r = 2, 3, \ldots$, and with $u = \frac{1}{2}x$ are

$$\begin{bmatrix} R_r(x) \\ R_{r+1}(x) \end{bmatrix} = \begin{bmatrix} -1 & \dfrac{r-1}{u} \\ -\dfrac{r}{u} & \dfrac{r(r-1)}{u^2} - 1 \end{bmatrix} \begin{bmatrix} -1 & \dfrac{r-3}{u} \\ -\dfrac{r-2}{u} & \dfrac{(r-2)(r-3)}{u^2} - 1 \end{bmatrix} \cdots \begin{bmatrix} -1 & \dfrac{1}{u} \\ -\dfrac{2}{u} & \dfrac{1 \cdot 2}{u^2} - 1 \end{bmatrix} \begin{bmatrix} R_0(x) \\ R_1(x) \end{bmatrix}$$

where $R_r(x)$ is $J_r(s)$ or $Y_r(x)$.

(a) Series representation of $J_r(x)$. With $u = \tfrac{1}{2}x$, $A = (\tfrac{1}{2}x)^2 = u^2$, $r = 0, 1, 2, \ldots$, and $\alpha = (-1)^{n+1}$,

$$J_0(x) = \left(1 - \frac{A}{1 \cdot 1}\left(1 - \frac{A}{2 \cdot 2}\left(1 - \frac{A}{3 \cdot 3}\left(1 - \cdots - \frac{A}{n \cdot n}\right)\right)\right)\right) - \alpha\epsilon_0$$

$$J_1(x) = u\left(1 - \frac{A}{1 \cdot 2}\left(1 - \frac{A}{2 \cdot 3}\left(1 - \frac{A}{3 \cdot 4}\left(1 - \cdots - \frac{A}{n(n+1)}\right)\right)\right)\right) - \alpha\epsilon_1$$

$$J_r(x) = \frac{u^r}{r!}\left(1 - \frac{A}{1(r+1)}\left(1 - \frac{A}{2(r+2)}\left(1 - \frac{A}{3(r+3)}\left(1 - \frac{A}{r(r+n)}\right)\right)\right)\right) - \alpha\epsilon_r$$

(b) Series representation of $Y_r(x)$. With u, A, r, α defined in (a),

$$Y_0(x) = \frac{1}{\pi}\left[B_0 + 2A\left(a_1 - \frac{A}{2 \cdot 2}\left(a_2 - \frac{A}{3 \cdot 3}\left(a_3 - \cdots - \frac{a_n A}{n \cdot n}\right)\right)\right) - \alpha\epsilon_0\right]$$

$$Y_1(x) = \frac{1}{\pi}\left[B_1 - u\left(b_1 - \frac{A}{1 \cdot 2}\left(b_2 - \frac{A}{2 \cdot 3}\left(b_3 - \cdots - \frac{b_n A}{n(n+1)}\right)\right)\right) - \alpha\epsilon_1\right]$$

$$Y_r(x) = \frac{1}{\pi}\left[B_r - \frac{u^r}{r!}\left(c_1 - \frac{A}{1(r+1)}\left(c_2 - \frac{A}{2(r+2)}\left(c_3 - \cdots - \frac{c_n A}{n(r+n)}\right)\right)\right) - \alpha\epsilon_r\right]$$

where $\quad a_1 = 1 \qquad\qquad b_1 = 1 \qquad\qquad c_1 = 1 + \frac{1}{2} + \frac{1}{3} + \cdots + \frac{1}{r}$

$$a_2 = a_1 + \frac{1}{2} \qquad b_2 = b_1 + 1 + \frac{1}{2} \qquad c_2 = 1 + c_1 + \frac{1}{r+1}$$

$$a_3 = a_2 + \frac{1}{3} \qquad b_3 = b_2 + \frac{1}{2} + \frac{1}{3} \qquad c_3 = \frac{1}{2} + c_2 + \frac{1}{r+2}$$

$$\cdots\cdots\cdots \qquad\qquad \cdots\cdots\cdots\cdots \qquad\qquad \cdots\cdots\cdots\cdots$$

$$a_n = a_{n-1} + \frac{1}{n} \quad b_n = b_{n-1} + \frac{1}{n-1} + \frac{1}{n} \quad c_n = \frac{1}{n-1} + c_{n-1} + \frac{1}{r+n-1}$$

and $\quad B_0 = 2(\gamma + \ln u)J_0(x) \qquad B_1 = 2(\gamma + \ln u)J_1(x) - \frac{1}{u}$

$$B_r = 2(\gamma + \ln u)J_r(x) - u^{-r}(r-1)!\left(1 + \frac{(r-2)A}{1}\left(1 + \frac{(r-3)A}{2}\left(1 + \cdots + \frac{(r-r)A}{r-1}\right)\right)\right)$$

(c) Graphs of $J_r(x)$ and $Y_r(x)$ for $r = 0, 1, 2$ are shown in $(10.9a)$ and $(10.10a)$, respectively. Selected special values of these functions are tabulated in $(10.9b)$ and $(10.10b)$. Also,

$$J_r(\infty) = 0 \qquad Y_r(\infty) = 0$$

(d) Numerical evaluation of $J_r(x)$ and $Y_r(x)$ is shown in (10.11). Numerical tables of $J_0(x), J_1(x)$ and $Y_0(x), Y_1(x)$ are given in $(A.16)$, $(A.17)$. The values of higher-order Bessel functions can be calculated in terms of lower-order values by the relations $(10.07d)$ as shown below.

(e) Example. If $J_0(4) = -0.397\,149\,810$ and $J_1(4) = -0.066\,043\,328$, then $u = \tfrac{4}{2}$, and

$$\begin{bmatrix} J_4(4) \\ J_5(4) \end{bmatrix} = \begin{bmatrix} -1 & \frac{3}{2} \\ -\frac{4}{2} & \frac{4 \cdot 3}{4} - 1 \end{bmatrix}\begin{bmatrix} -1 & \frac{1}{2} \\ -\frac{2}{2} & \frac{2 \cdot 1}{4} - 1 \end{bmatrix}\begin{bmatrix} J_0(4) \\ J_1(4) \end{bmatrix} = \begin{bmatrix} 0.281\,129\,065 \\ 0.132\,086\,656 \end{bmatrix}$$

(a) Graphs of $J_0(x)$ and $J_1(x)$

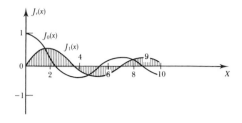

(b) Zeros and extreme values

x_{0k} = coordinate of $[J_0(x_{0k}) = 0]$ and of the extreme value of J_1 $(k > 0)$

x_{1k} = coordinate of $[J_1(x_{1k}) = 0]$ and of the extreme value of J_0 $(k > 0)$

k	x_{0k}	$J_0(x_{0k})$	$J_1(x_{0k})$	k	x_{1k}	$J_0(x_{1k})$	$J_1(x_{1k})$
1	2.404 83	0	0.519 15	1	3.831 71	−0.402 76	0
2	5.520 09	0	−0.340 26	2	7.015 59	0.300 12	0
3	8.653 73	0	0.271 42	3	10.173 47	−0.249 70	0
4	11.791 53	0	−0.232 46	4	13.323 69	0.218 36	0
5	14.930 92	0	0.206 66	5	16.470 63	−0.196 47	0
6	18.071 06	0	−0.187 73	6	19.615 86	0.180 06	0
7	21.211 64	0	0.173 27	7	22.760 08	−0.167 18	0
8	24.352 47	0	−0.161 70	8	25.903 67	0.156 72	0
9	27.493 48	0	0.152 18	9	29.046 23	−0.148 01	0
10	30.634 61	0	−0.144 17	10	32.189 68	0.140 61	0

Numerical values of $J_0(x)$ and $J_1(x)$ for selected argument x are tabulated in (A.16).

(c) Derivatives and integrals

$$\begin{bmatrix} J_0'(x) \\ J_1'(x) \end{bmatrix} = \begin{bmatrix} 0 & -1 \\ 1 & -\dfrac{1}{x} \end{bmatrix} \begin{bmatrix} J_0(x) \\ J_1(x) \end{bmatrix} \qquad \begin{bmatrix} J_0''(x) \\ J_1''(x) \end{bmatrix} = \begin{bmatrix} -1 & \dfrac{1}{x} \\ -\dfrac{1}{x} & \dfrac{2-x^2}{x^2} \end{bmatrix} \begin{bmatrix} J_0(x) \\ J_1(x) \end{bmatrix}$$

$$\begin{bmatrix} J_r'(x) \\ J_{r+1}'(x) \end{bmatrix} = \begin{bmatrix} \dfrac{r}{x} & -1 \\ 1 & -\dfrac{1+r}{x} \end{bmatrix} \begin{bmatrix} J_r(x) \\ J_{r+1}(x) \end{bmatrix} \qquad \begin{bmatrix} J_r''(x) \\ J_{r+1}''(x) \end{bmatrix} = \begin{bmatrix} \dfrac{r^2-r-x^2}{x^2} & \dfrac{1}{x} \\ -\dfrac{1}{x} & \dfrac{r^2+3r+2-x^2}{x^2} \end{bmatrix} \begin{bmatrix} J_r(x) \\ J_{r+1}(x) \end{bmatrix}$$

$\displaystyle \int J_0(x)\, dx = 2 \sum_{k=0}^{\infty} J_{2k+1}(x)$	$\displaystyle \int x J_0(x)\, dx = x J_1(x)$
$\displaystyle \int J_1(x)\, dx = -J_0(x)$	$\displaystyle \int x J_1(x)\, dx = -x J_0(x) + 2 \sum_{k=0}^{\infty} J_{2k+1}(x)$
$\displaystyle \int J_r(x)\, dx = 2 \sum_{k=0}^{\infty} J_{r+2k+1}(x)$	$\displaystyle \int \frac{1}{x} J_1(x)\, dx = -J_1(x) + 2 \sum_{k=0}^{\infty} J_{2k+1}(x)$
$\displaystyle \int x^r J_{r-1}(x)\, dx = x^r J_r(x)$	$\displaystyle \int x^{-r} J_{r+1}(x)\, dx = x^{-r} J_r(x)$

where the single primes and double primes indicate the first and second derivatives with respect to x.

(a) Graphs of $Y_0(x)$ and $Y_1(x)$

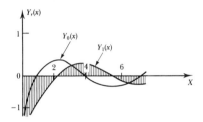

(b) Zeros and extreme values

x_{0k} = coordinate of $[Y_0(x_{0k}) = 0]$ and of the extreme value of Y_1 $(k > 0)$

x_{1k} = coordinate of $[Y_1(x_{1k}) = 0]$ and of the extreme value of Y_0 $(k > 0)$

k	x_{0k}	$Y_0(x_{0k})$	$Y_1(x_{0k})$	k	x_{1k}	$Y_0(x_{1k})$	$Y_1(x_{1k})$
1	0.893 58	0	N.A.	1	2.197 14	0.520 79	0
2	3.957 68	0	0.402 54	2	5.429 68	−0.340 32	0
3	7.086 05	0	−0.300 10	3	8.596 01	0.271 46	0
4	10.222 35	0	0.249 70	4	11.749 15	−0.232 46	0
5	13.361 10	0	−0.218 36	5	14.897 44	0.206 55	0
6	16.500 92	0	0.196 46	6	18.043 40	−0.187 73	0
7	19.641 31	0	0.180 06	7	21.188 07	0.173 27	0
8	22.782 03	0	0.167 18	8	24.331 94	−0.161 70	0
9	25.922 96	0	−0.156 73	9	27.475 29	0.152 18	0
10	29.064 03	0	0.148 01	10	30.618 39	−0.144 17	0

Numerical values of $Y_0(x)$ and $Y_1(x)$ for selected argument x are tabulated in (A.17).

(c) Derivatives and integrals

$$\begin{bmatrix} Y_0'(x) \\ Y_1'(x) \end{bmatrix} = \begin{bmatrix} 0 & -1 \\ 1 & -\dfrac{1}{x} \end{bmatrix}\begin{bmatrix} Y_0(x) \\ Y_1(x) \end{bmatrix} \qquad \begin{bmatrix} Y_0''(x) \\ Y_1''(x) \end{bmatrix} = \begin{bmatrix} -1 & \dfrac{1}{x} \\ -\dfrac{1}{x} & \dfrac{2-x^2}{x^2} \end{bmatrix}\begin{bmatrix} Y_0(x) \\ Y_1(x) \end{bmatrix}$$

$$\begin{bmatrix} Y_r'(x) \\ Y_{r+1}'(x) \end{bmatrix} = \begin{bmatrix} \dfrac{r}{x} & -1 \\ 1 & -\dfrac{1+r}{x} \end{bmatrix}\begin{bmatrix} Y_r(x) \\ Y_{r+1}(x) \end{bmatrix} \qquad \begin{bmatrix} Y_r''(x) \\ Y_{r+1}''(x) \end{bmatrix} = \begin{bmatrix} \dfrac{r^2-r-x^2}{x^2} & \dfrac{1}{x} \\ -\dfrac{1}{x} & \dfrac{r^2+3r+2-x^2}{x^2} \end{bmatrix}\begin{bmatrix} Y_r(x) \\ Y_{r+1}(x) \end{bmatrix}$$

$\displaystyle\int Y_0(x)\,dx = 2\sum_{k=0}^{\infty} Y_{2k+1}(x)$	$\displaystyle\int xY_0(x)\,dx = xY_1(x)$
$\displaystyle\int Y_1(x)\,dx = -Y_0(x)$	$\displaystyle\int xY_1(x)\,dx = -xY_0(x) + 2\sum_{k=0}^{\infty} Y_{2k+1}(x)$
$\displaystyle\int Y_r(x)\,dx = 2\sum_{k=0}^{\infty} Y_{r+2k+1}(x)$	$\displaystyle\int \frac{1}{x} Y_1(x)\,dx = -Y_1(x) + 2\sum_{k=0}^{\infty} Y_{2k+1}(x)$
$\displaystyle\int x^r Y_{r-1}(x)\,dx = x^r Y_r(x)$	$\displaystyle\int x^{-r} Y_{r+1}(x)\,dx = x^{-r} Y_r(x)$

where the single primes and double primes indicate the first and second derivatives with respect to x.

(a) Small argument. In $-3 \leq x \leq 3$, the series representing $J_r(x)$ in (10.08a) and $Y_r(x)$ in (10.08b) converge rapidly, and with $n = 7$, $|\epsilon_r| \leq 5 \times 10^{-7}$.

(b) Examples. For $x = 3$, $A = (x/2)^2 = 2.25$, and with $n = 7$,

$$J_0(3) = 1 - A\left(1 - \frac{A}{4}\left(1 - \frac{A}{9}\left(1 - \frac{A}{16}\left(1 - \cdots - \frac{A}{49}\right)\right)\right)\right) + \epsilon_0$$

$$= -0.260\,052\,35 + 4 \times 10^{-7}$$

$$Y_0(3) = \frac{1}{\pi}\left[B_0 + 2A\left(a_1 - \frac{A}{4}\left(a_2 - \frac{A}{9}\left(a_3 - \frac{A}{16}\left(a_4 - \cdots - \frac{a_7 A}{49}\right)\right)\right)\right)\right] + \bar{\epsilon}_0$$

$$= 0.376\,850\,44 - 4 \times 10^{-7}$$

where

$$B_0 = 2(0.577\,215\,67 + \ln 1.5)\, J_0(3) = -0.511\,096\,89$$

$$a_1 = 1 \qquad a_2 = a_1 + \tfrac{1}{2} \qquad a_3 = a_2 + \tfrac{1}{3} \qquad a_4 = a_3 + \tfrac{1}{4}$$

$$a_5 = a_4 + \tfrac{1}{5} \qquad a_6 = a_5 + \tfrac{1}{6} \qquad a_7 = a_6 + \tfrac{1}{7}$$

(c) Large argument. In $3 \leq x < \infty$, the following polynomial approximations introduced by E. E. Allen (Refs. 10.02, 10.03) yield good results.

$$J_0(x) = \frac{a}{\sqrt{x}}\cos b \qquad J_1(x) = \frac{c}{\sqrt{x}}\cos d \qquad Y_0(x) = \frac{a}{\sqrt{x}}\sin b \qquad Y_1(x) = \frac{c}{\sqrt{x}}\sin d$$

where with $A = 3/x$ and in terms of a_k, b_k, c_k, d_k tabulated in (d),

$$a = a_1 + A(a_2 + A(a_3 + A(a_4 + \cdots + a_7 A))) + \epsilon_a$$

$$b = b_1 + A(b_2 + A(b_3 + A(b_4 + \cdots + b_7 A))) + x + \epsilon_b$$

$$c = c_1 + A(c_2 + A(c_3 + A(c_4 + \cdots + c_7 A))) + \epsilon_c$$

$$d = d_1 + A(d_2 + A(d_3 + A(d_4 + \cdots + d_7 A))) + x + \epsilon_d$$

(d) Factors a_k, b_k, c_k, d_k

k	a_k	b_k	c_k	d_k
1	0.797 884 56	−0.785 398 16	0.797 884 56	−2.356 194 49
2	−0.000 000 77	−0.041 663 97	0.000 001 56	0.124 996 12
3	−0.005 527 40	−0.000 039 54	0.016 596 67	0.000 056 50
4	−0.000 095 12	0.002 625 73	0.000 171 05	−0.006 378 79
5	0.001 372 37	−0.000 541 25	−0.002 495 11	0.000 743 48
6	−0.000 728 05	−0.000 293 33	0.001 136 53	0.000 798 24
7	0.000 144 76	0.000 135 58	−0.000 200 33	−0.000 291 66
$|\epsilon|$	$\leq 1.6 \times 10^{-8}$	$\leq 7 \times 10^{-8}$	$\leq 4 \times 10^{-8}$	$\leq 9 \times 10^{-8}$

(e) Examples. For $x = 6$, $A = \tfrac{3}{6}$, and with

$a = 0.796\,555\,72$ $\qquad\qquad\qquad$ $b = 5.194\,047\,31$

$J_0(6) = 0.150\,645\,26 + 1 \times 10^{-9}$ $\qquad$ $Y_0(6) = -0.288\,194\,68 - 1 \times 10^{-9}$

$c = 0.801\,932\,33$ $\qquad\qquad\qquad$ $d = 3.705\,587\,20$

$J_1(6) = -0.276\,683\,86 - 1 \times 10^{-9}$ $\qquad$ $Y_1(6) = -0.175\,010\,34 - 1 \times 10^{-9}$

(a) Origin and definitions. General integral of spherical Bessel's differential equation

$$x^2 \frac{d^2 y}{dx^2} + 2x \frac{dy}{dx} + [ax^2 - r(r+1)]x = 0 \qquad (r \geq 0, a > 0)$$

is $y = C_1 j_x(ax) + C_2 y_r(ax)$ where for $a = 1$,

$$j_r(x) = \sqrt{\frac{\pi}{2x}} \, J_{r+1/2}(x) \quad = \text{spherical Bessel function of the first kind of order } r$$

$$y_r(x) = \sqrt{\frac{\pi}{2x}} \, Y_{r+1/2}(x) \quad = \text{spherical Bessel function of the second kind of order } r$$

$$J_{r+1/2}(x), Y_{r+1/2}(x) \qquad = \text{ordinary Bessel functions of order } r + \tfrac{1}{2} \ (10.07)$$

and $C_1, C_2 = $ constants of integration.

(b) Graphs of $j_0(x)$ and $j_1(x)$

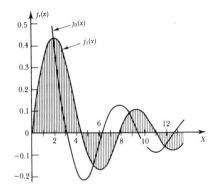

(c) Graphs of $y_0(x)$ and $y_1(x)$

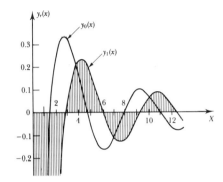

(d) Spherical Bessel functions for $r = 0, 1, 2, \ldots$ can be expressed in terms of elementary functions.

$j_0(x) = \dfrac{\sin x}{x}$	$y_0(x) = -\dfrac{\cos x}{x}$
$j_1(x) = \dfrac{\sin x}{x^2} - \dfrac{\cos x}{x}$	$y_1(x) = -\dfrac{\cos x}{x^2} - \dfrac{\sin x}{x}$
$j_2(x) = \left(\dfrac{3}{x^3} - \dfrac{1}{x}\right)\sin x - \dfrac{3}{x^2}\cos x$	$y_2(x) = -\left(\dfrac{3}{x^2} - \dfrac{1}{x}\right)\cos x - \dfrac{3}{x^2}\sin x$
$j_{-1}(x) = \dfrac{\cos x}{x}$	$y_{-1}(x) = \dfrac{\sin x}{x}$
$j_{-2}(x) = -\dfrac{\cos x}{x^2} - \dfrac{\sin x}{x}$	$y_{-2}(x) = -\dfrac{\sin x}{x^2} + \dfrac{\cos x}{x}$
$j_r(x) = \dfrac{2r-1}{x} j_{r-1}(x) - j_{r-2}(x)$	$y_r(x) = \dfrac{2r-1}{x} y_{r-1}(x) - y_{r-2}(x)$

(e) Example. For $r = 1$,

$$j_1(x) = \frac{1}{x} j_0(x) - j_{-1}(x) \qquad \text{or} \qquad j_{-1}(x) = \frac{1}{x}\left(\frac{\sin x}{x}\right) - \frac{\sin x}{x^2} + \frac{\cos x}{x} = \frac{\cos x}{x}$$

(a) Origin and definitions.
General integrals of the modified Bessel's differential equation

$$x^2 \frac{d^2y}{dx^2} + x \frac{dy}{dx} - (x^2 + r^2)x = 0 \qquad (r \geq 0)$$

are
$$y = \begin{cases} C_1 I_r(x) + C_2 I_{-r}(x) & r \neq 0, 1, 2, \ldots \\ C_1 I_r(x) + C_2 K_r(x) & r = \text{positive real number} \end{cases}$$

where $I_r(x)$ = modified Bessel function of the first kind of order
$K_r(x)$ = modified Bessel function of the second kind of order
C_1, C_2 = constants of integration

(b) Modified Bessel functions of the first kind
are defined with $u = \frac{1}{2}x$ as

$$I_r(x) = u^r \sum_{k=0}^{\infty} \frac{u^{2k}}{k!\Gamma(r+1+k)} \qquad I_{-r}(x) = u^{-r} \sum_{k=0}^{\infty} \frac{u^{2k}}{k!\Gamma(k+1-r)}$$

$$I_{-r}(x) = I_r(x) \qquad \text{for } r = 0, 1, 2, \ldots \text{ only}$$

where r = positive integer or positive fraction and $\Gamma(\ )$ = gamma function (2.28).

(c) Modified Bessel functions of the second kind
are defined in general as

$$K_r(x) = \begin{cases} \dfrac{\pi}{2} \dfrac{I_{-r}(x) - I_r(x)}{\sin r\pi} & r \neq 0, 1, 2, \ldots \\ \lim_{p \to r} \dfrac{\pi}{2} \dfrac{I_p(x) - I_p(x)}{\sin px} & r = 0, 1, 2, \ldots \end{cases}$$

For $r = 0, 1, 2, \ldots$, the limit must be evaluated by L'Hôpital's rule and yields with $u = \frac{1}{2}x$, $\alpha = (-1)^{r+1}$, $\beta = (-1)^{k+1}$,

$$K_r(x) = \alpha(\gamma + \ln u)I_r(x) - \frac{1}{2}u^{-r} \sum_{k=0}^{r-1} \frac{\beta(r-k-1)!u^{2k}}{k!}$$
$$- \frac{1}{2}\alpha u^r \sum_{k=0}^{\infty} \frac{[\psi(k) + \psi(r+k) + 2\gamma]u^{2k}}{k!(r+k)!}$$

$$K_{-r}(x) = K_r(x) \qquad \text{for } r = 0, 1, 2, \ldots, \text{ only}$$

where by (2.32d),

$$\gamma + \psi(r+k) = 1 + \frac{1}{2} + \frac{1}{3} + \cdots + \frac{1}{r+k} \qquad \gamma + \psi(k) = 1 + \frac{1}{2} + \frac{1}{3} + \cdots + \frac{1}{k}$$

$$\psi(0) = -\gamma \qquad \gamma = 0.577\,216\,665 \quad (2.18)$$

(d) Recurrent relations
for all r and with $u = \frac{1}{2}x$ are

$$\begin{bmatrix} I_r(x) \\ I_{r+1}(x) \end{bmatrix} = \begin{bmatrix} 1 & -\dfrac{r-1}{u} \\ -\dfrac{r}{u} & \dfrac{r(r-1)}{u^2} + 1 \end{bmatrix} \begin{bmatrix} 1 & -\dfrac{r-3}{u} \\ -\dfrac{r-2}{u} & \dfrac{(r-1)(r-3)}{u^2} + 1 \end{bmatrix} \cdots \begin{bmatrix} 1 & -\dfrac{1}{u} \\ -\dfrac{2}{u} & \dfrac{1 \cdot 2}{u^2} + 1 \end{bmatrix} \begin{bmatrix} I_0(x) \\ I_1(x) \end{bmatrix}$$

The function $K_r(x)$ satisfies the same relations with $I_r(x)$ replaced by $K_r(x)$ and by removing the minus sign in the off-diagonal terms as shown in (10.14e).

(a) Series representations of $I_r(x)$. With $u = \frac{1}{2}x$, $A = (\frac{1}{2}x)^2 = u^2$, and $r = 0, 1, 2, \ldots$,

$$I_0(x) = \left(1 + \frac{A}{1\cdot 1}\left(1 + \frac{A}{2\cdot 2}\left(1 + \frac{A}{3\cdot 3}\left(1 + \cdots + \frac{A}{n\cdot n}\right)\right)\right)\right) + \epsilon_0$$

$$I_1(x) = u\left(1 + \frac{A}{1\cdot 2}\left(1 + \frac{A}{2\cdot 3}\left(1 + \frac{A}{3\cdot 4}\left(1 + \cdots + \frac{A}{n(n+1)}\right)\right)\right)\right) + \epsilon_1$$

$$I_r(x) = \frac{u^r}{r!}\left(1 + \frac{A}{1(r+1)}\left(1 + \frac{A}{2(r+2)}\left(1 + \frac{A}{3(r+3)}\left(1 + \cdots + \frac{A}{n(r+n)}\right)\right)\right)\right) + \epsilon_r$$

(b) Series representation of $K_r(x)$. With $u = \frac{1}{2}x$, $A = (\frac{1}{2}x)^2$, $r = 0, 1, 2, \ldots$, and $\alpha = (-1)^{r+1}$,

$$K_0(x) = B_0 + A\left(a_1 + \frac{A}{2\cdot 2}\left(a_2 + \frac{A}{3\cdot 3}\left(a_3 + \frac{A}{4\cdot 4}\left(a_4 + \cdots + \frac{A}{n\cdot n}a_n\right)\right)\right)\right) + \epsilon_0$$

$$K_1(x) = B_1 - \frac{u}{2}\left(b_1 + \frac{A}{1\cdot 2}\left(b_2 + \frac{A}{2\cdot 3}\left(b_3 + \frac{A}{3\cdot 4}\left(b_4 + \cdots + \frac{A}{n(n+1)}b_n\right)\right)\right)\right) + \epsilon_1$$

$$K_r(x) = B_r - \left(\frac{\alpha}{2}\right)\frac{u^r}{r!}\left(c_1 + \frac{A}{1(r+1)}\left(c_2 + \frac{A}{2(r+2)}\left(c_3 + \frac{A}{3(r+3)}\left(c_4 + \cdots + \frac{A}{n(r+n)}c_n\right)\right)\right)\right) + \epsilon_r$$

where $a_1 = 1$ $b_1 = 1$ $c_1 = 1 + \dfrac{1}{2} + \dfrac{1}{3} + \cdots + \dfrac{1}{r}$

$$a_2 = a_1 + \frac{1}{2} \qquad b_2 = b_1 + 1 + \frac{1}{2} \qquad c_2 = 1 + c_1 + \frac{1}{r+1}$$

$$a_3 = a_2 + \frac{1}{3} \qquad b_3 = b_2 + \frac{1}{2} + \frac{1}{3} \qquad c_3 = \frac{1}{2} + c_2 + \frac{1}{r+2}$$

        $\cdots\cdots\cdots\cdots$ $\cdots\cdots\cdots\cdots$ $\cdots\cdots\cdots\cdots$

$$a_n = a_{n-1} + \frac{1}{n} \qquad b_n = {}_{\scriptstyle\bullet}b_{n-1} + \frac{1}{n-1} + \frac{1}{n} \qquad c_n = \frac{1}{n-1} + c_{n-1} + \frac{1}{r+n-1}$$

and $B_0 = -(\gamma + \ln u)I_0(x)$ $B_1 = (\gamma + \ln u)I_1(x) + \dfrac{1}{x}$

$$B_r = \alpha(\gamma + \ln u)I_r(x) + \frac{1}{2}u^{-r}(r-1)!\left(1 - \frac{(r-2)A}{1}\left(1 - \frac{(r-3)A}{2}\left(1 - \cdots - \frac{r-r}{r}A\right)\right)\right)$$

(c) Graphs of $I_r(x)$ and $K_r(x)$ for $r = 0, 1, 2$ are shown in (10.15a), (10.15b). Also,

$$I_r(\infty) = \qquad K_r(\infty) = 0$$

(d) Numerical evaluation of $I_r(x)$ and $K_r(x)$ is shown in (10.16), and numerical tables of $I_0(x)$, $I_1(x)$ and $K_0(x)$, $K_1(x)$ are given in (A.18), (A.19). The values of higher-order Bessel functions can be calculated in terms of lower-order values by relations (10.13d) as shown below.

(e) Example. $K_4(4)$ and $K_5(4)$ in terms of $K_0(4) = 0.011\,160$ and $K_1(4) = 0.012\,483$ given in (A.21) are calculated by the relations (10.13d) with $u = \frac{4}{2}$ as

$$\begin{bmatrix} K_4(4) \\ K_5(4) \end{bmatrix} = \begin{bmatrix} 1 & \dfrac{3}{2} \\ \dfrac{4}{2} & \dfrac{4\cdot 3}{4} + 1 \end{bmatrix}\begin{bmatrix} 1 & \dfrac{1}{2} \\ \dfrac{2}{2} & \dfrac{2\cdot 1}{4} + 1 \end{bmatrix}\begin{bmatrix} K_0(4) \\ K_1(4) \end{bmatrix} = \begin{bmatrix} 0.062\,229 \\ 0.154\,343 \end{bmatrix}$$

(a) Graphs of $I_0(x)$ and $I_1(x)$

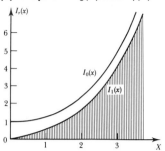

(b) Graphs of $K_0(x)$ and $K_1(x)$

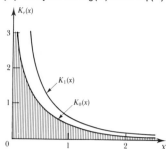

The numerical values of $I_0(x)$, $I_1(x)$, $K_0(x)$, $K_1(x)$ for selected arguments are tabulated in (A.18), (A.19).

(c) Derivatives and integrals

$$\begin{bmatrix} I_0'(x) \\ I_1'(x) \end{bmatrix} = \begin{bmatrix} 0 & 1 \\ 1 & -\dfrac{1}{x} \end{bmatrix} \begin{bmatrix} I_0(x) \\ I_1(x) \end{bmatrix} \qquad \begin{bmatrix} I_0''(x) \\ I_1''(x) \end{bmatrix} = \begin{bmatrix} 1 & -\dfrac{1}{x} \\ -\dfrac{1}{x} & \dfrac{2}{x^2}+1 \end{bmatrix} \begin{bmatrix} I_0(x) \\ I_1(x) \end{bmatrix}$$

$$\begin{bmatrix} I_r'(x) \\ I_{r+1}'(x) \end{bmatrix} = \begin{bmatrix} \dfrac{r}{x} & 1 \\ 1 & \dfrac{1+r}{x} \end{bmatrix} \begin{bmatrix} I_r(x) \\ I_{r+1}(x) \end{bmatrix} \qquad \begin{bmatrix} I_r''(x) \\ I_{r+1}''(x) \end{bmatrix} = \begin{bmatrix} \dfrac{r^2-r}{x^2}+1 & -\dfrac{1}{x} \\ -\dfrac{1}{x} & \dfrac{r^2+3r+2}{x^2}+1 \end{bmatrix} \begin{bmatrix} I_r(x) \\ I_{r+1}(x) \end{bmatrix}$$

$$\begin{bmatrix} K_0'(x) \\ K_1'(x) \end{bmatrix} = \begin{bmatrix} 0 & -1 \\ -1 & -\dfrac{1}{x} \end{bmatrix} \begin{bmatrix} K_0(x) \\ K_1(x) \end{bmatrix} \qquad \begin{bmatrix} K_0''(x) \\ K_1''(x) \end{bmatrix} = \begin{bmatrix} 1 & \dfrac{1}{x} \\ \dfrac{1}{x} & \dfrac{2}{x^2}+1 \end{bmatrix} \begin{bmatrix} K_0(x) \\ K_1(x) \end{bmatrix}$$

$$\begin{bmatrix} K_r'(x) \\ K_{r+1}' \end{bmatrix} = \begin{bmatrix} \dfrac{r}{x} & -1 \\ -1 & -\dfrac{r+1}{x} \end{bmatrix} \begin{bmatrix} K_r(x) \\ K_{r+1}(x) \end{bmatrix} \qquad \begin{bmatrix} K_r''(x) \\ K_{r+1}''(x) \end{bmatrix} = \begin{bmatrix} \dfrac{r^2-r}{x^2}+1 & \dfrac{1}{x} \\ \dfrac{1}{x} & \dfrac{r^2+3r+2}{x^2}+1 \end{bmatrix} \begin{bmatrix} K_r(x) \\ K_{r+1}(x) \end{bmatrix}$$

$\displaystyle\int I_0(x)\,dx = 2\sum_{k=0}^{\infty}(-1)^k I_{2k+1}(x)$	$\displaystyle\int K_0(x)\,dx = 2\sum_{k=0}^{\infty}(-1)^k K_{2k+1}(x)$
$\displaystyle\int I_1(x)\,dx = I_0(x)$	$\displaystyle\int K_1(x)\,dx = -K_0(x)$
$\displaystyle\int I_r(x)\,dx = 2\sum_{k=0}^{\infty}(-1)^k I_{r+2k+1}(x)$	$\displaystyle\int K_r(x)\,dx = 2\sum_{k=0}^{\infty}(-1)^k K_{r+2k+1}(x)$
$\displaystyle\int x^r I_{r-1}(x)\,dx = x^r I_r(x)$	$\displaystyle\int x^{-r} I_{r+1}(x)\,dx = x^{-r} I_r(x)$
$\displaystyle\int x^r K_{r-1}(x)\,dx = -x^r K_r(x)$	$\displaystyle\int x^{-r} K_{r+1}(x)\,dx = -x^{-r} K_r(x)$

where the single primes and double primes indicate the first and second derivatives with respect to x.

(a) Small argument. In $-4 < x < 4$, the series representing $I_r(x)$ in (10.14a) and $K_r(x)$ in (10.14b) converges rapidly and with $n = 10$, $|\epsilon_r| = 2 \times 10^{-7}$.

(b) Examples. For $x = 3.00$, $A = (\tfrac{1}{2}x)^2 = 2.25$, and with $n = 10$,

$$I_0(3) = 1 + A\left(1 + \frac{A}{4}\left(1 + \frac{A}{9}\left(1 + \frac{A}{16}\left(1 + \cdots + \frac{A}{100}\right)\right)\right)\right) + \epsilon_0$$

$$= 4.880\ 792\ 59 + 3 \times 10^{-9}$$

$$K_0(3) = B_0 + A\left(a_1 + \frac{A}{4}\left(a_2 + \frac{A}{9}\left(a_3 + \frac{A}{16}\left(a_4 + \cdots + \frac{A}{100}a_{10}\right)\right)\right)\right) + \bar{\epsilon}_0$$

$$= 0.034\ 739\ 40 + 1 \times 10^{-7}$$

where $B_0 = -(0.577\ 215\ 67 + \ln 1.5)I_0(3) = -4.796\ 260\ 65$

$$a_1 = 1 \qquad a_2 = a_1 + \tfrac{1}{2} \qquad a_3 = a_2 + \tfrac{1}{3} \qquad a_4 = a_3 + \tfrac{1}{4}$$

$$a_5 = a_4 + \tfrac{1}{5} \qquad a_6 = a_5 + \tfrac{1}{6} \qquad a_7 = a_6 + \tfrac{1}{7}$$

(c) Large argument. In $4 \le x < \infty$, the following polynomial approximations introduced by E. E. Allen (Refs. 10.02, 10.03) yield good results.

$$I_0(x) = \frac{e^x}{\sqrt{x}}\,a \qquad\qquad I_1(x) = \frac{e^x}{\sqrt{x}}\,b \qquad\qquad\qquad K_0(x) = \frac{e^{-x}}{\sqrt{x}}\,c \qquad\qquad K_1(x) = \frac{e^{-x}}{\sqrt{x}}\,d$$

where with $A = 2/x$ and in terms of a_k, b_k, c_k, d_k tabulated in (d),

$$a = a_1 + A(a_2 + A(a_3 + A(a_4 + \cdots + Aa_9))) + \epsilon_a$$

$$b = b_1 + A(b_2 + A(b_3 + A(b_4 + \cdots + Ab_9))) + \epsilon_b$$

$$c = c_1 + A(c_2 + A(c_3 + A(c_4 + \cdots + Ac_7))) + \epsilon_c$$

$$d = d_1 + A(d_2 + A(d_3 + A(d_4 + \cdots + Ad_7))) + \epsilon_d$$

(d) Factors a_k, b_k, c_k, d_k

k	a_k	b_k	c_k	d_k
1	0.398 942 28	0.398 942 28	1.253 314 14	1.253 314 14
2	0.024 911 10	−0.074 775 45	−0.078 323 58	0.234 986 19
3	0.007 921 37	−0.012 727 19	0.021 895 68	−0.036 556 20
4	−0.010 386 37	0.010 797 43	−0.010 624 46	0.015 042 68
5	0.113 248 84	−0.127 496 30	0.005 878 72	−0.007 803 53
6	−0.476 858 67	0 529 061 30	−0.002 515 40	0.003 256 14
7	1.145 187 90	−1.258 064 78	0.000 532 08	−0.000 682 45
8	−1.342 361 19	1.456 439 23		
9	0.599 395 83	−0.641 682 91		
$|\epsilon|$	$\le 1.9 \times 10^{-7}$	$\le 2.2 \times 10^{-7}$	$\le 1.9 \times 10^{-7}$	$\le 2.2 \times 10^{-7}$

(e) Examples. For $x = 4$, $A = \tfrac{2}{4}$, and with

$a = 0.414\ 003\ 85$ $b = 0.357\ 501\ 68$

$I_0(4) = 11.301\ 922\ 14 - 1.9 \times 10^{-7}$ $I_1(4) = 9.759\ 465\ 14 + 5 \times 10^{-9}$

$c = 1.218\ 595\ 34$ $d = 1.363\ 151\ 89$

$K_0(4) = 0.011\ 159\ 68 + 0$ $K_1(4) = 0.012\ 483\ 50 - 1 \times 10^{-7}$

(a) Origin and definitions. General integral of the modified spherical Bessel's differential equation

$$x^2 \frac{d^2y}{dx^2} + 2x \frac{dy}{dx} - [ax^2 + r(r+1)]x = 0 \qquad (r \geq 0, a > 0)$$

is $y = C_1 i_r(x) + C_2 k_r(x)$ where for $a = 1$

$$i_r(x) = \sqrt{\frac{\pi}{2x}} I_{r+1/2}(x) = \text{modified spherical Bessel function of the first kind of order } r$$

$$k_r(x) = \sqrt{\frac{\pi}{2x}} K_{r+1/2}(x) = \text{modified spherical Bessel functions of the second kind of order } r$$

$I_{r+1/2}(x), K_{r+1/2}(x) = $ modified Bessel functions of order $r + \frac{1}{2}$ (10.13)

and $C_1, C_2 = $ constants of integration. The graphs of $i_r(x)$ and $k_r(x)$ for $r = 0, 1$, are shown below.

(b) Graphs of $i_0(x)$ and $i_1(x)$

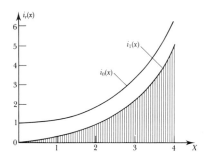

(c) Graphs of $k_0(x)$ and $k_1(x)$

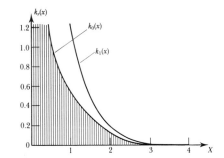

(d) Modified spherical Bessel functions for $r = 0, 1, 2, \ldots$ can be expressed in terms of elementary functions as

$i_0(x) = \dfrac{\sinh x}{x}$	$k_0(x) = \dfrac{\pi e^{-x}}{2x}$
$i_1(x) = \dfrac{\cosh x}{x} - \dfrac{\sinh x}{x^2}$	$k_1(x) = \dfrac{\pi e^{-x}}{2x}\left(1 + \dfrac{1}{x}\right)$
$i_2(x) = \left(\dfrac{3}{x^3} + \dfrac{1}{x}\right)\sinh x - \dfrac{3}{x^2}\cosh x$	$k_2(x) = \dfrac{\pi e^{-x}}{2x}\left(1 + \dfrac{3}{x} + \dfrac{3}{x^2}\right)$
$i_{-1}(x) = \dfrac{\cosh x}{x}$	$k_{-1}(x) = \dfrac{\pi e^{-x}}{2x}\left(1 - \dfrac{1}{x} - \dfrac{2}{x^2}\right)$
$i_{-2}(x) = \dfrac{\sinh x}{x} - \dfrac{\cosh x}{x}$	$k_{-2}(x) = \dfrac{\pi e^{-x}}{2x}\left(1 - \dfrac{1}{x} - \dfrac{2}{x^2}\right)$
$i_r(x) = -\dfrac{2r-1}{x}i_{r-1}(x) + i_{r-2}(x)$	$k_r(x) = -\dfrac{2r-1}{x}k_{r-1}(x) + k_{r-2}(x)$

(e) Example. For $r = 1$,

$$i_1(x) = -\frac{1}{x}i_0(x) + i_{-1}(x) \qquad \text{or} \qquad i_{-1}(x) = \frac{\sinh x}{x^2} - \frac{\sinh x}{x^2} + \frac{\cosh x}{x^2} = \frac{\cosh x}{x}$$

11
ORTHOGONAL POLYNOMIALS

(a) Notation. $k = 0, 1, 2, 3, \ldots$ $n = $ order of derivative

$D^k y = d^k y/dx^k = k$th-order derivative of $y = y(x)$ $c_k = $ real number

(b) Basic concept. The nth-order linear differential equation with variable coefficients,

$$\boxed{f_n(x)D^n y + f_{n-1}(x)D^{n-1}y + \cdots + f_2(x)D^2 y + f_1(x)Dy + f_0(x)y = g(x)}$$

can be solved under conditions defined in (c), (e) by one of the following power series.

(1) $\quad y = c_0 + c_1(x - x_0) + c_2(x - x_0)^2 + c_3(x - x_0)^3 + \cdots + c_k(x - x_0)^k + \cdots$

(2) $\quad y = (x - x_0)^m[c_0 - c_1(x + x_0) + c_2(x - x_0)^2 + c_3(x - x_0)^3 + \cdots + c_k(x - x_0)^k + \cdots]$

both of which may be of ascending or descending powers of x.

(c) Ordinary point. The real number x_0 in (1), (2) is said to be an ordinary point of the given differential equation if and only if

and $\quad \boxed{\begin{array}{l} f_n(x_0) \neq 0 \\[4pt] f_{n-1}(x), f_{n-2}(x), \ldots, f_2(x), f_1(x), f_0(x) \end{array}}$

possess a Taylor's series expansion (are analytic) in the neighborhood of x_0, which may be zero. At this point, the given differential equation possesses n linearly independent series solutions of the type (1) or (2). The radius of convergence of these series is the distance from x_0 to the nearest singular point.

(d) Singular point. A point which is not an ordinary point is called a singular point, and at such a point the functions $f_k(x)$ are not analytic.

(e) Regular singular point. If the conditions of (c) are not satisfied but the products

$$\boxed{(x - x_0)\frac{f_{n-1}(x)}{f_n(x)}, (x - x_0)^2\frac{f_{n-2}(x)}{f_n(x)}, \ldots, (x - x_0)^{n-1}\frac{f_1(x)}{f_n(x)}, (x - x_0)^n\frac{f_0(x)}{f_n(x)}}$$

are analytic at x_0, then x_0 is a regular singular point of the given differential equation, which has at least one of the series solutions of the type (1) or (2). The radius of convergence of the resulting series is again the distance from x_0 to the nearest singular point.

(f) Irregular singular point. A point which does not satisfy the conditions of (c) and (e) is called the irregular point, and no series solution exists at this point.

(g) Examples. To determine if $x_0 = 0$ is an ordinary point of the equation

$$2y'' - 4x^2 y' + 5xy = \sin 3x$$

the coefficient functions are examined first. Since $f_2(x) = 2$ for all values of x and $f_1(x) = -4x^2$, $f_0(x) = 5x$, $g(x) = \sin 3x$ are analytic for all values of $|x| < \infty$, the point x_0 which is in this interval is an ordinary point. In equation

$$x^2 y'' + x(x + 1)y' + 5y = x^2 \cos 4x$$

$x_0 = 0$ is a regular singular point, since after dividing the whole equation by x^2, $(x - 0)(x + 1)/x = x + 1$, $(x - 0)^2 5/x^2 = 5$, $[(x - 0)^2 \cos 4x]/x^2 = \cos 4x$ are all analytic at x_0.

(a) Notation. k, p, q, r = positive integers a_k, b_k, c_k = real numbers

$F(k)$ = function of k x_0 = particular point m = real number

(b) Change in limits. If

$$\sum_{k=p}^{q} F(k) = F(p) + F(p + 1) + F(p + 2) + \cdots + F(q)$$

where $q - p$ is a positive integer, then for $q = \infty$, $k = r - p$,

$$\boxed{\sum_{k=0}^{\infty} a_k F(k) x^{p+k} = \sum_{r=p}^{\infty} a_{r-p} F(r - p) x^r}$$

(c) Example. With $k = r - 1$,

$$\sum_{k=0}^{\infty} 2^k (2k + 1) x^{k+1} = \sum_{r=1}^{\infty} 2^{r-1}(2r - 1) x^r = x + 6x^2 + 20x^3 + 72x^4 + \cdots$$

(d) Equivalence. If $\boxed{\sum_{k=0}^{\infty} a_k x^k = \sum_{k=0}^{\infty} b_k x^k}$ then $a_k = b_k$

(e) Example. If

$$\sum_{k=0}^{\infty} k c_k x^k = b_0 x + b_1 x^2 + b_2 x^3 + b_3 x^4 + \cdots + b_k x^{k+1} + \cdots$$

then $c_1 = b_0, 2c_2 = b_1, 3c_3 = b_2, 4c_4 = b_3, \ldots, (k + 1)c_{k+1} = b_k, \ldots$

from which $b_1 = \dfrac{c_0}{1!}, b_2 = \dfrac{c_0}{2!}, b_3 = \dfrac{c_0}{3!}, \ldots, b_k = \dfrac{c_0}{k!}, \ldots$

and $\displaystyle\sum_{k=0}^{\infty} k c_k x^k = c_0 x \left(1 + \frac{x}{1!} + \frac{x^2}{2!} + \frac{x^3}{3!} + \cdots\right) = c_0 x \sum_{k=0}^{\infty} \frac{x^k}{k!}$

(f) Continuity. If a series of a continuous, single-valued function is uniformly convergent in a given interval, its sum is also a continuous function at least in the same interval. If the same series is also absolutely convergent in the said interval, its sum can be differentiated and integrated term by term in the same interval.

(g) Logarithmic derivative. If

$$f(m) = \frac{f_k(m)}{f_0(m)} \qquad \text{and} \qquad \ln\left[f(m)\right] = \ln\left[f_k(m)\right] - \ln\left[f_0(m)\right]$$

then

$$\frac{d}{dm}\left\{\ln\left[f(m)\right]\right\} = \frac{f'(m)}{f(m)} = \frac{f_k'(m)}{f_k(m)} - \frac{f_0'(m)}{f_0(m)}$$

and $\boxed{\dfrac{d}{dm}\left[f(m)\right] = f(m)\left[\dfrac{f_k'(m)}{f_k(m)} - \dfrac{f_0'(m)}{f_0(m)}\right]}$

(h) Example. If $f(m) = (m - a_1)(m - a_2)(m - a_3)/[(m - b_1)(m - b_2)(m - b_3)]$,

$$\frac{d}{dm}\left[f(m)\right] = \frac{(m - a_1)(m - a_2)(m - a_3)}{(m - b_1)(m - b_2)(m - b_3)}\left[\sum_{k=1}^{3} \frac{1}{m - a_k} - \sum_{k=1}^{3} \frac{1}{m - b_k}\right]$$

(a) General power series. The second-order ordinary differential equation with variable coefficients defined symbolically as

$$
f_2(x)\frac{d^2y}{dx^2} + f_1(x)\frac{dy}{dx} + f_0(x)y = g(x)
$$

where

$$
f_2(x) = \sum_{k=0}^{\infty} a_{2,k}x^k \qquad f_1(x) = \sum_{k=0}^{\infty} a_{1,k}x^k \qquad f_0(x) = \sum_{k=0}^{\infty} a_{0,k}x^k \qquad g(x) = \sum_{k=0}^{\infty} b_k x^k
$$

can be solved under certain conditions (11.01c) by the power series

$$
y = c_0 + c_1 x + c_2 x^2 + \cdots = \sum_{k=0}^{\infty} c_k x^k
$$

with derivatives

$$
\frac{dy}{dx} = \sum_{k=1}^{\infty} k c_k x^{k-1} \qquad \frac{dy^2}{dx^2} = \sum_{k=2}^{\infty} k(k-1)c_k x^{k-2}
$$

in which $c_0, c_1, c_2, \ldots$ are unknown coefficients.

(b) Coefficient equation. After y and its derivatives have been inserted in the given differential equation, the terms of like powers of x are combined and equated to the b coefficients of the same power of x. This leads to a system of infinitely many linear algebraic equations in which c coefficients are the unknowns.

$$
\begin{bmatrix}
A_{00} & A_{01} & A_{02} & & & \\
A_{10} & A_{11} & A_{12} & A_{13} & & \\
A_{20} & A_{21} & A_{22} & A_{23} & A_{24} & \\
\cdots & \cdots & \cdots & \cdots & \cdots & \cdots \\
A_{j0} & A_{j1} & A_{j2} & A_{j3} & A_{j4} & \cdots & A_{j,j+2} \\
\cdots & \cdots & \cdots & \cdots & \cdots & \cdots
\end{bmatrix}
\begin{bmatrix}
c_0 \\
c_1 \\
c_2 \\
\cdots \\
c_j \\
\cdots
\end{bmatrix}
=
\begin{bmatrix}
b_0 \\
b_1 \\
b_2 \\
\cdots \\
b_j \\
\cdots
\end{bmatrix}
$$

where

$$
A_{j,k} = k(k-1)a_{2,j-k+2} + ka_{1,j-k+1} + a_{0,j-k}
$$

as shown in (c) below.

(c) Coefficients $A_{j,k}$

k \ j	0	1	2	3	4	5
0	a_{00}	a_{10}	$2a_{20}$			
1	a_{01}	$a_{11}+a_{00}$	$2a_{21}+2a_{10}$	$2\cdot 3a_{20}$		
2	a_{02}	$a_{12}+a_{01}$	$2a_{22}+2a_{11}+a_{00}$	$2\cdot 3a_{21}+3a_{10}$	$3\cdot 4a_{20}$	
3	a_{03}	$a_{13}+a_{02}$	$2a_{23}+2a_{12}+a_{01}$	$2\cdot 3a_{22}+3a_{11}+a_{00}$	$3\cdot 4a_{21}+4a_{10}$	$4\cdot 5a_{20}$
4	a_{04}	$a_{14}+a_{03}$	$2a_{24}+2a_{13}+a_{02}$	$2\cdot 3a_{23}+3a_{12}+a_{01}$	$3\cdot 4a_{22}+4a_{11}+a_{00}$	$4\cdot 5a_{11}+5a_{10}$
$\cdots$	$\cdots$	$\cdots$	$\cdots$	$\cdots$	$\cdots$	$\cdots$

(d) Recurrent formula. The system of n equations thus obtained involves $n + 2$ c coefficients of which two remain unknown and are selected as the constants of integration. The successive elimination reveals the recurrent relation

$$c_{k+2} = -\frac{1}{A_{j,k+2}}\left(\sum_{k=0}^{j+1} A_{j,k}c_k - b_j\right)$$

which in many cases reduces to a very simple expression.

(e) Final solution of the given differential equation in (*a*) becomes

$$y = C_1 L_1(x) + C_2 L_2(x)$$

where the two independent shape functions are

$$L_1(x) = 1 + \bar{c}_2 x^2 + \bar{c}_4 x^4 + \cdots \qquad L_2(x) = x + \bar{c}_3 x^3 + \bar{c}_5 x^5 + \cdots$$

and $C_1 = c_0$, $C_2 = c_1$. The application of this procedure is shown in (11.04).

(f) Frobenius method. Many second-order equations do not lend themselves to the general solution by the power series (1) introduced in (11.01*b*). In such cases the series (2) in (11.01*b*) may yield the desired solution. Thus for

$$\frac{d^2y}{dt^2} + P(x)\frac{dy}{dx} + Q(x)y = R(x)$$

where

$$P(x) = \frac{f_1(x)}{f_2(x)} \qquad Q(x) = \frac{f_0(x)}{f_2(x)} \qquad R(x) = \frac{g(x)}{f_2(x)}$$

there exists at least one solution of the form

$$y = x^m \sum_{k=0}^{\infty} c_k x^k$$

in the neighborhood of x_0, if the condition stated in (11.01*e*) is satisfied.

(g) Indicial equation of this solution, obtained by equating to zero the coefficients of x^m, is with $R(x) = 0$,

$$m(m - 1) + mxP(x) + x^2 Q(x) = 0$$

and with $x = 0$ yields the roots m_1, m_2, which may be unequal, equal, or differ by an integer. Consequently, three types of solution may occur as shown in (11.05) to (11.07).

(h) Modified Frobenius method. At a regular singular point x_0 the solution must be taken in the form

$$y = (x - x_0)^m \sum_{k=0}^{\infty} c_k (x - x_0)^k$$

and the indicial equation in (*g*) must be changed to

$$m(m - 1) + m(x - x_0)P(x) + (x - x_0)^2 Q(x) = 0$$

which with $x = x_0$ yields the roots m_1, m_2, which may be of the three types defined in (*g*) and again three types of solution may occur.

(i) Two classes of solution are shown in the subsequent sections according to the type of point at which the series expansion is applied. They are designated as *ordinary point solutions* and *regular singular point solutions*.

(a) Example of solution about $x_0 = 0$. In

$$\frac{d^2y}{dx^2} + 4x\frac{dy}{dx} + 2y = 0$$

the selected point $x_0 = 0$ is an ordinary point, since by (11.01c),

$$f_2(x_0) = 1 \qquad \text{and} \qquad f_1(x) = 4x \qquad f_0(x) = 2$$

are analytic. Assuming the series (11.03a) to be the correct form of solution, the coefficients $A_{j,k}$ are tabulated first according to (11.03c),

k \\ j	0	1	2	3	4	5	6	7	8
0	2		$1 \cdot 2$						
1		$4 \cdot 1 + 2$		$2 \cdot 3$					
2			$4 \cdot 2 + 2$		$3 \cdot 4$				
3				$4 \cdot 3 + 2$		$4 \cdot 5$			
4					$4 \cdot 4 + 2$		$5 \cdot 6$		
5						$4 \cdot 5 + 2$		$6 \cdot 7$	
6							$4 \cdot 6 + 2$		$7 \cdot 8$
...								...	...

where $a_{00} = 2$, $a_{11} = 4$, $a_{20} = 1$ and all the remaining a coefficients are zero. From this table by the relations (11.03d) the coefficients of the series solution are related by

$$c_{k+2} = -\frac{4k + 2}{(k + 1)(k + 2)}c_k \qquad (k = 0, 1, 2, \ldots)$$

By selecting $C_1 = c_0$ and $C_2 = c_1$ as the constants of integration the c coefficients become

$$c_0 = C_1$$
$$c_2 = \left(-\frac{2}{1 \cdot 2}\right)C_1$$
$$c_4 = \left(-\frac{2}{1 \cdot 2}\right)\left(-\frac{10}{3 \cdot 4}\right)C_1$$
$$c_6 = \left(-\frac{2}{1 \cdot 2}\right)\left(-\frac{10}{3 \cdot 4}\right)\left(-\frac{18}{5 \cdot 6}\right)C_1$$
$$c_8 = \left(-\frac{2}{1 \cdot 2}\right)\left(-\frac{10}{3 \cdot 4}\right)\left(-\frac{18}{5 \cdot 6}\right)\left(-\frac{26}{7 \cdot 8}\right)C_1$$

$$c_1 = C_2$$
$$c_3 = \left(-\frac{6}{2 \cdot 3}\right)C_2$$
$$c_5 = \left(-\frac{6}{2 \cdot 3}\right)\left(-\frac{14}{4 \cdot 5}\right)C_2$$
$$c_7 = \left(-\frac{6}{2 \cdot 3}\right)\left(-\frac{14}{4 \cdot 5}\right)\left(-\frac{22}{6 \cdot 7}\right)C_2$$
$$c_9 = \left(-\frac{6}{2 \cdot 3}\right)\left(-\frac{14}{4 \cdot 5}\right)\left(-\frac{22}{6 \cdot 7}\right)\left(-\frac{30}{8 \cdot 9}\right)C_2$$

where the numerical factors in front of C_1 and C_2 are the scaled $\bar{c}$ coefficients introduced in (11.03e). Because of their form, the nested series solution offers a very compact representation shown below.

$$y = C_1\left(1 - x^2\left(1 - \frac{10}{3 \cdot 4}x^2\left(1 - \frac{18}{5 \cdot 6}x^2\left(1 - \frac{26}{7 \cdot 8}x^2\left(1 - \cdots\right)\right)\right)\right)\right)$$

$$+ C_2 x\left(1 - x^2\left(1 - \frac{14}{4 \cdot 5}x^2\left(1 - \frac{22}{6 \cdot 7}x^2\left(1 - \frac{30}{8 \cdot 9}x^2\left(1 - \cdots\right)\right)\right)\right)\right)$$

(b) Example of solution about $x_0 = a$. In

$$2\frac{d^2y}{dx^2} + (x + 2)\frac{dy}{dx} + 3y = 0$$

the selected point $x_0 = 2$ is an ordinary point, since by (11.01c),

$$f_2(x)_0 = 2 \qquad \text{and} \qquad f_1(x) = x + 2 \qquad f_0(x) = 3$$

are analytic. Using the substitution $u = x + 2$, the given equation reduces to

$$2\frac{d^2y}{du^2} + u\frac{dy}{du} + 3y = 0$$

where according to (11.03a), $a_{00} = 3$, $a_{11} = 1$, $a_{20} = 2$, and all the remaining a coefficients are zero. Assuming the series (1) in (11.01b) to be the correct form of solution, the coefficients $A_{j,k}$ are calculated by (11.03c) are

$A_{00} = 3$		$A_{02} = 1 \cdot 2 \cdot 2$					
	$A_{11} = 1 \cdot 1 + 3$		$A_{13} = 2 \cdot 3 \cdot 2$				
		$A_{22} = 2 \cdot 1 + 3$		$A_{24} = 3 \cdot 4 \cdot 2$			
			$A_{33} = 3 \cdot 1 + 3$		$A_{35} = 4 \cdot 5 \cdot 2$		
				$A_{44} = 4 \cdot 1 + 3$		$A_{16} = 5 \cdot 6 \cdot 2$	
					$\cdots$		

from which the recurrent relation (11.03d) becomes

$$c_k = -\frac{k + 1}{2k(k - 1)}c_{k-2} \qquad (k = 2, 3, 4, \ldots)$$

By selecting $C_1 = c_0$ and $C_2 = c_1$ as the constants of integration, the c coefficients are

$c_0 = C_1$	$c_6 = -\dfrac{3 \cdot 5 \cdot 7}{8(6!)}C_1$	$c_1 = C_2$	$c_7 = -\dfrac{4 \cdot 6 \cdot 8}{8(7!)}C_2$
$c_2 = -\dfrac{3}{2(2!)}C_1$	$c_8 = \dfrac{3 \cdot 5 \cdot 7 \cdot 9}{16(8!)}C_1$	$c_3 = -\dfrac{4}{2(3!)}C_2$	$c_9 = \dfrac{4 \cdot 6 \cdot 8 \cdot 10}{16(9!)}C_2$
$c_4 = \dfrac{3 \cdot 5}{4(4!)}C_1$	$\ldots\ldots\ldots\ldots$	$c_5 = \dfrac{4 \cdot 6}{4(5!)}C_2$	$\ldots\ldots\ldots\ldots$

As in (a), the numerical factors in front of C_1 and C_2 are the scaled $\bar{c}$ coefficients introduced in (11.03e). Because of the form of these factors, the resulting power series may be shown in nested form.

$$y = C_1\left(1 - \frac{3u^2}{2 \cdot 1 \cdot 2}\left(1 - \frac{5u^2}{2 \cdot 3 \cdot 4}\left(1 - \frac{7u^2}{2 \cdot 5 \cdot 6}\left(1 - \frac{9u^2}{2 \cdot 7 \cdot 8}(1 - \cdots)\right)\right)\right)\right)$$

$$+ C_2 u\left(1 - \frac{4u^2}{2 \cdot 2 \cdot 3}\left(1 - \frac{6u^2}{2 \cdot 4 \cdot 5}\left(1 - \frac{8u^2}{2 \cdot 6 \cdot 7}\left(1 - \frac{10u^2}{2 \cdot 8 \cdot 9}(1 - \cdots)\right)\right)\right)\right) \qquad (u = x + 2)$$

(a) General relations. If the unequal roots of the indicial equation in (11.03g) designated symbolically as

$$\boxed{m_1 = \alpha \qquad m_2 = \beta}$$

are inserted in the series of (11.03f), the final solution of the given differential equation becomes again

$$y = C_1 L_1(\alpha, x) + C_2 L_2(\beta, x)$$

where the two independent shape functions are

$$L_1(\alpha, x) = (1 + \bar{c}_{1,1}x + \bar{c}_{1,2}x^2 + \cdots)x^\alpha \qquad L_2(\beta, x) = (1 + \bar{c}_{2,1}x + \bar{c}_{2,2}x^2 + \cdots)x^\beta$$

and $C_1 = c_{1,0}$, $C_2 = c_{2,0}$, $\bar{c}_{1,k} = c_{1,k}/C_1$, $\bar{c}_{2,k} = c_{2,k}/C_2$ as shown in (c) below.

(b) Recurrent formulas for the calculation of $c_{1,k}$ and $c_{2,k}$ to be used in the final solution are

$$c_{1,k+2} = \frac{1}{A_{j,k+2}}\left(\sum_{k=0}^{j+1} A_{j,k}c_{1,k} - b_j\right) \qquad c_{2,k+2} = \frac{1}{B_{j,k+2}}\left(\sum_{k=0}^{j+1} B_{j,k}c_{2,k} - b_j\right)$$

in which $A_{j,k} = (k + \alpha)(k + \alpha - 1)a_{2,j-k+2} + (k + \alpha)a_{1,j-k+1} + a_{0,j-k}$

$$B_{j,k} = (k + \beta)(k + \beta - 1)a_{2,j-k+2} + (k + \beta)a_{1,j-k+1} + a_{0,j-k}$$

and j, k are the row number, column number, respectively, in the matrix equations formally identical to (11.03b).

(c) Example of solution about $x_0 = 0.$ In

$$4x\frac{d^2y}{dx^2} + 2\frac{dy}{dx} + 1 = 0$$

the selected point $x_0 = 0$ is a regular singular point, since by (11.03f),

$$xP(x) = \frac{1}{2} \qquad x^2Q(x) = \frac{x}{4}$$

are analytic. The indicial equation (11.03g)

$$m(m - 1) + \tfrac{1}{2} + \overset{0}{\cancel{x}} = 0$$

yields two unequal roots $m_1 = \alpha = 0$, $m_2 = \beta = \tfrac{1}{2}$ and indicates the solution given in (a) above. The recurrent formulas by (b) are

$c_{1,k+1} = -\dfrac{c_{1,k}}{2[\alpha(2\alpha - 1) + (4\alpha + 2k + 1)(k + 1)]}$	$= -\dfrac{c_{1,k}}{(2k + 1)(2k + 2)}$	$c_{1,1} = -\dfrac{c_{1,0}}{1 \cdot 2}$
$c_{2,k+1} = -\dfrac{c_{2,k}}{2[\beta(2\beta - 1) + (4\beta + 2k + 1)(k + 1)]}$	$= -\dfrac{c_{2,k}}{(2k + 2)(2k + 3)}$	$c_{2,1} = -\dfrac{c_{2,0}}{2 \cdot 3}$

and because of the form of these factors, the resulting power series may be shown in nested form as

$$y = C_1\left(1 - \frac{x}{1 \cdot 2}\left(1 - \frac{x}{3 \cdot 4}\left(1 - \frac{x}{5 \cdot 6}(1 - \cdots)\right)\right)\right) + C_2\left(1 - \frac{x}{2 \cdot 3}\left(1 - \frac{x}{4 \cdot 5}\left(1 - \frac{x}{6 \cdot 7}(1 - \cdots)\right)\right)\right)\sqrt{x}$$

$$= C_1 \cos\sqrt{x} + C_2 \sin\sqrt{x}$$

(a) General relations. If the roots of the indicial equation in (11.03g) are equal,

$$\boxed{m_1 = m_2 = m}$$

then the first shape function $L_1(m, x)$ is the same as in (11.05a) and the second shape function must be taken in the form of

$$L_2(m, x) = L_1(m, x) \ln (x - x_0) + \left.\frac{\partial L_1(m, x)}{\partial m}\right|_{m=0}$$

where $\partial L_1(m, x)/\partial m$ = partial derivative of $L_1(m, x)$ with respect to the root m. The final solution is

$$y = C_1 L_1(m, x) + C_2\left[L_1(m, x) \ln (x - x_0) + \left.\frac{\partial L_1(m, x)}{\partial m}\right|_{m=0}\right]$$

where C_1, C_2 are the constants of integration and in this case $c_{1,0} = 1, c_{2,0} = 0$.

(b) Example of solution about x_0 = 0. In

$$x\frac{d^2y}{dx^2} + \frac{dy}{dx} + xy = 0$$

the selected point $x_0 = 0$ is a regular singular point, since by (11.03f),

$$xP(x) = 1 \qquad x^2 Q(x) = x^2$$

are analytic. The indicial equation (11.03g)

$$m(m - 1) + mx\left(\frac{1}{x}\right) + x^{\nearrow 0} = 0$$

yields two equal roots $m_1 = m_2 = 0$ and indicates the solution given in (a) above. The first shape function with $m = 0$ and

$$\boxed{c_{1,k} = \frac{(-1)^k}{(2 \cdot 4 \cdot 6 \cdots 2k)^2}}$$

is $L_1(m, x) = 1 - \left(1 - \frac{x^2}{2^2}\left(1 - \frac{x^2}{4^2}\left(1 - \frac{x^2}{6^2}(1 - \cdots)\right)\right)\right)$

and the second shape function with $m = 0$ and

$$\boxed{c_{2,k} = \frac{\partial c_{1,k}}{\partial m} = \frac{(-1)^k s_k}{(2 \cdot 4 \cdot 6 \cdots 2k)^2}} \qquad \boxed{s_k = 1 + \frac{1}{2} + \frac{1}{3} + \cdots + \frac{1}{k}}$$

is $L_2(m, x) = L_1(m, x) \ln x + \frac{x^2}{4}\left(1 - \frac{s_2}{4^2 s_1}x^2\left(1 - \frac{s_3}{6^2 s_2}x^2\left(1 - \frac{s_4}{8^2 s_3}x^2(1 - \cdots)\right)\right)\right)$

where the partial derivative of $L_1(m, x)$ with respect to m is calculated by means of (11.02f). For example,

$$\frac{\partial c_{2,0}}{\partial m} = \left.\frac{\partial (1)}{\partial m}\right|_{m=0} = 0 \qquad \frac{\partial c_{2,1}}{\partial m} = -\left.\frac{\partial [1/(m + 2)^2]}{\partial m}\right|_{m=0} = \frac{1}{4}$$

$$\frac{\partial c_{2,2}}{\partial m} = \left.\frac{\partial [1/((m + 2)^2 (m + 4)^2)]}{\partial m}\right|_{m=0} = -\frac{1 + \frac{1}{2}}{2^2 4^2}$$

(a) General relations. If the roots of the indicial equation in (11.03g) differ by an integer,

$$\boxed{m_1 - m_2 = \alpha - \beta = r} \qquad (r = 1, 2, 3, \ldots)$$

then the first shape function $L_1(\alpha, x)$ is the same as in (11.05a) and the second shape function must be taken in the form of

$$L_2(\beta, x) = \frac{\partial}{\partial m}[(m - \beta)L_1(m, x)]_{m=\beta}$$

where the partial derivative is taken with respect to m which is a variable, whereas x is considered to be a constant. An alternative form of the second shape function is

$$L_2(\beta, x) = L_1(\alpha, x)\int \frac{e^{-t}}{[L_1(\alpha, x)]^2}\, dx \qquad \text{where } t = \int P(x)\, dx$$

The final solution is

$$y = C_1 L_1(\alpha, x) + C_2 L_2(\beta, x)$$

where C_1, C_2 are again the constants of integration.

(b) Example of solution about $x_0 = 0$. In

$$x^2 \frac{d^2y}{dx^2} + x\frac{dy}{dx} + (x^2 - 1)y = 0$$

the selected point $x_0 = 0$ is a regular singular point by (11.01f), and the indicial equation (11.03g)

$$m(m - 1) + mx\left(\frac{x}{x^2}\right) + x^2\frac{\overset{0}{\cancel{x^2}} - 1}{x^2} = 0$$

yields two roots differing by an integer; $m_1 = \alpha = 1$, $m_2 = \beta = -1$, $\alpha - \beta = 2$, and the first shape function with $\alpha = 1$ and

$$\boxed{c_{1,2k} = \frac{(-1)^k}{4^k k!(k + 1)!}}$$

is $\;L_1(\alpha, x) = x\left(1 - \frac{x^2}{4\cdot 1\cdot 2}\left(1 - \frac{x^2}{4\cdot 2\cdot 3}\left(1 - \frac{x^2}{4\cdot 3\cdot 4}(1 - \cdots)\right)\right)\right)$

and the second shape function with $\beta = -1$ and

$$\boxed{c_{2,2k} = (-1)^k\partial\left[\frac{1}{(m + 3)^2(m + 5)^2\cdots(m + 2k - 1)^2(m + 2k + 1)}\right]\Big/\partial m\Big|_{m=-1}}$$

is $\;L_2(\beta, x) = -\tfrac{1}{2}\ln x L_1(\alpha, x) + x^{-1}\left(1 + \frac{x^2}{4}\left(1 - \frac{5x^2}{16}\left(1 - \frac{x^2}{18}(1 - \cdots)\right)\right)\right)$

where the partial derivatives with respect to m are calculated by means of (11.02f). For example,

$$\frac{\partial[(m + 1)c_{1,2}]}{\partial m}\Big|_{m=-1} = \frac{\partial[-(m + 1)/((m + 1)(m + 3))]}{\partial m}\Big|_{m=-1} = \frac{1}{(m + 3)^2}\Big|_{m=-1} = \frac{1}{4}$$

(a) Characteristics. The polynomials $p_r(x)$ with $r = 0, 1, 2, \ldots$ designated by the names of their discoverers as Legendre, Chebyshev, Laguerre, and Hermite polynomials are characterized by four special conditions:

(1) *Two polynomials $p_m(x)$ and $p_n(x)$* of a given type are orthogonal in a specific interval with respect to their weight function $w(x)$.

(2) *Three polynomials $p_{r-1}(x), p_r(x), p_{r+1}(x)$* of a given type are related by their recurrence formula as

$$\boxed{p_{r+1}(x) = (\alpha_r x + \beta_r)p_r(x) - \gamma_r p_{r-1}(x)}$$

where $\alpha_r, \beta_r, \gamma_r$ are specific for each type.

(3) *Each polynomial* can be represented by its generalized Rodrigues formula as

$$p_r(x) = \frac{1}{h_r w(x)} \frac{d^r}{dx^r} \{w(x)[\phi(x)]^r\}$$

where h_r = orthogonality constant and $\phi(x)$ = polynomial in k of the first or second degree.

(b) Orthogonality. A set of polynomials

$$\boxed{\begin{aligned} p_r(x) &= \sum_{k=0}^{r} c_{r,k} x^k \\ &= c_{r,0} + c_{r,1}x + c_{r,2}x^2 + \cdots + c_{r,r-1}x^{r-1} + c_{r,r}x^r \end{aligned}}$$

of order r in x is said to be orthogonal in the interval $[a, b]$ with respect to the weight function $w(x)$ if and only if

$$\boxed{\int_a^b p_m(x)p_n(x)w(x)\,dx = \begin{cases} 0 & \text{for } m \neq n \\ h_r > 0 & \text{for } m = n = r \end{cases}}$$

where $m, n = 0, 1, 2, \ldots$.

(c) Coefficients of recurrence in $(a - 2)$ are

$$\alpha_r = \frac{k_{r+1}}{k_r} \qquad \beta_r = \frac{k_r k'_{r+1} - k_{r+1}k'_r}{k_r^2} \qquad \gamma_r = \frac{k_{r+1}k_{r-1}}{k_r^2}\frac{h_r}{h_{r-1}}$$

where $k_{r+1} = c_{r+1,r+1}$ $k_r = c_{r,r}$ $k_{r-1} = c_{r-1,r-1}$

$\qquad\qquad\quad k'_{r+1} = c_{r+1,r}$ $k'_r = c_{r,r-1}$ $k'_{r+1} = c_{r-1,r-2}$

and h_r, h_{r-1} are the orthogonality constants introduced in $(a\text{-}3)$ above.

(d) Origin. The orthogonal polynomials satisfy the differential equation

$$f_2(x)\frac{d^2y}{dx^2} + f_1(x)\frac{dy}{dx} + f_0 y = 0$$

where f_0 = constant independent of x and $f_1(x), f_2(x)$ are functions of x independent of f_0 and r. The nonvanishing a coefficients (11.03a) of these special equations are:

Differential equation	f_0	$f_1(x)$	$f_2(x)$
Legendre and Chebyshev	a_{00}	a_{11}	a_{20}, a_{22}
Laguerre	a_{00}	a_{10}, a_{11}	a_{21}
Hermite	a_{00}	a_{11}	a_{20}

(a) Origin and definitions. The general integral of the differential equation

$$(1 - x^2)\frac{d^2y}{dx^2} - 2x\frac{dy}{dx} + r(r + 1)y = 0 \qquad (-1 < x < 1)$$

is

$$y = \begin{cases} C_1\bar{P}_r(x) + C_2\bar{Q}_r(x) & \text{for } r = \text{real number} < |\infty| \\ C_1 P_r(x) + C_2 Q_r(x) & \text{for } r = 0, 1, 2, \ldots, n < \infty \end{cases}$$

where $\bar{P}_r(x)$ = Legendre function of the first kind of order r

$\bar{Q}_r(x)$ = Legendre function of the second kind of order r

$P_r(x)$ = Legendre polynomial of the first kind of degree r

$Q_r(x)$ = Legendre polynomial of the second kind of degree r

and C_1, C_2 = constants of integration.

(b) Region of ordinary points. Since by $(11.01c)$ in $(-1, 1)$,

$$f_2(x) = 1 - x^2 = 0 \qquad \text{and} \qquad f_1(x) = -2x \qquad f_0(x) = r(r + 1)$$

are analytic, all points in this region are ordinary points, and the solution of $(11.03a)$ but also the solution of $(11.05a)$ are applicable.

(c) Indicial equation of the second solution is

$$m(m - 1) = 0$$

whose roots are $m_1 = 0$ and $m_2 = 1$.

(d) Recurrence formulas for the calculation of $c_{1,k}$ and $c_{2,k}$ in $(11.05a)$ are

$$c_{1,k+2} = -\frac{(r - 2k)(r + 2k + 1)}{(2k + 1)(2k + 2)}c_{1,k} \qquad c_{2,k+2} = -\frac{(r - 2k - 1)(r + 2k + 2)}{(2k + 2)(2k + 3)}c_{2,k}$$

(e) Legendre functions of order r are defined in general with $A = x^2$ as

$$\bar{P}_r(x) = \bigwedge_{k=0}^{\infty}\left[1 - \frac{(r - 2k)(r + 2k + 1)}{(2k + 1)(2k + 2)}A\right]$$

$$= \left(1 - \frac{r(r + 1)}{1 \cdot 2}A\left(1 - \frac{(r - 2)(r + 3)}{3 \cdot 4}A\left(1 - \frac{(r - 4)(r + 5)}{5 \cdot 6}A(1 - \cdots)\right)\right)\right)$$

$$\bar{Q}_r(x) = x\bigwedge_{k=0}^{\infty}\left[1 - \frac{(r - 2k - 1)(r + 2k + 2)}{(2k + 2)(2k + 3)}A\right]$$

$$= x\left(1 - \frac{(r - 1)(r + 2)}{2 \cdot 3}A\left(1 - \frac{(r - 3)(r + 4)}{4 \cdot 5}A\left(1 - \frac{(r - 5)(r + 6)}{6 \cdot 7}A(1 - \cdots)\right)\right)\right)$$

where r = positive or negative number. For $x = 0$, $\bar{P}_r(0) = 1$, $\bar{Q}_r(0) = 0$, for all r.

(f) First derivatives of these functions are

$$\frac{d\bar{P}_r(x)}{dx} = -r(r + 1)x\bigwedge_{k=1}^{\infty}\left[1 - \frac{(r - 2k)(r + 2k + 1)}{(2k + 1)}A\right]$$

$$\frac{d\bar{Q}_r(x)}{dx} = \bigwedge_{k=0}^{\infty}\left[1 - \frac{(r - 2k - 1)(r + 2k + 3)}{(2k + 2)}A\right]$$

(a) Legendre polynomials. For $r = 0, 1, 2, \ldots$, Legendre functions in $(11.09e)$ reduce to finite polymomials defined symbolically as

$$P_r(x) = \frac{1}{2^r r!} \frac{d^r}{dx^r} [(x^2 - 1)^r]$$

$$Q_r(x) = LP_r(x) - W_{r-1}(x)$$

where

$$L = \ln \sqrt{\frac{1 + x}{1 - x}} \qquad W_{r-1}(x) = \frac{2r - 1}{r} \frac{P_{r-1}(x)}{1} + \frac{2r - 5}{r - 1} \frac{P_{r-3}(x)}{3} + \frac{2r - 9}{r - 2} \frac{P_{r-5}(x)}{5} + \cdots$$

and the series in $W_{r-1}(x)$ terminates with the term involving $P_0(x)$ if r is odd or with the term involving $P_1(x)$ if r is even. Special values are

$$P_r(0) = \begin{cases} (-1)^r \dfrac{1 \cdot 3 \cdot 5 \cdots (2r - 1)}{2 \cdot 4 \cdot 6 \cdots 2r} \\ 0 \end{cases} \qquad Q_r(0) = \begin{cases} 0 & \text{for } r \text{ even} \\ (-1)^{r+1} \dfrac{1 \cdot 3 \cdot 5 \cdots (2r - 1)}{2 \cdot 4 \cdot 6 \cdots 2r} & \text{for } r \text{ odd} \end{cases}$$

$$P_r(1) = 1 \qquad P_r(-1) = (-1)^r \qquad\qquad Q_r(1) = \infty \qquad Q_r(-1) = \infty(-1)^{r+1}$$

(b) Graphs $P_r(x)$

(c) Graphs of $Q_r(x)$

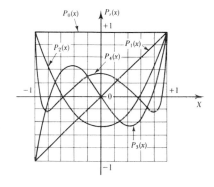

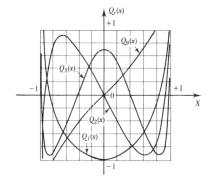

(d) Algebraic forms

r	$P_r(x)$	$Q_r(x)$
0	1	L
1	x	$LP_1(x) - 1$
2	$\frac{1}{2}(3x^2 - 1)$	$LP_2(x) - \frac{3}{2}x$
3	$\frac{1}{2}(5x^3 - 3x)$	$LP_3(x) - \frac{5}{2}x^2 + \frac{2}{3}$
4	$\frac{1}{8}(35x^4 - 30x^2 + 3)$	$LP_4(x) - \frac{35}{8}x^3 + \frac{55}{24}x$
5	$\frac{1}{8}(63x^5 - 70x^3 + 15x)$	$LP_5(x) - \frac{63}{8}x^4 + \frac{49}{8}x^2 - \frac{8}{15}$
6	$\frac{1}{16}(231x^6 - 315x^4 + 105x^2 - 5)$	$LP_6(x) - \frac{231}{16}x^5 + \frac{301}{24}x^3 - \frac{107}{80}x$
7	$\frac{1}{16}(429x^7 - 693x^5 + 315x^3 - 35x)$	$LP_7(x) - \frac{429}{16}x^6 + \frac{275}{8}x^4 - \frac{849}{80}x^2 + \frac{311}{560}$

$$L = \ln \sqrt{\frac{1 + x}{1 - x}} = \frac{x}{1} + \frac{x^3}{3} + \frac{x^5}{5} + \frac{x^7}{7} + \cdots$$

(e) Recurrence formulas

$$R_{r+1}(x) = \frac{2r+1}{r+1} xR_r(x) - \frac{r}{r+1} R_{r-1}(x) \qquad (r \geq 0)$$

$$R_r(x) = \frac{1}{2r+1} [R'_{r+1}(x) - R'_{r-1}(x)] \qquad (r > 0)$$

$$R'_r(x) = \frac{r}{x^2-1} [xR_r(x) - R_{r-1}(x)] \qquad (r > 0)$$

$$\int R_r(x)\,dx = \frac{1}{2r+1} [R_{r+1}(x) - R_{r-1}(x)] \qquad (r > 0)$$

where $R_r(x) = P_r(x)$ or $Q_r(x)$, and $R'_r(x) = dP_r(x)/dx$ or $dQ_r(x)/dx$.

(f) General relations.
In $-1 < x < 1$ and with the weight function $w(x) = 1$, the orthogonality condition is

$$\int_{-1}^{+1} P_m(x)P_n(x)\,dx = \begin{cases} 0 & \text{for } m \neq n \\ h_r & \text{for } m = n = r \end{cases}$$

where m, n, r are positive integers and the orthogonality coefficient is $h_r = 2/(2r+1)$.

$$\int_{-1}^{+1} x^r P_{r-m}(x)\,dx = \begin{cases} \dfrac{r(r!)}{(2r-m+1)!!\,m!!} & \text{for } m = 0, 2, 4, \ldots \\ 0 & \text{for } m = 1, 3, 5, \ldots \end{cases}$$

where ( )!! is the double factorial (2.30).

(g) Relations between $P_r(x)$ and $Q_r(x)$

$$P_r(x)Q_{r-1}(x) - P_{r-1}(x)Q_r(x) = \frac{1}{r} \qquad P_r(x)Q_{r-2}(x) - P_{r-2}(x)Q_r(x) = \frac{(2r-1)x}{r(r-1)}$$

11.11 TRIGONOMETRIC LEGENDRE POLYNOMIALS

(a) Origin and definitions.
The general integral of the differential equation

$$\boxed{\frac{d^2y}{d\phi^2} + \cot\phi \frac{dy}{d\phi} + r(r+1)y = 0 \qquad (-\infty < \phi < \infty)}$$

is $y = C_1 P_r(\cos\phi) + C_2 Q_r(\cos\phi)$

where $P_r(\cos\phi)$ = trigonometric Legendre polynomial of the first kind of degree r

$\qquad Q_r(\cos\phi)$ = trigonometric Legendre polynomial of the second kind of degree r

and C_1, C_2 = constants of integration.

(b) General expressions
of these polynomials for $r = 0, 1, 2, \ldots$ are

$$\boxed{\begin{aligned}
P_r(\cos\phi) &= \frac{(2r-1)!!}{2^{r-1}r!} \overset{s}{\underset{k=0}{\bigwedge}} \left[\cos(r-2k)\phi - \frac{(2k+1)(r-k)}{(k+1)(2r-2k-1)} \right] \\
Q_r(\cos\phi) &= -\ln(\tan\tfrac{1}{2}\phi) - W_{r-1}(\cos\phi)
\end{aligned}}$$

where $s = r/2$ if r is even and $s = (r-1)/2$ if r is odd and $W_{r-1}(\cos\phi)$ is the series in (11.10a) with $x = \cos\phi$ and terminating under the same conditions. The double factorial is again ( )!! as defined in (2.30).

(c) Graphs of $P_r(\cos \phi)$

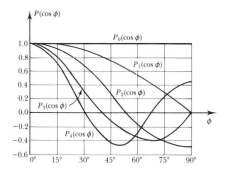

(d) Graphs of $Q_r(\cos \phi)$

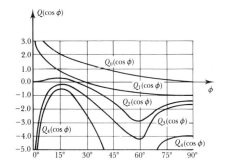

(e) Algebraic forms in terms $F_k = \cos k\phi$ with $k = 1, 2, 3, \ldots$ are

r	$P_r(\cos \phi)$	$Q_r(\cos \phi)$
0	1	L
1	F_1	$LP_1(F_1) - 1$
2	$\frac{1}{4}(3F_2 + 1)$	$LP_2(F_1) - \frac{3}{2}F_1$
3	$\frac{1}{8}(5F_3 + 3F_1)$	$LP_3(F_1) - \frac{5}{4}F_2 - \frac{7}{12}$
4	$\frac{1}{64}(35F_4 + 20F_2 + 9)$	$LP_4(F_1) - \frac{35}{32}F_3 - \frac{95}{96}F_1$
5	$\frac{1}{128}(63F_5 + 35F_3 + 30F_1)$	$LP_5(F_1) - \frac{63}{64}F_4 - \frac{14}{16}F_2 - \frac{317}{960}$
	$L = -\ln\left(\tan \frac{1}{2}\phi\right) - \dfrac{\cos \phi}{1} + \dfrac{\cos^3 \phi}{3} + \dfrac{\cos^5 \phi}{5} + \dfrac{\cos^7 \phi}{7} + \cdots$	

(f) Recurrence formulas

$$R_{r+1}(\cos \phi) = \frac{2r + 1}{r + 1} R_r(\cos \phi) - \frac{r}{r + 1} R_{r-1}(\cos \phi) \qquad (r \geq 0)$$

$$R_r'(\cos \phi) = \frac{r}{\sin^2 \phi} \left[\cos \phi R_r(\cos \phi) - R_{r-1}(\cos \phi)\right] \qquad (r > 0)$$

$$\int R_r(\cos \phi)\, d\phi = \frac{1}{r + 1} \left[R_{r+1}(\cos \phi) - R_{r-1}(\cos \phi)\right] \qquad (r > 0)$$

where $R_r(\cos \phi) = P_r(\cos \phi)$ or $Q_r(\cos \phi)$, and $R_r'(\cos \phi) = dP_r(\cos \phi)/d\phi$ or $dQ_r(\cos \phi)/d\phi$.

(g) Asymptotic approximations for large r in $\Delta < \phi < \pi - \Delta$ with $\Delta \ll \pi/10$ are

$$P_r(\cos \phi) = \sqrt{\frac{2}{r\pi \sin \phi}} \sin \left[(r + \tfrac{1}{2})\phi + \tfrac{1}{4}\pi\right] + \epsilon$$

$$Q_r(\cos \phi) = \sqrt{\frac{2}{r\pi \sin \phi}} \cos \left[(r + \tfrac{1}{2})\phi + \tfrac{1}{4}\pi\right] + \epsilon$$

where $\epsilon < 5(1/r)^{3/2}$.

(a) Origin and definitions. The general integral of the differential equation

$$(1 - x^2)\frac{d^2y}{dx^2} - 2x\frac{dy}{dx} + \left[p(r + 1) - \frac{p^2}{1 - x^2}\right]y = 0 \qquad (-1 < x < 1)$$

is $\;\; y = C_1 P_r^p(x) + C_2 Q_r^p(x) \qquad$ for $r, p = 0, 1, 2, \ldots, n < \infty$

where $P_r^p(x) = $ associated Legendre polynomial of the first kind of order p of degree r

$\qquad Q_r^p(x) = $ associated Legendre polynomial of the second kind of order p of degree r

and $C_1, C_2 = $ constants of integration. The region of regular singularity and the roots of the indicial equations are the same as in (11.09b) and (11.09c).

(b) General expressions of these polynomials for $r > p$ are

$$P_r^p(x) = (1 - x^2)^{p/2}\frac{d^p P_r(x)}{dx^p} = \frac{(1 - x^2)^{p/2}}{2^r r!}\frac{d^{p+r}}{dx^{p+r}}[(x^2 - 1)^r]$$

$$Q_r^p(x) = (1 - x^2)^{p/2}\frac{d^p Q_r(x)}{dx^p} = (1 - x^2)^{p/2}\frac{d^{p+r}}{dx^{p+r}}[LP_r(x) - W_{r-1}(x)]$$

where $P_r(x), Q_r(x), L, W_{r-1}(x)$ are defined in (11.10a). For $x = 0$ and for $r < p$, these polynomials are equal to zero, and

$$P_r^p(-x) = (-1)^{r-p}P_r^p(x) \qquad Q_r^p(-x) = (-1)^{r+p+1}Q_r^p(x) \qquad P_r^0(x) = P_r(x) \qquad Q_r^0(x) = Q_r(x)$$

(c) Recurrence formulas $(r \geq 1)$, $R_r^p(x) = P_r^p(x)$ or $Q_r^p(x)$.

$$R_r^{p+1}(x) = \frac{2px}{\sqrt{1 - x^2}}R_r^p(x) - (r - p - 1)(r + p)R_r^{p-1}(x) \qquad R_{r+1}^p(x) = \frac{2r + 1}{r + 1 - p}R_r^p(x) - \frac{r + p}{r + 1 - p}R_{r-1}^p(x)$$

$$\frac{d[R_r^p(x)]}{dx} = \frac{1}{x^2 - 1}[(r - p + 1)R_{r+1}^p(x) - (r + 1)xR_r^p(x)]$$

(d) Algebraic forms

$P_r'(x), P_r''(x), P_r'''(x) = $ first-, second-, third-order derivatives of $P_r(x)$ with respect to x

$$u = \sqrt{1 - x^2} \qquad\qquad L = \ln\sqrt{\frac{1 + x}{1 - x}} = \frac{x}{1} + \frac{x^3}{3} + \frac{x^5}{5} + \frac{x^7}{7} + \cdots \qquad\qquad L' = u^{-2}$$

p	r	$P_r^p(x)$	$Q_r^p(x)$
1	1	u	$u[L'P_1(x) - LP_1'(x)]$
	2	$3ux$	$u[L'P_2(x) + LP_2'(x)]$
	3	$3u(5x^2 - 1)/2$	$u[L'P_3(x) + LP_3'(x) - 5x]$
2	2	$3u^2$	$u^2[LP_2(x) + 2L'P_2'(x) + LP_2''(x)]$
	3	$15u^2x$	$u^2[LP_3(x) + 2L'P_3'(x) + LP_3''(x) - 5]$
	4	$15u^3$	$u^2[LP_4(x) + 2L'P_4'(x) + LP_4''(x) - 105x/4]$
3	3	$15u^3$	$u^3[L'P_3(x) + 3L'P_3'(x) + 3L'P_3''(x) + LP_3'''(x)]$
	4	$35u^3x$	$u^3[L'P_4(x) + 3L'P_4'(x) + 3L'P_4''(x) + LP_4'''(x) - 105x/4]$
	5	$15u^3(63x^2 - 7)/2$	$u^3[L'P_5(x) + 3L'P_5'(x) + 3L'P_5''(x) + LP_5'''(x) - 189x]$

(e) Orthogogonality condition stated in (11.10f) is valid for $P_m^p(x), P_n^p(x)$ with $w(x) = 1$ and the orthogonality constant (11.08b)

$$h_r^p = \frac{2}{2r + 1}\frac{(r + p)!}{(r - p)!}$$

(a) Origin and definitions. The general integral of the differential equation

$$\frac{d^2y}{d\phi^2} + \cot\phi\,\frac{dy}{d\phi} + \left[r(r+1) - \frac{p^2}{\sin^2\phi}\right]y = 0 \qquad (-\infty < \phi < \infty)$$

is $y = C_1 P_r^p(\cos\phi) + C_2 Q_r^p(\cos\phi)$ $(r, p = 0, 1, 2, \ldots, n < \infty)$

where $P_r^p(\cos\phi)$ = associated trigonometric Legendre polynomial of the first kind
             of order p of degree r

       $Q_r^p(\cos\phi)$ = associated trigonometric Legendre polynomial of the second kind
             of order p of degree r

and C_1, C_2 = constants of integration.

(b) General expressions of these polynomials for $r > p$ are

$$P_r^p(\cos\phi) = \sin^p\phi\,\frac{d^p P_r(\cos\phi)}{d(\cos\phi)^p} \qquad\qquad Q_r^p(\cos\phi) = \sin^p\phi\,\frac{d^p Q_r(\cos\phi)}{d(\cos\phi)^p}$$

where $P_r(\cos\phi)$, $Q_r(\cos\phi)$ are the trigonometric Legendre polynomials (11.11*e*).

(c) Recurrence formulas $(r \geq 1)$ with $A = \sin\phi$ and $B = \cos\phi$ are

$$R_r^{p+1}(B) = \frac{2pB}{A} R_r^p(B) - (r-p-1)(r+p)R_r^{p-1}(B) \qquad\qquad R_{r+1}^p(B) = \frac{2r+1}{r+1-p} R_r^p(B) - \frac{r+p}{r+1-p} R_{r-1}^p(B)$$

$$\frac{d[R_r^p(B)]}{dB} = -\frac{1}{A^2}[(r-p+1)R_{r+1}^p(B) - (r+1)BR_r^p(B)]$$

where $R_r^p(B) = P_r^p(B)$ or $Q_r^p(B)$.

(d) Algebraic forms

$P_r'(\cos\phi)$, $P_r''(\cos\phi)$, $P_r'''(\cos\phi)$ = first-, second-, third-order derivatives with respect to $\cos\phi$			
$A = \sin\phi$ $B = \cos\phi$ $L = -\ln(\tan\tfrac{1}{2}\phi) = \dfrac{B}{1} + \dfrac{B^3}{3} + \dfrac{B^5}{5} + \dfrac{B^7}{7} + \cdots$ $L' = B^{-2}$			

p	r	$P_r^p(\cos\phi)$	$Q_r^p(\cos\phi)$
1	1	A	$A[L'P_1(B) + LP_1'(B)]$
	2	$3AB$	$A[L'P_2(B) + LP_2'(B)]$
	3	$3A(5B^2 - 1)/2$	$A[L'P_3(B) + LP_3'(B) - 5B]$
2	2	$3A^2$	$A^2[LP_2(B) + 2L_2'P'(B) + LP_2''(B)]$
	3	$15A^2B$	$A^2[LP_3(B) + 2L'P_3'(B) + LP_3''(B) \quad 5]$
	4	$15A^3$	$A^2[LP_4(B) + 2L'P_4'(B) + LP_4''(B) - 105B/4]$
3	3	$15A^3$	$A^3[L'P_3(B) + 3LP_3'(B) + 3L'P_3''(B) + LP_3'''(B)]$
	4	$35A^3B$	$A^3[L'P_4(B) + 3LP_4'(B) + 3L'P_4''(B) + LP_4'''(B) - 105B/4]$
	5	$15A^3(63B^2 - 7)/2$	$A^3[L'P_5(B) + 3LP_5'(B) + 3L'P_5''(B) + LP_5'''(B) - 189B]$

(e) Asymptotic approximations for large r in $\Delta < \phi < \pi - \Delta$ with $\Delta \ll \pi/10$ are

$$P_r^p(\cos\phi) = (-r)^p\sqrt{\frac{2}{r\pi\sin\phi}}\,\sin\left[(r + \tfrac{1}{2})\phi + \tfrac{1}{4}\pi(1 + 2p)\right] + \epsilon$$

$$Q_r^p(\cos\phi) = (-r)^p\,\frac{2}{r\pi\sin\phi}\,\cos\left[(r + \tfrac{1}{2})\phi + \tfrac{1}{4}(1 + 2p)\right] + \epsilon \qquad \text{where } \epsilon < 5(1/r)^{3/2}.$$

(a) Origin and definitions. The general integral of the differential equation

$$(1 - x^2)\frac{d^2y}{dx^2} - x\frac{dy}{dx} + r^2y = 0 \qquad (-1 < x < 1)$$

is

$$y = \begin{cases} C_1\bar{T}_r(x) + C_2\bar{U}_r(x) & \text{for } r = \text{real number} < |\infty| \\ C_1 T_r(x) + C_2 U_r(x) & \text{for } r = 1, 2, 3, \ldots, n < \infty \\ C_1 + C_2 \arcsin x & \text{for } r = 0 \end{cases}$$

where $\bar{T}_r(x)$ = Chebyshev function of the first kind of order r

$\qquad \bar{U}_r(x)$ = Chebyshev function of the second kind of order r

$\qquad T_r(x)$ = Chebyshev polynomial of the first kind of degree r

$\qquad U_r(x)$ = Chebyshev polynomial of the second kind of degree r

and C_1, C_2 = constants of integration.

(b) Region of ordinary points. Since by $(11.01c)$ in $(-1, 1)$,

$$f_2(x) = 1 - x^2 \qquad \text{and} \qquad f_1(x) = -x \qquad f_0(x) = r^2$$

are analytic, all points of this region are ordinary points, and the solution of $(11.03a)$ but also the solution of $(11.05a)$ are applicable.

(c) Indicial equation of the second solution is

$$m(m - 1) = 0$$

whose roots are $m_1 = 0$ and $m_2 = 1$.

(d) Recurrent formulas for the calculation of $c_{1,k}$ and $c_{2,k}$ in $(11.05a)$ are

$$c_{1,k+2} = -\frac{r^2 - (2k)^2}{(2k + 1)(2k + 2)} c_{1,k} \qquad\qquad c_{2,k+2} = -\frac{r^2 - (2k + 1)^2}{(2k + 2)(2k + 3)} c_{2,k}$$

(e) Chebyshev functions of order r are defined in general with $A = x^2$ as

$$\bar{T}_r(x) = \bigwedge_{k=0}^{\infty}\left[1 - \frac{r^2 - (2k)^2}{(2k + 1)(2k + 2)} A\right]$$

$$= \left(1 - \frac{r^2}{1 \cdot 2}A\left(1 - \frac{r^2 - 4}{3 \cdot 4}A\left(1 - \frac{r^2 - 16}{5 \cdot 6}A(1 - \cdots)\right)\right)\right)$$

$$\bar{U}_r(x) = x\bigwedge_{k=0}^{\infty}\left[1 - \frac{r^2 - (2k + 1)^2}{(2k + 2)(2k + 3)} A\right]$$

$$= x\left(1 - \frac{r^2 - 1}{2 \cdot 3}A\left(1 - \frac{r^2 - 9}{4 \cdot 5}A\left(1 - \frac{r^2 - 25}{6 \cdot 7}A(1 - \cdots)\right)\right)\right)$$

where r = positive or negative number. For $x = 0$, $\bar{T}_r(0) = 1$, $\bar{U}_r(0) = 0$, for all r.

(f) First derivatives of these functions are

$$\frac{d\bar{T}_r(x)}{dx} = -r^2\bigwedge_{k=1}^{\infty}\left[1 - \frac{r^2 - (2k)^2}{(2k + 3)} A\right] \qquad\qquad \frac{d\bar{U}_r(x)}{dx} = \bigwedge_{k=0}^{\infty}\left[1 - \frac{r^2 - (2k + 1)^2}{(2k + 2)} A\right]$$

(a) Chebyshev polynomials. For $r = 1, 2, 3, \ldots$, Chebyshev functions in (11.14e) reduce to finite polynomials defined symbolically as

$$T_r(x) = \frac{(-2)^r r!}{(2r)!} \sqrt{1 - x^2} \, \frac{d^r}{dx^r} [(1 - x^2)^{r-1/2}] \qquad U_r(x) = \frac{\sqrt{1 - x^2}}{r} \frac{dT_r(x)}{dx}$$

Special values are $\quad T_r(0) = \begin{cases} (-1)^r \\ 0 \end{cases} \qquad\qquad U_r(0) = \begin{cases} 0 & \text{for } r \text{ even} \\ (-1)^r & \text{for } r \text{ odd} \end{cases}$

$$T_r(1) = 1 \qquad T_r(-1) = (-1)^r \qquad U_r(1) = 0 \qquad U_r(-1) = 0$$

(b) Graphs of $T_r(x)$

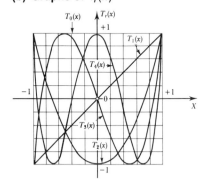

(c) Graphs of $U_r(x)$

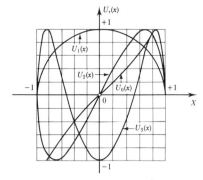

(d) Algebraic forms with $U = \sqrt{1 - x^2}$ are $T_0(x) = 1$, $U_0(x) = \sin^{-1} x$, and:

r	$T_r(x)$	$U_r(x)$
1	x	U
2	$2x^2 - 1$	$U(2x)$
3	$4x^3 - 3x$	$U(4x^2 - 1)$
4	$8x^4 - 8x^2 + 1$	$U(8x^3 - 4x)$
5	$16x^5 - 20x^3 + 5x$	$U(16x^4 - 12x^2 + 1)$
6	$32x^6 - 48x^4 + 18x^2 - 1$	$U(32x^5 - 32x^3 + 6x)$
7	$64x^7 - 112x^5 + 56x^3 - 7x$	$U(64x^6 - 80x^4 + 24x^2 - 1)$
8	$128x^8 - 256x^6 + 160x^4 - 32x^2 + 1$	$U(128x^7 - 192x^5 + 80x^3 - 8x)$
9	$256x^9 - 576x^7 + 432x^5 - 120x^3 + 9x$	$U(256x^8 - 448x^6 + 240x^4 - 40x^2 + 1)$
10	$512x^{10} - 1280x^8 + 1120x^6 - 400x^4 + 50x^2 - 1$	$U(512x^9 - 1024x^7 + 672x^5 - 160x^3 + 10x)$

(e) Recurrence formulas

$$T_{r+1}(x) = 2xT_r(x) - T_{r-1}(x) \qquad\qquad U_{r+1}(x) = 2xU_r(x) - U_{r-1}(x)$$

(f) Orthogonality condition defined in (11.08b) is applicable with $w(x) = \sqrt{1 - x^2}$ and the orthogonality constants

$$h_0 = \frac{1}{2\pi} \qquad h_r = \tfrac{1}{2}\pi \qquad (r = 1, 2, 3, \ldots)$$

(g) Trigonometric forms. With $x = \cos \phi$ in $-\infty < \phi < \infty$,

$$T_r(x) = \cos (r \arccos x) \qquad\qquad U_r(x) = \sin (r \arccos x)$$
$$T_r(\cos \phi) = \cos r\phi \qquad\qquad U_r(\cos \phi) = \sin r\phi$$

(a) Origin and definitions. The general integral of the differential

$$x\frac{d^2y}{dx^2} + (1 - x)\frac{dy}{dx} + ry = 0 \qquad (-\infty < x < \infty)$$

is $y = \begin{cases} C_1\bar{L}_r(x) + C_2\bar{L}_r^*(x) & \text{for } r = \text{real number} < |\infty| \\ C_1 L_r(x) + C_2 L_r^*(x) & \text{for } r = 0, 1, 2, \ldots, n < \infty \end{cases}$

where $\bar{L}_r(x)$ = Laguerre function of the first kind of order r

$\bar{L}_r^*(x)$ = Laguerre function of the second kind of order r

$L_r(x)$ = Laguerre polynomial of the first kind of degree r

$L_r^*(x)$ = Laguerre polynomial of the second kind of degree r

and C_1, C_2 = constants of integration.

(b) Laguerre functions of order r are defined in general with $A = 1/x$ as

$$\bar{L}_r(x) = (-x)^r \bigwedge_{k=0}^{\infty} \left[1 - \frac{(r - k)^2}{k + 1}A\right] \qquad \bar{L}_r^*(x) = \frac{e^x}{x^{r+1}} \bigwedge_{k=0}^{\infty} \left[1 + \frac{(r + k + 1)^2}{k + 1}A\right]$$

where r = positive or negative real number.

(c) Laguerre polynomials. For $r = 0, 1, 2, \ldots$, a Laguerre function of the first kind reduces to a finite polynomial and the function of the second kind must be replaced by a compatible power series as shown below, where $W_{r-1}(x)$ is derived by the procedure described in (11.06a).

$$L_r(x) = e^x\frac{d^r}{dx^r}[x^r e^{-x}] = r!\left(1 - \frac{rx}{1 \cdot 1}\left(1 - \frac{(r - 1)x}{2 \cdot 2}\left(1 - \frac{(r - 2)x}{3 \cdot 3}(1 - \cdots)\right)\right)\right)$$

$$L_r^*(x) = L_r(x)\int\frac{e^x\,dx}{x[L_r(x)]^2} = L_r(x)\ln x - W_{r-1}(x)$$

(d) Algebraic forms

$L_0(x) = 1$	$L_3(x) = 6 - 18x + 9x^2 - x^3$
$L_1(x) = 1 - x$	$L_4(x) = 24 - 96x + 72x^2 - 16x^3 + x^4$
$L_2(x) = 2 - 4x + x^2$	$L_5(x) = 120 - 600x + 600x^2 - 200x^3 + 25x^4 - x^5$

$L_6(x) = 720 - 4320x + 5400x^2 - 2400x^3 + 450x^4 - 36x^5 + x^6$

$L_7(x) = 5040 - 35\,280x + 52\,920x^2 + 29\,400x^3 + 7350x^4 - 882x^5 + 49x^6 - x^7$

$L_8(x) = 40\,320 - 322\,560x + 564\,480x^2 - 376\,320x^3 + 117\,600x^4 - 18\,816x^5 + 1568x^6 - 64x^7 + x^8$

(e) Recurrence formulas

$$L_{r+1}(x) = (2r + 1 - x)L_r(x) - r^2 L_{r-1}(x) \qquad \frac{d[L_{r+1}(x)]}{dx} = \frac{r + 1}{x}[L_{r+1}(x) - (r + 1)L_r(x)]$$

(f) Orthogonality condition defined in (11.08b) is applicable with $w(x) = e^{-x}$ and the orthogonality constant

$$h_r = (r!)^2$$

(a) Origin and definitions. The general integral of the differential equation

$$\frac{d^2y}{dx^2} - 2x\frac{dy}{dx} + 2ry = 0 \qquad (-\infty < x < \infty)$$

is

$$y = \begin{cases} C_1\bar{H}_r(x) + C_2\bar{H}_r^*(x) & \text{for } r = \text{real number} < |\infty| \\ C_1 H_r(x) + C_2 H_r^*(x) & \text{for } r = 0, 1, 2, \ldots, n < \infty \end{cases}$$

where $\bar{H}_r(x)$ = Hermite function of the first kind of order r

$\qquad \bar{H}_r^*(x)$ = Hermite function of the second kind of order r

$\qquad H_r(x)$ = Hermite polynomial of the first kind of degree r

$\qquad H_r^*(x)$ = Hermite polynomial of the second kind of parameter r

and C_1, C_2 = constants of integration.

(b) Hermite functions of order r are defined in general with $A = x^2$ as

$$\bar{H}_r(x) = \bigwedge_{k=0}^{\infty}\left[1 - \frac{2(r - 2k)A}{(1 + 2k)(2 + 2k)}\right] \qquad \bar{H}_r^*(x) = x\bigwedge_{k=0}^{\infty}\left[1 - \frac{2(r - 1 - 2k)A}{(2 + 2k)(3 + 2k)}\right]$$

where r = positive or negative real number.

(c) Hermite polynomials. For $r = 0, 1, 2, \ldots$, with $A = 1/(4x^2)$, a Hermite function of the first kind reduces to a finite polynomial and the function of the second kind must be replaced by a compatible function as shown below.

$$H_r(x) = (-1)^r e^{x^2}\frac{d^r}{dx^r}[e^{-x^2}] = (2x)^r\left(1 - \frac{r(r-1)A}{1}\left(1 - \frac{(r-2)(r-3)A}{2}\left(1 - \frac{(r-4)(r-5)A}{5}(1 - \cdots)\right)\right)\right)$$

$$H_r^*(x) = H_r(x)\int\frac{e^{x^2}\,dx}{[H_r(x)]^2} = H_r(x)\ln x - W_{r-1}(x)$$

where $W_{r-1}(x)$ is derived by the procedure described in (11.06a).

(d) Algebraic forms

$H_0(x) = 1$	$H_5(x) = 32x^5 - 160x^3 + 120x$
$H_1(x) = 2x$	$H_6(x) = 64x^6 - 480x^4 + 720x^2 - 120$
$H_2(x) = 4x^2 - 2$	$H_7(x) = 128x^7 - 1344x^5 + 3360x^3 - 1680$
$H_3(x) = 8x^3 - 12x$	$H_8(x) = 256x^8 - 3584x^4 - 13\,440x^2 + 1680$
$H_4(x) = 16x^4 - 48x^2 + 12$	$H_9(x) = 512x^9 - 9216x^7 + 48\,384x^5 - 80\,640x^3 + 30\,240x$

(e) Recurrence formulas

$$H_{r+1}(x) = 2xH_r(x) - 2rH_{r-1}(x) \qquad \frac{d[H_{r+1}(x)]}{dx} = 2(r + 1)H_r(x)$$

(f) Orthogonality condition defined in (11.08b) is applicable with $w(x) = e^{x^2}$ and the orthogonality constant

$$h_r = 2^r r!\sqrt{\pi}$$

(a) Origin and definitions. The general integral of the differential equation

$$x(1 - x)\frac{d^2y}{dx^2} + [(c - (a + b + 1)x]\frac{dy}{dx} - aby = 0$$

where a, b, c are real numbers, is

$$y = C_1\phi_1 F(\alpha_1, \beta_1, \gamma_1, u_1) + C_2\phi_2 F(\alpha_2, \beta_2, \gamma_2, u_2) \qquad r = 1, 2,$$

$$F(\alpha_r, \beta_r, \gamma_r, u_r) = \overset{\infty}{\underset{k=0}{\Lambda}}\left[1 + \frac{(\alpha_r + 1)(\beta_r + k)}{(1 + k)(\gamma_k + k)}u_r\right]$$

$$= \left(1 + \frac{\alpha_r\beta_r}{\gamma_r}u_r\left(1 + \frac{(\alpha_r + 1)(\beta_r + 1)}{2(\gamma_r + 1)}u_r\left(1 + \frac{(\alpha_r + 2)(\beta_r + 2)}{3(\gamma_r + 2)}u_r(1 + \cdots)\right)\right)\right)$$

is the ordinary hypergeometric function (series), u_r = function in x, ϕ_r = factor function in x, and C_1, C_2 = constants of integration. ϕ_r, u_r and $\alpha_r, \beta_r, \gamma_r$ take on specific values and forms depending on the parameters a, b, c.

(b) Regular singular points in this differential equation are $x = 0$, $x = 1$, and $x = \infty$. Consequently, there are three kinds of general integrals.

(c) General integrals of the first kind. If $|x| < 1$ and $c \neq 0, 1, 2, \ldots$, then

$\phi_1 = 1$	$\alpha_1 = a$	$\beta_1 = b$	$\gamma_1 = c$	$u_1 = x$
$\phi_2 = x^{1-c}$	$\alpha_2 = a - c + 1$	$\beta_2 = b - c + 1$	$\gamma_2 = 2 - c$	$u_2 = x$

If at least one of the parameters a, b is a negative integer, the hypergeometric function reduces to a finite polynomial

(d) General integrals of the second kind. If $|x - 1| < 1$ and $a + b - c \neq 0, 1, 2, \ldots$, then

$\phi_1 = 1$	$\alpha_2 = a$	$\beta_1 = b$	$\gamma_1 = a + b - c + 1$	$u_1 = 1 - x$
$\phi_2 = (1 - x)^{c-a-b}$	$\alpha_2 = c - b$	$\beta_2 = c - a$	$\gamma_2 = c - a - b + 1$	$u_2 = 1 - x$

If $c - a - b > 0$ and $x = 1$, then $\quad F(a, b, c, 1) = \dfrac{\Gamma(c)\Gamma(c - b - a)}{\Gamma(c - a)\Gamma(c - b)}$

where $\Gamma(\ )$ = gamma function (2.28).

(e) General integrals of the third kind. If $|x| > 1$ and $a - b \neq 0, 1, 2, \ldots$, then

$\phi_1 = x^{-a}$	$\alpha_1 = a$	$\beta_1 = a - c + 1$	$\gamma_1 = a - b + 1$	$u_1 = \dfrac{1}{x}$
$\phi_2 = x^{-b}$	$\alpha_2 = b$	$\beta_2 = b - c + 1$	$\gamma_2 = b - a + 1$	$u_2 = \dfrac{1}{x}$

If at least one of the parameters a, b is a negative integer, the hypergeometric function reduces to a finite polynomial.

(f) Geometric series. If $|x| < 1$ and $a = 1$, $b = c$, and $c = 0, 1, 2, \ldots$, the hypergeometric series reduces to an ordinary geometric series

$$F(1, c, c, x) = 1 + x + x^2 + x^3 + \cdots = \frac{1}{1 - x}$$

12

LEAST-SQUARES APPROXIMATIONS

(a) Statement of problem. In practice two types of problems occur in which the application of the collocation polynomials of difference type (3.07) to the approximation of a given function is not efficient.

(1) *Type I.* The given function is known only at a few points by values which are not well-established, and thus it may be required that the substitute function only approaches the given values.

(2) *Type II.* The given function is defined analytically for all values of x in an extended interval, and thus it may be required that all these values are included as much as possible in the substitute function.

In such cases, instead of using a high-degree–exact-fit polynomial, a low-degree–hit-miss polynomial may be used to minimize either the weighted mean-square error or the maximum absolute mean-square error in the selected interval. The process of finding this polynomial (substitute function) based on this concept is called the *method of least squares*.

(b) Notation

$y(x)$ = function to be approximated $w(x)$ = nonnegative weight function

$\bar{y}(x)$ = substitute function $\phi_j(x)$ = chosen function

m = number of known values n = number of chosen functions $(n \leq m)$

(c) Gauss' method is the oldest form of least-squares polynomial approximation over a set of discrete points, and (because of computational difficulties) it is applicable only in cases where low-degree polynomial approximation is required. If the values $y_0, y_1, y_2, \ldots, y_m$ are known at $x_0, x_1, x_2, \ldots, x_m$ and a substitute function

$$\bar{y}(x) = a_0 + a_1 x + a_2 x^2 + \cdots + a_n x^n = \sum_{j=0}^{n} a_j x^j$$

fits these values as close as possible, then the mean-square error

$$\epsilon^2 = \sum_{k=0}^{m} [\bar{y}_k(x_k) - y_k]^2$$

is a minimum if

$$\frac{\partial(\epsilon^2)}{\partial a_0} = 0, \frac{\partial(\epsilon^2)}{a_1} = 0, \frac{\partial(\epsilon^2)}{\partial a_2} = 0, \ldots, \frac{\partial(\epsilon^2)}{a_n} = 0$$

These n conditions yield

$$\begin{bmatrix} s_0 & s_1 & s_2 & \cdots & s_n \\ s_1 & s_2 & s_3 & \cdots & s_{n+1} \\ \cdots\cdots\cdots\cdots\cdots\cdots\cdots \\ s_n & s_{n+1} & s_{n+2} & \cdots & s_{2n} \end{bmatrix} \begin{bmatrix} a_0 \\ a_1 \\ \cdots \\ a_n \end{bmatrix} = \begin{bmatrix} t_0 \\ t_1 \\ \cdots \\ t_n \end{bmatrix}$$

where

$$s_k = \sum_{i=0}^{m} x_i^k \qquad t_k = \sum_{i=0}^{m} y_i x_i^k$$

This system of linear equations determines uniquely the desired coefficients, and the resulting polynomial $\bar{y}(x)$ possesses the minimum possible value of ϵ^2. The applications of this process are shown in (12.02), (12.03).

(d) Least-squares polynomial approximation over a set of points. If the values of $y(x)$ are known only at $m + 1$ points such as $y(x_0), y(x_1), y(x_2), \ldots, y(x_m)$, then the minimized mean-square error of the substitute function

$$\bar{y}(x) = a_0\phi_0(x) + a_1\phi_1(x) + a_2\phi_2(x) + \cdots + a_n\phi_n(x) = \sum_{j=0}^{n} a_j\phi_j(x)$$

in the prescribed interval (a, b) and in terms of chosen weights $w_0, w_1, w_2, \ldots, w_m$ is

$$\epsilon^2 = \sum_{k=0}^{m} w_k[\bar{y}(x_k) - y(x_k)]^2$$

If the functions $\phi_j(x)$ are polynomials of degree j and if they are mutually orthogonal so that

$$\sum_{k=0}^{m} w_k\phi_i(x_k)\phi_j(x_k) = 0 \qquad (i \neq j)$$

then the desired coefficients in $\bar{y}(x)$ are

$$a_j = \frac{\sum_{k=0}^{m} w_k y(x_k)\phi_j(x_k)}{\sum_{k=0}^{m} w_k[\phi_j(x_k)]^2}$$

The resulting substitute function $\bar{y}(x)$ has the striking advantage that the improvement in the approximation through addition of extra terms does not affect the previously computed coefficients. The applications of this process are shown in (12.04).

(e) Least-squares polynomial approximation over a given interval. If a given function $y(x)$ is to be approximated by

$$\bar{y}(x) = a_0\phi_0(x) + a_1\phi_1(x) + a_2\phi_2(x) + \cdots + a_n\phi_n(x) = \sum_{j=0}^{n} a_j\phi_j(x)$$

so that the minimized weighted mean-square error in the interval (a, b) is

$$\epsilon^2 = \int_a^b w(x)[\bar{y}(x) - y(x)]^2 \, dx$$

and the functions $\phi_j(x)$ are polynomials of degree j and if they are mutually orthogonal so that

$$\int_a^b w(x)\phi_i(x)\phi_j(x) \, dx = 0 \qquad (i \neq j)$$

then the desired coefficients in $\bar{y}(x)$ are

$$a_j = \frac{\int_a^b w(x)y(x)\phi_j(x) \, dx}{\int_a^b w(x)[\phi_j(x)]^2 \, dx}$$

and $\bar{y}(x)$ is the least-squares polynomial approximation of $y(x)$. The advantage of $\bar{y}(x)$ is the same as in (d). The applications of this process are shown in (12.05) to (12.11).

(a) General relations. If a set of values $y_0, y_1, \ldots, y_m$ in (12.01c) is to be approximated by a straight line

$$\bar{y}(x) = a_0 + a_1 x$$

the coefficients s_0, s_1, s_2, t_0, t_1 are determined first as

$$s_0 = m, s_1 = x_0 + x_1 + \cdots + x_m, s_2 = x_0^2 + x_1^2 + \cdots + x_m^2$$

$$t_0 = y_0 + y_1 + \cdots + y_m, t_1 = x_0 y_0 + x_1 y_1 + \cdots + x_m y_m$$

(b) Coefficient matrix equation (12.01c) in terms of these values becomes

$$\begin{bmatrix} s_0 & s_1 \\ s_1 & s_2 \end{bmatrix} \begin{bmatrix} a_0 \\ a_1 \end{bmatrix} = \begin{bmatrix} t_0 \\ t_1 \end{bmatrix}$$

from which

$$\begin{bmatrix} a_0 \\ a_1 \end{bmatrix} = \frac{1}{s_0 s_2 - s_1^2} \begin{bmatrix} s_2 & -s_1 \\ -s_1 & s_0 \end{bmatrix} \begin{bmatrix} t_0 \\ t_1 \end{bmatrix}$$

are the desired coefficients of the least-squares line in (a).

(c) Example. The following data,

x_y	2	4	6	8	10	12	14
y_k	2	6	4	10	6	12	14

should be fitted by a least-squares line $\bar{y}(x) = a_0 + a_1 x$. The matrix coefficients in (a), s_0, s_1, s_2 and t_0, t_1, are calculated below.

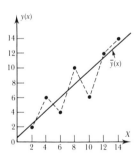

k	x_k^0	x_k	x_k^2	y_k	$x_k y_k$	$\bar{y}_k$
0	1	2	4	2	4	2.357
1	1	4	16	6	24	4.142
2	1	6	36	4	24	5.928
3	1	8	64	10	80	7.714
4	1	10	100	6	60	9.500
5	1	12	144	12	144	11.285
6	1	14	196	14	196	13.071
Σ	7	56	560	54	532	
	s_0	s_1	s_2	t_0	t_1	

The coefficient matrix in (b) yields

$$\begin{bmatrix} a_0 \\ a_1 \end{bmatrix} = \frac{1}{7(560) - 56^2} \begin{bmatrix} 560 & -56 \\ -56 & 7 \end{bmatrix} \begin{bmatrix} 54 \\ 532 \end{bmatrix} = \begin{bmatrix} 0.571 \\ 0.893 \end{bmatrix}$$

and the least-squares line is

$$\bar{y}(x) = 0.571 + 0.893x$$

The values of $\bar{y}_k$ are given in the last column of the table, and their fit is shown in the graph adjacent to the table.

(a) General relations. If a set of values $y_0, y_1, y_2, \ldots, y_m$ in (12.01c) is to be approximated by a second-degree parabola

$$\bar{y}(x) = a_0 + a_1 x + a_2 x^2$$

the coefficients $s_0, s_1, s_2, s_3, s_4, t_0, t_1, t_2$ are determined first as

$$
\begin{array}{lll}
s_0 = m & s_1 = x_0 + x_1 + \cdots + x_m & t_0 = y_0 + y_1 + \cdots + y_m \\[4pt]
& s_2 = x_0^2 + x_1^2 + \cdots + x_m^2 & t_1 = x_0 y_0 + x_1 y_1 + \cdots + x_m y_m \\[4pt]
& s_3 = x_0^3 + x_1^3 + \cdots + x_m^3 & t_2 = x_0^2 y_0 + x_1^2 y_1 + \cdots + x_m^2 y_m \\[4pt]
& s_4 = x_0^4 + x_1^4 + \cdots + x_m^4 &
\end{array}
$$

(b) Coefficient matrix equation (12.01c) in terms of these values becomes

$$
\begin{bmatrix} s_0 & s_1 & s_2 \\ s_1 & s_2 & s_3 \\ s_2 & s_3 & s_4 \end{bmatrix}
\begin{bmatrix} a_0 \\ a_1 \\ a_2 \end{bmatrix}
=
\begin{bmatrix} t_0 \\ t_1 \\ t_2 \end{bmatrix}
$$

from which, with $D = s_0 s_2 s_4 + 2 s_1 s_2 s_3 - s_1^2 s_4 - s_2^3 - s_3^2 s_0$,

$$
\begin{bmatrix} a_0 \\ a_1 \\ a_2 \end{bmatrix}
= \frac{1}{D}
\begin{bmatrix}
(s_2 s_4 - s_3^2) & -(s_1 s_4 - s_2 s_3) & (s_1 s_3 - s_2^2) \\
-(s_1 s_4 - s_2 s_3) & (s_0 s_4 - s_2^2) & -(s_0 s_3 - s_1 s_2) \\
(s_1 s_3 - s_2^2) & -(s_0 s_3 - s_1 s_2) & (s_0 s_2 - s_1^2)
\end{bmatrix}
\begin{bmatrix} t_0 \\ t_1 \\ t_2 \end{bmatrix}
$$

are the desired coefficients of the least-squares parabola in (a).

(c) Example. The following data,

x_k	1.00	1.05	1.15	1.32	1.51	1.68	1.92
y_k	0.52	0.73	1.08	1.44	1.39	1.46	1.58

should be fitted by a least-squares parabola defined in (a). The coefficients $s_1, s_2, s_3, s_4, t_0, t_1, t_2$ are calculated below and $s_0 = m = 7$.

k	x_k	x_k^2	x_k^3	x_k^4	y_k	$x_k y_k$	$x_k^2 y_k$	$\bar{y}_k$
0	1.00	1.00	1.00	1.00	0.52	0.52	0.52	0.909
1	1.05	1.10	1.16	1.22	0.73	0.77	0.80	0.914
2	1.15	1.32	1.52	1.75	1.08	1.24	1.43	0.941
3	1.32	1.74	2.30	3.04	1.44	1.90	2.51	1.032
4	1.51	2.28	3.44	5.20	1.39	2.10	3.17	1.206
5	1.68	2.82	4.74	7.97	1.46	2.45	4.12	1.424
6	1.92	3.68	7.08	13.59	1.58	3.03	5.81	1.835
Σ	9.63	13.94	21.24	33.77	8.20	12.01	18.36	
	s_1	s_2	s_3	s_4	t_0	t_1	t_2	

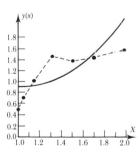

The coefficient matrix in (b) yields a_0, a_1, a_2 in terms of which the least-squares parabola is

$$\bar{y}(x) = 1.888 - 2.013x + 1.034x^2$$

The values of $\bar{y}_k$ are listed in the last column of the table, and their fit is shown in the graph adjacent to the table.

(a) Graham polynomial. If $y_0, y_1, y_2, \ldots, y_m$ are evenly spaced in the interval (a, b) and $m = 2M$ is an even integer, then with

$$x_m = \frac{a + b}{2} + \frac{k(b - a)}{2M} \qquad t = \frac{4x - 2(a + b)}{b - a} \qquad (k = 0, \pm 1, \pm 2, \ldots, \pm M)$$

the polynomial $\phi_k(x)$ in (12.01d) may be taken as

$$\phi_k(t, 2M) = 1 - \frac{k(k + 1)(M + t)}{(1!)^2(2M)} + \frac{(k - 1)(k)(k + 1)(k + 2)(M + t)(M + t - 1)}{(2!)^2(2M)(2M - 1)}$$
$$- \frac{(k - 2)(k)(k + 1)(k + 2)(k + 3)(M + t)(M + t - 1)(M + t - 2)}{(3!)^2(2M)(2M - 1)(2M - 2)} + \cdots$$

and the series terminates with the term containing $k - k + 1$ in the numerator. These polynomials are orthogonal and are very useful in the data-smoothing process.

(b) Five-point least-squares approximation. If only five values of y_k are used so that $M = 2$, then by (a), $t = -2, -1, 0, 1, 2$ and

$$\phi_0(t) = 1 \qquad \phi_1(t) = \tfrac{1}{2}t \qquad \phi_2(t) = \tfrac{1}{2}(t^2 - 2) \qquad \phi_3(t) = \tfrac{1}{6}(5t^3 - 17t)$$

t	$\phi_0(t)$	$\phi_1(t)$	$\phi_2(t)$	$\phi_3(t)$
-2	1	-1	1	-1
-1	1	$-\tfrac{1}{2}$	$-\tfrac{1}{2}$	2
0	1	0	-1	0
1	1	$\tfrac{1}{2}$	$-\tfrac{1}{2}$	-2
2	1	1	1	1

from which, by (12.01d),

$$a_0 = \frac{y_0 + y_1 + y_2 + y_3 + y_4}{(1)^2 + (1)^2 + (1)^2 + (1)^2 + (1)^2} \qquad a_1 = \frac{-y_0 - \tfrac{1}{2}y_1 + \tfrac{1}{2}y_3 + y_4}{(-1)^2 + (-\tfrac{1}{2})^2 + (0)^2 + (\tfrac{1}{2})^2 + (1)^2}$$

$$a_2 = \frac{y_0 - \tfrac{1}{2}y_1 - y_2 - \tfrac{1}{2}y_3 + y_4}{(1)^2 + (-\tfrac{1}{2})^2 + (-1)^2 + (\tfrac{1}{2})^2 + (1)^2} \qquad a_3 = \frac{-y_0 + 2y_1 - 2y_3 + y_4}{(-1)^2 + (2)^2 + (-2)^2 + (1)^2}$$

(c) Example. If the empirical data are given as

x_k	0.0	0.2	0.4	0.6	0.8
y_k	1.1	1.8	2.4	4.2	5.8
t	-2	-1	0	1	2
$\bar{y}_k$	1.386	1.6457	2.6314	4.0457	5.8386

where the last row gives the least-square approximations based on

$$\bar{y}(t) = a_0\phi_0(t) + a_1\phi_1(t) + a_2\phi_2(t) + a_3\phi_3(t)$$

in which, by (b),

$$a_0 = 3.0600 \qquad a_1 = 2.3600 \qquad a_2 = 0.4286 \qquad a_3 = -0.0100$$

(a) Substitute function $\bar{y}(x)$ for $y(x)$ in (a, b) transformed by $x = \frac{1}{2}[(a + b) + (b - a)u]$ to $\bar{y}(u)$ in $(-1, 1)$ is, according to $(12.01e)$,

$$\bar{y}(u) = \sum_{r=0}^{n} a_r P_r(u) \qquad (r = 0, 1, 2, \ldots, n)$$

where $P_r(u)$ = Legendre polynomial of the first kind of degree r (11.10) and

$$a_r = \frac{2r + 1}{2} \int_{-1}^{1} y(u) P_r(u)\, du$$

(b) Integrals used in the evaluation of a_r are

$$\int_{-1}^{1} f(u) P_r(u)\, du = \frac{1}{r!} \int_{-1}^{1} \left(\frac{1 - u^2}{2}\right)^r \frac{d^r}{du^r}[f(u)]\, du$$

(c) Square minimum error of this approximation is

$$|\epsilon|^2 = \int_{-1}^{1} [y(u)]^2\, du - \sum_{r=0}^{n} \frac{2a_r^2}{2r + 1}$$

(d) Legendre polynomials $P_r(u)$

$P_0(u) = 1$	$P_4(u) = \frac{1}{8}(35u^4 - 39u^2 + 3)$
$P_1(u) = u$	$P_5(u) = \frac{1}{8}(63u^5 - 70u^3 + 15u)$
$P_2(u) = \frac{1}{2}(3u^2 - 1)$	$P_6(u) = \frac{1}{16}(23u^6 - 315u^4 + 105u^2 - 5)$
$P_3(u) = \frac{1}{2}(5u^3 - 3u)$	$P_7(u) = \frac{1}{16}(429u' - 693u^5 + 315u^3 - 35u)$
$P_8(u) = \frac{1}{128}(6\,435u^8 - 12\,012u^6 + 6\,930u^4 - 1\,260u^2 + 35$	

and in general,

$$P_r(u) = \frac{1}{2r!} \frac{d^r}{du^r}[(u^2 - 1)^r] = \frac{(2r - 1)u}{r} P_{r-1}(u) - \frac{r - 1}{r} P_{r-2}(u)$$

(e) Special values

$$P_r(0) = \begin{cases} (-1)^{r/2} \dfrac{(r - 1)!!}{r!!} & r \text{ even} \\ 0 & r \text{ odd} \end{cases} \qquad\qquad \begin{aligned} P_r(1) &= 1 \\ P_r(-1) &= (-1)^r \end{aligned}$$

(f) Powers of u in terms of $P_r(u) = P_r$ are

$u = P_1$	$u^5 = \frac{1}{63}(270_1 + 28P_3 + 8P_5)$
$u^2 = \frac{1}{3}(1 + 2P_2)$	$u^6 = \frac{1}{231}(33 + 110P_2 + 72P_4 + 16P_6)$
$u^3 = \frac{1}{5}(3P_1 + 2P_3)$	$u^7 = \frac{1}{429}(143P_1 + 182P_3 + 88P_5 + 16P_7)$
$u^4 = \frac{1}{35}(7 + 20P_2 + 8P_4)$	$u^8 = \frac{1}{6435}(715 + 2600P_2 + 2160P_4 + 832P_6 + 128P_8)$

(a) Substitute function. When the errors near the ends of the interval (a, b) are to be considered, the polynomial approximation of $y(x)$ in terms of the Chebyshev weight function

$$w(x) = \frac{1}{\sqrt{(x-a)/(b-x)}}$$

transformed as in (12.05a) to

$$w(x) = \frac{1}{\sqrt{1-u^2}} \qquad (-1 < u < 1)$$

becomes

$$\bar{y}(u) = \frac{a_0}{\pi\sqrt{1-u^2}} + \frac{2}{\pi\sqrt{1-u^2}} \sum_{r=1}^{n} a_r T_r(u)$$

where $T_r(u) =$ Chebyshev polynomial of the first kind of degree r (11.15) and

$$a_0 = \int_{-1}^{1} y(u)\, du \qquad a_r = \int_{-1}^{1} y(u) T_r(u)\, du$$

(b) Square minimum error of this approximation is

$$|\epsilon|^2 = \int_{-1}^{1} \sqrt{1-u^2} \left[y(u) - \frac{1}{\sqrt{1-u^2}} \sum_{r=0}^{n} a_r T_r(u) \right]^2 du$$

(c) Chebyshev polynomials $T_r(u)$ are $T_0(u) = 1$, $T_1(u) = u$, and in general (11.15d),

$$T_r(u) = \frac{(-2)^r r!}{(2r)!} \sqrt{1-u^2}\, \frac{d^r}{du^r} [(1-u^2)^{r-1/2}] = 2u T_{r-1}(u) - T_{r-2}(u)$$

(d) Special values

$$T_r(0) = \begin{cases} (-1)^{r/2} & r \text{ even} \\ 0 & r \text{ odd} \end{cases} \qquad \begin{matrix} T_r(1) = 1 \\ T_r(-1) = (-1)^r \end{matrix}$$

(e) Derivatives

$$\frac{d[T_r(u)]}{du} = \begin{cases} 2r[T_{r-1}(u) + T_{r-3}(u) + \cdots + T_2(u) + \frac{1}{2}] & r \text{ odd} \\ 2r[T_{r-1}(u) + T_{r-3}(u) + \cdots + T_3(u) + T_1(u)] & r \text{ even} \end{cases}$$

(f) Indefinite integrals

$$\int T_0(u)\, du = T_1(u) + C \qquad \int T_1(u)\, du = -\tfrac{1}{4} T_2(u) + C$$

and in general

$$\int T_r(u)\, du = \frac{1}{2} \left[\frac{T_{r+1}(u)}{r+1} - \frac{T_{r-1}(u)}{r-1} \right] + C \qquad (r > 1)$$

where $C =$ constant of integration.

(g) Definite integrals used in the evaluation of a_r in (a) are

$$
\int_{-1}^{1} u^m T_r(u)\, du =
\begin{cases}
0 & (m + r)\ \text{odd} \\[2mm]
2(m!)\left[\dfrac{T_r(1)}{(m+1)!} - \dfrac{T'_r(1)}{(m+2)!} + \dfrac{T''_r(1)}{(m+3)!} - \cdots\right] & (m + r)\ \text{even}
\end{cases}
$$

$$
\int_{-1}^{1} y(u)\, T_r(u)\, du = [I_1 T_r(u) - I_2 T'_r(u) + I_3 T''_r(u) - \cdots]_{-1}^{1}
$$

where $m = 1, 2, 3, \ldots$, $T'_r(u) = d[T_r(u)]/du$, $T''_r(u) = d^2[T_r(u)]/du^2, \ldots$, and

$$
I_1 = \int y(u)\, du, \quad I_2 = \int\int y(u)\, du\, du, \ldots
$$

(h) Powers of u in terms of $T_r(u)$ are

$u = T_1$	$u^6 = \frac{1}{32}(10 + 15T_2 + 6T_4 + T_6)$
$u^2 = \frac{1}{2}(1 + T_2)$	$u^7 = \frac{1}{64}(35T_1 + 21T_3 + 7T_5 + T_7)$
$u^3 = \frac{1}{4}(3T_1 + T_3)$	$u^8 = \frac{1}{128}(35 + 56T_2 + 28T_4 + 8T_6 + T_8)$
$u^4 = \frac{1}{8}(3 + 4T_2 + T_4)$	$u^9 = \frac{1}{256}(126T_1 + 84T_3 + 36T_5 + 9T_7 + T_9)$
$u^5 = \frac{1}{16}(10T_1 + 5T_3 + T_5)$	$u^{10} = \frac{1}{512}(126 + 210T_2 + 120T_4 + 45T_6 + 10T_8 + T_{10})$

$$
u^{11} = \frac{1}{1024}(462T_1 + 330T_3 + 165T_5 + 55T_7 + 11T_9 + T_{11})
$$

$$
u^{12} = \frac{1}{2048}(462 + 792T_2 + 495T_4 + 220T_6 + 66T_8 + 12T_{10} + T_{12})
$$

and in general, with $p = \frac{1}{2}r$ and $q - \frac{1}{2}(r - 1)$,

$$
u^r =
\begin{cases}
\dfrac{\binom{r}{0}T_r + \binom{r}{1}T_{r-2} + \binom{r}{2}T_{r-4} + \cdots + \binom{r}{q}T_1}{2^{r-1}} & r\ \text{odd} \\[5mm]
\dfrac{\binom{r}{0}T_r + \binom{r}{1}T_{r-2} + \binom{r}{2}T_{r-4} + \cdots + \frac{1}{2}\binom{r}{p}T_0}{2^{r-1}} & r\ \text{even}
\end{cases}
$$

(i) Chebshev criterion states that of all polynomials of degree r having the leading coefficient equal to 1, the polynomial

$$
p_r(u) = \frac{T_r(u)}{2^{r-1}}
$$

has the smallest least upper bound for its absolute value in the interval $-1 \leq u \leq 1$. Since the upper bound of the Chebyshev polynomials is 1,

$$
\boxed{\text{Upper bound of } |p_r(u)| = \frac{1}{2^{r-1}}}
$$

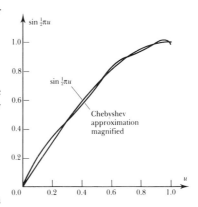

and the Chebyshev approximation confines the maximum error to a minimum. The adjacent graph for $\sin \frac{1}{2}\pi u$ illustrates the meaning of this statement.

(a) Change in interval. In many practical problems the approximation in $(0, 1)$ is more convenient than in $(-1, 1)$. In such cases the shifted Chebyshev polynomials are used.

$$\boxed{T_r^*(u) = T_r(2u - 1)}$$　　　　　　$$\boxed{T_0^*(u) = 1}$$

(b) Shifted polynomials

$T_1^*(u) = 2u - 1$	$T_3^*(u) = 32^3 - 48u^2 + 18u - 1$
$T_2^*(u) = 8u^2 - 8u + 1$	$T_4^*(u) = 128u^4 - 256u^3 + 160u^2 - 32u + 1$

$T_5^*(u) = 512u^5 - 1280u^4 + 1120u^3 - 400u^2 + 500 - 1$

$T_6^*(u) = 2048u^6 - 6144u^5 + 6144u^4 - 3584u^3 + 840u^2 - 72u + 1$

$T_7^*(u) = 8192u^7 - 28\,672u^6 + 39\,424u^5 - 26\,880u^4 + 9408u^3 - 1568u^2 + 98u - 1$

$T_8^*(u) = 32\,768u^8 - 131\,072u^7 + 213\,992u^6 - 180\,224u^5 + 84\,480u^4$
$$- 21\,504u^3 + 2688u^2 - 128u + 1$$

and in general,

$$\boxed{T_r^*(u) = (4u - 2)T_{r-1}^*(u) - T_{r-2}^*(u)}$$

or in terms of binomial coefficients,

$$T_r^*(u) = \sum_{k=0}^{r} A_k u^k \qquad A_k = (-1)^{r+k} 2^{2k-1} \left[2\binom{r + k}{r - k} - \binom{r + k - 1}{r - k} \right]$$

(c) Special values

$$T_r^*(0) = 1 \quad \text{if } r \text{ even} \qquad T_r^*(0) = -1 \quad \text{if } r \text{ odd} \qquad T_r^*(1) = 1 \quad \text{for all } r$$

(d) Powers of u in terms of $T_r^*(u) = T_r^*$ are

$$
\begin{bmatrix} u \\ u^2 \\ u^3 \\ u^4 \\ u^5 \\ u^6 \\ u^7 \\ u^8 \end{bmatrix}
=
\begin{bmatrix}
\frac{1}{2} & \frac{1}{2} & & & & & & & \\
\frac{3}{8} & \frac{4}{8} & \frac{1}{8} & & & & & & \\
\frac{10}{32} & \frac{15}{32} & \frac{6}{32} & \frac{1}{32} & & & & & \\
\frac{35}{128} & \frac{56}{128} & \frac{28}{128} & \frac{8}{128} & \frac{1}{128} & & & & \\
\frac{126}{512} & \frac{210}{512} & \frac{120}{512} & \frac{45}{512} & \frac{10}{512} & \frac{1}{512} & & & \\
\frac{462}{2048} & \frac{792}{2048} & \frac{495}{2048} & \frac{220}{2048} & \frac{66}{2048} & \frac{12}{2048} & \frac{1}{2048} & & \\
\frac{1716}{8192} & \frac{3003}{8192} & \frac{2002}{8192} & \frac{1001}{8192} & \frac{364}{8192} & \frac{91}{8192} & \frac{14}{8192} & \frac{1}{8192} & \\
\frac{6\,435}{32\,768} & \frac{11\,440}{32\,768} & \frac{8008}{32\,768} & \frac{4368}{32\,768} & \frac{1820}{32\,768} & \frac{560}{32\,768} & \frac{120}{32\,768} & \frac{16}{32\,768} & \frac{1}{32\,768}
\end{bmatrix}
\begin{bmatrix} T_0^* \\ T_1^* \\ T_2^* \\ T_3^* \\ T_4^* \\ T_5^* \\ T_6^* \\ T_7^* \\ T_8^* \end{bmatrix}
$$

and in general,

$$\boxed{u^r = \frac{1}{2^{2r-1}} \left[\frac{1}{2}\binom{2r}{r} T_0^* + \binom{2r}{r-1} T_1^* + \binom{2r}{r-2} T_2^* + \cdots + \binom{2r}{1} T_{r-1}^* + T_r^* \right]}$$

(e) Substitute function in $(0, 1)$ is then

$$\bar{y}^*(u) = \frac{a_0^*}{\pi\sqrt{1-u^2}} + \frac{2}{\pi\sqrt{1-u^2}} \sum_{r=1}^{n} a_r^* T_r^*(u)$$

where $T_r^*(u) = $ shifted Chebyshev polynomial of the first kind of order r, defined in (b), and

$$a_0^* = \int_0^1 y^*(u)\, du \qquad a_r^* = \int_0^1 y^*(u)\, T_r^*(u)\, du$$

(f) Derivatives

$$\frac{d[T_0^*(u)]}{du} = 0 \qquad\qquad \frac{d[T_1^*(u)]}{du} = 2T_0$$

$$\frac{d[T_2^*(u)]}{du} = 8T_1^*(u) \qquad\qquad \frac{d[T_3^*(u)]}{du} = 12[T^*(u) + \tfrac{1}{2}]$$

$$\frac{d[T_4^*(u)]}{du} = 16[T_3^*(u) + T_1^*(u)] \qquad\qquad \frac{d[T_5^*(u)]}{du} = 20[T_4^*(u) + T_2^*(u) + \tfrac{1}{2}]$$

and in general, for $r > 1$,

$$\frac{d[T_r^*(u)]}{du} = \begin{cases} 4r[T_{r-1}^*(u) + T_{r-3}^*(u) + \cdots + T_2^*(u) + \tfrac{1}{2}] & r \text{ odd} \\ 4r[T_{r-1}^*(u) + T_{r-3}^*(u) + \cdots + T_3^*(u) + T_1^*(u)] & r \text{ even} \end{cases}$$

(g) Indefinite integrals

$$\int T_0^*(u)\, du = \tfrac{1}{2}[T_1^*(u)] + C \qquad\qquad \int T_1^*(u)\, du = \tfrac{1}{8}[T_2^*(u)] + C$$

and in general, for $r > 1$,

$$\int T_r^*(u)\, du = \frac{1}{4}\left[\frac{T_{r+1}^*(u)}{r+1} - \frac{T_{r-1}^*(u)}{r-1} \right] + C$$

where $C = $ constant of integration.

(h) Definite integrals used in the evaluation of a_r^* in (e) above are with $m = 1, 2, 3, \ldots,$

$$\int_0^1 u^m T_r^*(u)\, du = \begin{cases} 0 & (m+r) \text{ odd} \\ (m!)\left[\dfrac{T_r^*(1)}{(m+1)!} - \dfrac{d[T_r^*(1)]/du}{(m+2)!} + \dfrac{d^2[T_r^*(u)]/du^2}{(m+3)!} + \cdots \right] & (m+r) \text{ even} \end{cases}$$

$$\int_0^1 y^*(u)\, T_r^*(u)\, du = [I_1^* D_0^* - I_2^* D_1^* + I_3^* D_2^* - \cdots]_0^1$$

where $\quad D_0^* = T_r^*(u),\ D_1^* = \dfrac{d[T_r^*(u)]}{du},\ D_2^* = \dfrac{d^2[T_r^*(u)]}{du^2},\ldots$

$$I_1^* = \int y^*(u)\, du,\ I_2^* = \iint y^*(u)\, du\, du,\ I_3^* = \iiint y^*(u)\, du\, du\, du, \ldots$$

and the series terminates with the term containing $D_n^* = 0$.

(a) Concept. Chebyshev polynomials provide a good approximation of a function with a small error, but the evaluation of the coefficients a_r in (12.06) and of a_r^* in (12.07) gives rise to rather complicated integrals. If, however, a truncated power series of the function $y(u)$ is readily available, the degree of the power series can be reduced with little loss of accuracy by expressing the powers of u in terms of $T_r(u)$ or $T_r^*(u)$ as shown in (b) below. This series of Chebyshev polynomials can be rearranged in an ordinary polynomial in u after stopping at the selected term containing $T_n(u)$ or $T_n^*(u)$ with a small coefficient. This process developed by Lanczos (Ref. 12.04) is called the *telescoping (economization)* of power series expansion. The resulting $\bar{y}(u)$ is not identical with the true least-squares approximation, but the method is a convenient tool of practical calculations. The following example shows the procedure of application.

(b) Economization of power series. The truncated Taylor's series

$$e^{-u} \cong 1 - \frac{u}{1!} + \frac{u^2}{2!} - \frac{u^3}{3!} + \frac{u^4}{4!} - \cdots \frac{u^{12}}{12!}$$

is expressed in terms of the relations (12.06h) abbreviated as $T_r(u) = T_r$,

$$e^{-u} \cong T_0 - T_1 + \frac{1}{2(2!)}(T_0 + T_2) - \frac{1}{4(3!)}(3T_1 + T_3) + \frac{1}{8(4!)}(3T_0 + 4T_2 + T_4)$$

$$- \frac{1}{16(5!)}(10T_1 + 5T_3 + T_5) + \frac{1}{32(6!)}(10T_0 + 15T_2 + 6T_4 + T_6)$$

$$- \frac{1}{64(7!)}(35T_1 + 21T_3 + 7T_5 + T_7) + \frac{1}{128(8!)}(35 + 56T_2 + 28T_4 + 8T_6 + T_8)$$

$$- \cdots \frac{1}{2048(12!)}(462 + 792T_2 + 495T_4 + 220T_6 + 66T_8 + 12T_{10} + T_{12})$$

After collecting factors of identical polynomials,

$$e^{-u} \cong + 1.266\,066T_0 - 1.130\,318T_1 + 0.271\,495T_2 - 0.044\,337T_3$$
$$+ 0.005\,474T_4 - 0.000\,543T_5 + 0.000\,500T_6 - 0.000\,003T_7$$
$$+ 0.000\,000T_8$$

The substitution of the relations (12.06c) for T_0, T_1, T_2, T_3, T_4, T_5 yields the final form of the economized series:

$$e^{-u} \cong 1.000\,045 - 1.000\,022u + 0.499\,199u^2$$
$$- 0.166\,488u^3 + 0.043\,793u^4$$
$$- 0.008\,446u^5 + 0.000\,045T_6$$

Since the largest value of T_6 is 1, the approximation error $\epsilon_c \leq 4.5(10)^{-5}$, whereas the error of the initial Taylor's series of the same degree (terminating with $u^5/5!$) is $\epsilon_T \leq 1.4(10)^{-3}$. The adjacent graph shows the relationships of these two errors.

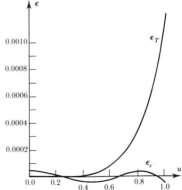

13
FOURIER
APPROXIMATIONS

(a) Concept. Whereas the polynomial approximations are in general well-suited for the approximation of a continuous function $y(x)$, when this function is *periodic*, it may be preferable to resort to an approximate function, which is periodic in the same interval. The most convenient form of such an approximation is a set of sine and cosine functions known as the *Fourier series*.

(b) Basic form. Any single-valued function $y(\theta)$ that is continuous except for a finite number of discontinuities in an interval $(-\pi \le \theta \le \pi)$ and has a finite number of maxima and minima in this interval may be represented by a convergent Fourier series as

$$y(\theta) = \tfrac{1}{2}A_0 + A_1 \cos \theta + A_2 \cos 2\theta + A_3 \cos 3\theta + \cdots$$

$$+ B_1 \sin \theta + B_2 \sin 2\theta + B_3 \sin 3\theta + \cdots$$

$$= \tfrac{1}{2}A_0 + \sum_{k=1}^{\infty} (A_k \cos k\theta + B_k \sin k\theta)$$

and approximated by $2n + 1$ terms of this series as

$$\boxed{\bar{y}(\theta) = \tfrac{1}{2}A_0 + \sum_{k=1}^{n} (A_k \cos k\theta + B_k \sin k\theta)}$$

(c) Principal characteristic of this series is its *orthogonality* over any period interval. In particular, in $(-\pi, \pi)$ with $j, k = 1, 2, 3, \ldots, n$,

$$\int_{-\pi}^{\pi} \cos j\theta \cos k\theta \, d\theta = 0 \quad (j \ne k) \qquad \int_{-\pi}^{\pi} \cos j\theta \cos k\theta \, d\theta = \pi \quad (j = k)$$

$$\int_{-\pi}^{\pi} \sin j\theta \sin k\theta \, d\theta = 0 \quad (j \ne k) \qquad \int_{-\pi}^{\pi} \sin j\theta \sin k\theta \, d\theta = \pi \quad (j = k)$$

$$\int_{-\pi}^{\pi} \cos j\theta \sin k\theta \, d\theta = 0 \quad (\text{for all } j, k)$$

(d) Coefficients of the truncated series, derived from the minimum error condition

$$\int_{-\pi}^{\pi} [y(\theta) - \tfrac{1}{2}A_0 - \sum_{k=1}^{n} (A_k \cos k\theta + B_k \sin k\theta)]^2 \, d\theta = \epsilon^2$$

are

$$\boxed{\tfrac{1}{2}A_0 = \frac{1}{2\pi} \int_{-\pi}^{\pi} y(\theta) \, d\theta \qquad A_k = \frac{1}{\pi} \int_{-\pi}^{\pi} y(\theta) \cos k\theta \, d\theta \qquad B_k = \frac{1}{\pi} \int_{-\pi}^{\pi} y(\theta) \sin k\theta \, d\theta}$$

The evaluation of these integrals is facilitated by the tables (13.03), and their particular forms corresponding to a selected set of functions are listed in (13.04) to (13.11).

(e) Nonperiodic functions that satisfy the conditions of (*b*) in a selected interval $(-\pi, \pi)$ can also be approximated by the truncated Fourier series, but only in this interval. Outside this interval, the series continues to be periodic regardless of the function to be approximated.

(a) Change in variable. In the development of Fourier series the variable may be changed as shown below, where L, T are positive numbers and θ, x, t are variables.

$$\theta = \frac{\pi x}{L} \qquad (-L < x < L)$$

$$\tfrac{1}{2}\bar{A}_0 = \frac{1}{2L} \int_{-L}^{L} y(x)\, dx$$

$$\bar{A}_k = \frac{1}{L} \int_{-L}^{L} y(x) \cos \frac{k\pi x}{L}\, dx$$

$$\bar{y}(x) = \tfrac{1}{2}\bar{A}_0 + \sum_{k=1}^{n} \left(\bar{A}_k \sin \frac{k\pi x}{L} + \bar{B}_k \sin \frac{k\pi x}{L} \right)$$

$$\bar{B}_k = \frac{1}{L} \int_{-L}^{L} y(x) \sin \frac{k\pi x}{L}\, dx$$

$$\theta = \frac{2\pi t}{T} \qquad (-\tfrac{1}{2}T < t < \tfrac{1}{2}T)$$

$$\tfrac{1}{2}A_0^* = \frac{1}{T} \int_{-T/2}^{T/2} y(t)\, dt$$

$$A_k^* = \frac{2}{T} \int_{-T/2}^{T/2} y(t) \cos \frac{2k\pi t}{T}\, dt$$

$$\bar{y}(t) = \tfrac{1}{2}A_0^* + \sum_{k=1}^{n} \left(A_k^* \cos \frac{2k\pi t}{T} + B_k^* \sin \frac{2k\pi t}{T} \right)$$

$$B_k^* = \frac{2}{t} \int_{-T/2}^{T/2} y(t) \sin \frac{2k\pi t}{T}\, dt$$

(b) Change in interval. In the development of Fourier series the interval can be shifted as shown below, where L, T, θ, x, t have the same meaning as in (a) and ω, C, K are signed numbers.

	$-2\pi < \theta < 0$	$\omega < \theta < \omega + 2\pi$	$0 < \theta < 2\pi$
A_k	$\dfrac{1}{\pi} \displaystyle\int_{-2\pi}^{0} y(\theta) \cos k\theta\, d\theta$	$\dfrac{1}{\pi} \displaystyle\int_{\omega}^{\omega+2\pi} y(\theta) \cos k\theta\, d\theta$	$\dfrac{1}{\pi} \displaystyle\int_{0}^{2\pi} y(\theta) \cos k\theta\, d\theta$
B_k	$\dfrac{1}{\pi} \displaystyle\int_{-2\pi}^{0} y(\theta) \sin k\theta\, d\theta$	$\dfrac{1}{\pi} \displaystyle\int_{\omega}^{\omega+2\pi} y(\theta) \sin k\theta\, d\theta$	$\dfrac{1}{\pi} \displaystyle\int_{0}^{2\pi} y(\theta) \sin k\theta\, d\theta$
	$-2L < x < 0$	$K < x < K + 2L$	$0 < x < 2L$
$\bar{A}_k$	$\dfrac{1}{L} \displaystyle\int_{-2L}^{0} y(x) \cos \frac{k\pi x}{L}\, dx$	$\dfrac{1}{L} \displaystyle\int_{K}^{K+2L} y(x) \cos \frac{k\pi x}{L}\, dx$	$\dfrac{1}{L} \displaystyle\int_{0}^{2L} y(x) \cos \frac{k\pi x}{L}\, dx$
$\bar{B}_k$	$\dfrac{1}{L} \displaystyle\int_{-2L}^{0} y(x) \sin \frac{k\pi x}{L}\, dx$	$\dfrac{1}{L} \displaystyle\int_{K}^{K+2L} y(x) \sin \frac{k\pi x}{L}\, dx$	$\dfrac{1}{L} \displaystyle\int_{0}^{2L} y(x) \sin \frac{k\pi x}{L}\, dx$
	$-T < t < 0$	$C < t < C + T$	$0 < t < T$
A_k^*	$\dfrac{2}{T} \displaystyle\int_{-T}^{0} y(t) \cos \frac{2k\pi t}{T}\, dt$	$\dfrac{2}{T} \displaystyle\int_{C}^{C+T} y(t) \cos \frac{2k\pi t}{T}\, dt$	$\dfrac{2}{T} \displaystyle\int_{0}^{T} y(t) \cos \frac{2k\pi t}{T}\, dt$
B_k^*	$\dfrac{2}{T} \displaystyle\int_{-T}^{0} y(t) \sin \frac{2k\pi t}{T}\, dt$	$\dfrac{2}{T} \displaystyle\int_{C}^{C+T} y(t) \sin \frac{2k\pi t}{T}\, dt$	$\dfrac{2}{T} \displaystyle\int_{0}^{T} y(t) \sin \frac{2k\pi t}{T}\, dt$

(c) Phase angles. The cosine and sine terms of the Fourier series may be combined in a single cosine or sine series with the phase angle ϕ_k or ψ_k, respectively.

$$\bar{y}(\theta) = \tfrac{1}{2}A_0 + \sum_{k=1}^{n} R_k \cos(k\theta + \phi_k)$$

$$\bar{y}(\theta) = \tfrac{1}{2}A_0 + \sum_{k=1}^{n} R_k \sin(k\theta + \psi_k)$$

$$R_k = \sqrt{A_k^2 + B_k^2} \qquad \phi_k = \tan^{-1}\left(-\frac{B_k}{A_k}\right)$$

$$R_k = \sqrt{A_k^2 + B_k^2} \qquad \psi_k = \tan^{-1}\frac{A_k}{B_k}$$

(a) Notation. The most common definite integrals used in the evaluation of Fourier coefficients A_k, B_k are tabulated below in terms of the following symbols.

$$k, m, n, r = \text{integers} \qquad a, b = \text{positive fractions} < 1 \qquad \eta = (-1)^k$$

$$u = x/L = \text{dimensionless variable} \qquad b = 1 - a \qquad \omega = \text{signed number} \neq 0$$

$$\lambda = k\pi \qquad C_k = \cos \lambda u \qquad S_k = \sin \lambda u \qquad \omega = c\pi \qquad C_c = \cos \omega u \qquad S_c = \sin \omega u$$

$$M_1 = (-1)^k \sum_{m=0}^{r} \frac{(-1)^m}{(2r - 2m - 1)!\lambda^{2m}} \qquad M_b = \frac{(-1)^r \sin \lambda a}{\lambda^{2r-1}} + (-1)^k \sum_{m=0}^{r} \frac{(-1)^m b^{2r-2m-1}}{(2r - 2m - 1)\lambda^{2m}}$$

$$N_1 = (-1)^k \sum_{m=0}^{r} \frac{(-1)^m}{(2r - 2m)!\lambda^m} \qquad N_b = \frac{(-1)^r \sin \lambda a}{\lambda^{2r}} + (-1)^k \sum_{m=0}^{r} \frac{(-1)^m b^{2r-2m}}{(2r - 2m)\lambda^{2m}}$$

$$P_1 = (-1)^k \sum_{m=0}^{r} \frac{(-1)^m}{(2r - 2m + 1)!\lambda^{2m}} \qquad P_b = \frac{(-1)^r \sin \lambda a}{\lambda^{2r-1}} - (-1)^k \sum_{m=0}^{r} \frac{(-1)^m b^{2r-2m+1}}{(2r - 2m + 1)!\lambda^{2m}}$$

(b) Trigonometric identities $(k = 0, 1, 2, \ldots)$

	$k = 0$	$k > 0$	k odd	k even
$\cos k\pi$	$+1$	$(-1)^k$	-1	$+1$
$\sin k\pi$	0	0	0	0

(c) Range $(-1 \leq u \leq 1)$

(1)	$\displaystyle\int_{-1}^{1} C_k\, du = 0$	(9)	$\displaystyle\int_{-1}^{1} S_k\, du = 0$
(2)	$\displaystyle\int_{-1}^{1} u C_k\, du = 0$	(10)	$\displaystyle\int_{-1}^{1} u S_k\, du = -\frac{2\eta}{\lambda}$
(3)	$\displaystyle\int_{-1}^{1} u^2 C_k\, du = \frac{4\eta}{\lambda^2}$	(11)	$\displaystyle\int_{-1}^{1} u^2 S_k\, du = 0$
(4)	$\displaystyle\int_{-1}^{1} u^{2r} C_k\, du = \frac{2(2r)!}{\lambda^2} M_1$	(12)	$\displaystyle\int_{-1}^{1} u^{2r} S_k\, du = 0$
(5)	$\displaystyle\int_{-1}^{1} u^{2r+1} C_k\, du = 0$	(13)	$\displaystyle\int_{-1}^{1} u^{2r+1} S_k\, du = -\frac{2(2r + 1)!}{\lambda} P_1$
(6)	$\displaystyle\int_{-1}^{1} C_c C_k\, du = \frac{2\omega\eta \sin \omega}{\omega^2 - \lambda^2}$	(14)	$\displaystyle\int_{-1}^{1} C_c S_k\, du = 0$
(7)	$\displaystyle\int_{-1}^{1} S_c C_k\, du = 0$	(15)	$\displaystyle\int_{-1}^{1} S_c S_k\, du = \frac{2\lambda\eta \sin \omega}{\omega^2 - \lambda^2}$
(8)	$\displaystyle\int_{-1}^{1} e^{\omega u} C_k\, du = \frac{2\omega\eta \sinh \omega}{\omega^2 + \lambda^2}$	(16)	$\displaystyle\int_{-1}^{1} e^{\omega u} S_k\, du = -\frac{2\lambda\eta \sinh \omega}{\omega^2 + \lambda^2}$

(d) Range $(0 \le u \le 1)$

(1)	$\displaystyle\int_0^1 C_k \, du = 0$		(9)	$\displaystyle\int_0^1 S_k \, du = \dfrac{1-\eta}{\lambda}$
(2)	$\displaystyle\int_0^1 u C_k \, du = \dfrac{\eta-1}{\lambda^2}$		(10)	$\displaystyle\int_0^1 u S_k \, du = -\dfrac{\eta}{\lambda}$
(3)	$\displaystyle\int_0^1 u^2 C_k \, du = \dfrac{2\eta}{\lambda^2}$		(11)	$\displaystyle\int_0^1 u^2 S_k \, du = \dfrac{2(\eta-1)}{\lambda^3} - \dfrac{\eta}{\lambda}$
(4)	$\displaystyle\int_0^1 u^{2r} C_k \, du = \dfrac{(2r)!}{\lambda^2} M_1$		(12)	$\displaystyle\int_0^1 u^{2r} S_k \, du = -\dfrac{(2r)!}{\lambda} N_1$
(5)	$\displaystyle\int_0^1 u^{2r+1} C_k \, du = \dfrac{(2r+1)!}{\lambda^2} N_1$		(13)	$\displaystyle\int_0^1 u^{2r+1} S_k \, du = -\dfrac{(2r+1)!}{\lambda} P_1$
(6)	$\displaystyle\int_0^1 C_c C_k \, du = \dfrac{\omega\eta \sin \omega}{\omega^2 - \lambda^2}$		(14)	$\displaystyle\int_0^1 C_c S_k \, du = \dfrac{\lambda(\eta \cos \omega - 1)}{\omega^2 - \lambda^2}$
(7)	$\displaystyle\int_0^1 S_c C_k \, du = \dfrac{\omega(1 - \eta \cos \omega)}{\omega^2 - \lambda^2}$		(15)	$\displaystyle\int_0^1 S_c S_k \, du = \dfrac{\lambda\eta \sin \omega}{\omega^2 - \lambda^2}$
(8)	$\displaystyle\int_0^1 e^{\omega u} C_k \, du = \dfrac{\omega(\eta e^{\omega} - 1)}{\omega^2 + \lambda^2}$		(16)	$\displaystyle\int_0^1 e^{\omega u} S_k \, du = \dfrac{\lambda(1 - \eta e^{\omega})}{\omega^2 + \lambda^2}$

(e) Range $(a \le u \le 1)$

(1)	$\displaystyle\int_a^1 C_k \, du = -\dfrac{\sin \lambda a}{\lambda}$		(6)	$\displaystyle\int_a^1 S_k \, du = \dfrac{\cos \lambda a - \eta}{\lambda}$
(2)	$\displaystyle\int_a^1 (u-a) C_k \, du = \dfrac{\eta - \cos \lambda a}{\lambda^2}$		(7)	$\displaystyle\int_a^1 (u-a) S_k \, du = \dfrac{\eta\lambda b - \sin \lambda a}{\lambda^2}$
(3)	$\displaystyle\int_a^1 (u-a)^2 C_k \, du = \dfrac{2(\eta b\lambda + \sin \lambda a)}{\lambda^3}$		(8)	$\displaystyle\int_a^1 (u-a)^2 S_k \, du = \dfrac{\eta(2 - b^2\lambda^2) - 2\cos \lambda a}{\lambda^3}$
(4)	$\displaystyle\int_a^1 (u-a)^{2r} C_k \, du = \dfrac{(2r)!}{\lambda^2} M_b$		(9)	$\displaystyle\int_a^1 (u-a)^{2r} S_k \, du = -\dfrac{(2r)!}{\lambda} N_b$
(5)	$\displaystyle\int_a^1 (u-a)^{2r+1} C_k \, du = \dfrac{(2r+1)!}{\lambda^2} N_b$		(10)	$\displaystyle\int_a^1 (u-a)^{2r+1} S_k \, du = -\dfrac{(2r+1)!}{\lambda} P_b$

(f) Range $(0 \le u \le a)$. By superposition of the respective cases (d) and (e),

$$\int_0^a y(u) C_k \, du = \int_0^1 y(u) C_k \, du - \int_a^1 y(u) C_k \, du$$

$$\int_0^a y(u) S_k \, du = \int_0^1 y(u) S_k \, du - \int_a^1 y(u) S_k \, du$$

(a) Notation

a, b, h, L = positive numbers $\eta = (-1)^k$ k, m, n, r = positive integers

$u = x/L$ $\lambda = k\pi$ $a + b = 1$ $C_k = \cos k\pi u$ $S_k = \sin k\pi u$

(b) Even function.
A function $y(x)$ is said to be even (symmetrical) if

$$y(-x) = y(x) \qquad (-L < x < L)$$

The Fourier approximation of an even function is

$$\bar{y}(x) = \tfrac{1}{2}\bar{A}_0 + \sum_{k=1}^{n} \bar{A}_k C_k$$

where

$$\tfrac{1}{2}\bar{A}_0 = \int_0^1 y(u)\, du \qquad \bar{A}_k = 2\int_0^1 y(u) C_k\, du$$

are the *Fourier coefficients* evaluated by the integral tables (13.03) and given for selected cases below.

(c) Rectangle (13.03d-1)

$$y(x) = y(uL) = h \qquad (-1 < u < 1)$$

$$\tfrac{1}{2}\bar{A}_0 = h \int_0^1 du = h$$

$$\bar{A}_k = 2h \int_0^1 C_k\, du = 0$$

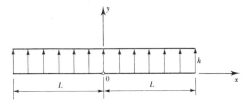

(d) Triangle (13.03d-2)

$$y(x) = \frac{hx}{L} = hu \qquad (-1 < u < 1)$$

$$\tfrac{1}{2}\bar{A}_0 = h \int_0^1 u\, du = \tfrac{1}{2}h$$

$$\bar{A}_k = 2h \int_0^1 u C_k\, du = 2h(\eta - 1)\lambda^{-2}$$

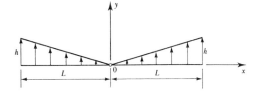

(e) Parabola of 2rth degree (13.03d-4)

$$y(x) = h\left(\frac{x}{L}\right)^{2r} = hu^{2r} \qquad (-1 < u < 1)$$

$$\tfrac{1}{2}\bar{A}_0 = h \int_0^1 u^{2r}\, du = h/(2r + 1)$$

$$\bar{A}_k = 2h \int_0^1 u^{2r} C_k\, du = 2h(2r)! M_1 \lambda^{-2}$$

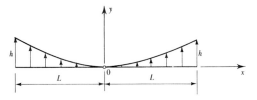

(f) Superposition

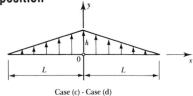

Case (c) - Case (d)

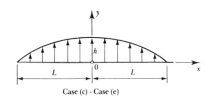

Case (c) - Case (e)

(g) Rectified sine curve (13.03*d*-7)

$$y(x) = h \left| \sin \frac{\pi x}{L} \right| = h \left| \sin \pi u \right| \qquad (-1 < u < 1)$$

$$\tfrac{1}{2}\bar{A}_0 = h \int_0^1 |\sin \pi u| \, du = 2h/\pi$$

$$\bar{A}_k = 2h \int_0^1 |\sin \pi u| \, C_k \, du = \begin{cases} 0 & k \text{ odd} \\[2mm] -\dfrac{4h}{(k^2 - 1)\pi} & k \text{ even} \end{cases}$$

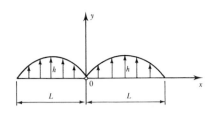

(h) Rectified cosine curve (Ref. 1.20, sec. 19.52)

$$y(x) = h \left| \cos \frac{\pi x}{L} \right| = h \left| \cos \pi u \right| \qquad (-1 < u < 1)$$

$$\tfrac{1}{2}\bar{A}_0 = h \int_0^1 |\cos \pi u| \, du = \frac{2h}{\pi}$$

$$\bar{A}_k = 2h \int_0^1 |\cos \pi u| \, C_k \, du = \begin{cases} 0 & k \text{ odd} \\[2mm] -\dfrac{4\eta h}{(k^2 - 1)\pi} & k \text{ even} \end{cases}$$

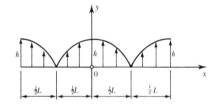

(i) Shifted rectangle (13.03*e*-1)

$$y(x) = y(u) = h \qquad (-1 < u < -a, a < u < 1)$$

$$\tfrac{1}{2}\bar{A}_0 = h \int_a^1 du = bh \qquad \bar{A}_k = 2h \int_a^1 C_k \, du = \frac{2h \sin \lambda a}{\lambda}$$

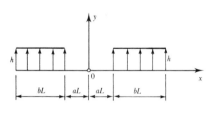

(j) Shifted triangle (13.03*e*-2)

$$y(x) = \left| \frac{h}{b_1}(u - a) \right| \qquad (-1 < u < -a, a < u < 1)$$

$$\tfrac{1}{2}\bar{A}_0 = \frac{h}{b}\int_a^1 (u - a)\, du = \tfrac{1}{2}bh$$

$$\bar{A}_k = \frac{2h}{b}\int_a^1 (u - a) C_k \, du = \frac{2h}{b}[\eta - \cos \lambda a]\lambda^{-2}$$

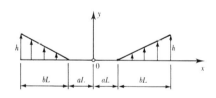

(k) Shifted parabola of 2*r*th degree (13.03*e*-4)

$$y(x) = \left| \frac{h}{b^{2r}}(u - a)^{2r} \right| \qquad (-1 < u < -a, a < u < 1)$$

$$\tfrac{1}{2}\bar{A}_0 = \frac{h}{b^{2r}}\int_a^1 (u - a)^{2r}\, du = \frac{bh}{2r + 1} \qquad \bar{A}_k = \frac{2h}{b^{2r}}\int_a^1 (u - a)^{2r} C_k \, du = \frac{2h(2r)!}{b^{2r}\lambda^2} M_b$$

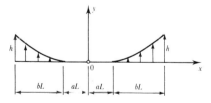

(l) Superposition

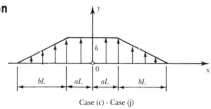

Case (c) - Case (j)

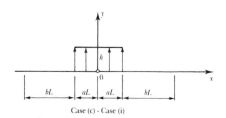

Case (c) - Case (i)

(a) Notation

a, b, h, L = positive numbers $\eta = (-1)^k$ k, m, n, r = positive integers

$u = x/L$ $\lambda = k\pi$ $a + b = 1$ $C_k = \cos k\pi u$ $S_k = \sin k\pi u$

(b) Odd function. A function $y(x)$ is said to be odd (antisymmetrical) if

$$y(-x) = -y(x) \qquad (-L < x < L)$$

The Fourier approximation of an odd function is

$$\bar{y}(x) = \sum_{k=1}^{n} \bar{B}_k S_k$$

where

$$\bar{B}_k = 2\int_0^1 y(u) S_k \, du$$

are the *Fourier coefficients* evaluated by the integral tables (13.03) and given for selected cases below.

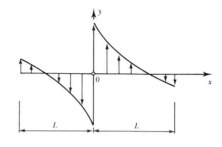

(c) Rectangle (13.03d-9)

$$y(x) = y(uL) = \begin{cases} -h & (-1 < u < 0) \\ h & (\ 0 < u < 1) \end{cases}$$

$$\bar{B}_k = 2h\int_0^1 S_k \, du = \frac{2h(1 - \eta)}{\lambda}$$

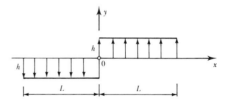

(d) Triangle (13.03d-10)

$$y(x) = \frac{hx}{L} = hu \qquad (-1 < u < 1)$$

$$\bar{B}_k = 2h\int_0^1 u S_k \, du = -\frac{2h\eta}{\lambda}$$

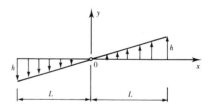

(e) Parabola of (2r + 1)th degree (13.03d-13)

$$y(x) = h\left(\frac{x}{L}\right)^{2r+1} = hu^{2r+1} \qquad (-1 < u < 1)$$

$$\bar{B}_k = 2h\int_0^1 u^{2r+1} S_k \, du = -\frac{(2r + 1)!}{\lambda} P_1$$

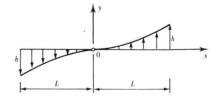

(f) Superposition

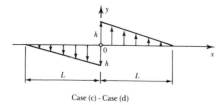

Case (c) - Case (d)

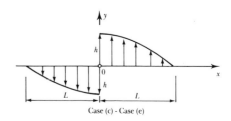

Case (c) - Case (e)

(g) Quadratic function $(13.03d\text{-}11,12)$

$$y(x) = y(uL) = \begin{cases} 4hu(1 + u) & (-1 < u < 0) \\ 4hu(1 - u) & (0 < u < 1) \end{cases}$$

$$\bar{B}_k = 8h \int_0^1 u(1 - u)S_k \, du = \frac{16h(1 - \eta)}{\lambda^3}$$

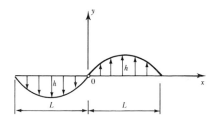

(h) Cubic function $(13.03d\text{-}10,13)$

$$y(x) = y(uL) = \frac{8hu(1 - u^2)}{3} \qquad (-1 < u < 1)$$

$$\bar{B}_k = \frac{8h}{3} \int_0^1 u(1 - u^2)S_k \, du = \frac{32\eta h}{\lambda^3}$$

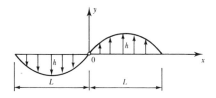

(i) Shifted rectangle $(13.03e\text{-}6)$

$$y(x) = y(uL) = \begin{cases} -h & (-1 < u < -a) \\ h & (a < u < 1) \end{cases}$$

$$\bar{B}_k = 2h \int_a^1 S_k \, du = \frac{2(\cos \lambda a - \eta)h}{\lambda}$$

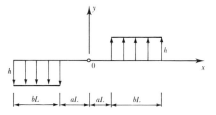

(j) Shifted triangle $(13.03e\text{-}7)$

$$y(c) = y(uL) = \begin{cases} \dfrac{h}{b}(u + a) & (-1 < u < -a) \\ \dfrac{h}{b}(u - a) & (a < u < 1) \end{cases}$$

$$\bar{B}_k = \frac{2h}{b} \int_a^1 (u - a)S_k \, du = \frac{2(\eta \lambda b - \sin \lambda a)h}{\lambda^2 b}$$

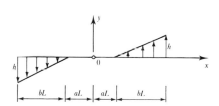

(k) Shifted parabola of $(2r + 1)$th degree $(13.03e\text{-}10)$

$$y(x) = y(uL) = \begin{cases} h\left(\dfrac{u + a}{b}\right)^{2r+1} & (-1 < u < -a) \\ h\left(\dfrac{u - a}{b}\right)^{2r+1} & (a < u < 1) \end{cases}$$

$$\bar{B}_k = \frac{2h}{b^{2r+1}} \int_a^1 (u - a)^{2r+1}S_k \, du = \frac{2h(2r + 1)!}{\lambda b^{2r+1}} P_b$$

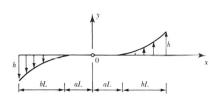

(l) Superposition

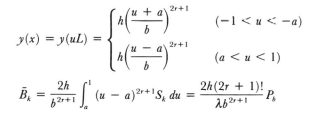

Case (c) - Case (j)

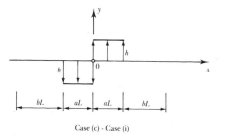

Case (c) - Case (i)

a, b = positive fractions	h, L = positive numbers	$k = 1, 2, 3, \ldots, n$
$u = x/L$ $\qquad \alpha = a\pi$	$\beta = b\pi$ $\qquad \theta = u\pi$	$j = 1, 3, 5, \ldots, s$

(1)

$a \leq 1$

$$\bar{y}(x) = ah + \frac{2h}{\pi} \sum_{k=1}^{n} \frac{\sin k\alpha \cos k\theta}{k}$$

(2)

$a \leq \frac{1}{2}$

$$\bar{y}(x) = 2ah + \frac{4h}{\pi} \sum_{k=1}^{n} \frac{\sin 2k\alpha \cos 2k\theta}{2k}$$

(3)

$a < 1$

$$\bar{y}(x) = (1 - a)h - \frac{2h}{\pi} \sum_{k=1}^{n} \frac{\sin k\alpha \cos k\theta}{k}$$

(4)

$b \geq a \leq \frac{1}{2}$

$$\bar{y}(x) = 2ah + \frac{4h}{\pi} \sum_{k=1}^{n} \frac{\sin k\alpha \cos k\theta}{k}$$

(5)

$b \geq a \leq \frac{1}{2}$

$$\bar{y}(x) = 2ah - \frac{4h}{\pi} \sum_{j=1}^{s} \left[\frac{\sin j\alpha \cos j\beta \cos j\theta}{j} - \frac{\sin j\alpha \sin j\beta \sin j\theta}{j} \right]$$

(6)

$b \geq a \leq \frac{1}{2}$

$$\bar{y}(x) = ah + \frac{2h}{\pi} \sum_{k=1}^{n} \left[\frac{\sin k\alpha \cos k\beta \cos k\theta}{k} - \frac{\sin k\alpha \sin k\beta \sin k\theta}{k} \right]$$

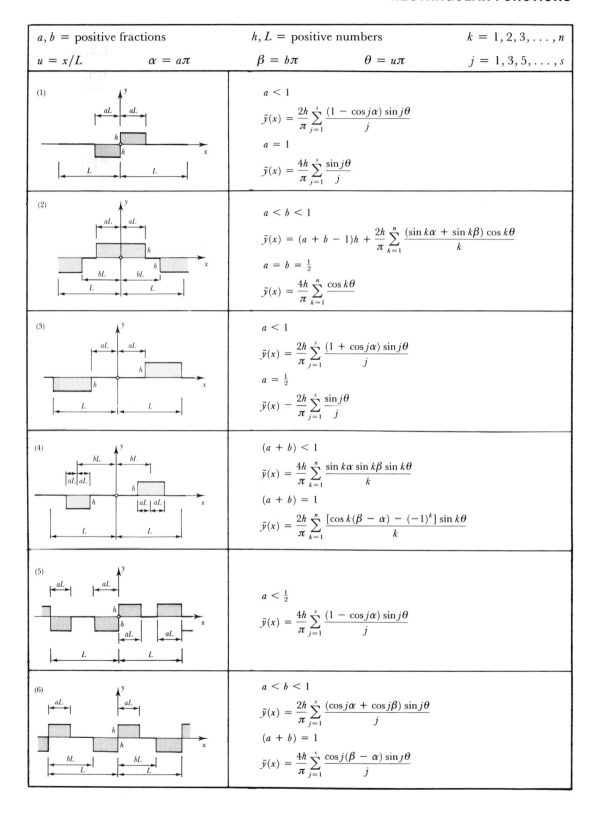

a, b = positive fractions h, L = positive numbers $k = 1, 2, 3, \ldots, n$

$u = x/L$ $\alpha = a\pi$ $\beta = b\pi$ $\theta = u\pi$ $j = 1, 3, 5, \ldots, s$

---

(1)

$a < 1$

$$\bar{y}(x) = \frac{2h}{\pi} \sum_{j=1}^{s} \frac{(1 - \cos j\alpha)\sin j\theta}{j}$$

$a = 1$

$$\bar{y}(x) = \frac{4h}{\pi} \sum_{j=1}^{s} \frac{\sin j\theta}{j}$$

---

(2)

$a < b < 1$

$$\bar{y}(x) = (a + b - 1)h + \frac{2h}{\pi} \sum_{k=1}^{n} \frac{(\sin k\alpha + \sin k\beta)\cos k\theta}{k}$$

$a = b = \frac{1}{2}$

$$\bar{y}(x) = \frac{4h}{\pi} \sum_{k=1}^{n} \frac{\cos k\theta}{k}$$

---

(3)

$a < 1$

$$\bar{y}(x) = \frac{2h}{\pi} \sum_{j=1}^{s} \frac{(1 + \cos j\alpha)\sin j\theta}{j}$$

$a = \frac{1}{2}$

$$\bar{y}(x) - \frac{2h}{\pi} \sum_{j=1}^{s} \frac{\sin j\theta}{j}$$

---

(4)

$(a + b) < 1$

$$\bar{y}(x) = \frac{4h}{\pi} \sum_{k=1}^{n} \frac{\sin k\alpha \sin k\beta \sin k\theta}{k}$$

$(a + b) = 1$

$$\bar{y}(x) = \frac{2h}{\pi} \sum_{k=1}^{n} \frac{[\cos k(\beta - \alpha) - (-1)^{k}]\sin k\theta}{k}$$

---

(5)

$a < \frac{1}{2}$

$$\bar{y}(x) = \frac{4h}{\pi} \sum_{j=1}^{s} \frac{(1 - \cos j\alpha)\sin j\theta}{j}$$

---

(6)

$a < b < 1$

$$\bar{y}(x) = \frac{2h}{\pi} \sum_{j=1}^{s} \frac{(\cos j\alpha + \cos j\beta)\sin j\theta}{j}$$

$(a + b) = 1$

$$\bar{y}(x) = \frac{4h}{\pi} \sum_{j=1}^{s} \frac{\cos j(\beta - \alpha)\sin j\theta}{j}$$

a, b = positive fractions	h, L = positive numbers	$k = 1, 2, 3, \ldots, n$
$u = x/L$ $\alpha = a\pi$	$\beta = b\pi$ $\theta = u\pi$ $\eta = (-1)^k$	$j = 1, 3, 5, \ldots, s$

(1)
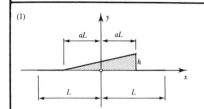

$a \le 1$

$$\bar{y}(x) = \frac{ah}{2} + \frac{h}{\pi}\sum_{k=1}^{n}\frac{\sin k\alpha \cos k\theta}{k} - \frac{h}{\pi}\sum_{k=1}^{n}\left(\frac{\cos k\alpha}{k} - \frac{\sin k\alpha}{ak^2}\right)\sin k\theta$$

(2)
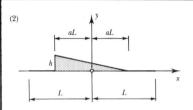

$a \le 1$

$$\bar{y}(x) = \frac{ah}{2} + \frac{h}{\pi}\sum_{k=1}^{n}\frac{\sin k\alpha \cos k\theta}{k} + \frac{h}{\pi}\sum_{k=1}^{n}\left(\frac{\cos k\alpha}{k} - \frac{\sin k\alpha}{ak^2}\right)\sin k\theta$$

(3)
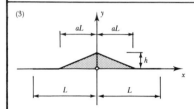

$a < 1$

$$\bar{y}(x) = \frac{ah}{2} + \frac{2h}{a\pi^2}\sum_{k=1}^{n}\frac{(1 - \cos k\alpha)\cos k\theta}{k^2}$$

$a = 1$

$$\bar{y}(x) = \frac{h}{2} + \frac{4h}{\pi^2}\sum_{j=1}^{s}\frac{\cos j\theta}{j^2}$$

(4)
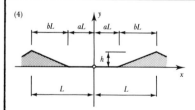

$a < 1$

$$\bar{y}(x) = \frac{bh}{2} + \frac{2h}{b\pi^2}\sum_{k=1}^{n}\frac{(\eta - \cos k\alpha)\cos k\theta}{k^2}$$

$a = 0$

$$\bar{y}(x) = \frac{h}{2} - \frac{4h}{\pi^2}\sum_{j=1}^{s}\frac{\cos j\theta}{j^2}$$

(5)
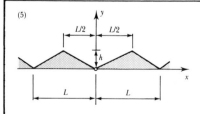

$$\bar{y}(x) = \frac{h}{2} - \frac{8h}{\pi^2}\sum_{j=1}^{s}\frac{\cos j\theta}{j^2}$$

(6)
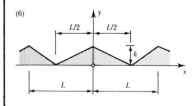

$$\bar{y}(x) = \frac{h}{2} + \frac{8h}{\pi^2}\sum_{j=1}^{s}\frac{\cos j\theta}{j^2}$$

h, L = positive numbers		$k = 1, 2, 3, \ldots, n$
$u = x/L$	$\theta = u\pi$	$j = 1, 3, 5, \ldots, s$

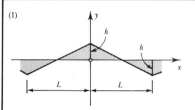

(1)

$$\bar{y}(x) = \frac{8h}{\pi^2} \sum_{j=1}^{s} \frac{\cos j\theta}{j^2}$$

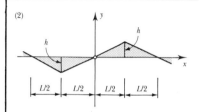

(2)

$$\bar{y}(x) = \frac{8h}{\pi^2} \sum_{j=1}^{s} \frac{\sin j\theta}{j^2}$$

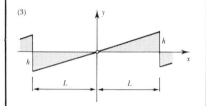

(3)

$$\bar{y}(x) = \frac{2h}{\pi} \sum_{k=1}^{n} \frac{(-1)^{k+1} \sin k\theta}{k}$$

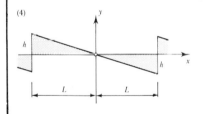

(4)

$$\bar{y}(x) = \frac{2h}{\pi} \sum_{k=1}^{n} \frac{(-1)^{k} \sin k\theta}{k}$$

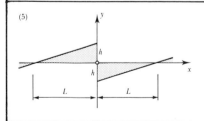

(5)

$$\bar{y}(x) = -\frac{2h}{\pi} \sum_{k=1}^{n} \frac{\sin k\theta}{k}$$

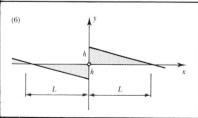

(6)

$$\bar{y}(x) = \frac{2h}{\pi} \sum_{k=1}^{n} \frac{\sin k\theta}{k}$$

a, b = positive fractions	h, L = positive numbers	$k = 1, 2, 3, \ldots, n$
$u = x/L$ $\quad \alpha = a\pi$	$\beta = b\pi$ $\quad \theta = u\pi$ $\quad \eta = (-1)^k$	$j = 1, 3, 5, \ldots, s$

(1)

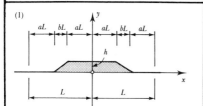

$$(2a + b) = 1$$

$$\bar{y}(x) = \frac{h}{2} + \frac{4h}{b\pi^2} \sum_{j=1}^{s} \frac{\cos j\alpha \cos j\theta}{j^2}$$

(2)

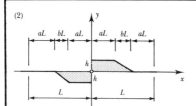

$$(2a + b) = 1$$

$$\bar{y}(x) = \frac{4h}{\pi} \sum_{j=1}^{s} \frac{\sin j\theta}{j} + \frac{2h\eta}{\pi} \sum_{k=1}^{n} \left[1 + \frac{(1 + \eta) \sin k\alpha}{k\pi b} \right] \frac{\sin k\theta}{k}$$

(3)

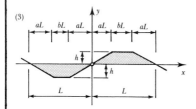

$$(2a + b) = 1$$

$$\bar{y}(x) = \frac{4h}{a\pi^2} \sum_{j=1}^{s} \frac{\sin j\alpha \sin j\theta}{j^2}$$

(4)

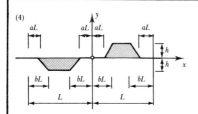

$$(a + b) \leq \tfrac{1}{2}$$

$$\bar{y}(x) = \frac{4h}{(b - a)\pi^2} \sum_{j=1}^{s} \frac{(\sin j\beta - \sin j\alpha) \sin j\theta}{j^2}$$

(5)

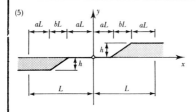

$$(2a + b) = 1$$

$$\bar{y}(x) = -\frac{2h\eta}{\pi} \sum_{k=1}^{n} \left(1 + \frac{\sin k\beta}{k\pi a} \right) \frac{\sin k\theta}{k}$$

(6)

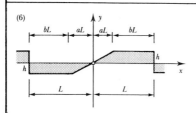

$$(a + b) = 1$$

$$\bar{y}(x) = -\frac{2h\eta}{\pi} \sum_{k=1}^{n} \left[1 + \frac{(1 + \eta) \sin k\alpha}{k\pi b} \right] \frac{\sin k\theta}{k}$$

a, b = positive fractions	h, L = positive numbers	$k = 1, 2, 3, \ldots, n$
$u = x/L$ $\theta = u\pi$ $\eta = (-1)^k$	c = noninteger number	$j = 1, 2, 5, \ldots, s$

(1)

$$y(x) = h \, |\sin \theta|$$

$$\bar{y}(x) = \frac{2h}{\pi} - \frac{4h}{\pi} \sum_{k=1}^{n} \frac{\cos 2k\theta}{(2k-1)(2k+1)}$$

(2)

$$y(x) = h \, |\cos \theta|$$

$$\bar{y}(x) = \frac{2h}{\pi} - \frac{4h}{\pi} \sum_{k=1}^{n} \frac{\cos 2k\theta}{(2k-1)(2k+1)}$$

(3)

$$-L < x < -L/2, \qquad L/2 < x < L, \qquad y(x) = 0$$

$$-L/2 < x < L/2, \qquad y(x) = h \cos \theta$$

$$\bar{y}(x) = \frac{h}{\pi} + \frac{h}{2} \cos \theta - \frac{2h\eta}{\pi} \sum_{k=1}^{n} \frac{\cos 2k\theta}{(2k-1)(2k+1)}$$

(4)

$$-L < x < 0, \qquad y(x) = 0$$

$$0 < x < L, \qquad y(x) = h \sin \theta$$

$$\bar{y}(x) = \frac{h}{\pi} + \frac{h}{2} \sin \theta - \frac{2h}{\pi} \sum_{k=1}^{n} \frac{\cos 2k\theta}{(2k-1)(2k+1)}$$

(5)

$$-L < x < L, \qquad y(x) = h \sin c\theta$$

$$\bar{y}(x) = -\frac{2h \sin c\pi}{\pi} \sum_{k=1}^{n} \frac{\eta k \sin k\theta}{k^2 - c^2}$$

(6)

$$-L < x < L, \qquad y(x) = h \cos c\theta$$

$$\bar{y}(x) = \frac{2ch \sin c\pi}{\pi} \left(\frac{1}{2c^2} + \sum_{k=1}^{n} \frac{\eta \cos k\theta}{c^2 - k^2} \right)$$

(a) Concept. Since the exponential and trigonometric functions are connected by the Euler's formulas as

$$\cos\theta = \frac{e^{i\theta} + e^{-i\theta}}{2} \qquad \sin\theta = \frac{e^{i\theta} - e^{-i\theta}}{2i} \qquad (i = \sqrt{-1})$$

the Fourier series expansion in (13.01d) becomes, with $k = 0, \pm1, \pm2, \dots$ and $\lambda u = k\pi x/L$,

$$y(x) = \tfrac{1}{2}\bar{A}_0 + \frac{1}{2}\sum_{k=1}^{\infty}\bar{A}_k(e^{i\lambda u} + e^{-i\lambda u}) - \tfrac{1}{2}i\sum_{k=1}^{\infty}\bar{B}_k(e^{i\lambda u} - e^{-i\lambda u}) = X_0 + \sum_{k=1}^{\infty}X_k e^{i\lambda u} + \sum_{k=1}^{\infty}Y_k e^{-i\lambda u}$$

where $X_0 = \tfrac{1}{2}\bar{A}_0$ $X_k = \tfrac{1}{2}(\bar{A}_k - i\bar{B}_k)$ $Y_k = \tfrac{1}{2}(\bar{A}_k + i\bar{B}_k)$

(b) Exponential form introduced above can be approximated by $2n + 1$ terms as

$$\boxed{\bar{y}(x) = \sum_{k=-n}^{n} X_k e^{i\lambda u}} \qquad (k = 0, \pm1, \pm2, \dots, \pm n)$$

(c) Coefficients of the truncated series are given by the general formula

$$\boxed{X_k = \frac{1}{2}\int_{-1}^{1} y(uL)e^{-i\lambda u}\, du}$$

and the set $X_{-n}, \dots, X_{-2}, X_{-1}, X_0, X_1, X_2, \dots, X_n$ is called the *spectrum* of $\bar{y}(x)$.

(d) Advantage of the exponential form is the compactness useful in the evaluation of derivatives and integrals. The *disadvantage* is its incapability to offer a simplification in cases of even and odd functions.

(e) Coefficients for selected particular functions are tabulated below.

$$X_k = \begin{cases} \dfrac{2h}{|k|\pi} & (k = \pm1, \pm5, \pm9, \dots) \\[2mm] \dfrac{-2h}{|k|\pi} & (k = \pm3, \pm7, \pm11, \dots) \\[2mm] 0 & (k = \pm2, \pm4, \pm6, \dots) \end{cases}$$

$$X_k = \begin{cases} \dfrac{4h}{k^2\pi^2} & (k = \pm1, \pm3, \pm5, \dots) \\[2mm] 0 & (k = \pm0, \pm2, \pm4, \dots) \end{cases}$$

$$X_k = \begin{cases} 0 & (k = \pm1, \pm3, \pm5, \dots) \\[2mm] \dfrac{2h}{(1-k^2)\pi} & (k = \pm0, \pm2, \pm4, \dots) \end{cases}$$

$$X_k = \begin{cases} \dfrac{-ikh}{4} & (k = \pm1) \\[2mm] 0 & (k = \pm3, \pm5, \pm7, \dots) \\[2mm] \dfrac{h}{(1-k^2)\pi} & (k = \pm0, \pm2, \pm4, \dots) \end{cases}$$

(a) Concept. The remarkable property of the sines and cosines is their orthogonality over both the continuous range and a set of equally spaced values in that range. Since in many engineering problems equally spaced values of the function are given, the approximation of this function by the Fourier series based on these discrete values is of practical importance. If the values are known at $2N + 1$ equally spaced points in $[0, L]$, as shown in the adjacent figure, so that their abscissas are

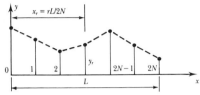

$$x_r = 0, \frac{L}{2N}, \frac{2L}{2N}, \dots, \frac{(2N-1)L}{2N}, L$$

then the finite Fourier approximation fitting the values $y_0, y_1, y_2, \dots, y_{2N-1}, y_{2N}$ is

$$\bar{y}(x) = \tfrac{1}{2}\bar{A}_0 + \sum_{k=1}^{N-1} \left(\bar{A}_k \cos \frac{2\pi kx}{L} + \bar{B}_k \sin \frac{2\pi kx}{L} \right) + \tfrac{1}{2}\bar{A}_n \cos \frac{2\pi kx}{L}$$

(b) Coefficients of the finite series obtained from the minimum error condition (13.01d) are

$$A_k = \frac{1}{N} \left(\tfrac{1}{2}y_0 + \sum_{r=1}^{2N-1} y_r \cos \frac{2\pi kx_r}{L} + \tfrac{1}{2}y_{2N} \right) \qquad (k = 0, 1, 2, \dots, N)$$

$$\bar{B}_k = \frac{1}{N} \sum_{r=1}^{2N-1} y_r \sin \frac{2\pi kx_r}{L} \qquad (k = 1, 2, 3, \dots, N-1)$$

(c) Coefficient matrices. The numerical evaluation of $\bar{A}_k$ and $\bar{B}_k$ can be made by two matrix equations, a typical example of which for $L = 30$, $2N = 6$ in terms of the given values $y_0, y_1, y_2, \dots, y_6$ is given below.

$$
\begin{bmatrix} \bar{A}_0 \\ \bar{A}_1 \\ \bar{A}_2 \\ \bar{A}_3 \\ \bar{A}_4 \\ \bar{A}_5 \\ \bar{A}_6 \end{bmatrix}
=
\begin{bmatrix}
1 & 1 & 1 & 1 & 1 & 1 & 1 \\
1 & C_1 & C_2 & C_3 & C_4 & C_5 & C_6 \\
1 & C_2 & C_4 & C_6 & C_8 & C_{10} & C_{12} \\
1 & C_3 & C_6 & C_9 & C_{12} & C_{15} & C_{18} \\
1 & C_4 & C_8 & C_{12} & C_{16} & C_{20} & C_{24} \\
1 & C_5 & C_{10} & C_{15} & C_{20} & C_{25} & C_{30} \\
1 & C_6 & C_{12} & C_{18} & C_{24} & C_{30} & C_{36}
\end{bmatrix}
\begin{bmatrix} y_0/6 \\ y_1/3 \\ y_2/3 \\ y_3/3 \\ y_4/3 \\ y_5/3 \\ y_6/6 \end{bmatrix}
$$

$$
\begin{bmatrix} \bar{B}_1 \\ \bar{B}_2 \\ \bar{B}_3 \\ \bar{B}_4 \\ \bar{B}_5 \end{bmatrix}
=
\begin{bmatrix}
S_1 & S_2 & S_3 & S_4 & S_5 \\
S_2 & S_4 & S_6 & S_8 & S_{10} \\
S_3 & S_6 & S_9 & S_{12} & S_{15} \\
S_4 & S_8 & S_{12} & S_{16} & S_{20} \\
S_5 & S_{10} & S_{15} & S_{20} & S_{25}
\end{bmatrix}
\begin{bmatrix} y_1/3 \\ y_2/3 \\ y_3/3 \\ y_4/3 \\ y_5/3 \end{bmatrix}
$$

where with $\alpha = = \pi/3$,

$$C_1 = \cos \alpha, \ C_2 = \cos 2\alpha, \dots, C_{36} = \cos 36\alpha$$

$$S_1 = \sin \alpha, \ S_2 = \sin 2\alpha, \dots, S_{25} = \sin 25\alpha$$

Both matrices are *symmetrical* and can be constructed for any value of L divided into any number of $2N$ segments with $\alpha = \pi/N$.

(a) Two Fourier series of the type $(13.01b)$

$$\bar{y}_1(\theta) = \tfrac{1}{2}A_{1,0} + \sum_{k=1}^{n} (A_{1,k}\cos k\theta + B_{1,k}\sin k\theta)$$

$$\bar{y}_2(\theta) = \tfrac{1}{2}A_{2,0} + \sum_{k=1}^{n} (A_{2,k}\cos k\theta + B_{2,k}\sin k\theta)$$

with known coefficients are considered in the operations defined below, where k, m, n, r are positive integers.

(b) Sum of two functions given by their respective Fourier series is

$$\bar{y}_1(\theta) \pm \bar{y}_2(\theta) = \tfrac{1}{2}C_0 + \sum_{k=1}^{n} (C_k\cos k\theta + D_k\sin k\theta)$$

where

$$C_0 = A_{1,0} \pm A_{2,0} \qquad C_k = A_{1,k} \pm A_{2,k} \qquad D_k = B_{1,k} \pm B_{2,k}$$

(c) Product of two functions given by their respective Fourier series is

$$\bar{y}_1(\theta) \cdot \bar{y}(\theta) = \tfrac{1}{2}F_0 + \sum_{k=1}^{n} (F_k\cos k\theta + G_k\sin k\theta)$$

where with $F_{-m} = F_m$ and $G_{-m} = -G_m$,

$$F_0 = \tfrac{1}{2}A_{1,0}A_{2,0} + \frac{1}{2}\sum_{r=1}^{n} (A_{1,r}A_{2,r} + B_{1,r}B_{2,r})$$

$$F_k = \tfrac{1}{2}A_{1,0}A_{2,k} + \frac{1}{2}\sum_{r=1}^{n} [(A_{2,r+k} + A_{2,r-k})A_{1,r} + (B_{2,r+k} + B_{2,r-k})B_{1,r}]$$

$$G_k = \tfrac{1}{2}A_{1,0}B_{2,k} + \frac{1}{2}\sum_{r=1}^{n} [(A_{2,r+k} - A_{2,r-k})B_{1,r} - (B_{2,r+k} - B_{2,r-k})A_{1,r}]$$

These relations are valid if and only if the product $y_1(\theta) \cdot y_2(\theta)$ is an integrable function.

(d) Derivative. If $y(\theta)$ is the function defined in $(13.01b)$ with $y(-\pi) = y(\pi)$ and its first derivative is sectionally continuous in $(-\pi, \pi)$, then its Fourier expansion is differentiable at each point where $y''(\theta)$ exists.

$$\frac{d\bar{y}(\theta)}{d\theta} = \sum_{k=1}^{n} k(-A_k\sin k\theta + B_k\cos k\theta)$$

(e) Integral. If $y(\theta)$ is the same as in (d) above, then

$$\int_{-\pi}^{\theta} \bar{y}(\theta)\, d\theta = \tfrac{1}{2}(\pi + \theta)A_0 + \sum_{k=1}^{n} \frac{1}{k}[A_k\sin k\theta - B_k(\cos k\theta - \cos k\pi)$$

(f) Higher derivatives and multiple integrals of the same functions can be derived by means of tables in Ref. 1.20, sec. 7.07-3, 9.10, 9.12, and 9.13.

14

LAPLACE TRANSFORMS

(a) Definition. If $F(t)$ is a real piecewise-continuous function of the real variable $t \geq 0$, then the *Laplace transform* of $F(t)$ is

$$\mathcal{L}\{F(t)\} = \int_0^\infty e^{-st} F(t)\, dt = f(s)$$

where s = real or complex and $e = 2.71828\ldots$, (2.18). The variable t can be x, y, z, or any other variable.

(b) Piecewise continuity. A function $F(t)$ is said to be piecewise-continuous (sectionally continuous) in an interval $a \leq t \leq b$ if the interval can be subdivided into a finite number of segments, in each of which the function is continuous and has a finite limit at both ends of each segment. The adjacent graph is an example of such function.

(c) Exponential order. A function $F(t)$ is said to be of exponential order N if with the real constant $M > 0$ and $t \to \infty$,

$$|F(t)| < Me^{Nt}$$

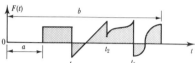

(d) Existence theorem. If $F(t)$ is piecewise-continuous in every segment of the interval $0 \leq t \leq R$ and of exponential order N for $t > R$, then its Laplace transform exists for all $s > N$. These conditions are sufficient but not necessary, since if they are not satisfied the transform may or may not exist.

(e) Laplace transforms of elementary functions derived by (a) in (f) to (i) are tabulated below, where a, ω = signed numbers, $s > 0$, and $n = 1, 2, 3, \ldots$.

	$F(t)$	$f(s)$		$F(t)$	$f(s)$
(1)	1	$\dfrac{1}{s}$	(7)	$\sin \omega t$	$\dfrac{\omega}{s^2 + \omega^2}$
(2)	t^n	$\dfrac{n!}{s^{n+1}}$	(8)	$\cos \omega t$	$\dfrac{s}{s^2 + \omega^2}$
(3)	e^{at}	$\dfrac{1}{s - a}$	(9)	$\sinh \omega t$	$\dfrac{\omega}{s^2 - \omega^2}$
(4)	e^{-at}	$\dfrac{1}{s + a}$	(10)	$\cosh \omega t$	$\dfrac{s}{s^2 - \omega^2}$
(5)	$e^{at} \sin \omega t$	$\dfrac{\omega}{(s - a)^2 + \omega^2}$	(11)	$e^{at} \sinh \omega t$	$\dfrac{\omega}{(s - a)^2 - \omega^2}$
(6)	$e^{at} \cos \omega t$	$\dfrac{s - a}{(s - a)^2 + \omega^2}$	(12)	$e^{at} \cosh \omega t$	$\dfrac{s - a}{(s - a)^2 - \omega^2}$

(f) Algebraic polynomial (Ref. 1.20, sec. 19.71)

$$\mathcal{L}\{t^n\} = \int_0^\infty t^n e^{-st}\, dt = \lim_{c \to \infty} \int_0^c t^n e^{-st}\, dt$$

$$= \lim_{c \to \infty} \left[-e^{-st}\left(t^n + \frac{n t^{n-1}}{s} + \frac{n(n-1) t^{n-2}}{s^2} + \cdots + \frac{n!}{s^{n+1}} \right) \right]_0^c = \frac{n!}{s^{n+1}}$$

(g) Exponential function (Ref. 1.20, sec. 19.71)

$$\mathcal{L}\{e^{at}\} = \int_0^\infty e^{at}e^{-st}\,dt = \lim_{c\to\infty}\int_0^c e^{(a-s)t}\,dt = \lim_{c\to\infty}\left(\frac{e^{(a-s)t}}{a-s}\right)_0^c = \frac{1}{s-a}$$

and similarly, $\mathcal{L}\{e^{-at}\} = \dfrac{1}{s+a}$

(h) Trigonometric functions (Ref. 1.20, sec. 9.06)

$$\mathcal{L}\{\sin \omega t\} = \int_0^\infty e^{-st}\sin \omega t\,dt = \lim_{c\to\infty}\int_0^c e^{-st}\sin \omega t\,dt$$

$$= \lim_{c\to\infty}\left[\frac{-e^{-st}}{s^2+\omega^2}(s\sin \omega t + \omega \cos \omega t)\right]_0^c = \frac{\omega}{s^2+\omega^2}$$

$$\mathcal{L}\{\cos \omega t\} = \int_0^\infty e^{-st}\cos \omega t\,dt = \lim_{c\to\infty}\int_0^c e^{-st}\cos \omega t\,dt$$

$$= \lim_{c\to\infty}\left[\frac{-e^{-st}}{a^2+\omega^2}(s\cos \omega t - \omega \sin \tau t)\right]_0^c = \frac{s}{s^2+\omega^2}$$

(i) Hyperbolic functions (Ref. 1.20, sec. 19.71)

$$\mathcal{L}\{\sinh \omega t\} = \int_0^\infty e^{-st}\sinh \omega t\,dt = \lim_{c\to\infty}\int_0^c \frac{e^{(\omega-s)t} - e^{-(\omega+s)t}}{2}\,dt$$

$$= \lim_{c\to\infty}\left[\frac{e^{(\omega-s)t}}{2(\omega-s)} + \frac{e^{-(\omega+s)t}}{2(\omega+s)}\right]_0^c = \frac{\omega}{s^2-\omega^2}$$

$$\mathcal{L}\{\cosh \omega t\} = \int_0^\infty e^{-st}\cosh \omega t\,dt = \lim_{c\to\infty}\int_0^c \frac{e^{(\omega-s)t} + e^{-(\omega+s)t}}{2}\,dt$$

$$= \lim_{c\to\infty}\left[\frac{e^{(\omega-s)t}}{2(\omega-s)} - \frac{e^{-(\omega+s)t}}{2(\omega+s)}\right] = \frac{s}{s^2-\omega^2}$$

(j) Winkler's functions (Ref. 1.20, sec. 9.06)

$$\mathcal{L}\{e^{at}\sin \omega t\} = \int_0^\infty e^{(a-s)t}\sin \omega t\,dt = \lim_{c\to\infty}\int_0^c e^{(a-s)t}\sin \omega t\,dt$$

$$= \lim_{c\to\infty}\left[\frac{e^{(a-s)t}}{(a-s)^2+\omega^2}((a-s)\sin \omega t - \omega \cos \omega t)\right]_0^c = \frac{\omega}{(s-a)^2+\omega^2}$$

$$\mathcal{L}\{e^{at}\cos \omega t\} = \int_0^\infty e^{(a-s)t}\cos \omega t\,dt = \lim_{c\to\infty}\int_0^c e^{(a-s)t}\cos \omega t\,dt$$

$$= \lim_{c\to\infty}\left[\frac{e^{(a-s)t}}{(a-s)^2+\omega^2}((a-s)\cos \omega t + \omega \sin \omega t)\right]_0^c = \frac{s-a}{(s-a)^2+\omega^2}$$

and similarly, $\mathcal{L}\{e^{-at}\sin \omega t\} = \dfrac{\omega}{(s+a)^2+\omega^2}$ $\mathcal{L}\{e^{-at}\cos \omega t\} = \dfrac{s+a}{(s+a)^2+\omega^2}$

(a) Notation

$F_k(t)$ = original function $\qquad\qquad$ $f_k(s)$ = Laplace transform of $F_k(x)$

a, b = signed numbers $\qquad$ s = parameter > 0 $\qquad$ $k, n = 1, 2, 3, \ldots$

(b) Linearity property. If $\mathscr{L}\{F_1(t)\} = f_1(s), \mathscr{L}\{F_2(t)\} = f_2(s), \ldots$, then

$$\mathscr{L}\{aF_1(t) \pm bF_2(t) \pm \cdots\} = a\mathscr{L}\{F_1(t)\} \pm b\mathscr{L}\{F_2(t)\} \pm \cdots = af_1(s) \pm bf_2(s) \pm \cdots$$

(c) Example (14.01e-7,8).

$$\mathscr{L}\{a \sin \omega t + b \cos \omega t\} = a\frac{\omega}{s^2 + \omega^2} + b\frac{s}{s^2 + \omega^2} = \frac{a\omega + bs}{s^2 + \omega^2}$$

(d) First shifting property. If $\mathscr{L}\{F(t)\} = f(s)$, then

$$\mathscr{L}\{e^{at}F(t)\} = \int_0^\infty e^{at}e^{-st}F(t)\, dt = f(s - a) \qquad \mathscr{L}\{e^{-at}F(t)\} = \int_0^\infty e^{-at}e^{-st}F(t)\, dt = f(s + a)$$

(e) Examples (14.01e-2,9,10).

$$\mathscr{L}\{t^n e^{at}\} = \int_0^\infty t^n e^{at}e^{-st}\, dt = \frac{n!}{(s - a)^{n+1}} \qquad \mathscr{L}\{t^n e^{-at}\} = \int_0^\infty t^n e^{-at}e^{-st}\, dt = \frac{n!}{(s + a)^{n+1}}$$

$$\mathscr{L}\{e^{-at} \cosh \omega t\} = \int_0^\infty e^{-at}e^{-st} \cosh \omega t\, dt = \frac{s + a}{(s + a)^2 - \omega^2}$$

(f) Second shifting property. If

$$\mathscr{L}\{F(t)\} = f(s) \qquad \text{and} \qquad G(t) = \begin{cases} 0 & (0 \le t \le a) \\ F(t - a) & (t > a) \end{cases}$$

then $\quad \mathscr{L}\{G(t)\} = \int_0^\infty e^{-st}G(t)\, dt = \int_0^a e^{-st}(0)\, dt + \int_a^\infty e^{-st}F(t - a)\, dt$

and with $u = t - a$,

$$\mathscr{L}\{G(t)\} = \int_0^\infty e^{-s(u+a)}F(u)\, du = e^{-as}\int_0^\infty e^{-su}F(u)\, du = e^{-as}f(s)$$

(g) Example (14.01e-2). If $F(t) = t^2$ and $G(t) = (t - a)^2$ as shown in the adjacent graphs, then

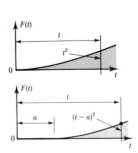

$$\mathscr{L}\{G(t)\} = \int_0^\infty (t - a)^2 e^{-st}\, dt = e^{-as}\frac{2}{s^3}$$

Similarly, if $F(t) = \sin \omega t$ and $G(t) = \sin[\omega(t - a)]$, then

$$\mathscr{L}\{G(t)\} = \int_0^\infty e^{-st} \sin[\omega(t - a)]\, dt = e^{-as}\frac{\omega}{s^2 + \omega^2}$$

(h) Scaling

$\mathscr{L}\{F(at)\} = \displaystyle\int_0^\infty e^{-st}F(at)\,dt$	$\mathscr{L}\left\{F\left(\dfrac{t}{a}\right)\right\} = \displaystyle\int_0^\infty e^{-st}F\left(\dfrac{t}{a}\right)dt$
With $u = at$, $dt = du/a$,	With $v = t/a$, $dt = a\,dv$,
$\mathscr{L}\{F(at)\} = \dfrac{1}{a}\displaystyle\int_0^\infty e^{-su/a}F(u)\,du = \dfrac{1}{a}f\left(\dfrac{s}{a}\right)$	$\mathscr{L}\left\{F\left(\dfrac{t}{a}\right)\right\} = a\displaystyle\int_0^\infty e^{-sav}F(v)\,dv = af(as)$

(i) Example (14.01*e*-5)

$$\mathscr{L}\left\{e^{at/b}\sin\frac{\omega t}{b}\right\} = \frac{\omega}{b[(s/b - a)^2 + \omega^2]}$$

(j) Multiplication by t^n. If $\mathscr{L}\{F(t)\} = f(s)$, then

$$\frac{df(s)}{ds} = \frac{d}{ds}\left[\int_0^\infty e^{-st}F(t)\,dt\right] = \int_0^\infty [-te^{-st}F(t)]\,dt$$

$$= -\int_0^\infty e^{-st}[tF(t)]\,dt = -\mathscr{L}\{tF(t)\}$$

and in general,

$$\boxed{\mathscr{L}\{t^n F(t)\} = (-1)^n \frac{d^n f(s)}{ds^n}}$$

(k) Example (14.01*e*-3,7)

$$\mathscr{L}\{t\sin\omega t\} = (-1)\frac{d}{ds}\left(\frac{\omega}{s^2 + \omega^2}\right) = \frac{2\omega s}{(s^2 + \omega^2)^2}$$

$$\mathscr{L}\{t^2 e^{at}\} = (-1)^2\frac{d^2}{ds^2}\left(\frac{1}{s + a}\right) = \frac{2}{(s + a)^3}$$

(l) Division by t. If

$$F(t) = tG(t) \qquad \mathscr{L}\{F(t)\} = f(s) \qquad \mathscr{L}\{G(t)\} = g(s)$$

then by (*j*), $\quad \mathscr{L}\{tG(t)\} = (-1)\dfrac{dg(s)}{ds} = f(s)$

and by integration,

$$g(s) = -\int_\infty^s f(s)\,ds = \int_s^\infty f(s)\,ds$$

or $\quad \boxed{\mathscr{L}\left\{\dfrac{F(t)}{t}\right\} = \displaystyle\int_s^\infty f(s)\,ds}$

(m) Examples (14.01*e*-7)

$$\left\{\frac{\sin\omega t}{t}\right\} = \int_s^\infty \frac{\omega}{s^2 + \omega^2}\,ds = \tan^{-1}\frac{s}{\omega}\Big|_s^\infty = \frac{\pi}{2} - \tan^{-1}\frac{s}{\omega} = \cot^{-1}\frac{s}{\omega}$$

(a) Notation

$F(t)$ = original function $f(s)$ = Laplace transform

$F^{(n)}(t)$ = nth-order derivative of $F(t)$ $(n = 1, 2, 3, \ldots)$

$F(+0), F^{(1)}(+0), F^{(2)}(+0), \ldots, F^{(n)}(+0)$ = initial values

$u_1, u_2, u_3, \ldots, u_n$ = functions of t

(b) First-order derivative. If $\mathscr{L}\{F(t)\} = f(s)$ and $dF(t)/dt = F'(t)$, then by parts,

$$\mathscr{L}\{F'(t)\} = \int_0^\infty e^{-st}F'(t)\, dt = \lim_{c \to \infty} \int_0^c e^{-st}F'(t)\, dt$$

$$= \lim_{c \to \infty} \left[e^{-st}F(t)\,|_0^c - \int_0^c (-se^{-st})F(t)\, dt \right]$$

and $$\boxed{\mathscr{L}\{F'(t)\} = sf(s) - F(+0)}$$

(c) Example. If $F'(t) = d(\sin \omega t)/(\omega\, dt) = \cos \omega t$ and $\mathscr{L}\{(\sin \omega t)/\omega\} = 1/(s^2 + \omega^2)$, then

$$\mathscr{L}\{\cos \omega t\} = s\mathscr{L}\left\{\frac{\sin \omega t}{\omega}\right\} + \frac{\sin \omega(+0)}{\omega} = \frac{s}{s^2 + \omega^2}$$

(d) Second-order derivative. If $\mathscr{L}\{F(t)\} = f(s)$ and $d^2F(t)/dt^2 = F''(t)$, then by parts,

$$\mathscr{L}\{F''(t)\} = \int_0^\infty e^{-st}F''(t)\, dt = \lim_{c \to \infty} \int_0^c e^{-st}F''(t)\, dt$$

$$= \lim_{c \to \infty} \left\{ [e^{-st}F'(t) + se^{-st}F(t)]_0^c - \int_0^c (-s^2 e^{-st}F(t)\, dt \right\}$$

and $$\boxed{\mathscr{L}\{F''(t)\} = s^2 f(s) - sF(+0) - F'(+0)}$$

(e) Example. If

$F(t) = \sinh \omega t$ $F'(t) = \omega \cosh \omega t$ $F''(t) = \omega^2 \sinh \omega t$

and $F(+0) = 0$ $F'(+0) = \omega$ $F''(+0) = 0$

then

$$\mathscr{L}\{\omega^2 \sinh \omega t\} = s^2 \mathscr{L}\{\sinh \omega t\} - s(0) - \omega = \frac{\omega s^2}{s^2 - \omega^2} - \omega = \frac{\omega^3}{s^2 - \omega^2}$$

(f) nth-order derivative. The transform of an nth-order derivative is by deduction of (b) and (d),

$$\boxed{\mathscr{L}\left\{\frac{d^n F(t)}{dt^n}\right\} = s^n f(s) - s^{n-1}F(+0) - s^{n-2}F^{(1)}(+0) - \cdots - sF^{(n-2)}(+0) - F^{(n-1)}(+0)}$$

where the lower derivatives must be continuous and the highest derivative $F^{(n)}(t)$ must be sectionally continuous in every finite interval of its application.

(g) Single integral. If

$$\mathscr{L}\{F(t)\} = f(s) \qquad G(t) = \int_0^t F(t)\,dt \qquad G(0) = 0 \qquad \frac{dG(t)}{dt} = F(t)$$

then by (14.02*b*) $\quad \mathscr{L}\left\{\dfrac{dG(t)}{dt}\right\} = s\mathscr{L}\{G(t)\} - G(+0) = f(s)$

from which
$$\boxed{\mathscr{L}\left\{\int_0^t F(t)\,dt\right\} = \frac{f(s)}{s}}$$

(h) Example (14.02*m*)

$$\mathscr{L}\left\{\int_0^t \frac{\sin \omega t}{t}\,dt\right\} = \frac{\mathscr{L}\{(\sin \omega t)/t\}}{s} = \frac{1}{s}\cot^{-1}\frac{s}{\omega}$$

(i) Multiple integrals. For the same function,

$$\boxed{\mathscr{L}\left\{\int_0^t\!\!\int F(t)\,dt\,dt\right\} = \frac{f(s)}{s^2}, \quad \mathscr{L}\left\{\int_0^t\!\!\int\!\!\int F(t)\,dt\,dt\,dt\right\} = \frac{f(s)}{s^3},\dots, \quad \mathscr{L}\left\{\int_0^t\!\!\int\cdots\int_0^t F(t)\,(dt)^n\right\} = \frac{f(s)}{s^n}}$$

(j) Example. If $F(t) = \cosh \omega t \cos \omega t$ and

$$\{\cosh \omega t \cos \omega t\} = \frac{s^3}{s^4 + 4\omega^4}$$

then $\quad \mathscr{L}\left\{\displaystyle\int_0^t\!\!\int \cosh \omega t \cosh \omega t\,dt\,dt\right\} = \dfrac{s}{s^4 + 4\omega^4}$

(k) Convolutions. If $\mathscr{L}\{F_1(t)\} = f_1(s)$, $\mathscr{L}\{F_2(t)\} = f_2(s),\dots$, then

$$\boxed{\begin{aligned}\mathscr{L}\left\{\int_0^t F_1(t - \tau)F_2(\tau)\,d\tau\right\} \\[2pt] \mathscr{L}\left\{\int_0^t F_1(\tau)F_2(t - \tau)\,d\tau\right\}\end{aligned}\Bigg\} = f_1(s)f_2(s)}$$

(l) Reduction. By (*k*) above,

$$\mathscr{L}\left\{\int_0^t (t - \tau)F(\tau)\,d\tau\right\} = \mathscr{L}\{t\}\cdot\mathscr{L}\{F(t)\} = \frac{f(s)}{s^2}$$

but by (*i*),

$$\frac{f(s)}{s^2} = \mathscr{L}\left\{\int_0^t\!\!\int F(t)\,dt\,dt\right\}$$

then in general,

$$\boxed{\int_0^t\!\!\int\cdots\int_0^t F(t)\,(dt)^n = \frac{1}{(n - 1)!}\int_0^t (t - \tau)^{n-1}F(\tau)\,d\tau}$$

(a) Three basic functions which play a useful role in the solution of engineering problems are:

    (1) Unit step function $\mathscr{U}(t - a)$
    (2) Unit impulse function $\mathscr{P}(t - a)$
    (3) Unit doublet function $\mathscr{Q}(t - a)$

Their analytical definitions and their Laplace transforms are given below in notation introduced in (14.01), (14.02). All Laplace transforms are equal to zero for $t < a$.

(b) Unit step function, also called Heaviside's unit function, shown in the adjacent graph, is

$$\mathscr{U}(t) = \begin{cases} 0 & (t < a) \\ 1 & (t \geq a) \end{cases}$$

and

$$\mathscr{L}\{\mathscr{U}(t)\} = \frac{1}{s} e^{-as}$$

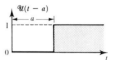

(c) Unit impulse function, also called Dirac Delta function, shown in the adjacent graph, is

$$\mathscr{P}(t - a) = \lim_{b \to 0} \left[\frac{\mathscr{U}(t - a)}{b} - \frac{\mathscr{U}(t - a - b)}{b} \right]$$

Other designations of this function are $\delta(t - a)$ or $\mathscr{D}(t - a)$, and the function is not a function in the true mathematical sense but a definition of distribution. Some of its properties are

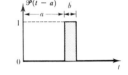

$$\int_0^\infty \mathscr{P}(t)\, dt = 1 \qquad\qquad \int_0^\infty \mathscr{P}(t - a)\, dt = 1$$

$$\int_0^\infty \mathscr{P}(t)F(t)\, dt = F(0) \qquad \int_0^\infty \mathscr{P}(t - a)\dot{F}(t)\, dt = F(t)$$

where $F(t)$ is a continuous function. The Laplace transforms of $\mathscr{P}(t)$ and $\mathscr{P}(t - a)$ are

$$\mathscr{L}\{\mathscr{P}(t)\} = 1 \qquad\qquad \mathscr{L}\{\mathscr{P}(t - a)\} = e^{-as}$$

$$\mathscr{L}\{\mathscr{P}(t)F(t)\} = f(s) \qquad \mathscr{L}\{\mathscr{P}(t - a)F(t)\} = e^{-as}f(s)$$

(d) Unit doublet function, shown in the adjacent graph, is

$$\mathscr{Q}(t - a) = \lim_{b \to 0} \frac{\mathscr{U}(t - a) - 2\mathscr{U}(t - a - b) + \mathscr{U}(t - a - 2b)}{b^2}$$

This function is not a function in the true mathematical sense but is again a definition of distribution. The Laplace transforms of $\mathscr{Q}(t)$ and $\mathscr{Q}(t - a)$ are

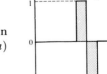

$$\mathscr{L}\{\mathscr{Q}(t)\} = s \qquad\qquad \mathscr{L}\{\mathscr{Q}(t - a)\} = se^{-as}$$

$$\mathscr{L}\{\mathscr{Q}(t)F(t)\} = sf(s) \qquad \mathscr{L}\{\mathscr{Q}(t - a)F(t)\} = se^{-as}f(s)$$

where $F(t)$ is the same as in (c).

(a) Three ramp functions which again (as their basic counterparts) play a useful role in engineering applications are:

   (1) Ramp function of first degree, $R_1(t - a)$
   (2) Ramp function of second degree, $R_2(t - a)$
   (3) Ramp function of nth degree, $R_n(t - a)$

Their analytical definitions and their Laplace transforms are given below in notation introduced in (14.01), (14.02). All Laplace transforms are equal to zero for $t < a$.

(b) Ramp function of first degree, shown in the adjacent figure, is

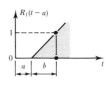

$$R_1(t - a) = \begin{cases} 0 & (t < a) \\ \dfrac{t - a}{b} & (t > a) \end{cases} \qquad \mathcal{L}\{R_1(t - a)\} = \frac{e^{-as}}{bs^2}$$

(c) Ramp function of second degree, shown in the adjacent figure, is

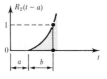

$$R_2(t - a) = \begin{cases} 0 & (t < a) \\ \dfrac{(t - a)^2}{b^2} & (t > a) \end{cases} \qquad \mathcal{L}\{R_2(t - a)\} = \frac{2e^{-as}}{b^2 s^3}$$

(d) Ramp function of nth degree, shown in the adjacent figure, is

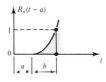

$$R_n(t - a) = \begin{cases} 0 & (t < a) \\ \dfrac{(t - a)^n}{b^n} & (t > a) \end{cases} \qquad \mathcal{L}\{R_n(t - a)\} = \frac{n!\,e^{-as}}{b^n s^{n+1}}$$

(e) Superposition of two rectangles (h = constant)

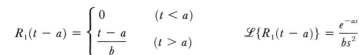

From the figures above,

$$\mathcal{L}\{F(t)\} = \frac{he^{-as}}{s} \quad (a < t < b) \qquad \mathcal{L}\{F(t)\} = \frac{he^{-as}}{s} - \frac{he^{-bs}}{s} \quad (t > b)$$

(f) Superposition of rectangle and two triangles (h = constant)

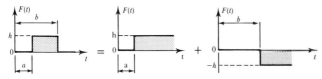

From the figures above,

$$\mathcal{L}\{F(t)\} = \frac{he^{-as}}{s} - \frac{he^{-as}}{(b - a)s} \quad (a < t < b) \qquad \mathcal{L}\{F(t)\} = \frac{he^{-as}}{s} - \frac{he^{-as}}{(b - a)s} + \frac{he^{-bs}}{(b - a)s} \quad (t > b)$$

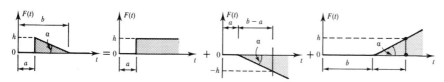

(a) Laplace transform of a periodic function,

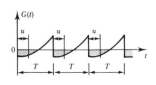

$$G(t + T) = G(t)$$

of period $T > 0$ shown in the adjacent graph is

$$\mathcal{L}\{G(t)\} = \int_0^\infty e^{-st} G(t)\, dt$$

and with $t_0 = u,\ t_1 = T + u,\ t_2 = 2T + u, \ldots,$

$$\mathcal{L}\{G(t)\} = \int_0^T e^{-su} G(u)\, du + \int_0^T e^{-s(T+u)} G(u)\, du + \int_0^T e^{-s(2T+2u)} G(u)\, du + \cdots$$

$$= (1 + e^{-sT} + e^{-2sT} + \cdots) \int_0^T e^{-su} G(u)\, du = \frac{\displaystyle\int_0^T e^{-su} G(u)\, du}{1 - e^{-sT}} = g(s)$$

(b) Rectangular function. If

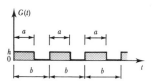

$$G(t) = \begin{cases} h & (0 < u < a) \\ 0 & (a < u < b) \end{cases}$$

and b is the period, then

$$g(s) = \frac{\displaystyle\int_0^a h e^{-su}\, du}{1 - e^{-bs}} = \frac{(1 - e^{-as})h}{(1 - e^{-bs})s}$$

(c) Triangular function. If

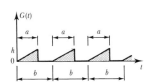

$$G(t) = \begin{cases} \dfrac{hu}{a} & (0 < u < a) \\ 0 & (a < u < b) \end{cases}$$

and b is the period, then by the geometric superposition $(14.05e)$, $(14.05f)$,

$$g(s) = \frac{\left(\dfrac{1}{as^2} - \dfrac{e^{-as}}{s} - \dfrac{e^{-as}}{as^2}\right)h}{1 - e^{-bs}} = \frac{[1 - (as + 1)e^{-as}]h}{a(1 - e^{-bs})s^2}$$

(d) Rectified sine half-wave. If

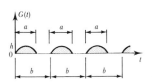

$$G(t) = \begin{cases} h \sin\left(\dfrac{\pi}{a} u\right) & (0 < u < a) \\ 0 & (a < u < b) \end{cases}$$

and $b = 2a$ is the period, then

$$g(s) = \frac{\displaystyle\int_0^a h e^{-su} \sin\left(\dfrac{\pi}{a} u\right) du}{1 - e^{-bs}} = \frac{a\pi h}{(\pi^2 + a^2 s^2)(1 - e^{-as})}$$

(a) Laplace transform of an antiperiodic function,

$$H(t + T) = -H(t)$$

of period $T > 0$ shown in the adjacent graph is

$$\mathcal{L}\{H(t)\} = \int_0^\infty e^{-st}H(t)\,dt$$

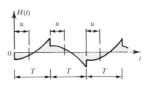

and with $t_0 = u,\ t_1 = T + u,\ t_2 = 2T + u, \ldots,$

$$\mathcal{L}\{H(t)\} = \int_0^T e^{-su}H(u)\,du - \int_0^T e^{-s(T+u)}H(u)\,du + \int_0^T e^{-s(2T+u)}H(u)\,du - \cdots$$

$$= (1 - e^{-sT} + e^{-2sT} - \cdots)\int_0^T e^{-su}H(t)\,du = \frac{\displaystyle\int_0^T e^{-su}H(u)\,du}{1 + e^{-sT}} = h(s)$$

(b) Triangular function. If with $k = 0, 1, 2, \ldots,$

$$H(t) = \begin{cases} (-1)^k \dfrac{h}{a}u & (0 < u < a) \\[2mm] (-1)^k \dfrac{h}{a}(b - u) & (a < u < b) \end{cases}$$

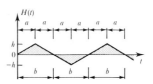

and $b = 2a$ is the period, then by the geometric superposition $(14.05e),\ (14.05f),$

$$h(s) = \frac{\dfrac{h}{as^2}(1 - 2e^{-as} + e^{-bs})}{1 + e^{-bs}} = \frac{h(1 - e^{-as})^2}{a(1 + e^{-bs})s^2}$$

(c) Alternating triangular function. If with $k = 0, 1, 2, \ldots,$

$$H(t) = \begin{cases} (-1)^k \dfrac{hu}{a} & (0 < u < a) \\[2mm] 0 & (a < u < b) \end{cases}$$

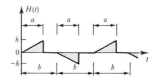

and b is the period, then by the relation between $g(s)$ and $h(s)$ defined in (a) in terms of $(14.06c),$

$$h(t) = \frac{[1 - (as + 1)e^{-as}]h}{a(1 + e^{-bs})s^2}$$

(d) Alternating rectified sine half-wave. If with $k = 0, 1, 2, \ldots,$

$$H(t) = \begin{cases} (-1)^k h \sin\left(\dfrac{\pi}{a}u\right) & (0 < u < a) \\[2mm] 0 & (a < u < b) \end{cases}$$

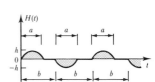

and $b = 2a$ is the period, then

$$h(s) = \frac{a\pi(1 - e^{-bs})h}{(\pi^2 + a^2s^2)(1 - e^{-as})(1 + e^{-bs})} = \frac{a\pi(1 + e^{-as})h}{(\pi^2 + a^2s^2)(1 + e^{-bs})}$$

(a) Definition. If the Laplace transform of a function $F(t)$ is $f(s)$ defined in $(14.01a)$, then $F(t)$ is the inverse of $f(s)$. Symbolically,

$$\boxed{\mathscr{L}^{-1}\{f(s)\} = F(t)}$$

Although in the strict sense of the uniqueness of functions the Laplace transform is not unique, in engineering applications this transform is essentially unique.

(b) Procedure of finding this inverse is usually simplified by using tables of Laplace transform pairs. Extensive tables are available in the literature (Refs. 1.01, 1.20, 14.01–14.06). In addition, certain properties of inverse Laplace transforms useful in finding the inverse are defined below.

(c) Inverse properties based on (14.02) and (14.03) are tabulated below.

(1) $\mathscr{L}^{-1}\left\{\dfrac{a}{s}\right\} = a$	(2) $\mathscr{L}^{-1}\left\{\dfrac{a}{s^2}\right\} = at$	(3) $\mathscr{L}^{-1}\left\{\dfrac{a}{s^n}\right\} = \dfrac{at^{n-1}}{(n-1)!}$
(4) $\mathscr{L}^{-1}\{f(s+a)\} = e^{-at}F(t)$		(5) $\mathscr{L}^{-1}\{f(s-a)\} = e^{at}F(t)$
(6) $\mathscr{L}^{-1}\{f(as)\} = \dfrac{1}{a}F\left(\dfrac{t}{a}\right)$		(7) $\mathscr{L}^{-1}\left\{f\left(\dfrac{s}{a}\right)\right\} = aF(at)$
(8) $\mathscr{L}^{-1}\{e^{as}f(s)\} = F(t+a)$		(9) $\mathscr{L}^{-1}\{e^{-as}f(s)\} = F(t-a)$
(10) $\mathscr{L}^{-1}\{s^n f(s)\} = \dfrac{d^n F(t)}{dt^n}$ if $F(0) = F'(0) = \cdots = F^{(n-1)}(0) = 0$		
(11) $\mathscr{L}^{-1}\left\{\dfrac{f(s)}{s}\right\} = \displaystyle\int_0^t F(t)\,dt$		(12) $\mathscr{L}^{-1}\left\{\dfrac{f(s)}{s^n}\right\} = \displaystyle\int_0^t\cdots\int_0^t F(t)\,(dt)^n$
(13) $\mathscr{L}^{-1}\left\{\dfrac{df(s)}{ds}\right\} = -tF(t)$		(14) $\mathscr{L}^{-1}\left\{\dfrac{d^n f(s)}{ds^n}\right\} = (-1)^n t^n F(t)$
(15) $\mathscr{L}^{-1}\{f_1(s)\cdot f_2(s)\} = \displaystyle\int_0^t F_1(t-\tau)F_2(\tau)\,d\tau = \int_0^t F_1(\tau)F_2(t-\tau)\,d\tau$		

(d) Examples. If

$$\mathscr{L}\left\{\frac{\sinh \omega t + \sin \omega t}{2\omega}\right\} = \frac{s^2}{s^4 - \omega^4} \qquad \mathscr{L}\{\sin t\} = \frac{1}{s^2 + 1}$$

then by $(c\text{-}10)$ and $(c\text{-}12)$,

$$\mathscr{L}^{-1}\left\{\frac{s^3}{s^4 - \omega^4}\right\} = \frac{d}{dt}\left(\frac{\sinh \omega t + \sin \omega t}{2\omega}\right) = \frac{\cosh \omega t + \cos \omega t}{2}$$

$$\mathscr{L}^{-1}\left\{\frac{s}{s^4 - \omega^4}\right\} = \int_0^t \frac{\sinh \omega t + \sin \omega t}{2\omega}\,dt = \frac{\cosh \omega t - \cos \omega t}{2\omega^2}$$

and by $(c\text{-}3)$,

$$\mathscr{L}^{-1}\left\{\frac{d}{ds}\left(\frac{1}{s^2 + 1}\right)\right\} = \mathscr{L}^{-1}\left\{\frac{-2s}{(s^2 + 1)^2}\right\} = -t \sin t$$

(a) Rational fraction. If a rational Laplace transform with $m \leq n - 1$,

$$f(s) = \frac{P(s)}{Q(s)} = \frac{b_0 + b_1 s + b_2 s^2 + \cdots + b_m s^m}{a_0 + a_1 s + a_2 s^2 + \cdots + b_n s^n}$$

then by the fundamental theorem of algebra, the denominator

$$Q(s) = (s - q_1)(s - q_2) \cdots (s - q_{n-1})(s - q_n)$$

where $q_1, q_2, \ldots, q_n$ are the roots, which according to (6.01d) can be real and distinct, or real and repeated, or complex conjugate and distinct, or complex conjugate and repeated. The roots of $Q(s)$ are calculated by one of the methods introduced in Chap. 6.

(b) Heaviside's formula, distinct roots. If q_k $(k = 1, 2, \ldots, n)$ are distinct, the inverse of $f(s)$ in (a) is

$$\mathcal{L}^{-1}\left\{\frac{P(s)}{Q(s)}\right\} = \sum_{k=1}^{n} \frac{P(q_k)}{Q'(q_k)} e^{q_k t}$$

where $\quad P(q_k) = b_0 + b_1 q_k + b_2 q_k^2 + \cdots + b_m q_k^m$

$$Q'(q_k) = \frac{d}{ds}[Q(s)]_{s=q_k}$$

$$= a_1 + 2a_2 q_k + 3a_3 q_k^2 + \cdots + n a_n q_k^{n-1}$$

(c) Example. If

$$f(s) = \frac{P(s)}{Q(s)} = \frac{2s + 1}{s^3 - 2s^2 - s + 2} = \frac{2s + 1}{(s - 2)(s - 1)(s + 1)}$$

the roots of $Q(s)$ are $q_1 = 2$, $q_2 = 1$, $q_3 = -1$, from which $P(2) = 5$, $P(1) = 3$, $P(-1) = -1$, $Q'(2) = 3$, $Q'(1) = -2$, $Q'(-1) = 6$, and

$$\mathcal{L}^{-1}\{f(s)\} = \mathcal{L}^{-1}\left\{\frac{\frac{5}{3}}{s - 2} - \frac{\frac{3}{2}}{s - 1} - \frac{\frac{1}{6}}{s + 1}\right\} = \tfrac{5}{3}e^{2t} - \tfrac{3}{2}e^{t} - \tfrac{1}{6}e^{-t}$$

(d) Heaviside's formula, repeated roots. If the roots of $Q(s) = 0$ are $q_1 = q_2 = \cdots = q_n = q$, the inverse of $f(s)$ is

$$\mathcal{L}^{-1}\left\{\frac{P(s)}{Q(s)}\right\} = \left[\sum_{k=1}^{n} \frac{A_k t^{n-k-1}}{(n - k - 1)!}\right] e^{qt}$$

where $\quad A_k = \frac{1}{(k - 1)!} \frac{d^{k-1}}{ds}[(s - q)^n f(s)]_{s=q_k} \quad (k = 1, 2, \ldots, n)$

(e) Heaviside's formula, distinct and repeated roots. If the roots of $Q(s) = 0$ are $q_1 = q_2 = \cdots = q_r = q$ and $q_{r+1}, q_{r+2}, \ldots, q_n$, the inverse of $f(s)$ is

$$\mathcal{L}^{-1}\left\{\frac{P(s)}{Q(s)}\right\} = \left[\sum_{k=1}^{r} \frac{A_k t^{r-k-1}}{(r - k - 1)!}\right] e^{qt} + \sum_{k=r+1}^{n} \frac{P(q_k)}{Q'(q_k)} e^{q_k t}$$

which is a combination of (d) and (b).

(a) Simple poles. In system engineering the roots of $Q(s) = 0$ in (14.09a) are called poles, and if

$$q_1 \neq q_2 \neq \cdots \neq q_n$$

the poles are called simple poles. The alternative solution of (14.09a) can be expressed as

$$\frac{P(s)}{Q(s)} = \frac{b_1}{s - q_1} + \frac{b_2}{s - q_2} + \cdots + \frac{b_n}{s - q_n}$$

where $\quad b_k = [(s - q_k)f(s)]_{s=q_k} \qquad (k = 1, 2, \ldots, n)$

and $\quad \mathscr{L}^{-1}\left\{\sum_{k=1}^{n} \frac{b_k}{s - q_k}\right\} = \sum_{k=1}^{n} b_k e^{q_k t}$

For simple poles this method is superior to the procedure (14.09b).

(b) Double poles. If

$$f(s) = \frac{b_0 + b_1 s}{(s - q)^2} \qquad (q_1 = q_2 = q)$$

then by (14.08c-5,10),

$$\mathscr{L}^{-1}\{f(s)\} = b_0 \mathscr{L}^{-1}\left\{\frac{1}{(s - q)^2}\right\} + b_1 \mathscr{L}^{-1}\left\{\frac{s}{(s - q)^2}\right\} = b_0 t e^{qt} + b_1 \frac{d}{dt}(t e^{qt})$$

and $\quad \mathscr{L}^{-1}\left\{\frac{b_0 + b_1 s}{(s - q)^2}\right\} = [(b_0 + b_1 q)t + b_1]e^{qt}$

(c) Triple poles. If

$$f(s) = \frac{b_0 + b_1 s + b_2 s^2}{(s - q)^3} \qquad (q_1 = q_2 = q_3 = q)$$

then by (14.08c-5,10)

$$\mathscr{L}^{-1}\left\{\frac{b_0 + b_1 s + b_2 s^2}{(s - q)^3}\right\} = b_0 \frac{t^2 e^{qt}}{2} + b_1 \frac{d}{dt}\left(\frac{t^2 e^{qt}}{2}\right) + b_2 \frac{d^2}{dt^2}\left(\frac{t^2 e^{qt}}{2}\right)$$

(d) Multiple poles. If

$$f(s) = \frac{b_0 + b_1 s + b_2 s^2 + \cdots + b_{n-1} s^{n-1}}{(s - q)^n} \qquad (q_1 = q_2 = \cdots = q_n)$$

then by (14.08c-5,10)

$$\mathscr{L}^{-1}\left\{\frac{b_0 + b_1 s + b_2 s^2 + \cdots + b_{n-1} s^{n-1}}{(s - q)^n}\right\} = \frac{b_0 t^{n-1}}{(n - 1)!}e^{qt} + \sum_{k=1}^{n-1} b_k \frac{d^k}{dt^k}\left[\frac{t^{n-1}}{(n - 1)!}e^{qt}\right]$$

(a) Simple factor. If with α, β = real signed numbers,

$$f(s) = \frac{b_0 + b_1 s}{(s - \alpha)^2 \pm \beta^2}$$

$$q_1 = \begin{cases} \alpha + \beta i \\ \alpha + \beta \end{cases} \qquad q_2 = \begin{cases} \alpha - \beta i \\ \alpha - \beta \end{cases}$$

then by (14.08c-5,10)

$$\mathscr{L}^{-1}\{f(s)\} = b_0 \mathscr{L}^{-1}\left\{\frac{1}{(s - \alpha)^2 \pm \beta^2}\right\} + b_1 \mathscr{L}^{-1}\left\{\frac{s}{(s - \alpha^2) \pm \beta^2}\right\}$$

and

$$\mathscr{L}^{-1}\left\{\frac{b_0 + b_1 s}{(s - \alpha)^2 + \beta^2}\right\} = \frac{b_0}{\beta} e^{\alpha t} \sin \beta t + \frac{b_1}{\beta} \frac{d}{dt}(e^{\alpha t} \sin \beta t)$$

$$\mathscr{L}^{-1}\left\{\frac{b_0 + b_1 s}{(s - \alpha)^2 - \beta^2}\right\} = \frac{b_0}{\beta} e^{\alpha t} \sinh \beta t + \frac{b_1}{\beta} \frac{d}{dt}(e^{\alpha t} \cosh \beta t)$$

(b) Double factors. If with α, β defined in (a),

$$f(s) = \frac{b_0 + b_1 s + b_2 s^2 + b_3 s^3}{[(s - \alpha)^2 \pm \beta^2]^2}$$

$$q_{1,2} = \begin{cases} \alpha + \beta i \\ \alpha + \beta \end{cases} \qquad q_{3,4} = \begin{cases} \alpha - \beta i \\ \alpha - \beta \end{cases}$$

then by (a),

$$\mathscr{L}^{-1}\left\{\frac{1}{(s - \alpha)^2 \pm \beta^2}\right\} = \frac{e^{\alpha t}}{\beta}\begin{cases} \sin \beta t \\ \sinh \beta t \end{cases}$$

and by convolution (14.08c-15)

$$\mathscr{L}^{-1}\left\{\frac{1}{(s - \alpha)^2 + \beta^2} \cdot \frac{1}{(s - \alpha)^2 + \beta^2}\right\} = \frac{1}{\beta^2}\int_0^t e^{\alpha(t-\tau)} \sin \beta(t - \tau) e^{\alpha \tau} \sin \beta\tau \, d\tau$$

$$= \frac{1}{\beta^2}\int_0^t \frac{e^{\alpha t}}{2}[\cos \beta(t - 2\tau) - \cos \beta\tau] \, d\tau = \frac{e^{\alpha t}}{2\beta^3}(\sin \beta t - \beta t \cos \beta t) = \phi_1(t)$$

Similarly,

$$\mathscr{L}^{-1}\left\{\frac{1}{(s - \alpha)^2 - \beta^2} \cdot \frac{1}{(s - \alpha)^2 - \beta^2}\right\} = \frac{e^{\alpha t}}{2\beta^3}(\beta t \cosh \beta t - \sinh \beta t) = \psi_1(t)$$

with these relations,

$$\mathscr{L}^{-1}\left\{\frac{b_0 + b_1 s + b_2 s^2 + b_3 s^3}{[(s - \alpha)^2 + \beta^2]^2}\right\} = b_0\phi_1(t) + b_1\phi_1'(t) + b_2\phi_1''(t) + b_3\phi_1'''(t)$$

$$\mathscr{L}^{-1}\left\{\frac{b_0 + b_1 s + b_2 s^2 + b_3 s^3}{[(s - \alpha)^2 - \beta^2]^2}\right\} = b_0\psi_1(t) + b_1\psi_1'(t) + b_2\psi_1''(t) + b_3\psi_1'''(t)$$

where $\phi_1^{(k)}(t) = d^k\phi_1(t)/dt^k$ and $\psi_1^{(k)}(t) = d^k\psi_1(t)/dt^k$.

(c) Multiple factors. If

$$f(s) = \frac{P(s)}{[(s - \alpha)^2 \pm \beta^2]^n}$$

the inverse is found by a successful application of the convolution.

(a) Notation

a, b = positive constants $\neq 0$ $k = 0, 1, 2, \ldots,$ $r = 1, 2, 3, \ldots$

$(\ )!!$ = double factorial (2.30) $\Gamma(\ )$ = gamma function (2.28)

$\exp(\ )$ = exponential function (9.05) $\mathrm{erf}(\ )$ = error function (10.01)

$J_k(\ )$ = Bessel function (10.07) $I_k(\ )$ = modified Bessel function (10.13)

(b) Irrational algebraic functions

$f(s)$	$F(t)$	$f(s)$	$F(t)$
$\dfrac{1}{\sqrt{s}}$	$\dfrac{1}{\sqrt{\pi t}}$	$\dfrac{1}{\sqrt{s^{2r+1}}}$	$\dfrac{(2t)^r}{(2r-1)!!\sqrt{\pi t}}$
$\dfrac{1}{\sqrt{s+a}}$	$\dfrac{1}{\sqrt{\pi t}}e^{-at}$	$\dfrac{1}{\sqrt{s-a}}$	$\dfrac{1}{\sqrt{\pi t}}e^{at}$
$\dfrac{1}{s\sqrt{s+a}}$	$\dfrac{\mathrm{erf}\sqrt{at}}{\sqrt{a}}$	$\dfrac{1}{(s-a)\sqrt{s}}$	$e^{at}\dfrac{\mathrm{erf}\sqrt{at}}{\sqrt{a}}$
$\dfrac{1}{\sqrt{(s+a)^{2r+1}}}$	$\dfrac{(2t)^r e^{-at}}{(2r-1)!!\sqrt{\pi t}}$	$\dfrac{1}{\sqrt{(s-a)^{2r+1}}}$	$\dfrac{(2t)^r e^{at}}{(2r-1)!!\sqrt{\pi t}}$
$\dfrac{1}{\sqrt{s^2+a^2}}$	$J_0(at)$	$\dfrac{1}{\sqrt{(s^2-a^2)}}$	$I_0(at)$
$\dfrac{1}{s\sqrt{s^2+a^2}}$	$\int_0^t J_0(at)\,dt$	$\dfrac{1}{s\sqrt{s^2-a^2}}$	$\int_0^t I_0(at)\,dt$
$\dfrac{1}{\sqrt{(s^2+a^2)^3}}$	$\dfrac{tJ_1(at)}{a}$	$\dfrac{1}{\sqrt{(s^2-a^2)^3}}$	$\dfrac{tI_1(at)}{a}$
$\dfrac{s}{\sqrt{(s^2+a^2)^3}}$	$tJ_0(at)$	$\dfrac{s}{\sqrt{(s^2-a^2)^3}}$	$tI_0(at)$
$\dfrac{s^2}{\sqrt{(s^2+a^2)^3}}$	$J_0(at)-atJ_1(at)$	$\dfrac{s^2}{\sqrt{(s^2-a^2)^3}}$	$I_0(at)+atI_1(at)$

(c) Exponential functions

$\dfrac{\exp(a/s)}{s}$	$I_0(2\sqrt{at})$	$\dfrac{\exp(-a/s)}{s}$	$J_0(2\sqrt{at})$
$\dfrac{\exp(a/s)}{s^{k+1}}$	$\sqrt{\left(\dfrac{t}{a}\right)^k}I_k(2\sqrt{at})$ $k>1$	$\dfrac{\exp(-a/s)}{s^{k+1}}$	$\sqrt{\left(\dfrac{t}{a}\right)^k}J_k(2\sqrt{at})$ $(k>1)$
$\dfrac{\exp(a/s)}{\sqrt{s}}$	$\dfrac{1}{\sqrt{\pi t}}\cosh(2\sqrt{at})$	$\dfrac{\exp(-a/s)}{\sqrt{s}}$	$\dfrac{1}{\sqrt{\pi t}}\cos(2\sqrt{at})$
$\dfrac{\exp(a/s)}{s\sqrt{s}}$	$\dfrac{1}{\sqrt{\pi t}}\sinh(2\sqrt{at})$	$\dfrac{\exp(-a/s)}{s\sqrt{s}}$	$\dfrac{1}{\sqrt{\pi t}}\sin(2\sqrt{at})$

15
ORDINARY
DIFFERENTIAL
EQUATIONS

(a) Definitions. A differential equation is an algebraic or transcendental equality involving differentials or derivatives. The *order* of a differential equation is the order of the highest derivative, and its *degree* is the degree of the highest derivative. The number of independent variables defines the differential equation as *ordinary differential equation* (single variable) or *partial differential equation* (two or several variables). A differential equation which is of the first degree in the dependent variable and in all its derivatives is called the *linear differential equation*. In general,

$$\sum_{r=0}^{n} a_r(x)\frac{d^r Y}{dx^r} = g(x) \qquad (r = 0, 1, 2, \ldots, n)$$

where Y = dependent variable, x = independent variable, n = order, $a_r(x)$ = variable coefficient (given function in x), and $g(x)$ = input or forcing function (given function in x).

(b) Special forms. The general form (a) is then designated as the nonhomogeneous ordinary linear differential equation with variable coefficients. If $g(x) = 0$, the equation (a) is called homogeneous, and if $a_r(x) = a_r$ (constant), the equation in (a) is designated with constant coefficients.

(c) General solution of the ordinary differential equation (a) is

$$Y = Y(x, C_1, C_2, \ldots, C_n)$$

where $C_1, C_2, \ldots, C_n$ are arbitrary and independent constants to be determined from the initial conditions (initial-value problem) or from the boundary conditions (boundary-value problem).

(d) Closed-form solutions are available only in a few simple cases but lead frequently to approximate solutions of more complex problems. Selected cases are tabulated in Ref. 1.20, pp. 176–183, and are not repeated here.

(e) Series solutions appear in the form of power series introduced in Chap. 11 and of Fourier series defined in Chap. 13 and applied in (15.02) below.

(f) Laplace transform solutions are tabulated for selected cases in Ref. 1.20, pp. 248–280. The derivations of these solutions and the evaluation of the respective convolution integrals are shown in the subsequent sections of this chapter.

(g) Numerical solutions shown in the last sections of this chapter are based either on the Taylor's series expansion or on the collocation polynomials, whose indeterminate coefficients result in the least error of the assumed approximation.

15.02 FOURIER SOLUTION, GENERAL CASE

(a) Linear differential equation with constant coefficients and an arbitrary input function has a Fourier solution if $g(x)$ can be approximated in the selected interval $(-L, L)$ by a truncated Fourier series

$$\bar{g}(x) = \tfrac{1}{2}\bar{A}_0 + \sum_{k=1}^{p} \bar{A}_k \cos\frac{k\pi x}{L} + \sum_{k=1}^{p} \bar{B}_k \sin\frac{k\pi x}{L} \qquad (k = 1, 2, \ldots, p)$$

where $\bar{A}_0, \bar{A}_k, \bar{B}_k$ are the Fourier coefficients defined in (13.03).

(b) Assumed solution of the differential equation (a) is then

$$\bar{Y} = \tfrac{1}{2}\bar{a}_0 + \sum_{k=1}^{p} \bar{a}_k \cos\frac{k\pi x}{L} + \sum_{k=1}^{p} \bar{b}_k \sin\frac{k\pi x}{L} + \sum_{r=1}^{n} \bar{C}_r K_r(x)$$

where the last sum is the complementary solution given for large number of cases in Ref. 1.20, pp. 179–183.

(c) Constants $\bar{a}_0, \bar{a}_k, \bar{b}_k$ are determined by inserting this solution in the differential equation and equating the coefficients of the respective trigonometric functions. The remaining constants $\bar{C}_r$ are obtained from r conditions (initial values or boundary values).

(d) Example. The motion of a given mass m (kg) suspended on an elastic spring of stiffness k (N/m) and acted on by a periodic force h of square-wave variation (N),

$$g(t) = \begin{cases} -h & -\tfrac{1}{2}T < t < 0 \\ h & 0 < t < \tfrac{1}{2}T \end{cases}$$

is governed by the differential equation

$$\frac{d^2 Y}{dt^2} + \omega^2 Y = \frac{g(t)}{m} \qquad \left(\omega = \sqrt{\frac{k}{m}}\right)$$

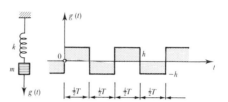

where T = period (s) and t = time (s). With the forcing frequency $\Omega = 2\pi/T$, this equation becomes, in terms of the Fourier expansion introduced in (13.07-1),

$$\frac{d^2 \bar{Y}}{dt^2} + \omega^2 \bar{Y} = \frac{4h}{\pi m}\sum_{k=1}^{p}\frac{\sin k\Omega t}{k} \qquad (k = 1, 3, 5, \ldots, p)$$

and its assumed solution is taken in the form of (b) as

$$\bar{Y} = \sum_{k=1}^{p} \bar{b}_k \sin k\Omega t + \bar{C}_1 \cos \omega t + \bar{C}_2 \sin \omega t$$

After inserting $\bar{Y}$ in the differential equation, the constants $\bar{b}_k$ are determined by (c) as

$$\bar{b}_k = \frac{4h}{k\pi(\omega^2 - k^2\Omega^2)m}$$

and the approximate solution becomes

$$\bar{Y} = \bar{C}_1 \cos \omega t + \bar{C}_2 \sin \omega t + \frac{4h}{\pi m}\sum_{k=1}^{p}\frac{\sin k\Omega t}{k(\omega^2 - k^2\Omega^2)}$$

If the initial conditions are given as $Y(0) = Y_0, dY(0)/dt = 0$,

$$Y(0) = \bar{C}_1 \qquad 0 = \omega\bar{C}_2 + \frac{4h\Omega}{\pi m}\sum_{k=1}^{p}\frac{1}{(\omega^2 - k^2\Omega^2)}$$

and the final approximation is

$$\bar{Y} = Y_0 \cos \omega t + \frac{4h}{\omega\pi m}\sum_{k=1}^{p}\frac{1}{k(\omega^2 - k^2\Omega^2)}(\omega \sin k\Omega t - k\Omega \sin \omega t)$$

(a) Linear differential equation with constant coefficients and an arbitrary input function defined as

$$\sum_{r=0}^{n} a_r \frac{d^r Y}{dx^r} = g(x)$$

has a Laplace transform solution (LTS) if $Y = Y(x)$ and $g(x)$ have Laplace transforms

$$\mathscr{L}\{Y(x)\} = f(s) \qquad \mathscr{L}\{g(x)\} = h(s)$$

and the initial conditions are

$$\left|\frac{d^r Y(x)}{dx^r}\right|_{x=0} = Y_0^{(r)} \qquad Y_0^{(0)} = Y_0$$

(b) Laplace transform of the differential equation in (a) is then, according to $(14.03f)$,

$$f(s)M(s) - N(s) = h(s)$$

from which

$$f(s) = \frac{N(s) + h(s)}{M(s)}$$

where $\quad N(s) = \sum_{r=0}^{n-1} N_r(s) \qquad N_r(s) = Y_0^{(r)} \sum_{k=r+1}^{n} a_k s^{k-r-1} \qquad M(s) = \sum_{k=0}^{n} a_k s^k$

(c) Inverse of this transform is the solution of the given differential equation represented symbolically as

$$Y = Y_0 L_1(x) + Y_0^{(1)} L_2(x) + Y_0^{(2)} L_2(x) + Y_0^{(3)} L_3(x) + \cdots + Y_0^{(n-1)} L_n(x) + G(x)$$

where

$$
\begin{bmatrix} L_1(x) \\ \\ L_2(x) \\ \\ L_3(x) \\ \cdots \\ L_n(x) \end{bmatrix}
=
\begin{bmatrix}
a_n & a_{n-1} & a_{n-2} & \cdots & a_3 & a_2 & a_1 \\
& a_n & a_{n-1} & \cdots & a_4 & a_3 & a_2 \\
& & a_n & \cdots & a_5 & a_4 & a_3 \\
& & & \cdots\cdots\cdots & & & \\
& & & & & & a_n
\end{bmatrix}
\begin{bmatrix}
\mathscr{L}^{-1}\left\{\dfrac{s^{n-1}}{M(s)}\right\} \\
\mathscr{L}^{-1}\left\{\dfrac{s^{n-2}}{M(s)}\right\} \\
\mathscr{L}^{-1}\left\{\dfrac{s^{n-3}}{M(s)}\right\} \\
\cdots \\
\mathscr{L}^{-1}\left\{\dfrac{1}{M(s)}\right\}
\end{bmatrix}
$$

are the shape functions and

$$G(x) = \mathscr{L}^{-1}\left\{\frac{h(s)}{M(s)}\right\}$$

(d) Convolution integral in (c),

$$G(x) = \int_0^x L_n(x - \tau)g(\tau)\, d\tau = \int_0^x L_n(\tau)g(x - \tau)\, d\tau$$

can be evaluated by Goursat's formula $(4.17b)$. Particular cases of this integral are tabulated in (15.04).

(e) Transport matrix equation formed by the assembly of $1, Y, Y^{(1)}, Y^{(2)}, \ldots, Y^{(n-1)}$,

$$
\begin{bmatrix} 1 \\ Y \\ Y^{(1)} \\ Y^{(2)} \\ Y^{(3)} \\ \cdots \\ Y^{(n-1)} \end{bmatrix}
=
\left[\begin{array}{c|ccccc}
1 & 0 & 0 & 0 & \cdots & 0 \\ \hline
G(x) & L_1(x) & L_2(x) & L_3(x) & \cdots & L_n(x) \\
G^{(1)}(x) & L_1^{(1)}(x) & L_2^{(1)}(x) & L_3^{(1)}(x) & \cdots & L_n^{(1)}(x) \\
G^{(2)}(x) & L_1^{(2)}(x) & L_2^{(2)}(x) & L_3^{(2)}(x) & \cdots & L_n^{(2)}(x) \\
G^{(3)}(x) & L_1^{(3)}(x) & L_2^{(3)}(x) & L_3^{(3)}(x) & \cdots & L_n^{(3)}(x) \\
\cdots & & & & & \\
G^{(n-1)}(x) & L_1^{(n-1)}(x) & L_2^{(n-1)}(x) & L_3^{(n-1)}(x) & \cdots & L_n^{(n-1)}(x)
\end{array}\right]
\begin{bmatrix} 1 \\ Y_0 \\ Y_0^{(1)} \\ Y^{(2)} \\ Y^{(3)} \\ \cdots \\ Y_0^{(n=1)} \end{bmatrix}
$$

$\mathbf{H}_R \qquad\qquad\qquad \mathbf{T}_{RL}(x) \qquad\qquad\qquad \mathbf{H}_L$

relates the state vector of the right end $\mathbf{H}_R$ in the interval $[0, x]$ to the state vector of the left end $\mathbf{H}_L$ of the same interval by

$$\boxed{\ \mathbf{T}_{RL}(x) = \text{transport matrix } [(n + 1) \times (n + 1)]\ }$$

(f) Transport functions in (e) are related by the recurrent relations

$$\frac{dL_j(x)}{dx} = L_j^{(1)}(x), \frac{dL_j^{(1)}(x)}{dx} = L_j^{(2)}(x), \ldots \qquad (j = 1, 2, \ldots, n)$$

which shows that only the first n functions are required for the construction of the $\mathbf{L}(x)$ submatrix.

(g) Load functions in (c) are related by the recurrent relations

$$\frac{dG(x)}{dx} = G^{(1)}(x), \frac{dG^{(1)}(x)}{dx} = G^{(2)}(x), \ldots$$

which shows that only the first load function (convolution integral) is required for the construction of the submatrix $\mathbf{G}(x)$.

(h) Transport chain. Once the transport matrix is available for single segment, the extension to a multisegment domain is accomplished by matrix multiplication. Considering the domain given in the adjacent figure by segments 01, 12, 23 of lengths l_1, l_2, l_3, respectively, the state vectors at 1, 2, 3 are

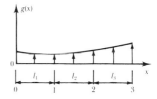

$$\mathbf{H}_1 = \mathbf{T}_{10}\mathbf{H}_0$$

$$\mathbf{H}_2 = \mathbf{T}_{21}\mathbf{T}_{10}\mathbf{H}_0 \qquad\qquad \mathbf{H}_2 = \mathbf{T}_{20}\mathbf{H}_0$$

$$\mathbf{H}_3 = \mathbf{T}_{32}\mathbf{T}_{21}\mathbf{T}_{10}\mathbf{H}_0 \qquad\qquad \mathbf{H}_3 = \mathbf{T}_{30}\mathbf{H}_0$$

where $\mathbf{T}_{20} = \mathbf{T}_{21}\mathbf{T}_{10}$ and $\mathbf{T}_{30} = \mathbf{T}_{32}\mathbf{T}_{21}\mathbf{T}_{10}$. These relations show that the product of two, three, and any number of segmental matrices is always a *new transport matrix*.

(a) Convolution integrals (15.03*d*) are evaluated below for selected cases of $g(x)$ in terms of the following equivalents. The infinite series in (3) to (5) may have a closed-form sum in some cases.

w = signed number $k = 1, 2, \ldots, n$ a, b, c, α, β = positive constants

$$L_{n+1}(x) = \int_0^x L_n(x)\,dx \qquad \beta = \pi/b \qquad \phi = \pi/2 \qquad L_{n+k}(x) = \int_0^x L_{n+k-1}(x)\,dx$$

Input function $g(x)$	Convolution integral $G(x)$
(1)	$wL_{n+1}(x)$
(2)	$\dfrac{k!w}{b^k} L_{n+k+1}(x)$
(3)	$w\left[\displaystyle\sum_{k=1}^{\infty} (-\beta)^{k-1} L_{n+k}(\tau) \sin\left[\beta(x-\tau) - (k-1)\phi\right]\right]_0^x$
(4)	$w\left[\displaystyle\sum_{k=1}^{\infty} (-\beta)^{k-1} L_{n+k}(\tau) \cos\left[(x-\tau) - (k-1)\phi\right]\right]_0^x$
(5)	$we^{-\alpha x}\left[\displaystyle\sum_{k=1}^{\infty} (-\alpha)^{k-1} L_{n+k}(\tau) e^{\alpha\tau}\right]_0^x$
(6)	$0 \hfill (0 \le x < b)$ $PL_n(x-b) \hfill (b \le x)$
(7)	$0 \hfill (0 \le x < b)$ $QL_{n-1}(x-b) \hfill (b \le x)$

(b) Shifting along the x axis. If $g(x)$ is shifted along x, then $G(x)$ may be expressed in terms of the results in (*a*) by replacing x by $x - a$ or $x - b$ and $L_j(x)$ by $A_j = L_j(x - a)$ or $L_j(x - b) = B_j$, respectively.

Input function $g(x)$	Convolution integral $G(x)$	
(1)	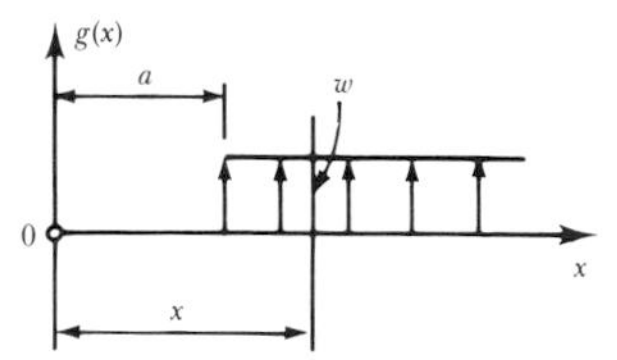 0	$(0 \leq x < a)$
	wA_{n+1}	$(a \leq x)$
(2)	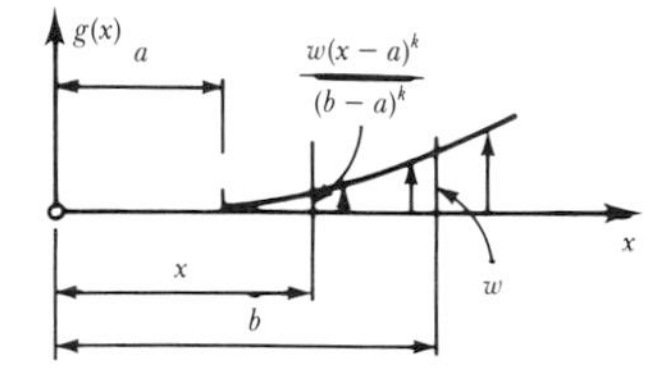 0	$(0 \leq x < a)$
	$\dfrac{k!w}{(b-a)^k} A_{n+k+1}$	$(k = 1, 2, \ldots \quad a \leq x)$

(c) Geometric superposition of convolution integrals tabulated in (*a*), (*b*) may be used for evaluation of convolution integrals of piecewise-continuous functions as shown below.

Input function $g(x)$	Convolution integral $G(x)$	
(1)	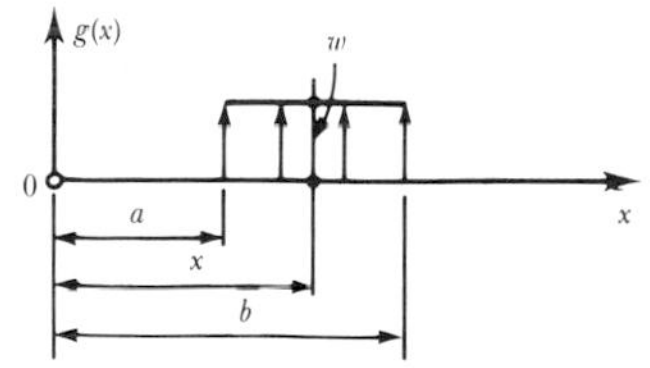 0	$(0 \leq x < a)$
	wA_{n+1}	$(a \leq x < b)$
	$w(A_{n+1} - B_{n+1})$	$(b \leq x)$
(2)	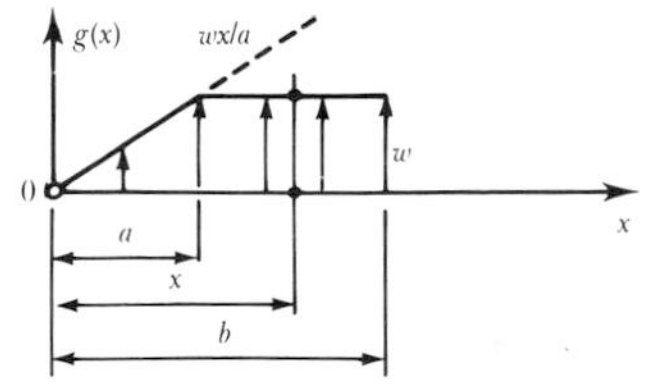 $\dfrac{w}{a} L_{n+2}$	$(0 \leq x < a)$
	$\dfrac{w}{a}(L_{n+2} - A_{n+2})$	$(a \leq x \leq b)$
	$\dfrac{w}{a}(L_{n+2} - A_{n+2} - aB_{n+1})$	$(b \leq x)$
(3)	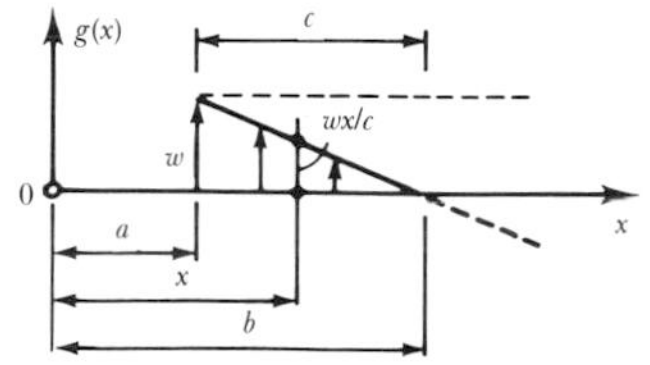 0	$(0 \leq x < a)$
	$w\left(A_{n+1} - \dfrac{1}{c}A_{n+2}\right)$	$(a \leq x < b)$
	$w\left(A_{n+1} - \dfrac{1}{c}A_{n+2} + \dfrac{1}{c}B_{n+2}\right)$	$(b \leq x)$

(a) Notation. Y = solution M, N = signed numbers $\neq 0$ $p = N/M$ $\lambda = |N/M|$

(b) Laplace transform of the first-order differential equation

$$\boxed{M\frac{dY}{dx} + NY = g(x)}$$ $$\boxed{Y_0 = \text{initial value}}$$

is by (15.03b) for $N > 0$, and for $N < 0$,

$$f(s) = Y_0\frac{1}{s + \lambda} + \frac{1}{M(s + \lambda)}h(s)$$ $$f(s) = Y_0\frac{1}{s - \lambda} + \frac{1}{M(s - \lambda)}h(s)$$

where $f(s) = \mathcal{L}\{Y\}$ and $h(s) = \mathcal{L}\{g(x)\}$.

(c) Inverse Laplace transform of (b) by (15.03c) is the Laplace tansform solution (LTS) of the differential equation given symbolically as

$$\boxed{Y = Y_0L_1(x) + G(x)}$$

where the shape functions (15.03c) are

$$L_1(x) = \begin{cases} \mathcal{L}^{-1}\left\{\dfrac{1}{s + \lambda}\right\} = e^{-\lambda x} & (N > 0) \\[2mm] \mathcal{L}^{-1}\left\{\dfrac{1}{s - \lambda}\right\} = e^{\lambda x} & (N < 0) \end{cases}$$

and the *convolution integral* (15.03d) in terms of the respective $L_1(x)$ is

$$\boxed{G(x) = \mathcal{L}^{-1}\left\{\frac{h(s)}{M(s \pm \lambda)}\right\} = \frac{1}{M}\int_0^x L_1(x - \tau)g(\tau)\,d\tau}$$

Particular cases of these integrals are tabulated in Ref. 1.20, sec. 16.36.

(d) Example. The Laplace transform solution of

$$10Y' + 20Y = 13\sin 3x \qquad Y_0 = 4$$

is by (c) for $N = 20 > 0$ and $\lambda = \frac{20}{10} = 2$,

$$Y = 4e^{-2x} + \frac{13}{10}\int_0^x e^{-2(x-\tau)}\sin 3\tau\,d\tau$$

where the convolution integral by Ref. 1.20, sec. 9.06, is

$$G(x) = \frac{13}{10}\int_0^x e^{-2(x-\tau)}\sin 3\tau\,d\tau = \frac{13e^{-2x}}{10}\left[\frac{e^{2\tau}}{2^2 + 3^2}(2\sin 3\tau - 3\cos 3\tau)\right]_0^x$$

$$= \tfrac{1}{10}[2\sin 3x - 3(\cos 3x - e^{-2x})]$$

or directly by Ref. 1.20, sec. 16.36, case 28, with $M = 10$, $N = 20$, $\lambda = 2$, $\beta = 3$, $w = 13$,

$$G(x) = \frac{w[\lambda \sin \beta x - \beta(\cos \beta x - L_1(x))]}{M(\lambda^2 + \beta^2)}$$

$$= \frac{13[2\sin 3x - 3(\cos 3x - e^{-2x})]}{10(2^2 + 3^2)} = \tfrac{1}{10}[2\sin 3x - 3(\cos 3x - e^{-2x})]$$

(a) Kelvin model consisting of an elastic spring of stiffness k (N/m) and a dashpot of viscosity η (N · s/m) is shown in the adjacent figure. The constitutive equations of the spring and the dashpot are, respectively,

$$F_S = k\Delta_S \qquad\qquad F_D = \eta\dot{\Delta}_D$$

where F_S, F_D = forces (N)

$\quad\Delta_S, \Delta_D$ = displacements (m)

$\quad\dot{\Delta}_S, \dot{\Delta}_D$ = velocities (m/s)

$\qquad t$ = time (s)

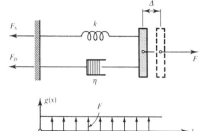

If the applied force F is given by the diagram shown above, then

$$F = F_S + F_D \qquad\qquad \Delta = \Delta_S = \Delta_D \qquad\qquad \dot{\Delta} = \dot{\Delta}_S = \dot{\Delta}_D$$

(b) Governing differential equation of this model is

$$\boxed{\eta\dot{\Delta} + k\Delta = F}$$

and its solution by (15.05c), with the initial condition $\Delta_0 = 0$, is

$$\boxed{\Delta = \Delta_0 e^{-\lambda t} + \frac{1}{\eta}\int_0^t e^{-\lambda(t-\tau)}F\,d\tau = \frac{F}{k}(1 - e^{-\lambda t})}$$

15.07 STANDARD LINEAR MODEL

(a) Standard linear model consisting of two elastic springs of stiffnesses k_1, k_2 and one dashpot of viscosity η_1 is shown in the adjacent figure. If the applied force is given by the diagram shown below and the same constitutive equations are valid as in (15.06a), then

$$F = F_1 + F_2$$

$$\Delta = \Delta_{S,1} + \Delta_{D,1} = \Delta_{S,2}$$

$$\dot{\Delta} = \dot{\Delta}_{S,1} + \dot{\Delta}_{D,1} + \dot{\Delta}_{S,2}$$

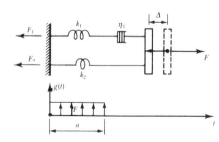

(b) Governing differential equation of this model in terms of

$$F_1 = F - F_2 \qquad\qquad \dot{F}_1 = \dot{F} - \dot{F}_2$$

$$F_2 = k_2\Delta \qquad\qquad \dot{F}_2 = k_2\dot{\Delta}$$

becomes

$$\boxed{\dot{\Delta} + \lambda\Delta = \frac{k_1 F}{\eta_1(k_1 + k_2)} + \frac{\dot{F}}{k_1(k_1 + k_2)}} \qquad\qquad \boxed{\lambda = \frac{k_1 k_2}{\eta_1(k_1 + k_2)}}$$

The solution by (15.05c) with $\Delta_0 = \alpha_0$ and with $t > a$ is then

$$\boxed{\Delta = \alpha_0 e^{-\lambda t} + \frac{F}{k_1}(1 - e^{-\lambda t}) - \frac{F}{k_2}[1 - e^{-\lambda(t-a)}]}$$

(a) Notation. Y = solution $\quad M, N$ = signed numbers $\neq 0 \quad p = N/M \quad \lambda = \sqrt{|N/M|}$

(b) Laplace transform of the second-order differential equation,

$$\boxed{M\frac{d^2Y}{dx^2} + NY = g(x)} \qquad \boxed{Y_0, Y_0' = \text{initial values}}$$

is by (15.03b) for $N > 0$, $\quad f(s) = Y_0\dfrac{s}{s^2 + \lambda^2} + Y_0'\dfrac{1}{s^2 + \lambda^2} + \dfrac{1}{M(s^2 + \lambda^2)}h(s)$

and for $N < 0$, $\quad f(s) = Y_0\dfrac{s}{s^2 - \lambda^2} + Y_0'\dfrac{1}{s^2 - \lambda^2} + \dfrac{1}{M(s^2 - \lambda^2)}h(s)$

where $f(s)$, $h(s)$ have the same meaning as in (15.05).

(c) Inverse Laplace transform of (b) according to (15.03c) is the Laplace transform solution (LTS) of the differential equation given symbolically as

$$\boxed{Y = Y_0 L_1(x) + Y_0' L_2(x) = G(x)}$$

where the shape functions (15.03c) are tabulated below.

$L_j(x)$	$N > 0$	$N < 0$
$L_1(x)$	$\mathscr{L}^{-1}\left\{\dfrac{s}{s^2 + \lambda^2}\right\} = \cos \lambda x$	$\mathscr{L}^{-1}\left\{\dfrac{s}{s^2 - \lambda^2}\right\} = \cosh \lambda x$
$L_2(x)$	$\mathscr{L}^{-1}\left\{\dfrac{1}{s^2 + \lambda^2}\right\} = \dfrac{\sin \lambda x}{\lambda}$	$\mathscr{L}^{-1}\left\{\dfrac{1}{s^2 - \lambda^2}\right\} = \dfrac{\sinh \lambda x}{\lambda}$

The convolution integral (15.03d) in terms of the respective $L_2(x)$ is

$$\boxed{G(x) \equiv \mathscr{L}^{-1}\left\{\frac{h(x)}{s^2 \pm \lambda^2}\right\} = \frac{1}{M}\int_0^x L_2(x - \tau)g(\tau)\,d\tau}$$

Particular cases of these integrals are tabulated in Ref. 1.20, sec. 16.38.

(d) Example. The Laplace transform solution of

$$10Y'' + 40Y = 6x^2 \qquad Y_0 = 4 \qquad Y_0' = 5$$

is by (c) for $N = 40 > 0$ and $\lambda = \sqrt{40/10} = 2$,

$$Y = 4\cos 2x + \tfrac{5}{2}\sin 2x + \frac{1}{10}\int_0^x \tfrac{1}{2}\sin[2(x - \tau)]6\tau^2\,d\tau$$

where the convolution integral by Ref. 1.20, sec. 9.04-2, is

$$G(x) = \frac{3}{10}\int_0^x \tau^2 \sin 2(x - \tau)\,d\tau$$

$$= \frac{3}{10}\left[\tau^2\frac{\cos 2(x - \tau)}{2} + 2\tau\frac{\sin 2(x - \tau)}{4} - 2\frac{\cos 2(x - \tau)}{8}\right]_0^x$$

$$= \tfrac{3}{40}(\cos 2x + 2x^2 - 1)$$

or directly by Ref. 1.20, sec. 16.38, case 3.

(a) Hookes' model of spring constant k (N/m) and mass m (kg) is acted on by a forcing function defined in the adjacent graph. The initial values of this motion are $Y_0 = \alpha_0$ and $\dot{Y}_0 = 0$.

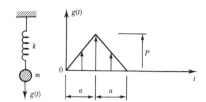

(b) Differential equation of this motion is

$$\boxed{m\ddot{Y} + kY = g(t)} \qquad \boxed{\omega = \sqrt{k/m}}$$

and $\quad Y = \alpha_0 \cos \omega t + G(t)$

where

$$G(t) = \begin{cases} \dfrac{P}{am} L_4 & (0 \le t \le a) \\[2ex] \dfrac{P}{am}(L_4 - 2A_4) & (a \le t < 2a) \\[2ex] \dfrac{P}{am}(L_4 - 2A_4 + B_4) & (2a < t) \end{cases}$$

$$L_4 = (\omega t - \sin \omega t)/\omega^3$$

$$A_4 = [\omega(t - a) - \sin \omega(t - a)]/\omega^3$$

$$B_4 = [\omega(t - 2a) - \sin \omega(t - 2a)]/\omega^3$$

The parameters of this motion are P = maximum force (N), a = time (s), ω = natural frequency (rad/s), Y = displacement (m), and $\dot{Y}$ = velocity (m/s).

15.10 BEAM DEFLECTION

(a) Cantilever beam of length l (m), moment of inertia I (in^4), elastic modulus E (Pa) is acted on by a triangular load of maximum intensity p (N/m), as shown in the adjacent figure.

(b) Deflection differential equation of this beam is

$$\boxed{\frac{d^2Y}{dx^2} = \frac{M}{EI}}$$

where Y = transverse deflection and the bending moment at x is

$$M = -px^3/(6l)$$

The end conditions are $Y(0) = Y_0$, $Y'(0) = Y'_0$ and $Y(l) = 0$, $Y'(l) = 0$.

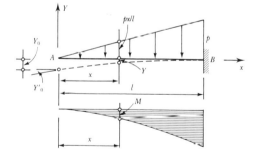

(c) Solution of (*b*) by (15.08*b*) is

$$Y = Y_0 = Y'_0 x + G(x) \qquad G(x) = -\frac{p}{EI}\int_0^x \frac{\tau^3}{6l}(x - \tau)\,d\tau = -\frac{px^5}{120EIl}$$

For $x = l$, the deflection and slope of the elastic curve at B are

$$Y(l) \overset{0}{=} Y_0 + Y'_0 l - \frac{pl^4}{120EI} \qquad Y'(l) \overset{0}{=} Y'_0 - \frac{pl^3}{24EI}$$

from which the end slope and deflection at A are

$$\boxed{Y'_0 = \frac{pl^3}{24EI} \qquad Y_0 = -\frac{pl^4}{30EI}}$$

(a) Notation. Y = solution $\qquad M, N, R$ = signed numbers $\neq 0$ $\qquad p = N/m$ $\qquad q = R/M$

$$\lambda = \tfrac{1}{2}p \qquad\qquad \Delta = p^2 - 4q \qquad\qquad \omega = \tfrac{1}{2}\sqrt{|\Delta|}$$

(b) Laplace transform of the second-order-differential equation

$$\boxed{M\frac{d^2Y}{dx^2} + N\frac{dY}{dx} + RY = g(x)} \qquad\qquad \boxed{Y_0,\, Y_0' = \text{initial values}}$$

is by (15.03b), $f(s) = Y_0\dfrac{s+p}{s^2+ps+q} + Y_0'\dfrac{1}{s^2+ps+q} + \dfrac{1}{M(s^2+ps+q)}h(s)$

where $f(s)$ and $h(s)$ have the same meaning as in (15.05b).

$$f(s) = \begin{cases} Y_0\dfrac{s+2\lambda}{(s+\lambda)^2+\omega^2} + Y_0'\dfrac{1}{(s+\lambda)^2+\omega^2} + \dfrac{h(s)}{M[(s+\lambda)^2+\omega^2]} & (\Delta < 0) \\[3mm] Y_0\dfrac{s+2\lambda}{(s+\lambda)^2} + Y_0'\dfrac{1}{(s+\lambda)^2} + \dfrac{h(s)}{M(s+\lambda)^2} & (\Delta = 0) \\[3mm] Y_0\dfrac{s+2\lambda}{(s+\lambda)^2-\omega^2} + Y_0'\dfrac{1}{(s+\lambda)^2-\omega^2} + \dfrac{h(s)}{M[(s+\lambda)^2-\omega^2]} & (\Delta > 0) \end{cases}$$

(c) Inverse Laplace transform of (b) by (15.03c) is the Laplace transform solution (LTS) of the differential equation given symbolically as

$$\boxed{Y = Y_0L_1(x) + Y_0'L_2(x) + G(x)}$$

where $L_1(x)$, $L_2(x)$ are the shape functions (15.03c) given in (d) below, and in terms of the respective $L_2(x)$,

$$\boxed{G(x) = \mathcal{L}^{-1}\left\{\frac{1}{M(s^2+ps+q)}\right\} = \frac{1}{M}\int_0^x L_2(x-\tau)g(\tau)\,d\tau}$$

is the *convolution integral* (15.03d). Particular cases are tabulated in Ref. 1.20, sec. 16.40.

(d) Shape functions

$\Delta < 0$	$L_1(x) = \mathcal{L}^{-1}\left\{\dfrac{s+\lambda+\lambda}{(s+\lambda)^2+\omega^2}\right\} = e^{-\lambda x}\left(\cos\omega x + \dfrac{\lambda}{\omega}\sin\omega x\right)$
	$L_2(x) = \mathcal{L}^{-1}\left\{\dfrac{1}{(s+\lambda)^2+\omega^2}\right\} = e^{-\lambda x}\dfrac{\sin\omega x}{\omega}$
$\Delta = 0$	$L_1(x) = \mathcal{L}^{-1}\left\{\dfrac{s+\lambda+\lambda}{(s+\lambda)^2}\right\} = e^{-\lambda x}(1+\lambda x)$
	$L_2(x) = \mathcal{L}^{-1}\left\{\dfrac{1}{(s+\lambda)^2}\right\} = e^{-\lambda x}x$
$\Delta > 0$	$L_1(x) = \mathcal{L}^{-1}\left\{\dfrac{s+\lambda+\lambda}{(s+\lambda)^2-\omega^2}\right\} = e^{-\lambda x}\left(\cosh\omega x + \dfrac{\lambda}{\omega}\sinh\omega x\right)$
	$L_2(x) = \mathcal{L}^{-1}\left\{\dfrac{1}{(s+\lambda)^2-\omega^2}\right\} = e^{-\lambda x}\dfrac{\sinh\omega x}{\omega}$

A complete set of shape functions in terms of the respective phase angles is tabulated in (15.12).

(a) Table of shape functions $(15.11d)$

(1) Case I: $\Delta < 0$, $\psi = \tan^{-1}\dfrac{\omega}{\lambda}$, $E = e^{-\lambda x}$, $r = \sqrt{\lambda^2 + \omega^2}$

$L_1(x)$	$\dfrac{Er}{2\omega}\sin(\omega x + \psi)$	$L_2(x)$	$\dfrac{E}{\omega}\sin \omega x$
$L_3(x)$	$-\dfrac{1}{\omega r}[E\sin(\omega x + \psi) - \sin\psi]$		
$L_4(x)$	$\dfrac{1}{\omega r^2}[E\sin(\omega x + 2\psi) + \omega x\sin\psi - \sin 2\psi]$		
$L_5(x)$	$-\dfrac{1}{\omega r^3}[E\sin(\omega x + 3\psi) - \tfrac{1}{2}\omega^2 x^2\sin\psi + \omega x\sin\psi - \sin 3\psi]$		

(2) Case II: $\Delta = 0$, $\psi = 0$, $E = e^{-\lambda x}$

$L_1(x)$	$(1 + \lambda x)E$	$L_2(x)$	xE
$L_3(x)$	$-\dfrac{1}{\lambda^2}[(1 + \lambda x)E - 1]$		
$L_4(x)$	$\dfrac{1}{\lambda^3}[(2 + \lambda x)E + \lambda x - 2]$		
$L_5(x)$	$-\dfrac{1}{\lambda^4}[(3 + \lambda x)E - \tfrac{1}{2}\lambda^2 x^2 + 2\lambda x - 3]$		

(3) Case III: $\Delta > 0$, $\psi = \tanh^{-1}\dfrac{\omega}{\lambda}$, $E = e^{-\lambda x}$, $r = \sqrt{\lambda^2 - \omega^2}$

$L_1(x)$	$\dfrac{Er}{\omega}\sinh(\omega x + \psi)$	$L_2(x)$	$\dfrac{E}{\omega}\sinh \omega x$
$L_3(x)$	$-\dfrac{1}{\omega r}[E\sinh(\omega x + \psi) - \sin\psi]$		
$L_4(x)$	$\dfrac{1}{\omega r^2}[E\sinh(\omega x + 2\psi) + \omega x\sinh\psi - \sinh 2\psi]$		
$L_5(x)$	$-\dfrac{1}{\omega r^3}[E\sinh(\omega x + 3\psi) - \tfrac{1}{2}\omega^2 x^2\sinh\psi + \omega x\sinh\psi - \sinh 3\psi]$		

(b) Derivatives and integrals of shape functions in all cases are

$$dL_1(x)/dx = -L_0(x) \qquad\qquad dL_2(x)/dx = L_1(x) - 2\lambda L_2(x)$$

$$dL_j(x)/dx = L_{j-1}(x) \qquad \text{for } j = 3, 4, \ldots$$

$$\int_0^x L_j(x)\,dx = L_{j+1}(x) \qquad \text{for } j = 2, 3, \ldots$$

(a) Notation. $m = 0, 1, 2, \ldots$ $\qquad\qquad\qquad$ b, w, α, β = signed numbers

$\qquad\qquad$ p, q: see (15.11a) $\qquad\qquad\qquad\qquad\qquad$ A, B, C, D = constants

(b) Algebraic input function. If $g(x) = w(x/b)^m$, then in (15.11c),

$$G(x) = \mathscr{L}^{-1}\left\{\frac{1}{M(s^2 + ps + q)}\frac{m!w}{s^{m+1}b^m}\right\} = \frac{w}{Mb^m}\int_0^x (x - \tau)^m L_2(\tau)\,d\tau$$

and by (4.17b),

$$\boxed{\begin{aligned}G(x) = \frac{w}{Mb^m}&[(x - \tau)^m L_3(\tau) + m(x - \tau)^{m-1}L_4(\tau) + m(m - 1)(x - \tau)^{m-2}L_5(\tau)\\ &+ \cdots + m!L_{m+3}(\tau)]_0^x = \frac{m!w}{Mb^m}L_{m+3}(x)\end{aligned}}$$

where $L_{m+3}(x)$ is the respective shape function defined in (15.12a), (15.12b).

(c) Harmonic input function. If $g(x) = w\sin\beta x$, then the convolution integral in (15.11c) becomes too involved and the solution by partial fractions is much simpler.

$$G(x) = \frac{w}{M}\mathscr{L}^{-1}\left\{\frac{1}{s^2 + ps + q}\frac{\beta}{s^2 + \beta^2}\right\} = \frac{w}{M}\mathscr{L}^{-1}\left\{\frac{As + B}{s^2 + ps + q} + \frac{Cs + D}{s^2 + \beta^2}\right\}$$

where with $u = q - \beta^2$, $v = p\beta$,

$$A = \frac{v}{u^2 + v^2} \qquad B = \frac{(p^2 - u)\beta}{u^2 + v^2} \qquad C = -\frac{v}{u^2 + v^2} \qquad D = \frac{u\beta}{u^2 + v^2}$$

In terms of these constants,

$$\boxed{\begin{aligned}G(x) &= \frac{w}{M(u^2 + v^2)}\mathscr{L}^{-1}\left\{\frac{vs + vp - u\beta}{s^2 + ps + q} - \frac{vs - u\beta}{s^2 + \beta^2}\right\}\\ &= \frac{uw[\sin\beta x - \beta L_2(x)] - vw[\cos\beta x - L_1(x)]}{M(u^2 + v^2)}\end{aligned}}$$

(d) Exponential input function. If $g(x) = we^{-\alpha x}$, then the solution by partial fractions is even simpler.

$$G(x) = \frac{w}{M}\mathscr{L}^{-1}\left\{\frac{1}{s^2 + ps + q}\frac{1}{s + \alpha}\right\} = \frac{w}{M}\mathscr{L}^{-1}\left\{\frac{As + B}{s^2 + ps + q} + \frac{C}{s + \alpha}\right\}$$

where with $u = \alpha^2 - p\alpha + q$,

$$A = -\frac{1}{u} \qquad B = \frac{\alpha - p}{u} \qquad C = \frac{1}{u}$$

In terms of these constants,

$$\boxed{G(x) = \frac{w}{uM}\mathscr{L}^{-1}\left\{\frac{-s - p + \alpha}{s^2 + ps + q} + \frac{1}{s + \alpha}\right\} = \frac{w}{uM}[e^{-\alpha x} - L_1(x) + \alpha L_2(x)]}$$

(a) Notation. Y = solution M, N = signed numbers $p = N/M$

$$\lambda = \sqrt[4]{\frac{N}{4M}} \qquad\qquad \eta = \sqrt[4]{\frac{N}{M}}$$

(b) Laplace transform of the fourth-order differential equation

$$\boxed{M\frac{d^4Y}{dx^4} + NY = g(x)} \qquad\qquad \boxed{Y_0, Y_0', Y_0'', Y_0''' = \text{initial values}}$$

is, by $(15.03b)$,

$$f(s) = Y_0 \frac{s^3}{s^4 + p} + Y_0' \frac{s^2}{s^4 + p} + Y_0'' \frac{s}{s^4 + p} + Y_0''' \frac{1}{s^4 + p} + \frac{h(s)}{M(s^4 + p)}$$

where $f(s)$ and $h(s)$ have the same meaning as in $(15.05b)$.

$$f(s) = \begin{cases} Y_0 \dfrac{s^3}{s^4 + 4\lambda^4} + Y_0' \dfrac{s^2}{s^4 + 4\lambda^4} + Y_0'' \dfrac{s}{s^4 + 4\lambda^4} + Y_0''' \dfrac{1}{s^4 + 4\lambda^4} + \dfrac{h(s)}{M(s^4 + 4\lambda^4)} & (p > 0) \\[3mm] Y_0 \dfrac{s^3}{s^4 - \eta^4} + Y_0' \dfrac{s^2}{s^4 - \eta^4} + Y_0'' \dfrac{s}{s^4 - \eta^4} + Y_0''' \dfrac{1}{s^4 - \eta^4} + \dfrac{h(s)}{M(s^4 - \eta^4)} & (p < 0) \end{cases}$$

(c) Inverse Laplace transform of (b) by $(15.03c)$ is the Laplace transform solution (LTS) of the differential equation given symbolically as

$$\boxed{Y = Y_0 L_1(x) + Y_0' L_2(x) + Y_0'' L_3(x) + Y_0''' L_4(x) + G(x)}$$

where $L_1(x), L_2(x), L_3(x), L_4(x)$ are the shape functions $(15.03c)$ listed in $(15.15b)$, $(15.15c)$, and in terms of the respective $L_4(x)$,

$$\boxed{G(x) = \mathcal{L}^{-1}\left\{\frac{h(s)}{M(s^4 + p)}\right\} = \frac{1}{M}\int_0^x L_4(x - \tau)g(\tau)\,d\tau}$$

is the *convolution integral* $(15.03d)$. Particular cases of these integrals are tabulated in Ref. 1.20, sec. 16.44.

(d) Transport matrix equation introduced in $(15.03e)$ is, for all p,

$$\begin{bmatrix} 1 \\ Y(x) \\ Y'(x) \\ Y''(x) \\ Y'''(x) \end{bmatrix} = \begin{bmatrix} 1 & 0 & 0 & 0 & 0 \\ G(x) & L_1(x) & L_2(x) & L_3(x) & L_4(x) \\ G'(x) & -pL_4(x) & L_1(x) & L_2(x) & L_3(x) \\ G''(x) & -pL_3(x) & -pL_4(x) & L_1(x) & L_2(x) \\ G'''(x) & -pL_2(x) & -pL_3(x) & -pL_4(x) & L_1(x) \end{bmatrix} \begin{bmatrix} 1 \\ Y_0 \\ Y_0' \\ Y_0'' \\ Y_0''' \end{bmatrix}$$

where

$$dL_4(x)/dx = L_3(x) \qquad dL_3(x)/dx = L_2(x) \qquad dL_2(x)/dx = L_1(x)$$

but $dL_1(x)/dx = -pL_4(x)$ and $G'(x), G''(x), G'''(x)$ are the first-, second-, and third-order derivatives of $G(x)$.

(a) Equivalents

$$S_1 = \sin \lambda x \qquad C_1 = \cos \lambda x \qquad \bar{S}_1 = \sinh \lambda x \qquad \bar{C}_1 = \cosh \lambda x$$

$$S_2 = \sin \eta x \qquad C_2 = \cos \eta x \qquad \bar{S}_2 = \sinh \eta x \qquad \bar{C}_2 = \cosh \eta x$$

(b) Case I (15.14b), (15.14c) — $p > 0 \qquad \lambda = \sqrt[4]{\dfrac{N}{4EI}} \qquad r = 0, 1, 2, \ldots \qquad k = 5, 6, \ldots$

$L_1(x)$	$\mathscr{L}^{-1}\left\{\dfrac{s^3}{s^4 + 4\lambda^4}\right\}$	$\bar{C}_1 C_1$	$\displaystyle\sum_{r=0}^{\infty} (-4)^r \dfrac{(\lambda x)^{4r}}{(4r)!}$
$L_2(x)$	$\mathscr{L}^{-1}\left\{\dfrac{s^2}{s^4 + 4\lambda^4}\right\}$	$\dfrac{1}{2\lambda}(\bar{C}_1 S_1 + \bar{S}_1 C_1)$	$\dfrac{1}{\lambda}\displaystyle\sum_{r=0}^{\infty} (-4)^r \dfrac{(\lambda x)^{4r+1}}{(4r+1)!}$
$L_3(x)$	$\mathscr{L}^{-1}\left\{\dfrac{s}{s^4 + 4\lambda^4}\right\}$	$\dfrac{1}{2\lambda^2}\bar{S}_1 S_1$	$\dfrac{1}{\lambda^2}\displaystyle\sum_{r=0}^{\infty} (-4)^r \dfrac{(\lambda x)^{4r+2}}{(4r+2)!}$
$L_4(x)$	$\mathscr{L}^{-1}\left\{\dfrac{1}{s^4 + 4\lambda^4}\right\}$	$\dfrac{1}{4\lambda^3}(\bar{C}_1 S_1 - \bar{S}_1 C_1)$	$\dfrac{1}{\lambda^3}\displaystyle\sum_{r=0}^{\infty} (-4)^r \dfrac{(\lambda x)^{4r+3}}{(4r+3)!}$
$L_5(x)$	$\displaystyle\int_0^x L_4(x)\, dx$	$\dfrac{1}{4\lambda^4}[1 - L_1(x)]$	$\dfrac{1}{\lambda^4}\displaystyle\sum_{r=0}^{\infty} (-4)^r \dfrac{(\lambda x)^{4r+4}}{(4r+4)!}$
$L_6(x)$	$\displaystyle\int_0^x L_5(x)\, dx$	$\dfrac{1}{4\lambda^4}[x - L_2(x)]$	$\dfrac{1}{\lambda^5}\displaystyle\sum_{r=0}^{\infty} (-4)^r \dfrac{(\lambda x)^{4r+5}}{(4r+5)!}$
$L_k(x)$	$\displaystyle\int_0^x L_{k-1}(x)\, dx$	$\dfrac{1}{4\lambda^4}\left[\dfrac{x^{k-5}}{(k-5)!} - L_{k-4}(x)\right]$	$\dfrac{1}{\lambda^{k-1}}\displaystyle\sum_{r=0}^{\infty} (-4)^r \dfrac{(\lambda x)^{4r+k-1}}{(4r+k-1)!}$

(c) Case II (15.14b), (15.14c) — $p < 0 \qquad \eta = \sqrt[4]{\left|\dfrac{N}{M}\right|} \qquad r = 0, 1, 2, \ldots \qquad k = 5, 6, \ldots$

$L_1(x)$	$\mathscr{L}^{-1}\left\{\dfrac{s^3}{s^4 - 4\eta^4}\right\}$	$\dfrac{1}{2}(\bar{C}_2 + C_2)$	$\displaystyle\sum_{r=0}^{\infty} \dfrac{(\eta x)^{4r}}{(4r)!}$
$L_2(x)$	$\mathscr{L}^{-1}\left\{\dfrac{s^2}{s^4 - 4\eta^4}\right\}$	$\dfrac{1}{2\eta}(\bar{S}_1 + S_2)$	$\dfrac{1}{\eta}\displaystyle\sum_{r=0}^{\infty} \dfrac{(\eta x)^{4r+1}}{(4r+1)!}$
$L_3(x)$	$\mathscr{L}^{-1}\left\{\dfrac{s}{s^4 - 4\eta^4}\right\}$	$\dfrac{1}{2\eta^2}(\bar{C}_2 - C_2)$	$\dfrac{1}{\eta^2}\displaystyle\sum_{r=0}^{\infty} \dfrac{(\eta x)^{4r+2}}{(4r+2)!}$
$L_4(x)$	$\mathscr{L}^{-1}\left\{\dfrac{1}{s^4 - 4\eta^4}\right\}$	$\dfrac{1}{4\eta^3}(\bar{S}_2 - S_2)$	$\dfrac{1}{\eta^3}\displaystyle\sum_{r=0}^{\infty} \dfrac{(\eta x)^{4r+3}}{(4r+3)!}$
$L_5(x)$	$\displaystyle\int_0^x L_4(x)\, dx$	$\dfrac{1}{\eta^4}[L_1(x) - 1]$	$\dfrac{1}{\eta^4}\displaystyle\sum_{r=0}^{\infty} \dfrac{(\eta x)^{4r+4}}{(4r+4)!}$
$L_6(x)$	$\displaystyle\int_0^x L_5(x)\, dx$	$\dfrac{1}{\eta^4}[L_2(x) - x]$	$\dfrac{1}{\eta^5}\displaystyle\sum_{r=0}^{\infty} \dfrac{(\eta x)^{4r+5}}{(4r+5)!}$
$L_k(x)$	$\displaystyle\int_0^x [L_{k-1}(x)\, dx$	$\dfrac{1}{\eta^4}\left[L_{k-4}(x) - \dfrac{x^{k-5}}{(k-5)!}\right]$	$\dfrac{1}{\eta^{k-1}}\displaystyle\sum_{r=0}^{\infty} \dfrac{(\eta x)^{4r+k-1}}{(4r+k-1)!}$

(a) Notation $m = 0, 1, 2, \ldots$ $b, w, \alpha, \beta =$ signed numbers

             p: see $(15.14a)$ $A, B, C, D, E, F =$ constants

(b) Algebraic input function. If $g(x) = w(x/b)^m$, then in $(15.14c)$ by $(15.03d)$,

$$G(x) = \mathcal{L}^{-1}\left\{\frac{1}{M(s^4 + p)}\frac{m!w}{s^{m+1}b^m}\right\} = \frac{w}{Mb^m}\int_0^x (x - \tau)^m L_4(\tau)\, d\tau$$

and by $(4.17b)$,

$$G(x) = \frac{w}{Mb^m}\left[(x - \tau)^m L_5(\tau) + m(x - \tau)^{m-1}L_5(\tau) + m(m - 1)(x - \tau)^{m-2}L_7(\tau)\right.$$

$$\left. + \cdots + m!L_{m+5}(\tau)\right]_0^x = \frac{m!w}{Mb^m}L_{m+5}(x)$$

where $L_{m+5}(x)$ is the respective shape function in $(15.15b)$, $(15.15c)$.

(c) Harmonic input function. If $g(x) = w \sin \beta x$, then again the convolution integral in $(15.14c)$ becomes too involved and the solution by partial fractions is much simpler.

$$G(x) = \frac{w}{M}\mathcal{L}^{-1}\left\{\frac{1}{s^4 + p}\frac{\beta}{s^2 + \beta^2}\right\} = \frac{w}{M}\mathcal{L}^{-1}\left\{\frac{As^3 + Bs^2 + Cs + D}{s^4 + p} + \frac{Es + F}{s^2 + \beta^2}\right\}$$

where with $u = p + \beta^4$,

$$A = C = E = 0 \qquad B = -\beta/u \qquad D = \beta^3/u \qquad F = \beta/u$$

and

$$G(x) = \frac{w}{M(p + \beta^4)}\mathcal{L}^{-1}\left\{\frac{-\beta s^2 + \beta^3}{s^4 + p} + \frac{\beta}{s^2 + \beta^2}\right\}$$

$$= \frac{w}{M(p + \beta^4)}\left[\sin \beta x - \beta L_2(x) + \beta^3 L_4(x)\right]$$

(d) Exponential input function. If $g(x) = we^{-\alpha x}$, then the solution by partial fractions is even simpler.

$$G(x) = \frac{w}{M}\mathcal{L}^{-1}\left\{\frac{1}{s^4 + p}\frac{1}{s + \alpha}\right\} = \frac{w}{M}\mathcal{L}^{-1}\left\{\frac{As^3 + Bs^2 + Cs + D}{s^4 + p} + \frac{E}{s + \alpha}\right\}$$

where with $u = p + \alpha^4$,

$$A = -1/u \qquad B = \alpha/u \qquad C = -\alpha^2/u \qquad D = \alpha^3/u \qquad E = 1/u$$

and

$$G(x) = \frac{w}{M(p + \alpha^4)}\mathcal{L}^{-1}\left\{\frac{-s^3 + \alpha s^2 - \alpha^2 s + \alpha^3}{s^4 + p} + \frac{1}{s + \alpha}\right\}$$

$$= \frac{w}{M(p + \alpha^4)}\left[e^{-\alpha x} - L_1(x) + \alpha L_2(x) - \alpha^2 L_3(x) + \alpha^3 L_4(x)\right]$$

(a) System. A straight prismatic bar of moment of inertia $I = 500\,\text{in}^4$, elastic modulus $E = 3(10)^3\,\text{ksi}$ is free at L, fixed at R, supported along its entire length $l = 200$ in on an elastic foundation of modulus $k = 1.5\,\text{ksi}$ and acted on by a uniformly distributed load of intensity $w = 0.2\,\text{k/in}$ as shown in the adjacent figure. (1 ksi = 1000 lbf/in².)

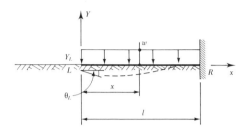

(b) Governing differential equation of this beam is

$$\frac{d^4Y}{dx^4} + \frac{k}{EI}Y = -\frac{w}{EI} \qquad\qquad Y = \text{deflection (in)}$$

(c) General solution of this equation by (15.14c) is

$$Y = Y_L L_1(x) + Y'_L L_2(x) + Y''_L L_3(x) + Y'''_L L_4(x) + G(x)$$

where $L_1(x)$, $L_2(x)$, $L_3(x)$, $L_4(x)$ are the shape functions given in (15.15b) in terms of $\lambda = 0.0236/\text{in}$, and by (15.16b) with $m = 0$, $b = 1$, $M = EI$,

$$G(x) = -\frac{w}{EI}L_5(x)$$

(d) End values of the beam are

Y_L = left-end deflection $\qquad\qquad Y'_L$ = left-end slope = θ_L

Y''_L = left-end bending moment$/EI = M_L/EI = 0$

$-Y'''_L$ = left-end shear$/EI = V_L/EI = 0$

Y_R = right-end deflection = $0 \qquad\qquad Y'_R$ = right-end slope = $\theta_R = 0$

Y''_R = right-end bending moment$/EI = M_R/EI$

$-Y'''_R$ = right-end shear$/EI = V_R/EI$

(e) Transport matrix equation (15.14d) with $p = k/EI = 10^{-6}\,\text{in}^{-4}$ and $l = 200$ in is

$$
\begin{bmatrix} 1 \\ Y_R \\ \theta_R \\ \hline M_R/EI \\ -V_R/EI \end{bmatrix}
=
\begin{bmatrix}
1 & 0 & 0 & 0 & 0 \\
G(l) & L_1(l) & L_2(l) & L_3(l) & L_4(l) \\
G'(l) & -pL_4(l) & L_1(l) & L_2(l) & L_3(l) \\
\hline
G''(l) & -pL_3(l) & -pL_4(l) & L_1(l) & L_2(l) \\
G'''(l) & -pL_2(l) & -pL_3(l) & -pL_4(l) & L_1(l)
\end{bmatrix}
\begin{bmatrix} 1 \\ Y_L \\ \theta_L \\ \hline M_L/EI \\ -V_L/EI \end{bmatrix}
$$

From the second and third row above,

$Y_L = -0.135$ in $\qquad\qquad \theta_L = -0.0001$ rad

and with these results, the fourth and fifth row yield

$M_R = -233.8\,\text{k} \cdot \text{in} \qquad\qquad V_R = 9.3\,\text{k}$

The foundation reaction is then $(0.2)(200) - 9.3 = 30.7\,\text{k}$.

(a) System. A straight prismatic bar of length $l = 200$ in, moment of inertia $I = 33.33$ in^4, elastic modulus $E = 30(10)^3$ ksi is simply supported and acted on by a uniformly distributed load of intensity $w = 0.2$ k/in and ponding load of intensity $g = 3.61(10)^{-3}$ ksi as shown in the adjacent figure.

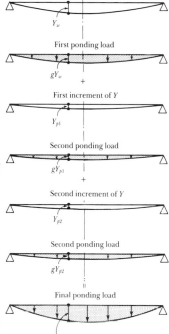

(b) Ponding process is a successive accumulation of rainwater on the slab supported by this beam and consists of the following. First, the beam deflects under the applied load and the slab forms a dry pond which during the first rain is filled with water forming the ponding load of intensity gY_w, where Y_w is the initial deflection due to applied loads. The first pond load produces the first increment in Y_w called Y_{p1}, which in turn provides space for additional water to form a second ponding load. Thus the initial deflection and all the incremental deflections due to ponding form the final elastic curve which is defined by the following equation.

(c) Governing differential equation of this system in terms of Y = deflection (in) is

$$\boxed{\dfrac{d^4Y}{dx^4} - \dfrac{g}{EI}Y = -\dfrac{w}{EI}}$$

(d) General solution of this differential equation is given in (15.14c) in terms of shape functions tabulated in (15.15c). The load function of this solution (convolution integral) with $\eta = 0.007\,75/$in and with $m = 0$, $b = 1$, $M = EI$ is by (15.16b),

$$\boxed{G(x) = -\dfrac{w}{EI}L_5(x)}$$

(e) End values of this beam are designated by the symbols introduced in (15.17d) on the opposite page, and their zeros are, in this case,

$$Y_L = 0 \qquad M_L = 0 \qquad Y_R = 0 \qquad M_R = 0$$

(f) Transport matrix equation (15.14d) is formally identical to (15.17e), but the parameter p is different: $p = -g/EI = 1.204(10)^{-7}$ in^{-4}. From the second and fourth row,

$$\theta_L = -0.070\,81 \text{ rad} \qquad V_L = -21 \text{ k}$$

and with these results, the third and fifth row yield,

$$\theta_R = 0.070\,81 \text{ rad} \qquad V_R = 21 \text{ k}$$

Since the applied load is 40 k, the ponding load is $-V_L + V_R - 40 = 2$ k.

(a) Notation. Y = solution M, N = signed numbers $\neq 0$ $p = N/M$

$$\lambda = \sqrt{\frac{N}{M}} \qquad\qquad \eta = \sqrt{\left|\frac{N}{M}\right|}$$

(b) Laplace transform of the fourth-order differential equation

$$M\frac{d^4 Y}{dx^4} + N\frac{d^2 Y}{dx^2} = f(x) \qquad\qquad Y_0, Y_0', Y_0'', Y_0''' = \text{initial values}$$

is, by (15.03b),

$$f(s) = Y_0\frac{1}{s} + Y_0'\frac{1}{s^2} + Y_0''\frac{1}{s(s^2 + p)} + Y_0'''\frac{1}{s^2(s^2 + p)} + \frac{h(s)}{Ms^2(s^2 + p)}$$

where $f(s)$ and $h(s)$ are the Laplace transforms of Y and $g(x)$, respectively.

$$f(s) = \begin{cases} Y_0\dfrac{1}{s} + Y_0'\dfrac{1}{s^2} + Y_0''\dfrac{1}{s(s^2 + \lambda^2)} + Y_0'''\dfrac{1}{s^2(s^2 + \lambda^2)} + \dfrac{h(s)}{Ms^2(s^2 + \lambda^2)} & (p > 0) \\[3mm] Y_0\dfrac{1}{s} + Y_0'\dfrac{1}{s^2} + Y_0''\dfrac{1}{s(s^2 - \eta^2)} + Y_0'''\dfrac{1}{s^2(s^2 - \eta^2)} + \dfrac{h(s)}{Ms^2(s^2 - \eta^2)} & (p < 0) \end{cases}$$

(c) Inverse Laplace transform of (b) by (15.03c) is the Laplace transform solution (LTS) of the differential equation given symbolically as

$$Y = Y_0 + Y_0'x + Y_0''L_3(x) + Y_0'''L_4(x) + G(x)$$

where $L_3(x)$, $L_4(x)$ are the shape functions (15.03c) listed in (15.20b), (15.20c), and in terms of $L_4(x)$,

$$G(x) = \mathscr{L}^{-1}\left\{\frac{h(s)}{Ms^2(s^2 + p)}\right\} = \frac{1}{M}\int_0^x L_4(x - \tau)g(\tau)\,d\tau$$

is the *convolution integral* (15.03d). Particular cases of these integrals are tabulated in Ref. 1.20, sec. 16.46. Three typical derivations are shown in (15.21).

(d) Transport matrix equation introduced in (15.03e) is, for all p,

$$\begin{bmatrix} 1 \\ Y(x) \\ Y'(x) \\ Y''(x) \\ Y'''(x) \end{bmatrix} = \left[\begin{array}{c|cccc} 1 & 0 & 0 & 0 & 0 \\ \hline G(x) & 1 & x & L_3(x) & L_4(x) \\ G'(x) & 0 & 1 & L_2(x) & L_3(x) \\ G''(x) & 0 & 0 & L_1(x) & L_2(x) \\ G'''(x) & 0 & 0 & L_0(x) & L_1(x) \end{array}\right]\begin{bmatrix} 1 \\ Y_0 \\ Y_0' \\ Y_0'' \\ Y_0''' \end{bmatrix}$$

where $dL_4(x)/dx = L_3(x)$ $dL_3(x)/dx = L_2(x)$

$dL_2(x)/dx = L_1(x)$ $dL_1(x)/dx = L_0(x)$

and $G'(x)$, $G''(x)$, $G'''(x)$ are the first-, second-, and third-order derivatives of $G(x)$, respectively. The application of this matrix equation is shown in (15.22).

(a) Equivalents.　$S = \sin \lambda x$　　　$C = \cos \lambda x$　　　$\bar{S} = \sinh \eta x$　　　$\bar{C} = \cosh \eta x$

(b) Case I $(15.19b)$, $(15.19c)$

		$p > 0 \qquad \lambda = \sqrt{\dfrac{N}{M}} \qquad r = 0, 1, 2, \ldots \qquad k = 5, 6, 7, \ldots$	
$L_0(x)$	$\dfrac{dL_1(x)}{dx}$	$-\lambda S$	$-\lambda \displaystyle\sum_{r=0}^{\infty} (-1)^r \dfrac{(\lambda x)^{2r+1}}{(2r+1)!}$
$L_1(x)$	$\mathscr{L}^{-1}\left\{\dfrac{s}{s^2 + \lambda^2}\right\}$	C	$\displaystyle\sum_{r=0}^{\infty} (-1)^r \dfrac{(\lambda x)^{2r}}{(2r)!}$
$L_2(x)$	$\mathscr{L}^{-1}\left\{\dfrac{1}{s^2 + \lambda^2}\right\}$	$\dfrac{1}{\lambda} S$	$\dfrac{1}{\lambda}\displaystyle\sum_{r=0}^{\infty} (-1)^r \dfrac{(\lambda x)^{2r+1}}{(2r+1)!}$
$L_3(x)$	$\mathscr{L}^{-1}\left\{\dfrac{1}{s(s^2 + \lambda^2)}\right\}$	$\dfrac{1}{\lambda^2}(1 - C)$	$\dfrac{1}{\lambda^2}\displaystyle\sum_{r=0}^{\infty} (-1)^r \dfrac{(\lambda x)^{2r+2}}{(2r+2)!}$
$L_4(x)$	$\mathscr{L}^{-1}\left\{\dfrac{1}{s^2(s^2 + \lambda^2)}\right\}$	$\dfrac{1}{\lambda^2}\left(x - \dfrac{S}{\lambda}\right)$	$\dfrac{1}{\lambda^3}\displaystyle\sum_{r=0}^{\infty} (-1)^r \dfrac{(\lambda x)^{2r+3}}{(2r+3)!}$
$L_5(x)$	$\displaystyle\int_0^x L_4(x)\,dx$	$\dfrac{1}{\lambda^2}\left[\dfrac{x^2}{2} - L_3(x)\right]$	$\dfrac{1}{\lambda^4}\displaystyle\sum_{r=0}^{\infty} (-1)^r \dfrac{(\lambda x)^{2r+4}}{(2r+4)!}$
$L_k(x)$	$\displaystyle\int_0^x L_{k-1}(x)\,dx$	$\dfrac{1}{\lambda^2}\left[\dfrac{x^{k-3}}{(k-3)!} - L_{k-2}(x)\right]$	$\dfrac{1}{\lambda^{k-1}}\displaystyle\sum_{r=0}^{\infty} (-1)^r \dfrac{(\lambda x)^{2r+k-1}}{(2r+k-1)!}$

(c) Case II $(15.19b)$, $(15.19c)$

		$p < 0 \qquad \eta = \sqrt{\left|\dfrac{N}{M}\right|} \qquad r = 0, 1, 2, \ldots \qquad k = 5, 6, 7, \ldots$	
$L_0(x)$	$\dfrac{dL_1(x)}{dx}$	$\eta \bar{S}$	$-\eta \displaystyle\sum_{r=0}^{\infty} \dfrac{(\eta x)^{2r+2}}{(2r+2)!}$
$L_1(x)$	$\mathscr{L}^{-1}\left\{\dfrac{s}{s^2 - \eta^2}\right\}$	$\bar{C}$	$\displaystyle\sum_{r=0}^{\infty} (-1)^r \dfrac{(\eta x)^{2r}}{(2r)!}$
$L_2(x)$	$\mathscr{L}^{-1}\left\{\dfrac{1}{s^2 - \eta^2}\right\}$	$\dfrac{1}{\eta} \bar{S}$	$\dfrac{1}{\eta}\displaystyle\sum_{r=0}^{\infty} \dfrac{(\eta x)^{2r+1}}{(2r+1)!}$
$L_3(x)$	$\mathscr{L}^{-1}\left\{\dfrac{1}{s(s^2 - \eta^2)}\right\}$	$\dfrac{1}{\eta^2}(\bar{C} - 1)$	$\dfrac{1}{\eta^2}\displaystyle\sum_{r=0}^{\infty} \dfrac{(\eta x)^{2r+2}}{(2r+2)!}$
$L_4(x)$	$\mathscr{L}^{-1}\left\{\dfrac{1}{s^2(s^2 - \eta^2)}\right\}$	$\dfrac{1}{\eta^2}\left(\dfrac{\bar{S}}{\eta} - x\right)$	$\dfrac{1}{\eta^3}\displaystyle\sum_{r=0}^{\infty} \dfrac{(\eta x)^{2r+3}}{(2r+3)!}$
$L_5(x)$	$\displaystyle\int_0^x L_4(x)\,dx$	$\dfrac{1}{\eta^2}\left[L_3(x) - \dfrac{x^2}{2}\right]$	$\dfrac{1}{\eta^4}\displaystyle\sum_{r=0}^{\infty} \dfrac{(\eta x)^{2r+4}}{(2r+4)!}$
$L_k(x)$	$\displaystyle\int_0^x L_{k-1}(x)\,dx$	$\dfrac{1}{\eta^2}\left[L_{k-2}(x) - \dfrac{x^{k-3}}{(k-3)!}\right]$	$\dfrac{1}{\eta^{k-1}}\displaystyle\sum_{r=0}^{\infty} \dfrac{(\eta x)^{2r+k-1}}{(2r+k-1)!}$

(a) Notation. $m = 0, 1, 2, \ldots$ $b, w, \alpha, \beta = $ signed numbers

p: see (15.19) $A, B, C, D, E, F = $ constants

(b) Algebraic input function. If $g(x) = w(x/n)^m$, then in (15.19c) by (15.03d),

$$G(x) = \mathscr{L}^{-1}\left\{\frac{1}{Ms^2(s^2 + p)}\frac{m!w}{s^{m+1}b^m}\right\}$$

$$= \frac{w}{Mb^m}\int_0^x (x - \tau)^m L_4(\tau)\, d\tau = \frac{m!w}{Mb^m}L_{m+5}(x)$$

where $L_{m+5}(x)$ is the respective shape function defined in (15.20b), (15.20c)

(c) Harmonic input function. If $g(x) = w\sin\beta x$, then again the convolution integral in (15.19c) becomes too involved and the solution by partial fractions is much simpler.

$$G(x) = \frac{w}{M}\mathscr{L}^{-1}\left\{\frac{1}{s^2(s^2 + p)}\frac{\beta}{s^2 + \beta^2}\right\}$$

$$= \frac{w}{Mu}\mathscr{L}^{-1}\left\{\frac{As^3 + Bs^2 + Cs + D}{s^2(s^2 + p)} + \frac{Es + F}{s^2 + \beta^2}\right\}$$

where with $u = (\beta^2 - p)\beta^2$,

$$A = C = E = 0 \qquad B = -\frac{1}{(\beta^2 - p)\beta} \qquad D = \frac{1}{\beta} \qquad F = \frac{1}{(\beta^2 - p)\beta}$$

and

$$G(x) = \frac{w}{Mu}\mathscr{L}^{-1}\left\{\frac{\beta}{s^2 + \beta^2} - \frac{\beta}{s^2} + \frac{\beta^3}{s^2(s^2 + p)}\right\}$$

$$= \frac{w}{M(\beta^2 - p)\beta^2}[\sin\beta x - \beta x + \beta^3 L_4(x)]$$

(d) Exponential input function. If $g(x) = we^{-\alpha x}$, then the solution by partial fractions is even simpler.

$$G(x) = \frac{w}{M}\mathscr{L}^{-1}\left\{\frac{1}{s^2(s^2 + p)}\frac{1}{s + \alpha}\right\}$$

$$= \frac{w}{M}\mathscr{L}^{-1}\left\{\frac{As^3 + Bs^2 + Cs + D}{s^2(s^2 + p)} + \frac{Es + F}{s + \alpha}\right\}$$

where with $u = (p + \alpha^2)\alpha^2$,

$$A = -F = -1/u \qquad B = \alpha/u \qquad C = -1/\alpha^2 \qquad D = 1/\alpha \qquad E = 0$$

and

$$G(x) = \frac{w}{Mu}\mathscr{L}^{-1}\left\{\frac{-s^3 + \alpha^2 s^2 - \alpha^2 us + \alpha^3 u}{s^2(s^2 + p)} + \frac{1}{s + \alpha}\right\}$$

$$= \frac{w}{Mu}[e^{-\alpha x} - 1 + \alpha x - \alpha^2 L_3(x) + \alpha^3 L_4(x)]$$

(a) System. A straight prismatic bar of moment of inertia $I = 200\ \text{in}^4$, elastic modulus $E = 3(10)^4\ \text{ksi}$ is simply supported at L, fixed at R, and acted on by an axial force $P = 150\ \text{k}$ and a transverse uniform load of intensity $w = 0.2\ \text{k/in}$ as shown in the adjacent figure, where $l = 480\ \text{in}$.

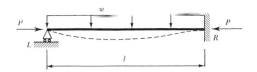

(b) Governing differential of this beam-column (Ref. 16.21) is

$$\frac{d^4Y}{dx^4} + \frac{P}{EI}\frac{d^2Y}{dx^2} = -\frac{w}{EI}$$

$$Y = \text{deflection (in)}$$

(c) General solution of this equation by (15.19c) is

$$Y = Y_L + Y_L' x + Y_L'' L_3(x) + Y_L''' L_4(x) + G(x)$$

where $L_3(x)$, $L_4(x)$ are the shape functions given in (15.20b) in terms of $\lambda = 0.005/\text{in}$, and by (15.21b) with $m = 0$, $b = 1$, $M = EI$,

$$G(x) = -\frac{w}{EI} L_5(x)$$

(d) End values of the beam-column are

$Y_L = \text{left-end deflection} = 0$ $\qquad$ $Y_L' = \text{left-end slope} = \theta_L$

$Y_L'' = \text{left-end bending moment}/EI = M_L/EI = 0$

$-Y_L''' = \text{left-end shear}/EI + \text{axial force times slope}/EI = (V_L + P\theta_L)/EI$

$Y_R = \text{right-end deflection} = 0$ $\qquad$ $Y_R' = \text{right-end slope} = \theta_R = 0$

$Y_R'' = \text{right-end bending moment}/EI = M_R/EI$

$-Y_R''' = \text{right-end shear}/EI + \text{axial force times slope}/EI = (V_R + P\theta_R)/EI$

(e) Transport matrix equation (15.19d) adjusted to the end conditions given in (d) above is

$$
\begin{bmatrix} 1 \\ \hline 0 \leftarrow Y_R \\ 0 \leftarrow \theta_R \\ \hline M_R/EI \\ -(V_R + P\theta_R)/EI \end{bmatrix}
-
\begin{bmatrix}
1 & 0 & 0 & 0 & 0 \\ \hline
G(l) & 1 & l & L_3(l) & L_4(l) \\
G'(l) & 0 & 1 & L_2(l) & L_3(l) \\ \hline
G''(l) & 0 & 0 & L_1(l) & L_2(l) \\
G'''(l) & 0 & 0 & L_0(l) & L_1(l)
\end{bmatrix}
\begin{bmatrix} 1 \\ \hline Y_L \rightarrow 0 \\ \theta_L \rightarrow 0 \\ \hline M_L/EI \\ -(V_L + P\theta_L)/EI \end{bmatrix}
$$

From the second and third row above,

$\qquad \theta_L = -0.1078\ \text{rad}$ $\qquad\qquad$ $V_L = -32.84\ \text{k}$

and with these results, the fourth and fifth row yield

$\qquad M_R = -7.277 \times 10^3\ \text{k} \cdot \text{in}$ $\qquad\qquad$ $V_R = 63.16\ \text{k}$

The magnification effect of the axial force on the bending moment at R is 1.267, since the propped-end beam moment with $P = 0$ is $-5.760 \times 10^3\ \text{k} \cdot \text{in}$.

(a) Notation. Y = solution M, N, R = signed numbers $\neq 0$

$$p = \frac{N}{M} \qquad q = \frac{R}{M} \qquad \Delta = p^2 - 4q$$

(b) Laplace transform of the fourth-order differential equation

$$M\frac{d^4Y}{dx^4} + N\frac{d^2Y}{dx^2} + RY = g(x)$$

$$Y_0, Y_0', Y_0'', Y_0''' = \text{initial values}$$

is, by (15.03*b*),

$$f(s) = Y_0[f_1(s) + pf_3(s)] + Y_0'[f_2(s) + pf_4(s)] + Y_0''f_3(s) + Y_0'''f_4(s) + f_4(s)h(s)$$

where

$$f_1(s) = \frac{s^3}{s^4 + ps^2 + q} \qquad f_2(s) = \frac{s^2}{s^4 + ps^2 + q} \qquad f_3(s) = \frac{s}{s^4 + ps^2 + q}$$

$$f_4(s) = \frac{1}{s^4 + ps^2 + q} \qquad\qquad h(s) = \frac{1}{M}\mathscr{L}\{g(x)\}$$

(c) Inverse Laplace transform of (*b*) by (15.03*c*) is the Laplace transform solution (LTS) of the differential equation given symbolically as

$$Y = Y_0[L_1(x) + pL_3(x)] + Y_0'[L_2(x) + pL_4(x)] + Y_0''L_3(x) + Y_0'''L_4(x) + G(x)$$

where $L_j(x) = \mathscr{L}^{-1}\{f_j(s)\}$ $(j = 1, 2, 3, 4)$

are the shape functions (15.03*d*) derived in (15.24) and

$$G(x) = \mathscr{L}^{-1}\{f_4(s)h(s)\} = \frac{1}{M}\int_0^x L_4(x - \tau)g(\tau)\,d\tau$$

is the *convolution integral* (15.03*d*). Particular forms of this integral may be derived by the procedures introduced in the preceding sections, and the closed-form evaluations are given in Ref. 1.20, sec. 16.49.

(d) Properties of shape functions. For $k = 1, 2, 3, \ldots,$

$$\frac{d^k L_4(x)}{dx^k} = L_{4-k}(x) \qquad \int_0^x\int \cdots \int_0^x L_4(x)\,(dx)^k = L_{4+k}(x)$$

(e) Transport matrix equation introduced in (15.03*e*) is in this case for all Δ,

$$\begin{bmatrix} 1 \\ Y(x) \\ Y'(x) \\ Y''(x) \\ Y'''(x) \end{bmatrix} = \begin{bmatrix} 1 & 0 & 0 & 0 & 0 \\ G(x) & L_1(x) + pL_3(x) & L_2(x) + pL_4(x) & L_3(x) & L_4(x) \\ G'(x) & L_0(x) + pL_2(x) & L_1(x) + pL_3(x) & L_2(x) & L_3(x) \\ G''(x) & L_{-1}(x) + pL_1(x) & L_0(x) + pL(x) & L_1(x) & L_2(x) \\ G'''(x) & L_{-2}(x) + pL_0(x) & L_{-1}(x) + pL_1(x) & L_0(x) & L_1(x) \end{bmatrix} \begin{bmatrix} 1 \\ Y_0 \\ Y_0' \\ Y_0'' \\ Y_0''' \end{bmatrix}$$

where $L_{-2}(x), L_{-1}(x), \ldots, L_3(x), L_4(x)$ are the shape functions defined in (*c*), (*d*) above and $G(x)$ is the convolution integral introduced in (*c*) above. Its first, second, and third derivatives are $G'(x), G''(x), G'''(x)$, respectively.

(a) Basic transform. As apparent from the Laplace transform of the fourth-order differential equation in (15.23b), the basic transform is

$$f_4(s) = \frac{1}{s^4 + ps^2 + q}$$

and its inverse is the highest shape function of LTS designated as $L_4(x)$. The remaining shape functions required in the solution are then the derivatives and the integrals of $L_4(x)$ as indicated in (15.23c). Since the denominator of $f_4(s)$ may be expressed in six distinct algebraic forms (depending on the magnitude and sign of p, q), six distinct forms of $L_4(x)$ are available (Ref. 1.20, sec. 16.48).

(b) Case I. If $p > 0$ and $p^2 - 4q = 0$, then with $\lambda = \sqrt{p/2}$, by (14.11b),

$$\mathcal{L}^{-1}\{f_4(s)\} = \mathcal{L}^{-1}\left\{\frac{1}{(s^2 + \lambda^2)(s^2 + \lambda^2)}\right\} = \frac{1}{2\lambda^3}(\sin \lambda x - \lambda x \cos \lambda x) = L_4(x)$$

(c) Case II. If $p < 0$ and $p^2 - 4q = 0$, then with $\eta = \sqrt{|p/2|}$, by (14.11b),

$$\mathcal{L}^{-1}\{f_4(s)\} = \mathcal{L}^{-1}\left\{\frac{1}{(s^2 - \eta^2)(s^2 - \eta^2)}\right\} = \frac{1}{2\eta^3}(\eta x \cosh \eta x - \sinh \eta x) = L_4(x)$$

(d) Case III. If $p > 0$ and $p^2 - 4q > 0$, then with $\omega = \sqrt{p^2 - 4q}$, $\lambda = \sqrt{(p - \omega)/2}$, $\eta = \sqrt{(p + \omega)/2}$, by (14.11$b$),

$$\mathcal{L}^{-1}\{f_4(s)\} = \mathcal{L}^{-1}\left\{\frac{1}{(s^2 + \lambda^2)(s^2 + \eta^2)}\right\} = \frac{1}{\omega}\left(\frac{\sin \lambda x}{\lambda} - \frac{\sin \eta x}{\eta}\right) = L_4(x)$$

(e) Case IV. If $p < 0$ and $p^2 - 4q > 0$, then with $\omega = \sqrt{p^2 - 4q}$, $\lambda = \sqrt{(|p| - \omega)/2}$, $\eta = \sqrt{(|p| + \omega)/2}$, by (14.11$b$),

$$\mathcal{L}^{-1}\{f_4(s)\} = \mathcal{L}^{-1}\left\{\frac{1}{(s^2 - \lambda^2)(s^2 - \eta^2)}\right\} = \frac{1}{\omega}\left(\frac{\sinh \eta x}{\eta} - \frac{\sinh \lambda x}{\lambda}\right) = L_4(x)$$

(f) Case V. If $p > 0$ and $p^2 - 4q < 0$, then with $\omega = \sqrt[4]{|q|}$, $\lambda = \frac{1}{2}\sqrt{2\omega^2 - p}$, $\eta = \frac{1}{2}\sqrt{2\omega^2 + p}$, $\phi = \tan^{-1}(\lambda/\eta)$, by (14.11$b$),

$$\mathcal{L}^{-1}\{f_4(s)\} = \mathcal{L}^{-1}\left\{\frac{1}{[(s - \lambda)^2 + \eta^2][(s + \lambda)^2 + \eta^2]}\right\}$$

$$= \frac{1}{\omega\lambda\eta}[e^{\lambda x}\sin(\eta x - \phi) + e^{-\lambda x}\sin(\eta x + \phi)] = L_4(x)$$

(g) Case VI. If $p < 0$ and $p^2 - 4q < 0$, then with $\omega = \sqrt[4]{|q|}$, $\lambda = \frac{1}{2}\sqrt{2\omega^2 - |p|}$, $\eta = \frac{1}{2}\sqrt{2\omega^2 + |p|}$, $\psi = \tan^{-1}(\eta/\lambda)$, by (14.11$b$),

$$\mathcal{L}^{-1}\{f_4(s)\} = \mathcal{L}^{-1}\left\{\frac{1}{[(s - \lambda)^2 - \eta^2][(s + \lambda)^2 - \eta^2]}\right\}$$

$$= \frac{1}{\omega\lambda\eta}[e^{\eta x}\sin(\lambda x - \psi) + e^{-\eta x}\sin(\lambda x + \psi)] = L_4(x)$$

(a) Classification of methods. If the explicit solution of an ordinary differential equation is too involved or cannot be constructed, the effort must be made to compute the values of Y at certain discrete points, usually selected at equidistant spacing h to facilitate the future interpolation. Although the literature of this topic is very voluminous, with the availability of digital computers, four classes of methods provide all the necessary tools for the solution of most engineering problems. They are:

(1) Euler's and Taylor's series methods
(2) Runge–Kutta methods
(3) Predictor–corrector methods
(4) Finite-difference methods

Their theoretical background and selected applications are shown in the remaining sections of this chapter.

(b) Numerical solution. The first step in the numerical solution of a given differential equation is the division of the interval of investigation $[a, b]$ into r segments along the axis of the argument x (or any other argument) by the points designated as

$$j = 0, 1, 2, \ldots, k - 1, k, k + 1, \ldots, r - 1, r$$

so that the spacing between these points is given by

$$h_k = x_k - x_{k-1} \qquad (k = 1, 2, 3, \ldots, r)$$

If an equal spacing is used as mentioned in (a), then

$$h_k = h = (b - a)/r$$

and

$$x_k = x_0 + kh = a + kh \qquad \text{or} \qquad x_k = x_r - (r - k)h = b - (r - k)h$$

The goal of the solution is to find the values Y_k corresponding to these arguments.

(c) Types of problems. In addition to the differential equation governing the problem, certain conditions must be known for the solution of the equation. Their number must equal the order of the differential equation, and according to their position in the interval $[a, b]$, the problem is designated as:

(1) *Initial-value problem* if all conditions are known at $x_0 = a$
(2) *Boundary-value problem* if some conditions are known at $x_0 = a$ and/or some other conditions are known at $x_r = b$.

These conditions may be the values of Y and/or of the derivatives of Y at the respective points. A special case of the boundary-value analysis is the problem, the unknown of which are not the values of Y, but certain characteristic values, called the *eigenvalues*. The concept of eigenvalues was introduced in (8.01) to (8.04), and their calculations from the given differential equation are shown in (15.39), (15.40).

(a) Concept.

The simplest and yet seldom-used method originally suggested by Euler is applicable to the solution of the first-order differential equation

$$\frac{dY}{dx} = f(x, Y)$$

$$Y(0) = Y_0 = \text{known condition}$$

If the function $f(x, Y)$ is continuous, sufficiently differentiable, and bounded over the interval of integration, then by using the forward difference representation of the first derivative, the values of Y becomes, by successive substitution,

$$Y_1 = Y_0 + hf_0, \; Y_2 = Y_1 + hf_1, \ldots, Y_{k+1} = Y_k + hf_k$$

where $f_0 = f(x_0, Y_0), f_1 = f(x_1, Y_1), \ldots, f_k = f(x_k, Y_k)$ and h is the equal spacing introduced in (15.25b). Geometrically the method replaces the desired solution (which is a continuous curve) by a string polygon whose slope of each side is the approximation of the first derivative of Y. Although the method is very simple, its results are so inaccurate that it is almost never used in engineering calculations. If combined with Richardson extrapolation (4.06f), however, it may yield good results as shown below.

(b) Example.

The first-order differential equation

$$dY/dx + Y = 1 + x$$

in $[0.0, 0.6]$ with the initial condition $Y_0 = 1$ is solved below in spacing $h = 0.05, 0.10$, and 0.20 by Euler's method. The results are then extrapolated and compared with the exact solution listed in the last column as

$$Y = x + e^{-x}$$

x	$h = 0.05$ A	$h = 0.10$ B	$h = 0.20$ C	D $2A - B$	E $2B - C$	F $\frac{1}{3}(4C - E)$	G Y_{exact}
0.00	1.000 000	1.000 00	1.000 00	1.000 00	1.000 00	1.000 00	1.000 00
0.05	1.000 00						1.001 23
0.10	1.002 50	1.000 00		1.005 00			1.004 84
0.15	1.007 35						1.010 71
0.20	1.014 51	1.010 00	1.000 00	1.019 20	1.020 00	1.018 93	1.018 73
0.25	1.023 79						1.028 80
0.30	1.035 09	1.029 00		1.041 18			1.040 82
0.35	1.048 34						1.054 69
0.40	1.063 42	1.056 10	1.040 00	1.070 74	1.072 20	1.070 25	1.070 32
0.45	1.080 25						1.087 63
0.50	1.098 74	1.090 49		1.106 99			1.106 53
0.55	1.118 80						1.126 95
0.60	1.140 36	1.131 44	1.112 00	1.149 28	1.150 88	1.148 75	1.148 81

(c) Local truncation error of Euler's method is

$$\epsilon_{T,k+1} \leq \frac{h^2}{2} \frac{d^2 Y}{dx^2}\bigg|_{x=\eta}$$

where $x_k < \eta < x_{k+1}$. The true truncation error is affected by the propagated errors of the preceding steps as well as by the local truncation error.

(a) Concept. If the function $f(x, Y)$ in (15.26a) has $n + 1$ continuous derivatives, then its Taylor's series expansion about x_k,

$$Y_{k+1} = Y_k + hY'_k + \frac{h^2}{2!} Y''_k + \frac{h^3}{3!} Y'''_k + \cdots + \frac{h^n}{n!} Y_k^{(n)} + \frac{h^{n+1}}{(n+1)!} Y^{(n+1)}(\eta)$$

$$= Y_k + h\Phi_n(x_k, Y_k) + \epsilon_{n+1} \qquad x_k < \eta < x_{k+1})$$

is the approximation of the solution Y at x_{k+1}. For the evaluation, x_k and Y_k are assumed to be known and the remaining values are calculated as

$$Y'_k = f(x_k, y_k) = f_k \qquad\qquad Y''_k = f_{kx} + f_k f_{ky}$$

$$Y'''_k = f_{kxx} + 2f_k f_{kxy} + f_k^2 f_{kyy} + f_{ky}(f_{kx} + f_k f_{ky})$$

. . . .

where $f_{kx} = \partial f_k/\partial x$, $f_{ky} = \partial f_k/\partial Y$, $f_{kxx} = \partial^2 f_k/d^2 x$, ... This approach is called the *Taylor's method of order n*, and, as apparent, Euler's method is a particular case of this method for $n = 1$. The local truncation error ϵ_{n+1} can be estimated by the last term of the expansion.

(b) Example. The first-order differential equation $dY/dx + Y = x + 1$ in $[0, 1]$ with the initial condition $Y_0 = 1$ and $h = 0.1$ is solved for $n = 2$ and $n = 4$. The derivatives of $f = Y' = -Y + x + 1$ are

$$f' = \frac{d}{dx}(-Y + x - 1) = -Y' + 1 = Y - x \qquad\qquad f'_k = Y_k - x_k$$

$$f'' = \frac{d}{dx}(Y - x) = Y' - 1 = -Y + x \qquad\qquad f''_k = -Y_k + x_k$$

$$f''' = \frac{d}{dx}(-Y + x) = -Y' + 1 = Y - x \qquad\qquad f'''_k = Y_k - x_k$$

and

$$\Phi_n(x_k, Y_k) = \left[1 - \frac{h}{2!} + \frac{h^2}{3!} - \frac{h^3}{4!} + \cdots + (-1)\frac{h^{n-1}}{n!}\right](x_k - Y_k) + 1$$

For $n = 2$, $Y_0 = 1$,

$$Y_{k+1} = 9.050\,00(10)^{-1}Y_k + 9.500\,00(10)^{-3}k + 0.1$$

For $n = 4$, $Y_0 = 1$

$$Y_{k+1} = 9.048\,38(10)^{-1}Y_k + 9.516\,35(10)^{-3}k + 0.1$$

Since the closed-form solution of the given differential equation is $Y = x + e^{-x}$, the accuracy of the approximations can be verified.

$Y_1 = 1.004\,84, 1.005\,00, 1.004\,84$ $Y_2 = 1.018\,73, 1.019\,03, 1.018\,73$

$Y_3 = 1.040\,82, 1.041\,22, 1.040\,82$ $Y_4 = 1.070\,32, 1.070\,84, 1.070\,32$

$Y_5 = 1.106\,53, 1.070\,84, 1.106\,53$ . . . .

where the first number is the exact value and the second and third numbers are approximations based on $n = 2$ and $n = 4$, respectively.

(a) Concept. Instead of computing higher derivatives in the truncated Taylor series, the Runge–Kutta method calculates only the first derivatives but at several points of the subinterval h. There is a whole family of Runge–Kutta methods of various orders for the solution of equation (15.26a). Three most commonly used methods are given below.

(1) *Heun method*

$$Y_{k+1} = Y_k + h(K_A + K_B)/2 \qquad K_A = f(x_k, Y_k) \qquad K_B = f(x_k + h, Y_k + hK_A)$$

(2) *Polygon method*

$$Y_{k+1} = Y_k + hK_B \qquad K_A = f(x_k, Y_k) \qquad K_B = f(x_k + \tfrac{1}{2}h, Y_k + \tfrac{1}{2}K_A)$$

(3) *Fourth-order method*

$$Y_{k+1} = Y_k + \frac{h}{6}(K_A + 2K_B + 2K_C + K_D)$$

$$K_A = f(x_k, Y_k) \qquad\qquad K_C = f(x_k + \tfrac{1}{2}h, Y_k + \tfrac{1}{2}hK_B)$$

$$K_B = f(x_k + \tfrac{1}{2}h, Y_k + \tfrac{1}{2}hK_A) \qquad K_D = f(x_k + h, Y_k + hK_C)$$

The estimates of errors are complicated and very often unsatisfactory. The spacing h in (3) is to be chosen so small that

$$K_R - K_C \leq \frac{5K_B}{100}$$

(b) Example. For the differential equation (15.27b), the fourth-order Runge–Kutta method based on relations (a-3) with $h = 0.1$ yields the following approximations of Y_1, Y_2, and Y_3.

	x_k	Y_k	$f(x, Y) = x - Y + 1$	Y_{k+1}
	$x_0 = 0$	$Y_0 = 1$	$K_A - 0$	
1	0.05	$Y_0 + 0.05K_A$	$K_B = 0.0500$	
	0.05	$Y_0 + 0.05K_B$	$K_C = 0.0475$	
	0.10	$Y_0 + 0.10K_C$	$K_D = 0.0952\,5$	$Y_1 = 1.0048\,4$
	$x_1 = 0.1$	$Y_1 = 1.00484$	$K_A = 0.0951\,6$	
2	0.15	$Y_1 + 0.05K_A$	$K_B = 0.1404\,0$	
	0.15	$Y_1 + 0.05K_B$	$K_C = 0.1381\,4$	
	0.20	$Y_1 + 0.01K_C$	$K_D = 0.1813\,5$	$Y_2 = 1.0187\,7$
	$x_2 = 0.2$	$Y_2 = 1.01877$	$K_A = 0.1812\,3$	
3	0.25	$Y_2 + 0.05K_A$	$K_B = 0.2221\,7$	
	0.25	$Y_2 + 0.05K_B$	$K_C = 0.2201\,2$	
	0.30	$Y_2 + 0.10K_C$	$K_D = 0.2592\,2$	$Y_3 = 1.0408\,5$

(a) Classification by steps. The numerical methods available for the solution of first-order differential equations generally fall into two categories:

(1) *Single-step methods* introduced in (15.26) to (15.28) requiring one initial value Y_0 at the left end of the interval of integration which is divided into r subintervals (usually of equal length). The calculation of Y_{k+1} involves only the preceding Y_k, x_k, and h.

(2) *Multistep methods* introduced in (c) to (e) below requiring several values Y_0, Y_1, ..., Y_k on the left of Y_{k+1} in the interval of integration, divided again into r subintervals (usually of equal spacing h). The calculation of Y_{k+1} involves the appropriate preceding values of Y, x, and h.

(b) Multistep methods are governed by the general formula

$$Y_{k+1} = \sum_{j=1}^{m} a_j Y_{k-j+1} + h \sum_{j=0}^{m} b_j f_{k-j+1}$$

where k is a given point in the interval of integration and m is the order of approximation. If $b_0 = 0$, the designation of explicit or predictor formula for Y_{k+1}^p is used; in turn, when $b_0 \neq 0$, the designation of implicit or corrector formula for Y_{k+1}^c is used. The recurrent application of these two formulas is called the predictor–corrector method described in (15.30). The most commonly used formulas are listed below. The most important aspect of these methods is their ability to estimate the truncation error.

(c) Milne's method

$$Y_{k+1}^p = Y_{k-3} + 4h(2f_k - f_{k-1} + 2f_{k-2})/6$$

$$Y_{k+1}^c = Y_{k-1} + h(f_{k+1}^p + 4f_k + f_{k-1})/3$$

$$Y_{k+1} = Y_{k+1}^c - (Y_{k+1}^c - Y_{k+1}^p)/29 = Y_{k+1}^c - \epsilon_{k+1}$$

(d) Adams–Bashforth–Moulton method

$$Y_{k+1}^p = Y_k + h(55f_k - 59f_{k-1} + 37f_{k-2} - 9f_{k-3})/24$$

$$Y_{k+1}^c = Y_k + h(9f_{k+1}^p + 19f_k - 5f_{k-1} + f_{k-2})/24$$

$$Y_{k+1} = Y_{k+1}^c - 19(Y_{k+1}^c - Y_{k+1}^p)/270 = Y_{k+1}^c - \epsilon_{k+1}$$

(e) Hamming's method

$$Y_{k+1}^p = Y_{k-3} + 4h(2f_k - f_{k-1} + 2f_{k-2})/3$$

$$Y_{k+1}^c = [9Y_k - Y_{k-2} + 3h(f_{k+1}^p + 2f_k - f_{k-1})]/8$$

$$Y_{k+1} = Y_{k+1}^c - 9(Y_{k+1}^c - Y_{k+1}^p)/121 = Y_{k+1}^c - \epsilon_{k+1}$$

(a) Predictor–corrector procedure. First, the starting values

$$Y_0 , Y_1 , Y_2 , \ldots , Y_k$$

and $f_0 = f(x_0 , Y_0), f_1 = f(x_1 , Y_1), f_2 = f(x_2 , Y_2), \ldots , f_k = f(x_k , Y_k)$

of the respective equation are calculated by one of the preceding methods (15.27), (15.28). Next the predictor-corrector set is selected from (15.29c) to (15.29e) and the following *iteration procedure* is applied.

Step P_0: $Y_{k+1}^{(0)}$ is calculated by the predictor formula in terms of the starting values.

Step F_0: $f_{k+1}^{(0)}$ is calculated by $f(x, Y)$ in terms of x_{k+1} and $Y_{k+1}^{(0)}$.

Step C_1: $Y_{k+1}^{(1)}$ is calculated by the corrector formula in terms of the appropriate starting values and x_{k+1}, $Y_{k+1}^{(0)}$, $f_{k+1}^{(0)}$.

Step F_1: $f_{k+1}^{(1)}$ is calculated by $f(x,y)$ in terms of x_{k+1} and $Y_{k+1}^{(1)}$.

Step C_2: $Y_{k+1}^{(2)}$ is calculated by the corrector formula in terms of the appropriate starting values and x_{k+1}, $Y_{k+1}^{(1)}$, $f_{k+1}^{(1)}$.

         . . . .

The iteration process for Y_{k+1} stops with $C_s - C_{s-1} \cong 0$. If Y_{k+2} is desired, the same procedure is repeated with Y_{k+1} used as another starting value.

(b) Example. The application of Hamming's method in the solution of $dY/dx = x - Y + 1$ with $x_0 = 0$, $Y_0 = 1$, and $h = 0.1$ is shown below. The starting values by the Runge–Kutta method (15.28b) are

$Y_0 = 1.000\,000$	$Y_1 = 1.004\,84$	$Y_2 = 1.018\,77$	$Y_3 = 1.040\,82$
$f_0 = 0$	$f_1 = 0.095\,16$	$f_2 = 0.181\,23$	$f_3 = 0.259\,18$

P_0: $Y_4^{(0)} = 1.000\,00 + 0.4(0.527\,45)/3 = 1.070\,33$

F_0: $f_4^{(0)} = 0.4 - 1.070\,33 + 1 = 0.329\,67$

C_1: $Y_4^{(1)} = [8.362\,54 + 0.3(0.666\,80)]/8 = 1.070\,32$

F_1: $f_4^{(1)} = 0.4 - 1.070\,32 + 1 = 0.329\,68$

C_2: $Y_4^{(2)} = [8.362\,54 + 0.3(0.666\,81)]/8 = 1.070\,32$

and the final value is

$$Y_4 = Y_4^{(2)} - 9(Y_4^{(2)} - Y_4^{(1)})/121 = 1.070\,32$$

which is also the true value $Y_4 = 0.4 + e^{-04}$. The calculation of Y_5 would require the starting values $Y_1 , Y_2 , Y_3 , Y_4 , f_1 , f_2 , f_3 , f_4$ and the repetition of the same procedure.

15.31 SYSTEM OF FIRST-ORDER DIFFERENTIAL EQUATIONS WITH CONSTANT COEFFICIENTS, TRANSPORT METHOD

(a) Reduction of order. Every differential equation of nth order can be replaced by a system of n simultaneous differential equations of first order. The solution of this system is a matrix series. Three typical reductions are shown below, where A_0, A_1, A_2, A_3, A_4 are constants and U, V, W, Y, $g(x)$ are functions of x.

(b) Second-order differential equation

$$A_2 \frac{d^2Y}{dx^2} + A_1 \frac{dY}{dx} + A_0 Y = g(x)$$

reduces to

$$A_2 \frac{dU}{dx} + A_1 U + A_0 Y = g(x) \qquad \frac{dY}{dx} = U$$

(c) Third-order differential equation

$$A_3 \frac{d^3Y}{dx^3} + A_2 \frac{d^2Y}{dx^2} + A_1 \frac{dY}{dx} + A_0 Y = g(x)$$

reduces to

$$A_3 \frac{dU}{dx} + A_2 U + A_1 V + A_0 Y = g(x) \qquad \frac{dY}{dx} = V \qquad \frac{dV}{dx} = U$$

(d) Fourth-order differential equation

$$A_4 \frac{d^4Y}{dx^4} + A_3 \frac{d^3Y}{dx^3} + A_2 \frac{d^2Y}{dx^2} + A_1 \frac{dY}{dx} + A_0 Y = g(x)$$

reduces to

$$A_4 \frac{dU}{dx} + A_3 U + A_2 V + A_1 W + A_0 Y = g(x) \qquad \frac{dY}{dx} = W \qquad \frac{dW}{dx} = V \qquad \frac{dV}{dx} = U$$

(e) Differential matrix equation constructed from these reductions becomes in case (d) with $B_k = A_k/A_4$.

$$\underbrace{\begin{bmatrix} dU/dx \\ dV/dx \\ dW/dx \\ dY/dx \end{bmatrix}}_{d\mathbf{H}/dx} = \underbrace{\begin{bmatrix} -B_3 & -B_2 & -B_1 & -B_0 \\ 0 & 1 & 0 & 0 \\ 0 & 0 & 1 & 0 \\ 0 & 0 & 0 & 1 \end{bmatrix}}_{\mathbf{D}} \underbrace{\begin{bmatrix} U \\ V \\ W \\ Y \end{bmatrix}}_{\mathbf{H}} + \underbrace{\begin{bmatrix} g(x)/A_4 \\ 0 \\ 0 \\ 0 \end{bmatrix}}_{\mathbf{L}(x)}$$

(f) Transport matrix equation. The solution of the differential matrix equation in (e) is

$$\boxed{\mathbf{H} = \exp(\mathbf{D}x)\,\mathbf{H}_0 + \exp(\mathbf{D}x)\int_0^x \exp(-\mathbf{D}\tau)\,\mathbf{L}(\tau)\,d\tau = \mathbf{T}(x)\mathbf{H}_0 + \mathbf{G}(x)}$$

where $\mathbf{H}$, $\mathbf{H}_0$ are the state vectors at $x > 0$, $x = 0$, respectively, $\mathbf{D}$ is the matrix of constant coefficients, $\mathbf{T}(x)$ is the transport matrix in x defined in (g) below, and $\mathbf{G}(x)$ is the convolution integral vector defined in (i) below.

(g) Transport matrix $\mathbf{T}(x)$ in (f) is, according to (15.03e),

$$
\mathbf{T}(x) = \exp\left(\mathbf{D}x\right) = \mathbf{I} + \mathbf{D}\frac{x}{1!} + \mathbf{D}^2\frac{x^2}{2!} + \mathbf{D}^3\frac{x^3}{3!} + \cdots
$$

and the sum of this series of matrix functions is the transport matrix,

$$
\mathbf{T}(x) = \begin{bmatrix}
T_{11}(x) & T_{12}(x) & T_{13}(x) & T_{14}(x) \\
T_{21}(x) & T_{23}(x) & T_{23}(x) & T_{24}(x) \\
T_{31}(x) & T_{32}(x) & T_{33}(x) & T_{34}(x) \\
T_{41}(x) & T_{42}(x) & T_{43}(x) & T_{44}(x)
\end{bmatrix}
$$

(h) Inverse of transport matrix $\mathbf{T}(x)$ is

$$
[\mathbf{T}(x)]^{-1} = \mathbf{T}(-x) = \exp\left(-\mathbf{D}x\right) = \mathbf{I} - \mathbf{D}\frac{x}{1!} + \mathbf{D}^2\frac{x^2}{2!} - \mathbf{D}^3\frac{x^3}{3!} + \cdots
$$

and the sum of this series of matrix functions is the same matrix as in (g) but the argument is $-x$.

(i) Convolution integral $\mathbf{G}(x)$ is then

$$
\mathbf{G}(x) = \frac{1}{A_4}\begin{bmatrix}
T_{11}(x) & T_{12}(x) & T_{13}(x) & T_{14}(x) \\
T_{21}(x) & T_{22}(x) & T_{23}(x) & T_{24}(x) \\
T_{31}(x) & T_{32}(x) & T_{33}(x) & T_{34}(x) \\
T_{41}(x) & T_{42}(x) & T_{43}(x) & T_{44}(x)
\end{bmatrix}\int_0^x\begin{bmatrix}
T_{11}(-\tau)g(\tau) \\
T_{21}(-\tau)g(\tau) \\
T_{31}(-\tau)g(\tau) \\
T_{41}(-\tau)g(\tau)
\end{bmatrix}dt = \begin{bmatrix}
G_{10}(x) \\
G_{20}(x) \\
G_{30}(x) \\
G_{40}(x)
\end{bmatrix}
$$

(j) Closed-form transport matrix equation written for the subinterval of length h is

$$
\underbrace{\begin{bmatrix}
1 \\
U_1 \\
V_1 \\
W_1 \\
Y_1
\end{bmatrix}}_{\hat{\mathbf{H}}_1} = \underbrace{\left[\begin{array}{c|cccc}
1 & 0 & 0 & 0 & 0 \\
\hline
G_{10}(h) & T_{11}(h) & T_{12}(h) & T_{13}(h) & T_{14}(h) \\
G_{20}(h) & T_{21}(h) & T_{22}(h) & T_{23}(h) & T_{24}(h) \\
G_{30}(h) & T_{31}(h) & T_{32}(h) & T_{33}(h) & T_{34}(h) \\
G_{40}(h) & T_{41}(h) & T_{42}(h) & T_{43}(h) & T_{44}(h)
\end{array}\right]}_{\hat{\mathbf{T}}_{10}}\underbrace{\begin{bmatrix}
1 \\
U_0 \\
V_0 \\
W_0 \\
Y_0
\end{bmatrix}}_{\hat{\mathbf{H}}_0}
$$

where $\hat{\mathbf{H}}_1$ is the state vector at $x = h$.

(k) Transport chain. The final solution of the differential equation (b) is

$$
\hat{\mathbf{H}}_1 = \hat{\mathbf{T}}_{10}\hat{\mathbf{H}}_0, \ \hat{\mathbf{H}}_2 = \hat{\mathbf{T}}_{21}\hat{\mathbf{H}}_1, \dots, \hat{\mathbf{H}}_n = \hat{\mathbf{T}}_{n,n-1}\hat{\mathbf{H}}_{n-1}
$$

where the T coefficients of matrices $\hat{\mathbf{T}}_{10}, \hat{\mathbf{T}}_{21}, \dots, \hat{\mathbf{T}}_{n,n-1}$ are the same numbers and the $\mathbf{G}$ function corresponds to the g functions of the respective subintervals. The numerical example of this method is shown in (15.32).

(a) Reduction. The second-order differential equation

$$\boxed{\frac{d^2Y}{dt^2} - 4\frac{dY}{dt} - 4Y = 0}$$

with $t_0 = 0$, $Y_0 = 1$, $Y'_0 = 2$, and $h = 0.1$ in $[0, 1]$ is reduced by $(15.31b)$ to

$$\frac{dU}{dt} + 4U - 4Y = 0 \qquad \frac{dY}{dt} = -U$$

(b) Differential matrix equation $(15.31e)$ is then

$$\underbrace{\begin{bmatrix} dU/dt \\ dY/dt \end{bmatrix}}_{d\mathbf{H}/dt} = \underbrace{\begin{bmatrix} -4 & 4 \\ -1 & 0 \end{bmatrix}}_{\mathbf{D}} \underbrace{\begin{bmatrix} U \\ Y \end{bmatrix}}_{\mathbf{H}}$$

(c) Transport matrix in series form $(15.31g)$ becomes

$$\mathbf{T}(t) \cong \begin{bmatrix} 1 & 0 \\ 0 & 1 \end{bmatrix} + \begin{bmatrix} -4 & 4 \\ -1 & 0 \end{bmatrix}\frac{t}{1!} + \begin{bmatrix} 12 & -16 \\ 4 & -4 \end{bmatrix}\frac{t^2}{2!} + \begin{bmatrix} -32 & 48 \\ -12 & 16 \end{bmatrix}\frac{t^3}{3!}$$

$$+ \begin{bmatrix} 80 & -128 \\ 32 & -48 \end{bmatrix}\frac{t^4}{4!} + \begin{bmatrix} -192 & 320 \\ -112 & 192 \end{bmatrix}\frac{t^5}{5!} = \begin{bmatrix} T_{11}(t) & T_{22}(t) \\ T_{21}(t) & T_{22}(t) \end{bmatrix}$$

from which for $t = h = 0.1$,

$$T_{11}(t) \cong 1 - 4t/1! + 12t^2/2! - 32t^3/3! + 80t^4/4! - 192t^5/5! = 0.654\,986$$

$$T_{12}(t) \cong 0 + 4t/1! - 16t^2/2! + 48t^3/3! - 128t^4/4! + 320t^5/5! = 0.327\,493$$

$$T_{21}(t) \cong 0 - t/1! + 4t^2/2! - 12t^3/3! + 32t^4/4! - 112t^5/5! = -0.081\,876$$

$$T_{22}(t) \cong 1 + 0/1! - 4t^2/2! + 16t^3/3! - 48t^4/4! + 192t^5/5! = 0.982\,483$$

Since this differential equation has a closed-form solution, the approximate values listed above can be compared to the exact values given below for $t = 0.1$.

$$T_{11}(t)\ (1 - 2t)e^{-2t} = 0.654\,985 \qquad T_{12}(t) = 4t3^{-2t} = 0.327\,492$$

$$T_{21}(t) = -te^{-2t} = -0.081\,873 \qquad T_{22}(t) = (1 + 2t)e^{-2t} = 0.982\,477$$

(d) State vectors at $1, 2, 3$ by $(15.31j)$ with $U_0 = -2$, $Y_0 = 1$ are

$$\begin{bmatrix} U_1 \\ Y_1 \end{bmatrix} \cong \begin{bmatrix} 0.654\,984 & 0.327\,493 \\ -0.081\,876 & 0.982\,483 \end{bmatrix}\begin{bmatrix} -2 \\ 1 \end{bmatrix} = \begin{bmatrix} -0.982\,475 \\ 1.146\,235 \end{bmatrix}$$

$$\begin{bmatrix} U_2 \\ Y_2 \end{bmatrix} \cong \begin{bmatrix} 0.654\,984 & 0.327\,493 \\ -0.081\,876 & 0.982\,483 \end{bmatrix}\begin{bmatrix} -0.982\,475 \\ 1.146\,235 \end{bmatrix} = \begin{bmatrix} -0.268\,121 \\ 1.206\,598 \end{bmatrix}$$

$$\begin{bmatrix} U_3 \\ Y_3 \end{bmatrix} \cong \begin{bmatrix} 0.654\,984 & 0.327\,493 \\ -0.081\,876 & 0.982\,483 \end{bmatrix}\begin{bmatrix} -0.268\,121 \\ 1.206\,598 \end{bmatrix} = \begin{bmatrix} 0.219\,537 \\ 1.207\,414 \end{bmatrix}$$

and similarly $\mathbf{H}_4 = \mathbf{T}\mathbf{H}_3$, $\mathbf{H}_5 = \mathbf{T}\mathbf{H}_4, \dots, \mathbf{H}_{10} = \mathbf{T}\mathbf{H}_9$. These results can be compared to the exact values which are $U_1 = -0.982\,477$, $U_2 = -0.268\,128$, $U_3 = 0.219\,525$, $Y_1 = 1.146\,223$, $Y_2 = 1.206\,576$, $Y_3 = 1.207\,385$.

(a) Differential matrix equation. If the coefficients A_0, A_1, A_2, A_3, A_4 in (15.31d) are functions of the argument x, then the differential equation has variable coefficients and the matrix equation (15.31e) has also variable coefficients in D.

$$\begin{bmatrix} dU/dx \\ dV/dx \\ dW/dx \\ dY/dx \end{bmatrix} = \begin{bmatrix} -B_3(x) & -B_2(x) & -B_1(x) & -B_0(x) \\ 0 & 1 & 0 & 0 \\ 0 & 0 & 1 & 0 \\ 0 & 0 & 0 & 1 \end{bmatrix} \begin{bmatrix} U \\ V \\ W \\ Y \end{bmatrix}$$

In such a case a substitute model with constant coefficients in each subinterval must be used or the solution of this matrix equation must be approximated by the Runge–Kutta method, which is well suited for this purpose.

(b) Runge–Kutta method applied in the solution of the differential equation shown in (a) divides the subinterval h into two segments, shown in the adjacent figure. The coordinates of their endpoints are x_{0i}, x_{0j}, x_{0k} and the D matrices at these points are

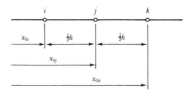

$$\mathbf{D}_i = \begin{bmatrix} -B_3(x_{0i}) & -B_2(x_{0i}) & -B_1(x_{0i}) & -B_0(x_{0i}) \\ 0 & 1 & 0 & 0 \\ 0 & 0 & 1 & 0 \\ 0 & 0 & 0 & 1 \end{bmatrix}$$

$$\mathbf{D}_j = \begin{bmatrix} -B_3(x_{0j}) & -B_2(x_{0j}) & -B_1(x_{0j}) & -B_0(x_{0j}) \\ 0 & 1 & 0 & 0 \\ 0 & 0 & 1 & 0 \\ 0 & 0 & 0 & 1 \end{bmatrix}$$

$$\mathbf{D}_k = \begin{bmatrix} -B_3(x_{0k}) & -B_2(x_{0k}) & -B_1(x_{0k}) & -B_0(x_{0k}) \\ 0 & 1 & 0 & 0 \\ 0 & 0 & 1 & 0 \\ 0 & 0 & 0 & 1 \end{bmatrix}$$

(c) Runge–Kutta solution in interval $ik = h$ is

$$\mathbf{H}_k = \underbrace{\left[\mathbf{I} + \mathbf{K}_A + \frac{h}{1!} + \mathbf{K}_B \frac{h^2}{2!} + \mathbf{K}_C \frac{h^3}{3!} + \mathbf{K}_D \frac{h^4}{4!} \right]}_{\mathbf{T}_{ki}} \mathbf{H}_i$$

where $\mathbf{I}$ is a unit matrix,

$$\mathbf{K}_A = \tfrac{1}{6}(\mathbf{D}_i + 4\mathbf{D}_j + \mathbf{D}_k) \qquad\qquad \mathbf{K}_C = \tfrac{1}{2}(\mathbf{D}_i \mathbf{D}_j^2 + \mathbf{D}_j^2 \mathbf{D}_k)$$

$$\mathbf{K}_B = \tfrac{1}{3}(\mathbf{D}_i \mathbf{D}_j + \mathbf{D}_j^2 + \mathbf{D}_j \mathbf{D}_k) \qquad\qquad \mathbf{K}_D = \mathbf{D}_i \mathbf{D}_j^2 \mathbf{D}_k$$

and $\mathbf{H}_i$, $\mathbf{H}_k$ are the state vectors related by the transport matrix $\mathbf{T}_{ki}$. The inclusion of the convolution integral follows the pattern introduced in (15.31i).

(a) Concept. The finite-diffrence method replaces the interval of integration $[a, b]$ by a selected set of points (joints, nodes, pivotal points) and approximates the differential equation and its initial-value equations or its boundary-value equations by the respective difference equations related to these points. This substitution leads to a set of algebraic equations written for all selected points, and the solution of this set yields the approximate values of Y at these points. The spacing of these points is usually equidistant, but uneven spacing may be used in special situations. Because of higher accuracy, only central differences $(3.02d)$, $(3.02e)$ are commonly used. Since odd-order differential equations are hard to handle by finite differences, the transformation to even-order equations is performed by differentiation or integration.

(b) Set of joints shown in the adjacent figure is defined by the respective coordinates as

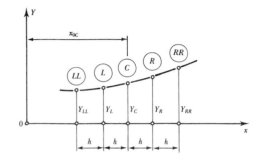

$$LL[x_{OC} - 2h, Y_{LL}]$$

$$L[x_{OC} - h, Y_L]$$

$$C[x_{OC}, Y_C]$$

$$R[x_{OC} + h, Y_R]$$

$$RR = [x_{OC} + 2h, Y_{RR}]$$

where h = constant spacing and LL, L, C, R, RR = designations of joints.

(c) Difference operators at C based on relations $(3.02d)$, $(3.02e)$ are, with $k = 1, 2, \ldots,$

$$\delta^{(0)} Y_C = Y_C \qquad\qquad \delta^{(2k+1)} Y_C = -\delta^{(2k)} Y_L + \delta^{(2k)} Y_R$$

$$\delta^{(1)} Y_C = -Y_L + Y_R \qquad\qquad \delta^{(2k+2)} Y_C = \delta^{(2k)} Y_L - 2\delta^{(2k)} Y_C + \delta^{(2k)} Y_R$$

$$\delta^{(2)} Y_C = Y_L - 2Y_C + Y_R$$

(d) Difference equations at C are

$$\left[\frac{dY}{dx}\right]_C \cong \frac{\delta^{(1)} Y_C}{2h} = \frac{-Y_L + Y_R}{2h} \qquad\qquad \left[\frac{d^3 Y}{dx^3}\right]_C \cong \frac{\delta^{(3)} Y_C}{2h^3} = \frac{-Y_{LL} + 2Y_L - 2Y_R + Y_{RR}}{2h^3}$$

$$\left[\frac{d^2 Y}{dx^2}\right]_C \cong \frac{\delta^{(2)} Y_C}{h^2} = \frac{Y_L - 2Y_C + Y_R}{h^2} \qquad\qquad \left[\frac{d^4 Y}{dx^4}\right]_C \cong \frac{\delta^{(4)} Y_C}{h^4} = \frac{Y_{LL} - 4Y_L + 6Y_C - 4Y_R + Y_{RR}}{h^4}$$

and in general,

$$\left[\frac{d^{2k+1} Y}{dx^{2k+1}}\right]_C \cong \frac{\delta^{(2k+1)} Y_C}{2h^{2k+1}} = \frac{-\delta^{(2k)} Y_L + \delta^{(2k)} Y_R}{2h^{2k+1}}$$

$$\left[\frac{d^{2k+2} Y}{dx^{2k+2}}\right]_C \cong \frac{\delta^{(2k+2)} Y_C}{h^{2k+2}} = \frac{\delta^{(2k+1)} Y_L - 2\delta^{(2k+1)} Y_C + \delta^{(2k+1)} Y_R}{h^{2k+2}}$$

where each equation defines the approximation of the respective derivative in terms of Y's and the constant spacing h.

(e) Joint input functions. Since the values of Y and its derivatives are related to LL, L, C, R, RR, the input function $g(x)$ must also be related to these joints by equivalents called the joint loads. For this purpose each segment h is treated as a simple beam whose reactions are the joint loads. The table (f) gives the most common cases of joint loads expressed in terms of

P = concentrated load $\qquad\qquad$ Q = applied couple

w = intensity of distributed load $\qquad$ G = joint load

α = string angle $\qquad\qquad$ h = segment length $\qquad$ $m + n = 1$ $\qquad$ m, n = fractions

(f) Table of joint loads

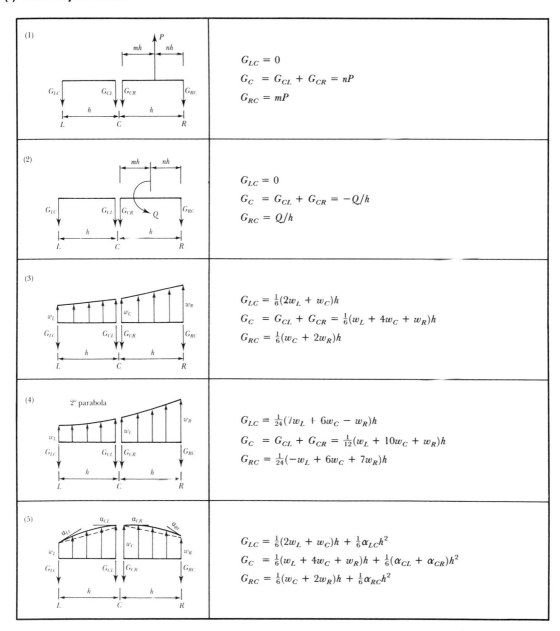

(1)
$$G_{LC} = 0$$
$$G_C = G_{CL} + G_{CR} = nP$$
$$G_{RC} = mP$$

(2)
$$G_{LC} = 0$$
$$G_C = G_{CL} + G_{CR} = -Q/h$$
$$G_{RC} = Q/h$$

(3)
$$G_{LC} = \tfrac{1}{6}(2w_L + w_C)h$$
$$G_C = G_{CL} + G_{CR} = \tfrac{1}{6}(w_L + 4w_C + w_R)h$$
$$G_{RC} = \tfrac{1}{6}(w_C + 2w_R)h$$

(4) 2° parabola
$$G_{LC} = \tfrac{1}{24}(7w_L + 6w_C - w_R)h$$
$$G_C = G_{CL} + G_{CR} = \tfrac{1}{12}(w_L + 10w_C + w_R)h$$
$$G_{RC} = \tfrac{1}{24}(-w_L + 6w_C + 7w_R)h$$

(5)
$$G_{LC} = \tfrac{1}{6}(2w_L + w_C)h + \tfrac{1}{6}\alpha_{LC}h^2$$
$$G_C = \tfrac{1}{6}(w_L + 4w_C + w_R)h + \tfrac{1}{6}(\alpha_{CL} + \alpha_{CR})h^2$$
$$G_{RC} = \tfrac{1}{6}(w_C + 2w_R)h + \tfrac{1}{6}\alpha_{RC}h^2$$

(a) Governing difference equation for the second-order differential equation with variable coefficients,

$$M(x)\frac{d^2Y}{dx^2} + N(x)\frac{dY}{dx} + R(x)Y = g(x)$$

is at C, $K_L Y_L - K_C Y_C + K_R Y_R = \bar{G}_C$

where with $M(x_{0C}) = M_C$, $N(x_{0C}) = N_C$, $R(x_{0C}) = R_C$,

$K_L = 1 - \dfrac{N_C}{2M_C}$	$K_C = 2 - \dfrac{R_C h^2}{M_C}$	$K_R = 1 + \dfrac{N_C}{2M_C}$

$\bar{G}_C = G_C h^2 / M_C =$ scaled joint load (15.34f)

This equation applies at all intermediate points in $[0, l]$, and for the endpoints $0, r$ the endpoint equations given in (c) must be used. The set of points subdividing the length l is shown in (b), where the spacing $h = l/r$ and $-1, r + 1$ are the fictitious points used in the endpoint equations.

(b) Subdivision of interval

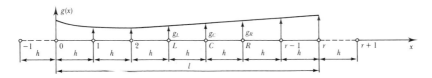

(c) Endpoint equations. Two endpoint equations are required in the solution of the second-order equation. They must be selected from the following set.

$Y(0) = \alpha_0$ $Y_0 = \alpha_0$ $Y(rh) = \beta_0$ $Y_r = \beta_0$

$Y'(0) = \alpha_1$ $Y'_0 = \frac{1}{2}(-Y_{-1} + Y_1)/h = \alpha_1$ $Y'(rh) = \beta_1$ $Y'_r = \frac{1}{2}(-Y_{r-1} + Y_{r+1})/h = \beta_1$

(d) Numerov's method. If $N(x) = 0$ in (a), then the differential equation may be written symbolically as

$$\frac{d^2Y}{dx^2} = f(x, y)$$

and its difference equation becomes

$$Y_L - 2Y_C + Y_R = -\frac{h^2}{12}\left(\frac{R_L}{M_L}Y_L + 10\frac{R_C}{M_C}Y_C + \frac{R_R}{M_R}Y_R\right) + \frac{G_C h^2}{M_C}$$

from which
$K_L = 1 + \dfrac{R_L h^2}{12M_L}$	$K_C = 2 - \dfrac{10R_C h^2}{12M_C}$	$K_R = 1 + \dfrac{R_R h^2}{12M_R}$

This method uses the second-order difference as the approximation of the second derivative of Y and the parabolic joint load formula for the approximation of the variation of Y. The better accuracy of Numerov's method is shown in (15.36b).

(a) Standard method. The boundary-value problem governed by

$$\boxed{\frac{d^2Y}{dx^2} - Y = 10x}$$

in $[0, 8]$ with the end conditions $Y(0) = 0$, $Y(8) = 0$, $r = 4$, and $h = 2$ is solved by the standard finite-difference method introduced in (15.35a). The governing difference equation with $M = 1$, $N = 0$, $R = -1$ and $K_L = 1$, $K_C = 2 + 4$, $K_R = 1$ becomes

$$Y_L - 6Y_C + Y_R = 40x_C$$

By the recurrent use,

$$\begin{bmatrix} -6 & 1 & 0 \\ 1 & -6 & 1 \\ 0 & 1 & -6 \end{bmatrix}\begin{bmatrix} Y_1 \\ Y_2 \\ Y_3 \end{bmatrix} = \begin{bmatrix} 80 \\ 160 \\ 240 \end{bmatrix}$$

from which

$$Y(0) = 0 \qquad Y(2) = Y_1 = -19.608 \qquad Y(4) = Y_2 = -37.647$$

$$Y(6) = Y_3 = -46.275 \qquad Y(8) = Y_4 = 0$$

If the number of segments is increased from 4 to 8, then $r = 8$, $h = 1$, and the finite-difference matrix equation becomes

$$\begin{bmatrix} -3 & 1 & & & & & \\ 1 & -3 & 1 & & & & \\ & 1 & -3 & 1 & & & \\ & & 1 & -3 & 1 & & \\ & & & 1 & -3 & 1 & \\ & & & & 1 & -3 & 1 \\ & & & & & 1 & -3 \end{bmatrix}\begin{bmatrix} Y_1 \\ Y_2 \\ Y_3 \\ Y_4 \\ Y_5 \\ Y_6 \\ Y_7 \end{bmatrix} = \begin{bmatrix} 10 \\ 20 \\ 30 \\ 40 \\ 50 \\ 60 \\ 70 \end{bmatrix}$$

and $Y(0) = Y_0 = 0$ $Y(5) = Y_5 = -45.543$ (-46.017)

$\qquad Y(1) = Y_1 = -9.920$ (-9.937) $Y(6) = Y_6 = -48.328$ (-49.173)

$\qquad Y(2) = Y_2 = -19.759$ (-19.805) $Y(7) = Y_7 = -39.443$ (-40.570)

$\qquad Y(3) = Y_3 = -29.352$ (-29.462) $Y(8) = Y_8 = 0$

$\qquad Y(4) = Y_4 = -38.298$ (-38.535)

The exact solution of the differential equation is $Y(x) = 80 \sinh x \, (\sinh 8)^{-1} - 10x$ and the exact values of $Y(x)$ are listed in parentheses above.

(b) Numerov's method applied to the same problem with $r = 4$, $h = 2$ yields after scaling

$$\begin{bmatrix} -8 & 1 & 0 \\ 1 & -8 & 1 \\ 0 & 1 & -8 \end{bmatrix}\begin{bmatrix} Y_1 \\ Y_2 \\ Y_3 \end{bmatrix} = \begin{bmatrix} 120 \\ 240 \\ 360 \end{bmatrix}$$

from which $Y(2) = Y_1 = -19.839$, $Y(4) = Y_2 = -38.710$, $Y(6) = Y_3 = -49.839$. The comparison of these results shows that Numerov's method is far superior to the standard method.

(a) Governing difference equation for the fourth-order differential equation

$$M(x)\frac{d^4Y}{dx^4} + N(x)\frac{d^2Y}{dx^2} + R(x)Y = g(x)$$

is at C,

$$K_{LL}Y_{LL} - K_L Y_L + K_C Y_C - K_R Y_R + K_{RR}Y_{RR} = \bar{G}_C$$

where with $\phi_j = (N_j h^2)/M_j$ and $\psi_j = (R_j h^4)/M_j$,

$K_{LL} = 1$	$K_L = 4 - \phi_L$	$K_C = 6 - 2\phi_C + \psi_C$	$K_R = 4 - \phi_R$	$K_{RR} = 1$

and $\bar{G}_C = G_C h^3/M_c$ = scaled joint load (15.34*f*). This equation applies at all intermediate points, and as in (15.35*a*) for the endpoints 0 and r the endpoint equations shown in (*b*) below must be used.

(b) Endpoint equations. Four such equations are required for the solution of the fourth-order equation. They must be selected from the following sets.

$$Y(0) = \alpha_0 \qquad Y_0 = \alpha_0 \qquad Y'(0) = \alpha_1 \qquad Y_0' = \frac{-Y_{-1} + Y_1}{2h} = \alpha_1$$

$$Y''(0) = \alpha_2 \qquad Y_0'' = \frac{Y_{-1} - 2Y_0 + Y_1}{h^2} = \alpha_2$$

$$Y'''(0) = \alpha_3 \qquad Y_0''' = \frac{-Y_{-2} + 2Y_{-1} - 2Y_1 + Y_2}{2h^3} = \alpha_3$$

$$Y(rh) = \beta_0 \qquad Y_r = \beta_0 \qquad Y'(rh) = \beta_1 \qquad Y_r' = \frac{-Y_{r-1} + Y_{r+1}}{2h} = \beta_1$$

$$Y''(rh) = \beta_2 \qquad Y_r'' = \frac{Y_{r-1} - 2Y_r + Y_{r+1}}{h^2} = \beta_2$$

$$Y'''(rh) = \beta_3 \qquad Y_r''' = \frac{-Y_{r-2} + 2Y_{r-1} - 2Y_{r+1} + Y_{r+2}}{2h^3} = \beta_3$$

(c) Stüssi's method. The differential equation is written symbolically as

$$\frac{d^4Y}{dx^4} = f(x, Y, Y'')$$

and its difference approximation is the same as in (*a*) but with $\phi_j = (N_j h^2/(12M_j)$ and $\psi_j = (R_j h^4)/(12M_j)$

$K_L = 4 - 8\phi_L - \psi_L$	$K_C = 6 - 18\phi_C + 10\psi_C$	$K_R = 4 - 8\phi_R - \psi_R$
$K_{LL} = 1 + \phi_{LL}$		$K_{RR} = 1 + \phi_{RR}$

This method uses the fourth-order difference expression as the approximation of the fourth derivative of Y and the parabolic load formula for the approximation Y'' and Y.

(a) Standard method. The boundary-value problem governed by

$$\boxed{\frac{d^4 Y}{dx^4} + 4y = 20}$$

in $[0, 8]$ with the end conditions $Y(0) = 0$, $Y'(0) = 0$, $Y(8) = 0$, $Y'(8) = 0$, $r = 4$, and $h = 2$ is solved by the method introduced in (15.37a). The governing difference equation with $M_C = 1$, $N_C = 0$, $R_C = 4$ and $K_{LL} = 1$, $K_L = 4$, $K_C = 70$, $K_R = 4$, $K_{RR} = 1$ becomes

$$Y_{LL} - 4Y_L + 70Y_C - 4Y_R + Y_{RR} = 100$$

and the endpoint equations are $Y_{-1} = Y_1$, $Y_0 = Y_4 = 0$, $Y_5 = Y_3$. The use of these endpoint equations and the recurrent use of the governing difference equation yields

$$\begin{bmatrix} 71 & -4 & 1 \\ -4 & 70 & -4 \\ 1 & -4 & 71 \end{bmatrix} \begin{bmatrix} Y_1 \\ Y_2 \\ Y_3 \end{bmatrix} = \begin{bmatrix} 160 \\ 160 \\ 160 \end{bmatrix}$$

from which $Y_1 = 2.364$, $Y_2 = 2.556$, $Y_3 = 2.364$. If the number of segments is increased to $r = 8$, then with $h = 1$, the governing difference equation becomes

$$Y_{LL} - 4Y_L + 10Y_C - 4Y_R + Y_{RR} = 10$$

and

$$Y(1) = Y_1 = 1.544 \quad (1.226)$$

$$Y(2) = Y_2 = 2.390 \quad (2.329)$$

$$Y(3) = Y_3 = 2.578 \quad (2.617)$$

$$Y(4) = Y_4 = 2.584 \quad (2.629)$$

$$Y(5) = Y_5 = 2.578 \quad (2.578)$$

$$Y(6) = Y_6 = 2.390 \quad (2.329)$$

$$Y(7) = Y_Y = 1.544 \quad (1.226)$$

The exact solution of the differential equation is given in parentheses above.

(b) Stüssi's method applied to the same problem with $r = 4$ and $h = 2$ yields, after scaliing by 12,

$$\begin{bmatrix} 724 & 16 & 12 \\ 16 & 712 & 16 \\ 12 & 16 & 724 \end{bmatrix} \begin{bmatrix} Y_1 \\ Y_2 \\ Y_3 \end{bmatrix} = \begin{bmatrix} 1920 \\ 1920 \\ 1920 \end{bmatrix}$$

from which $Y(2) = Y_1 = 2.553$, $Y(4) = Y_2 = 2.581$, $Y(6) = Y_3 = 2.553$. The comparison of these results with those of (a) shows that for $R/M > 1$, Stüssi's method yields poorer results compared to those obtained in (a).

(a) Standard characteristic determinant of the second-order differential equation in $[0, L]$,

$$\frac{d^2Y}{dx^2} + \lambda^2 Y = 0$$

$$Y(0) = 0, \; Y(L) = 0$$

is by the relation (15.35a), with $r = 4$, $h = L/4$, and $\phi = h^2\lambda^2$,

$$\begin{vmatrix} -(2 - \phi) & 1 & 0 \\ 1 & -(2 - \phi) & 1 \\ 0 & 1 & -(2 - \phi) \end{vmatrix} = 0$$

The expansion of this determinant yields the polynomial in λ^2,

$$-(2 - h^2\lambda^2)^3 + 2(2 - h^2\lambda^2) = 0$$

The roots of this equation are the eigenvalues

$$\lambda_1 = \frac{4}{L}\sqrt{2 - \sqrt{2}} = \frac{3.061}{L} \qquad \lambda_2 = \frac{4}{L}\sqrt{2} = \frac{5.657}{L} \qquad \lambda_3 = \frac{4}{L}\sqrt{2 + \sqrt{2}} = \frac{7.391}{L}$$

The exact eigenvalue equation is $\sin \lambda L = 0$, and the exact eigenvalues are

$$\lambda_1 = \frac{\pi}{L} = \frac{3.142}{L} \qquad \lambda_2 = \frac{2\pi}{L} = \frac{6.283}{L} \qquad \lambda_3 = \frac{3\pi}{L} = \frac{9.425}{L}$$

(b) Numerov's determinant of the same equation by the relation (15.35d) with $r = 4$, $h = L/4$, and $\phi = h^2\lambda^2$ as

$$\begin{vmatrix} -(24 - 10\phi) & (12 + \phi) & 0 \\ (12 + \phi) & -(24 - 10\phi) & (12 + \phi) \\ 0 & (12 + \phi) & -(24 - 10\phi) \end{vmatrix} = 0$$

As in (a), the expansion of this determinant is the characteristic equation

$$-(24 - 10h^2\lambda^2)^3 + 2(12 + h^2\lambda^2)^2(24 - 10h^2\lambda^2) = 0$$

The roots of this equation are the eigenvalues

$$\lambda_1 = \frac{4}{L}\sqrt{\frac{246 - \sqrt{41472}}{98}} = \frac{3.139}{L} \qquad \lambda_2 = \frac{4}{L}\sqrt{\frac{24}{10}} = 6.197 \qquad \lambda_3 = \frac{4}{L}\sqrt{\frac{264 + \sqrt{41472}}{98}} = \frac{9.254}{L}$$

(c) Matrix method given in (8.11a) yields the simplest and the most accurate procedure for the calculation of eigenvalues. The closed-form solution of the matrix equation in (a) is

$$\bar{\lambda}_k = r\sqrt{2 - 2\cos\frac{k\pi}{r}}$$

For $r = 10$, $\quad \bar{\lambda}_1 = 3.128\,87, \bar{\lambda}_2 = 6.180\,34, \bar{\lambda}_3 = 9.079\,81, \ldots$

$r = 20$, $\quad \bar{\lambda}_1 = 3.138\,36, \bar{\lambda}_2 = 6.257\,38, \bar{\lambda}_3 = 9.337\,81, \ldots$

$r = 30$, $\quad \bar{\lambda}_1 = 3.140\,16, \bar{\lambda}_2 = 6.271\,71, \bar{\lambda}_3 = 9.386\,07, \ldots$

$r = 50$, $\quad \bar{\lambda}_1 = 3.141\,08, \bar{\lambda}_2 = 6.279\,05, \bar{\lambda}_3 = 9.410\,83, \ldots$

$r = 100$, $\quad \bar{\lambda}_1 = 3.141\,46, \bar{\lambda}_2 = 6.282\,15, \bar{\lambda}_3 = 9.421\,29, \ldots$

$r = 1000$, $\quad \bar{\lambda}_1 = 3.141\,59, \bar{\lambda}_2 = 6.283\,18, \bar{\lambda}_3 = 9.424\,74, \ldots$

The accuracy of these results depends only on the computer capability in the evaluation of the cosine function for very small arguments.

(a) Standard characteristic determinant of the fourth-order differential equation in $[0, L]$,

$$\boxed{\frac{d^4Y}{dx^4} + \lambda^2\frac{d^2Y}{dx^2} = 0}$$

with boundary values

$$\boxed{Y(0) = 0 \qquad Y'(0) = 0 \qquad Y''(L) = 0 \qquad Y'''(L) + \lambda^2 Y'(L) = 0}$$

is constructed, with $r = 4$, $h = L/4$, $R = h^2\lambda^2$, from $(15.37b)$

$$Y_0 = 0 \qquad -Y_{-1} + Y_1 = 0 \qquad Y_3 - 2Y_4 + Y_5 = 0$$

$$-Y_2 + 2Y_3 - 2Y_5 + Y_6 + R(-Y_3 + Y_5) = 0$$

and from the relation $(15.37a)$ as

$$\begin{vmatrix} (7-2R) & -(4-R) & 1 & & & \\ -(4-R) & (6-R) & -(4-R) & & & \\ 1 & -(4-R) & (6-2R) & -(4-R) & & \\ & 1 & -(4-R) & (6-2R) & -(4-R) & 1 \\ & -1 & (2-R) & 0 & -(2-R) & 1 \\ & 1 & -2 & 1 & 0 \end{vmatrix} = 0$$

The first three roots of this determinant equation are the eigenvalues

$$\lambda_1 = \frac{1.092}{L} \qquad \lambda_2 = \frac{4.087}{L} \qquad \lambda_3 = \frac{6.648}{L}$$

The exact eigenvalue equation is

$$\cos \lambda L = 0$$

and the exact eigenvalues are

$$\lambda_1 = \frac{\pi}{2L} = \frac{1.571}{L} \qquad \lambda_2 = \frac{3\pi}{L} = \frac{4.712}{L} \qquad \lambda_3 = \frac{5\pi}{L} = \frac{7.854}{L}$$

(b) Stüssi's determinant of the same equation by the same endpoint equations and by the relation $(15.37c)$ with $r = 4$, $h = L/4$, and $\phi = h^2\lambda^2/12$ is

$$\begin{vmatrix} (7-19\phi) & -(4-8\phi) & (1-\phi) & & & \\ -(4-8\phi) & (6-18\phi) & -(4-8\phi) & (1-\phi) & & \\ (1-\phi) & -(4-8\phi) & (6-18\phi) & -(4-8\phi) & (1-\phi) & \\ & (1-\phi) & -(4-8\phi) & (6-18\phi) & -(4-8\phi) & 1-\phi \\ & 1 & (2-12\phi) & 0 & -(2-12\phi) & 1 \\ & 1 & -2 & 1 & 0 \end{vmatrix} = 0$$

The first three roots of this determinant equation are the eigenvalues

$$\lambda_1 = \frac{1.568}{L} \qquad \lambda_2 = \frac{4.669}{L} \qquad \lambda_3 = \frac{7.635}{L}$$

which shows again the superiority of Stüssi's method.

(a) Numerical methods for well-posed problems must satisfy the conditions of consistency, stability, and convergence.

(1) *Consistency.* If in the limit $h \to 0$ the substitute function and the true solution are equivalent, then the numerical solution is said to be consistent.

(2) *Stability.* If the errors at one stage of calculations remain finite all the time (do not cause larger errors as the computation continues), then the numerical method is said to be stable.

(3) *Convergence.* If the results of the numerical method approach the true values as h approaches zero, then the numerical method is said to be convergent.

(b) One-step methods such as Taylor's series and Runge–Kutta methods do not show numerical instability if the step sizes of integration are small. The Taylor's series method is the simplest method, and it can be easily adjusted to the differential equation and to the desired accuracy. The only difficulty arising in the application is the differentiation of the given differential equation. The higher derivatives can be conveniently calculated by the extensive tables given in Ref. 1.20. Runge–Kutta formulas are the most widely used procedures for the numerical solution of ordinary differential equations. Their primary advantages are their easy programming, good stability, and step-size changes without complications. Their disadvantages are the significantly longer required computer time and the lack of dependable error estimation.

(c) Multistep methods such as Adams–Bashforth–Moulton method and Hamming's method are strongly stable for a sufficiently small h, and their great advantage is their requirement of one derivative calculation. They can be easily adjusted to the differential equation and to the desired accuracy.

(d) Finite-difference methods provide a simple procedure for the solution of boundary-value problems, and if combined with the extrapolation procedure, they offer results of good accuracy. An extensive evaluation of finite-difference solutions of ordinary differential equations can be found in Ref. 15.03.

16
PARTIAL
DIFFERENTIAL
EQUATIONS

(a) Definitions. A partial differential equation of order n is a *functional equation* containing at least one partial derivative of the unknown function

$$\Phi = \Phi(x_1, x_2, \ldots, x_m)$$

of two or more variables $x_1, x_2, \ldots, x_m$. The order of a partial differential equation is the order of the highest derivative of Φ.

(b) Solution of a partial differential equation of order n given in general form as

$$F(x_1, x_2, \ldots, x_m, \Phi, \frac{\partial \Phi}{\partial x_1}, \frac{\partial \Phi}{\partial x_2}, \ldots, \frac{\partial \Phi}{\partial x_m}, \frac{\partial^2 \Phi}{\partial x_1^2}, \frac{\partial^2 \Phi}{\partial x_2^2}, \ldots, \frac{\partial^2 \Phi}{\partial x_m^2}, \ldots) = 0$$

is any function which satisfies the equation identically. The *general solution* is a solution which contains a number of arbitrary functions equal to the order of the equation. A *particular solution* (particular integral) is a special case of the general solution with specific functions substituted for the arbitrary functions. These specific functions are generated by the given conditions (initial conditions, boundary conditions). Many partial differential equations admit *singular solutions* (singular integrals) unrelated to the general solution.

(c) Linear partial differential equations and systems of such equations appear in the description of engineering and science problems. In this chapter, the classical and numerical solutions of the second-order partial differential equations (which dominate the applications) are presented. The *general linear partial differential equation of order two* in two independent variables x and y is of the form

$$A\Phi_{xx} + B\Phi_{xy} + C\Phi_{yy} + D\Phi_x + E\Phi_y + F\Phi = G$$

where $A, B, \ldots, G$ may depend on x, y but not on Φ, and Φ_{xx}, Φ_{xy}, Φ_{yy}, Φ_x, Φ_y are the partial derivatives of $\Phi = \Phi(x, y)$. This equation is homogeneous if $G = 0$, otherwise inhomogeneous. The solution of this equation represents a surface, called the integral surface. If on this surface exist curves across which Φ_{xx}, Φ_{xy}, Φ_{yy} are discontinuous or indeterminate, then these curves are called the *characteristics* of the respective equations.

(d) Classification. Because of the nature of their solution, the partial differential equations in (*c*) are designated as:

(1) *Elliptic equations* if $B^2 - 4AC < 0$

(2) *Hyperbolic equations* if $B^2 - 4AC > 0$

(3) *Parabolic equations* if $B^2 - 4AC = 0$

Elliptic equations are of a steady-state type associated with conditions which require that Φ or its normal derivatives, or both, take on the closed boundary some specific values and the closed boundary prescribes the region of application of Φ. Parabolic and hyperbolic equations are of the non-steady-state type (propagation type), and their solutions are allowed to propagate in the open domain at least in one variable.

(a) Types of solutions. There are many methods by which the partial differential equations can be solved. The most commonly used methods are:

(1) Separation-of-variables method   (4) Finite-difference method

(2) Laplace transform method   (5) Finite-element method

(3) Complex-variable method

The concepts of the first, second, and fourth method are introduced below.

(b) Separation-of-variables method. Although this method is not universally applicable, it offers a convenient and simple solution to many important equations in engineering and physical sciences. The basic idea is to assume the solution in the form of the product of two or several functions, each of which contains one variable only. The insertion of this product in the given differential equation separates this equation into two or several ordinary differential equations. The product of the solution of these equations is then the solution of the partial differential equation. Particular applications of this method are given in (16.03) to (16.12).

(c) Laplace transform method. The basic idea of this method is to obtain the Laplace transform of the respective partial differential equation and of the associated end conditions with respect to one variable. By this transformation, the partial differential equation is reduced to an ordinary differential equation whose solution is found by taking its Laplace transform inverse. The following relations are used in this method. If

$$\mathscr{L}\{F(x, t)\} = \int_0^\infty e^{-st}F(x, t)\, dt = f(x, s)$$

is the definition of the Laplace transform of $F(x, t)$, then the transforms of the derivatives of $F(x, t)$ are

$$\mathscr{L}\left\{\frac{\partial^n F(x, t)}{\partial t^n}\right\} = s^n f(x, s) - s^{n-1}F(x, +0) - s^{n-2}\frac{\partial F(x, +0)}{\partial t} - \cdots - \frac{\partial^{n-1}F(x, +0)}{\partial t^{n-1}}$$

$$\mathscr{L}\left\{\frac{\partial F(x, t)}{\partial x}\right\} = \frac{\partial f(x, s)}{\partial x} \qquad\qquad \mathscr{L}\left\{\frac{\partial^n F(x, s)}{\partial x^n}\right\} = \frac{\partial^n f(x, s)}{\partial x^n}$$

$$\mathscr{L}\left\{\frac{\partial^2 F(x, t)}{\partial x\, \partial y}\right\} = s\frac{\partial f(x, s)}{\partial x} - \frac{\partial F(x, +0)}{\partial x}$$

Applications of this method are shown in (16.13) to (16.15).

(d) Finite-difference method (as in the case of ordinary differential equations) replaces the region of the partial differential equation by a specific grid and approximates the governing equation and the end condition equations by the difference equations, in terms of the unknown values Φ_j, which are related to the nodal points of the selected grid. This substitution leads to a set of algebraic equations, and the solution of these equations yields the numerical approximations of Φ_j. Finite-difference molecules and their applications are tabulated in (16.17) to (16.31).

(a) Notation

A, B, C = integration constants α, β = separation constants

x, y = cartesian coordinates r, θ = polar coordinates

$U = U(x, y)$ = cartesian solution $V = V(r, \theta)$ = polar solution

(b) Cartesian Laplace's equation in two-dimensions,

$$\frac{\partial^2 U}{\partial x^2} + \frac{\partial^2 U}{\partial y^2} = 0$$

admits a solution $U = X(x)Y(x) = XY$, which when inserted in this equation yields

$$\frac{1}{X}\frac{d^2 X}{dx^2} = -\frac{1}{Y}\frac{d^2 Y}{dy^2}$$

Since each side of this equation is a function of one variable only, each side must be equal to the same constant, denoted as α^2 (which can be real or complex).

(c) Two ordinary differential equations resulting from this separation are

$$\frac{d^2 X}{dx^2} + \alpha^2 X = 0 \qquad\qquad \frac{d^2 Y}{dy^2} - \alpha^2 Y = 0$$

and their solutions are $X = A_1 e^{i\alpha x} + A_2 e^{-i\alpha x}$ $Y = B_1 e^{\alpha y} + B_2 e^{-\alpha y}$

(d) General solution is then

$$U = XY = \begin{cases} (A_1 e^{i\alpha x} + A_2 e^{-i\alpha x})(B_1 e^{\alpha y} + B_2 e^{-\alpha y}) & (\alpha \neq 0) \\ \\ (A_1 + A_2 x)(B_1 + B_2 y) & (\alpha = 0) \end{cases}$$

(e) Polar Laplace's equation in two-dimensions,

$$\frac{\partial^2 V}{\partial r^2} + \frac{1}{r}\frac{\partial V}{\partial r} + \frac{1}{r^2}\frac{\partial^2 V}{\partial \theta^2} = 0$$

admits a solution $V = R(r)Q(\theta) = RQ$, which when inserted in this equation yields after separation two ordinary differential equations

$$\frac{d^2 R}{dr^2} + \frac{1}{r}\frac{dR}{dr} - \frac{\beta^2 R}{r^2} = 0 \qquad\qquad \frac{d^2 Q}{d\theta^2} + \beta^2 Q = 0$$

and their solutions are $R = A_1 r^\beta + A_2 r^{-\beta}$ $Q = B_1 e^{i\beta\theta} + B_2 e^{-i\beta\theta}$

(f) General solution is then

$$V = RQ = \begin{cases} (A_1 r^\beta + A_2 r^{-\beta})(B_1 e^{i\beta\theta} + B_2 e^{-i\beta\theta}) & (\beta \neq 0) \\ \\ (A_1 + A_2 \ln r)(B_1 + B_2 \theta) & (\beta = 0) \end{cases}$$

(g) Constants of integration A_1, A_2 and B_1, B_2 are to be determined from the given conditions as shown in (16.04c).

(a) Temperature distribution $U(x,y)$ in a rectangular slab of sides a, b (shown in the adjacent figure) is governed by the Laplace's equation

$$\frac{\partial^2 U}{\partial x^2} + \frac{\partial^2 U}{\partial y^2} = 0$$

and the edge conditions are given as

$$U(0,y) = f_1(y),\ U(a,y) = f_2(y),\ U(x,b) = f_3(x),\ U(x,0) = f_4(x)$$

(b) Separation-of-variables solutions are, by (16.03c),

$$X(x) = A_1 e^{i\alpha x} + A_2 e^{-i\alpha x} = \bar{A}_1 \cos \alpha x \ + \bar{A}_2 \sin \alpha x$$

$$Y(y) = B_1 e^{\alpha y} + B_2 e^{-\alpha y} = \bar{B}_1 \cosh \alpha y + \bar{B}_2 \sinh \alpha y$$

or in terms of phase arguments x_0, y_0,

$$X(x) = \bar{C} \sin \alpha(x + x_0) \qquad Y(y) = \bar{D} \sinh \alpha(y - y_0)$$

(c) Particular edge conditions. If $f_1(y) = f_2(y) = f_4(x) = 0$ and $f_3(x) = f(x)$, then

(1) $U(0,y) = 0 = X(0)Y(y)$ (3) $U(x,0) = f(x) = X(x)Y(0)$

(2) $U(a,y) = 0 = X(a)Y(y)$ (4) $U(x,b) = 0 \quad = X(x)Y(b)$

From (1), $Y(y) \neq 0$, $X(0) = 0 = \bar{A}_1$ and from (2), $Y(y) \neq 0$, $X(a) = 0 = \bar{A}_2 \sin \alpha a$, where with $\bar{A}_2 \neq 0$, $\sin \alpha a = 0$, which implies

$$\alpha_k = k\pi/a \qquad (k = 1, 2, 3, \ldots)$$

and yields $X_k(x) = \bar{A}_{2,k} \sin \alpha_k x$.

From (4), $X(x) \neq 0$, $Y(b) = 0 = \bar{D}_k \sinh \alpha_k(b - y_0)$, which requires $y_0 = b$. Thus from (1), (2), and (4), with $\alpha_k = \pi/a, 2\pi/a, 3\pi/a, \ldots$,

$$U(x,y) = \sum_{k=1}^{\infty} \bar{A}_{2,k} \sin \alpha_k x \bar{D}_k \sinh \alpha_k(y - b)$$

$$= \sum_{k=1}^{\infty} \bar{E}_k \sin \alpha_k x \sinh \alpha_k(y - b)$$

which must satisfy (3) as

$$U(x,0) = \sum_{k=1}^{\infty} \bar{E}_k \sin \alpha_k x \sinh \alpha_k(-b) = f(x)$$

where $\bar{E}_k \sinh \alpha_k(-b)$ is the sine Fourier coefficient of $f(x)$, given in (13.02) as

$$\bar{E}_k = -\frac{2}{a \sinh \alpha_k b} \int_0^a f(x) \sin \alpha_k x\, dx$$

(d) Final solution for the particular case defined in (c) with $\bar{E}_k$ given above is

$$U(x,y) = \sum_{k=1}^{\infty} \bar{E}_k \sin \alpha_k x \sinh \alpha_k(y - b)$$

(a) Notation

A, B, C = integration constants t = time c = signed number

x, y = cartesian coordinates $\alpha, \beta, \lambda, \omega$ = separation constants

$U = U(x, y)$ = cartesian solution $V = V(x, y, t)$ = cartesian solution in two-dimensions
in one-dimension

(b) Cartesian wave equation in one dimension,

$$\frac{\partial^2 U}{\partial x^2} - \frac{1}{c^2}\frac{\partial^2 U}{\partial t^2} = 0$$

admits a solution $U = X(x)T(t) = XT$, which when inserted in this equation yields

$$\frac{1}{X}\frac{d^2 X}{dx^2} = \frac{1}{c^2 T}\frac{d^2 T}{dt^2}$$

As in (16.03b), each side must be equal to the same constant, denoted as α^2 (which again can be real or complex).

(c) Two ordinary differential equations resulting from this separation are

$$\frac{d^2 X}{dx^2} + \alpha^2 X = 0 \qquad\qquad \frac{d^2 T}{dt^2} + (\alpha c)^2 T = 0$$

and their solutions are $X = A_1 e^{i\alpha x} + A_2 e^{-i\alpha x}$ $Y = B_1 e^{i\omega t} + B_2 e^{-i\omega t}$ $(\omega = \alpha c)$

(d) General solution in one dimension is then

$$U = XT = \begin{cases} (A_1 e^{i\alpha x} + A_2 e^{-i\alpha x})(B_1 e^{i\omega t} + B_2 e^{-i\omega t}) & (\alpha \neq 0) \\ A_1 + A_2 x & (\alpha = 0) \end{cases}$$

(e) Cartesian wave equation in two dimensions

$$\frac{\partial^2 V}{\partial x^2} + \frac{\partial^2 V}{\partial y^2} - \frac{1}{c^2}\frac{\partial^2 V}{\partial t^2} = 0$$

admits a solution $V = X(x)Y(y)T(t) = XYT$, which when inserted in this equation yields after separation three ordinary differential equations

$$\frac{d^2 X}{dx^2} + \alpha^2 X = 0 \qquad\qquad \frac{d^2 Y}{dy^2} + \beta^2 Y = 0 \qquad\qquad \frac{d^2 T}{dt^2} + \omega^2 T = 0$$

and, with $\omega^2 = c^2(\alpha^2 + \beta^2)$, their solutions are

$$X = A_1 e^{i\alpha x} + A_2 e^{-i\alpha x} \qquad Y = B_1 e^{i\beta y} + B_2 e^{-i\beta y} \qquad T = C_1 e^{i\omega t} + C_2 e^{-i\omega t}$$

(f) General solution in two-dimensions is then

$$V = XYT = \begin{cases} (A_1 e^{i\alpha x} + A_2 e^{-i\alpha x})(B_1 e^{i\beta y} + B_2 e^{-i\beta y})(C_1 e^{i\omega t} + C_2 e^{-i\omega t}) & (\alpha \neq 0, \beta \neq 0) \\ (A_1 + A_2 x)(B_1 e^{i\lambda y} + B_2 e^{-i\lambda y})(C_1 e^{i\omega t} + C_2 e^{-i\omega t}) & (\alpha = 0, \beta = \lambda) \end{cases}$$

where $A_1, A_2, \ldots$ have the same meaning as in (16.03g).

(a) Vibration of tightly stretched string of length $a(m)$ fixed at both
ends and subjected to a uniform tension S (shown in the adjacent
figure) is governed by the displacement equation

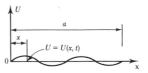

$$\frac{\partial^2 U}{\partial x^2} - \frac{1}{c^2}\frac{\partial^2 U}{\partial t^2} = 0$$

and the end conditions

$$U(0, t) = 0, \, U(a, t) = 0, \, dU(x, 0)/dt = 0, \, U(x, 0) = f(x)$$

where $U = U(x, t) =$ displacement (m), $S =$ tension (N), $g =$ gravitational acceleration (m/s²),
$w =$ weight of string per unit length (N/m), $f(x) =$ shape of string at $t = 0$ (m), and $c^2 = Sg/w$.

(b) Separation-of-variables solutions by $(16.05c)$ are

$$X(x) = A_1 e^{i\alpha x} + A_2 e^{-i\alpha x} = \bar{A}_1 \cos \alpha x + \bar{A}_2 \sin \alpha x$$

$$T(t) = B_1 e^{i\omega t} + B_2 e^{-i\omega t} = \bar{B}_1 \cos \omega t + \bar{B}_2 \sin \omega t \qquad (\omega = \alpha c)$$

(c) Particular conditions

(1)　$U(0, t) = 0 = X(0) T(t)$　　　　　　(3)　$dU(x, 0)/dt = 0 = X(x) \, dT(0)/dt$

(2)　$U(a, t) = 0 = X(a) T(t)$　　　　　　(4)　$U(x, 0) = f(x) = X(x) T(0)$

From (1), $T(t) \neq 0$, $X(0) = 0 = \bar{A}_1$ and from (2), $T(t) \neq 0$, $X(a) = 0 = \bar{A}_2 \sin \alpha a$, where with
$\bar{A}_2 \neq 0$, $\sin \alpha a = 0$, which implies

$$\alpha_k = k\pi/a \qquad (k = 1, 2, 3, \ldots)$$

and yields $X_k(x) = \bar{A}_{2,k} \sin \alpha_k x$.
　　From (3), $X(x) \neq 0$, $dT(0)/dt = 0$, $\bar{B}_2 = 0$ and yields

$$T_k(t) = \bar{B}_{1,k} \cos \omega_k t \qquad\qquad \omega_k = \alpha_k c = ck\pi/a$$

Then

$$U(x, t) = \sum_{k=1}^{\infty} X_k(x) T_k(t) = \sum_{k=1}^{\infty} \bar{A}_{2,k} \sin \alpha_k x \bar{B}_{1,1} \cos \omega_k t$$

$$= \sum_{k=1}^{\infty} \bar{C}_k \sin \alpha_k x \cos \omega_k t$$

Finally, from (4),

$$U(x, 0) = \sum_{k=1}^{\infty} \bar{C}_k \sin \alpha_k x = f(x)$$

where $\bar{C}_k$ is the sine Fourier coefficient of $f(x)$, given in (13.02) as

$$\bar{C}_k = \frac{2}{a} \int_0^a f(x) \sin \alpha_k x \, dx$$

(d) Final solution for the particular case defined in (c) with C_k given above is

$$U(x, t) = \sum_{k=1}^{\infty} \bar{C}_k \sin \alpha_k x \cos \omega_k t$$

(a) Notation

A, B, C = integration constants $\qquad$ t = time $\qquad$ c = signed number

x, y = cartesian coordinates $\qquad$ $\alpha, \beta, \lambda, \omega$ = separation constants

$U = U(x, t)$ = cartesian solution $\qquad$ $V = V(x, y, t)$ = cartesian solution in two-dimensions
  in one-dimension

(b) Cartesian diffusion equation in one-dimension,

$$\frac{\partial^2 U}{\partial x^2} - \frac{1}{c^2}\frac{\partial U}{\partial t} = 0$$

admits a solution $U = X(x)T(t) = XT$, which when inserted in this equation yields

$$\frac{1}{X}\frac{d^2 X}{dx^2} = \frac{1}{c^2 T}\frac{dT}{dt}$$

As in (16.04b), each side must be equal to the same constant, designated as α^2 (which again can be real or complex).

(c) Two ordinary differential equations resulting from this separation are

$$\frac{d^2 X}{dx^2} + \alpha^2 X = 0 \qquad\qquad \frac{dT}{dt} + (\alpha c)^2 T = 0$$

and their solutions are $\quad X = A_1 e^{i\alpha x} + A_2 e^{-i\alpha x} \qquad T = B_1 e^{-\omega t} \qquad (\omega = \alpha^2 c^2)$

(d) General solution in one-dimension is then

$$U = XT = \begin{cases} (A_1 e^{i\alpha x} + A_2 e^{-i\alpha x})B_1 e^{-\omega t} & (\alpha \neq 0) \\ (A_1 + A_2 x)B_1 e^{-\omega t} & (\alpha = 0) \end{cases}$$

(e) Cartesian diffusion equation in two-dimensions,

$$\frac{\partial^2 V}{\partial x^2} + \frac{\partial^2 V}{\partial y^2} - \frac{1}{c^2}\frac{\partial V}{\partial t} = 0$$

admits a solution $V = X(x)Y(y)T(t) = XYT$, which when inserted in this equation yields after separation three ordinary differential equations

$\dfrac{d^2 X}{dx^2} + \alpha^2 X = 0$	$\dfrac{d^2 Y}{y^2} + \beta^2 Y = 0$	$\dfrac{dT}{dt} + (\alpha c)^2 T = 0$

and, with $\omega = c^2(\alpha^2 + \beta^2)$, their solutions are

$$X = A_1 e^{i\alpha x} + A_2 e^{-i\alpha x} \qquad Y = B_1 e^{-\beta y} + B_2 e^{-i\beta y} \qquad T = C_1 e^{-\omega t}$$

(f) General solution in two-dimensions is then

$$V = XYT = \begin{cases} (A_1 e^{i\alpha x} + A_2 e^{-i\alpha x})(B_1 e^{i\beta y} + B_2 e^{-i\beta y})C_1 e^{-\omega t} & (\alpha \neq 0, \beta \neq 0) \\ (A_1 + A_2 x)(B_1 e^{i\lambda y} + B_2 e^{-i\lambda y})C_1 e^{-\omega t} & (\alpha = 0, \beta = \lambda) \end{cases}$$

where $A_1, A_2, \ldots$ have the same meaning as in (16.03g).

(a) Variation of voltage in submarine cable of length a (m) with grounded ends and with initial voltage distribution $f(x)$ is governed by

$$\frac{\partial^2 V}{\partial x^2} - RC\frac{\partial V}{\partial t} = 0$$

and the conditions

$$V(0, t) = 0 \qquad\qquad V(a, t) = 0 \qquad\qquad V(x, 0) = f(x)$$

where R = resistance (Ω), C = capacitance (F), and $V = V(x, t)$ = voltage (V).

(b) Separation-of-variables solutions by (16.07c) are

$$X(x) = A_1 e^{i\alpha x} + A_2 e^{-i\alpha x} = \bar{A}_1 \cos \alpha x + \bar{A}_2 \sin \alpha x$$

$$T(t) = B_1 e^{-\omega t} \qquad (\omega = \alpha^2/(RC))$$

(c) Particular conditions

(1) $V(0, t) = 0 = X(0)T(t)$ (3) $V(x, 0) = f(x) = X(x)T(0)$

(2) $V(a, t) = 0 = X(a)T(t)$

From (1), $T(t) \neq 0$, $X(0) = 0 = \bar{A}_1$ and from (2), $T(t) \neq 0$, $X(a) = 0 = \bar{A}_2 \sin \alpha a$, where with $\bar{A}_2 \neq 0$, $\sin \alpha a = 0$, which implies

$$\alpha_k = k\pi/a \qquad (k = 1, 2, 3, \ldots)$$

and yields, $X_k(x) = \bar{A}_{2,k} \sin \alpha_k x$. The sum of all solutions is then

$$V(x, t) = \sum_{k=1}^{\infty} \bar{A}_{2,k} \sin \alpha_k x B_{1,k} e^{-\omega_k t} = \sum_{k=1}^{\infty} \bar{C}_k e^{-\omega_k t} \sin \alpha_k x$$

Finally, from (3),

$$V(x, 0) = \sum_{k=1}^{\infty} \bar{C}_k \sin \alpha_k x = f(x)$$

where C_k is the sine Fourier series coefficient of $f(x)$ given in (13.02) as

$$\bar{C}_k = \frac{2}{a} \int_0^a f(x) \sin \alpha_k x \, dx$$

(d) Final solution for the particular case defined in (c) is, with $\bar{C}_k$ defined above,

$$V(x, t) = \sum_{k=1}^{\infty} \bar{C}_k e^{-\omega_k t} \sin \alpha_k x$$

(e) Special case for $f(x) = V_0$ yields with

$$\bar{C}_{2k} = 0 \qquad\qquad \bar{C}_{2k-1} = \frac{4V_0}{(2k - 1)\pi}$$

$$V(x, t) = \frac{4}{\pi} V_0 \sum_{k=1}^{\infty} \frac{1}{2k - 1} e^{-\lambda_k t} \sin \beta_k x$$

where $\beta_k = (2k - 1)\pi/a$ and $\lambda_k = \beta_k^2/RC$.

(a) Notation

E = modulus of elasticity (Pa)

x, y = cartesian coordinates

$V = V(x, t)$ = deflection (m)

t = time

I = moment of inertia (m^4)

m = mass per unit length (kg/m)

ω = angular frequency (1/s)

(b) Governing differential equation of free, transverse vibration without damping of a straight beam of constant section and length a (m).

$$EI\frac{\partial^4 V}{\partial x^4} + m\frac{\partial^2 V}{\partial t^2} = 0$$

admits a general solution $V = X(x)T(t) = XT$, which when inserted in this equation yields

$$\frac{EI}{m}\frac{d^4 X}{dx^4} = -\frac{1}{T}\frac{d^2 T}{dt^2}$$

As in (16.03b), each side must be equal to the same constant designated as ω^2.

(c) Two ordinary differential equations resulting from this separation are

$$\frac{d^4 X}{dx^4} - \lambda^4 X = 0 \qquad \frac{d^2 T}{dt^2} + \omega^2 T = 0$$

where $\lambda = \sqrt[4]{m\omega^2/EI}$.

(d) Laplace transform solutions of these ordinary differential equations are the *eigenfunction*

$$X = A_1 L_1(x) + A_2 L_2(x) + A_3 L_3(x) + A_4 L_4(x)$$

and the *time function* $T = B_1 K_1(t) + B_2 K_2(t)$

where

$L_1(x) = \frac{1}{2}(\cosh \lambda x + \cos \lambda x)$	$L_2(x) = \frac{1}{2}(\sinh \lambda x + \sin \lambda x)/\lambda$
$L_3(x) = \frac{1}{2}(\cosh \lambda x - \cos \lambda x)/\lambda^2$	$L_4(x) = \frac{1}{2}(\sinh \lambda x - \sin \lambda x)/\lambda$
$K_1(t) = \cos \omega t$	$K_2(t) = (\sin \omega t)/\omega$

(e) Integration constants in (d) are

$$A_1 = X(+0) \qquad A_2 = X'(+0) \qquad A_3 = X''(+0) \qquad A_4 = X'''(+0)$$

$$B_1 = \frac{m}{\Delta}\int_0^a V(x, 0)X(x)\, dx \qquad\qquad B_2 = \frac{m}{\omega\Delta}\int_0^a \dot{V}(x, 0)X(x)\, dx$$

where the primes indicate the derivatives with respect to x, the overdot the derivative with respect to t, and

$$\Delta = \int_0^a mX^2(x)\, dx$$

is called the *general mass*.

(f) General solution as shown in (16.10) on the opposite page is then a sum of products $X_k T_k$, each corresponding to a particular angular frequency ω_k.

(a) Transverse vibration without damping of a simple beam is governed by the differential equation given in (16.09*b*) and by the conditions

$$V(0, t) = 0 \qquad V(x, 0) = \frac{5wa^2x(a - x)}{96EI} = f(x) \qquad V(a, 0) = 0$$

$$V''(0, t) = 0 \qquad \dot{V}(x, 0) = 0 \qquad V''(a, t) = 0$$

which indicates that the initial deflection curve is due to the weight of the beam, specified by the unit length weight w (N/m), the initial velocity of all sections of the beam is zero, and the end deflections and end moments are zero during the motion.

(b) Separation-of-variables solutions are given in (16.09*d*) with the eigenvalue ω to be determined below.

(c) Particular conditions

(1) $V(0, t) = 0 = X(0)T(t)$ (2) $V''(0, t) = 0 = X''(0)T(t)$
(3) $V(a, t) = 0 = X(a)T(T)$ (4) $V''(a, t) = 0 = X''(a)T(t)$
(5) $\dot{V}(x, 0) = 0 = X(x)\dot{T}(0)$ (6) $V(x, 0) = f(x) = X(x)T(0)$

According to (16.09*e*) and by (1) to (4), $A_1 = A_3 = 0$, $A_2 = X'(0)$, $A_4 = X'''(0)$, and the transport matrix (15.03*e*) reduces for $x = a$ to

$$\begin{bmatrix} L_2(a) & L_4(a) \\ \lambda^4 L_4(a) & L_2(a) \end{bmatrix} \begin{bmatrix} X'(0) \\ X'''(0) \end{bmatrix} = \begin{bmatrix} 0 \\ 0 \end{bmatrix}$$

The determinant of this matrix equation yields the *frequency equation*

$$\sin \lambda a = 0 \qquad \lambda_k = k\pi/a \qquad (k = 1, 2, 3, \ldots)$$

and the *eigenfunction* becomes

$$X_k(x) = C_k \sin \lambda_k x$$

From (5), $\dot{T}(0) = 0 = B_2$, and with $\omega_k = EIk^2\pi^2/ma^2$, the *time function* becomes

$$T_k(t) = B_{1,k} \cos \omega_k t$$

The sum of all solutions is then

$$\boxed{V(x, t) = \sum_{k=1}^{\infty} C_k \sin \lambda_k x B_{1,k} \cos \omega_k t = \sum_{k=1}^{\infty} D_k \sin \lambda_k x \cos \omega_k t}$$

Finally, from (6),

$$V(x, 0) = \sum_{k=1}^{\infty} D_k \sin \lambda_k x = \frac{5wa^2x(a - x)}{96EI} = f(x)$$

where D_k is the sine Fourier series coefficient given in (13.02), which in this case by (13.03, cases 10, 11) reduces to

$$\boxed{D_k = \frac{5wa^4[1 - (-1)^k]}{24(k\pi)^3 EI} = d_k \frac{wa^4}{EI}}$$

so that $d_2 = d_4 = d_6 = \cdots = 0$ and the odd coefficients become

$$d_1 = 1.342\,81(10)^{-2} \qquad\qquad d_3 = 4.977\,09(10)^{-4} \qquad\qquad d_5 = 1.075\,05(10)^{-4}$$

$$d_7 = 3.917\,21(10)^{-5} \qquad\qquad d_9 = 1.843\,37(10)^{-5} \qquad\qquad d_{11} = 1.009\,63(10)^{-5}$$

(a) Forcing function. The general case of the linear partial differential equation of second order introduced in (16.01c) admits a separation-of-variables solution if the forcing function G is one of the following types:

(1) $G = g(x)$ \
(2) $G = g(y)$

(3) $G = Y(y)g(x)$ \
(4) $G = X(x)g(y)$

and the differential equation has the form

$$M_2\Phi_{xx} + N_2\Phi_{yy} + M_1\Phi_x + N_1\Phi_y + (M_0 + N_0)\Phi = G$$

where $M_2, M_1, M_0, N_2, N_1, N_0$ are constants.

(b) Solution of the first kind with $G = g(x)$ assumes

$$\Phi_1 = X(x)Y(y) \qquad \Phi_2 = Z(x)$$

so that Φ_1 is the solution of the homogeneous equation and

$$\Phi = \Phi_1 + \Phi_2$$

is the solution of the nonhomogeneous equation. After separation,

$$M_2\frac{d^2X}{dx^2} + M_1\frac{dX}{dx} + (M_0 - \lambda^2)X = 0$$

$$N_2\frac{d^2Y}{dy^2} + N_1\frac{dY}{dy} + (N_0 + \lambda^2)Y = 0$$

$$M_2\frac{d^2Z}{dx^2} + M_1\frac{dZ}{dx} + (M_0 + N_0)Z = g(x)$$

where λ is the separation constant and X, Y, Z are the integrals of the respective differential equations.

(c) Solution of the second kind with $G = g(y)$ assumes

$$\Phi_1 = X(x)Y(y) \qquad \Phi_2 = Z(y)$$

so that Φ_1 is the solution of the homogeneous equation and

$$\Phi = \Phi_1 + \Phi_2$$

is the solution of the nonhomogeneous equation. After separation,

$$M_2\frac{d^2X}{dx^2} + M_1\frac{dX}{dx} + (M_0 - \lambda^2)X = 0$$

$$N_2\frac{d^2Y}{dy^2} + N_1\frac{dY}{dy} + (N_0 + \lambda^2)Y = 0$$

$$N_2\frac{d^2Z}{dy^2} + N_1\frac{dZ}{dy} + (M_0 + N_0)Z = g(y)$$

where λ and X, Y, Z have the same meaning as in (b).

(d) Solution of the third kind with $G = Y(y)g(x)$ assumes

$$\Phi = X(x)Y(y)$$

to be the complete solution and after separation

$$M_2 \frac{d^2X}{dx^2} + M_1 \frac{dX}{dx} + (M_0 - \lambda^2)X = g(x)$$

$$N_2 \frac{d^2Y}{dy^2} + N_1 \frac{dY}{dy} + (N_0 + \lambda^2)Y = 0$$

where λ and X, Y have the same meaning as in (*b*).

(e) Solution of the fourth kind with $G = X(x)g(y)$ assumes

$$\Phi = X(x)Y(y)$$

to be the complete solution and after separation

$$M_2 \frac{d^2X}{dx^2} + M_1 \frac{dX}{dx} + (M_0 - \lambda^2)X = 0$$

$$N_2 \frac{d^2Y}{dy^2} + N_2 \frac{dY}{dy} + (N_0 + \lambda^2)Y = g(y)$$

where λ and X, Y have the same meaning as in (*b*).

(f) Example. The forced vibration of the elastic string in (16.06*a*) is governed by

$$\frac{\partial^2 U}{\partial x^2} - \frac{1}{c^2}\frac{\partial^2 U}{\partial t^2} = -bx \qquad (b = \text{constant})$$

If $U(0, t) = 0$, $U(a, t) = 0$, $U(x, 0) = 0$, $\dot{U}(x, 0) = 0$, and with

$$U = X(x)T(t) + Z(x)$$

the separation yields with $\omega = \alpha c$,

$$\frac{d^2X}{dx^2} + \alpha^2 X = 0 \qquad \frac{d^2T}{dt^2} + \omega^2 T = 0 \qquad \frac{d^2Z}{dx^2} = -bx$$

The solution of the first two equations is given in (16.06*c*), and the integral of the third equation is

$$Z = -bx^3/6$$

The complete solution is then with $\lambda_k = k\pi/a$ and $\omega_k = ck\pi/a$,

$$U(x, t) = \sum_{k=1}^{\infty} (\bar{A}_{1,k} \cos \lambda_k x + \bar{A}_{2,k} \sin \lambda_k x)(\bar{B}_{1,k} \cos \omega_k t + \bar{B}_{2,k} \sin \omega_k t) - \frac{bx^3}{6}$$

where $\bar{A}_{1,k} = 0$ $\qquad \bar{A}_{2,k}\bar{B}_{1,k} = \dfrac{ba^3}{6 \sin \lambda_k a}$ $\qquad \bar{A}_{2,k}\bar{B}_{2,k} = \dfrac{ba^3 \omega_k}{6 \sin \lambda_k a}$

or in simpler form,

$$U(x, t) = \frac{ba^3}{6} \sum_{k=1}^{\infty} \left[\frac{\sin \lambda_k x}{\sin \lambda_k a}\left(\cos \omega_k t + \frac{\sin \omega_k t}{\omega} \right) \right] - \frac{bx^3}{6}$$

(a) Concept and basic relations of LTM are given in (16.02c), and their applications in the solution of partial differential equations are introduced in (16.13) to (16.15). A table of Laplace transform pairs used in these solutions is shown below.

(b) Table of Laplace transform pairs

a, c, x = constants > 0 $\qquad \alpha = k\pi/a$	$\beta = (2k-1)\pi/(2a)$ $\qquad \eta = (-1)^k$
$u(x, s)$	$U(x, t)$
(1) $\dfrac{\sinh (sx/c)}{s \sinh (sa/c)}$	$\dfrac{x}{a} + \dfrac{2}{a}\sum\limits_{k=1}^{\infty} \dfrac{\eta}{\alpha} \sin \alpha x \cos \alpha ct$
(2) $\dfrac{\sinh (sx/c)}{s^2 \sinh (sa/c)}$	$\dfrac{xt}{a} + \dfrac{2}{ac}\sum\limits_{k=1}^{\infty} \dfrac{\eta}{\alpha^2} \sin \alpha x \cos \alpha ct$
(3) $\dfrac{\sinh (sx/c)}{s \cosh (sa/c)}$	$\dfrac{2}{a}\sum\limits_{k=1}^{\infty} \dfrac{\eta}{\beta} \sin \beta x \sin \beta ct$
(4) $\dfrac{\sinh (sx/c)}{s^2 \cosh (sa/c)}$	$\dfrac{x}{c} + \dfrac{2}{ac}\sum\limits_{k=1}^{\infty} \dfrac{\eta}{\beta^2} \sin \beta x \cos \beta ct$
(5) $\dfrac{\cosh (sx/c)}{s \sinh (sa/c)}$	$\dfrac{ct}{a} + \dfrac{2}{a}\sum\limits_{k=1}^{\infty} \dfrac{\eta}{\alpha} \sin \alpha x \sin \alpha ct$
(6) $\dfrac{\cosh (sx/c)}{s^2 \sinh (sa/c)}$	$\dfrac{ct^2}{2a} + \dfrac{2}{ac}\sum\limits_{k=1}^{\infty} \dfrac{\eta}{\alpha^2} \cos \alpha x \,(1 - \cos \alpha ct)$
(7) $\dfrac{\cosh (sx/c)}{s \cosh (sa/c)}$	$1 + \dfrac{2}{a}\sum\limits_{k=1}^{\infty} \dfrac{\eta}{\beta} \cos \beta x \cos \beta ct$
(8) $\dfrac{\cosh (sx/c)}{s^2 \cosh (sa/c)}$	$t + \dfrac{2}{ac}\sum\limits_{k=1}^{\infty} \dfrac{\eta}{\beta^2} \cos \beta x \sin \beta ct$
(9) $\dfrac{\sinh (x\sqrt{s}/c)}{\sinh (a\sqrt{s}/c)}$	$\dfrac{2c^2}{a}\sum\limits_{k=1}^{\infty} \eta\alpha \exp (-\alpha^2 c^2 t) \sin \alpha x$
(10) $\dfrac{\sinh (x\sqrt{s}/c)}{\sqrt{s} \cosh (a\sqrt{s}/c)}$	$\dfrac{2c}{a}\sum\limits_{k=1}^{\infty} \eta \exp (-\beta^2 c^2 t) \sin \beta x$
(11) $\dfrac{\sinh (x\sqrt{s}/c)}{s \sinh (a\sqrt{s}/c)}$	$\dfrac{x}{a} + \dfrac{2}{a}\sum\limits_{k=1}^{\infty} \dfrac{\eta}{\alpha} \exp (-\beta^2 c^2 t) \sin \beta x$
(12) $\dfrac{\sinh (x\sqrt{s}/c)}{s^2 \sinh (a\sqrt{s}/c)}$	$\dfrac{xt}{a} + \dfrac{2}{ac^2}\sum\limits_{k=1}^{\infty} \dfrac{\eta}{\alpha^3} [1 - \exp (-\alpha^2 c^2 t] \sin \alpha x$
(13) $\dfrac{\cosh (x\sqrt{s}/c)}{\cosh (a\sqrt{s}/c)}$	$\dfrac{2c^2}{a}\sum\limits_{k=1}^{\infty} \eta\beta \exp (-\beta^2 c^2 t) \cos \beta x$
(14) $\dfrac{\cosh (x\sqrt{s}/c)}{\sqrt{s} \sinh (a\sqrt{s}/c)}$	$\dfrac{c}{a} + \dfrac{2c}{a}\sum\limits_{k=1}^{\infty} \eta \exp (-\alpha^2 c^2 t) \cos \alpha x$
(15) $\dfrac{\cosh (x\sqrt{s}/c)}{s \cosh (a\sqrt{s}/c)}$	$1 + \dfrac{2}{a}\sum\limits_{k=1}^{\infty} \dfrac{\eta}{\beta} \exp (-\beta^2 c^2 t) \cos \beta x$
(16) $\dfrac{\cosh (x\sqrt{s}/c)}{s^2 \cosh (a\sqrt{s}/c)}$	$\dfrac{x^2 - a^2}{2c^2} + t - \dfrac{2}{ac^2}\sum\limits_{k=1}^{\infty} \dfrac{\eta}{\beta^3} \exp (-\beta^2 c^2 t) \cos \beta x$

(a) Notation (16.05a)

$U(x, t)$ = displacement

$U(0, t)$ = left-end displacement

a, c, U_0 = positive constants

$U(x, 0)$ = initial displacement ($t = 0$)

$U(a, t)$ = right-end displacement

(b) Cartesian wave equation (16.05b) in one dimension,

$$\frac{\partial^2 U(x, t)}{\partial x^2} - \frac{1}{c^2}\frac{\partial^2 U(x, t)}{\partial t^2} = 0 \qquad (0 \le x \le a \qquad t \ge 0)$$

with boundary conditions

(1) $U(0, t) = 0$

(2) $U(a, t) = 0$

and initial conditions

(3) $U(x, 0) = U_0 \sin (\pi x/a)$

(4) $dU(x, 0)/dx = 0$

is transformed with respect to t by (16.03c) as

$$\mathcal{L}\left\{\frac{\partial^2 U(x, t)}{\partial x^2}\right\} = \frac{1}{c^2}\mathcal{L}\left\{\frac{\partial^2 U(x, t)}{\partial t^2}\right\}$$

and becomes $\dfrac{d^2 u(x, s)}{dx^2} = \dfrac{1}{c^2}[s^2 u(x, s) - sU(x, 0) - U'(x, 0)]$

where $u(x, s) = \displaystyle\int_0^\infty e^{-st} U(x, t)\, dt$

is the transform of $U(x, t)$ and by (3), (4) above, $U(x, 0) = U_0 \sin (\pi x/a)$, $U'(x, 0) = 0$.

(c) Transformed equation is

$$\frac{d^2 u(x, s)}{dx^2} - \frac{s^2}{c^2} u(x, s) = -U_0 \frac{s^2}{c^2} \sin \frac{\pi x}{a}$$

and its general solution (Ref. 1.20, p. 180) is

$$u(x, s) = A \cosh \frac{sx}{c} + B \sinh \frac{sx}{c} + U_0 \frac{s}{c^2}\frac{\sin (\pi x/a)}{s^2/c^2 + \pi^2/s^2}$$

(d) Integration constants are from (1) and (2) in (b), $A = 0$, $B = 0$, and the final form of the integral of the transformed differential equation is

$$u(x, s) = U_0 \frac{s \sin (\pi x/a)}{s^2 + c^2\pi^2/a^2}$$

(e) Inverse Laplace transform of this equation taken from (14.01e-8) is the solution of the wave equation in (b).

$$U(x, t) = U_0 \sin (\pi x/a) \cos (c\pi t/a)$$

(f) Alternative solution. There is always the possibility to take the Laplace transform of the given differential equation with respect to x rather than t as done above. This choice would, however, lead to certain difficulties as would become apparent upon the completion of this alternative approach. In practice, it may be desirable to take the transform with respect to each variable and compare which of these variables offers the best simplification.

(a) Notation $(16.07a)$

$U(x, t)$ = temperature

$U(0, t)$ = left end temperature

a, c, U_0 = positive constants

$U(x, 0)$ = initial temperature $(t = 0)$

$U(a, t)$ = right end temperature

(b) Cartesian heat equation $(16.07b)$ in one-dimension,

$$\frac{\partial^2 U(x, t)}{\partial x^2} - \frac{1}{c^2} \frac{\partial U(x, t)}{\partial t} = 0 \qquad (0 \le x \le a \qquad t \ge 0)$$

with boundary conditions

(1) $U(0, t) = 0$

and initial value

(2) $U(a, t) = U_0$

(3) $U(x, 0) = 0$

is transformed with respect to t by $(16.03c)$ as

$$\mathcal{L}\left\{\frac{\partial^2 U(x, t)}{\partial x^2}\right\} = \frac{1}{c^2} \mathcal{L}\left\{\frac{\partial U(x, t)}{\partial t}\right\}$$

and becomes

$$\frac{d^2 u(x, s)}{dx^2} = \frac{1}{c^2}[su(x, s) - U(x, 0)]$$

where $u(x, s) = \displaystyle\int_0^\infty e^{-st} U(x, t)\, dt$

is the transform of $U(x, t)$ and by (3), $U(x, 0) = 0$.

(c) Transformed equation is

$$\frac{d^2 u(x, s)}{dx^2} - \frac{s}{c^2} u(x, s) = 0$$

and its general solution (Ref. 1.20, p. 180) is

$$u(x, s) = A \cosh\left(x\sqrt{s}/c\right) + B \sinh\left(x\sqrt{s}/c\right)$$

(d) Integration constants are from (1) and (2) in (b),

$$A = 0 \qquad B = U_0 \frac{1}{s \sinh\left(a\sqrt{s}/c\right)}$$

and the final form of the integral of the transformed differential equation is

$$u(x, s) = U_0 \frac{\sinh\left(x\sqrt{s}/c\right)}{s \sinh\left(a\sqrt{s}/c\right)}$$

(e) Inverse Laplace transform of this equation taken from $(16.12b\text{-}11)$ is the solution of the heat equation in (b).

$$U(x, t) = U_0\left[\frac{x}{a} + \frac{2}{a}\sum_{k=1}^{\infty} \frac{\eta}{\alpha} \exp\left(-\alpha^2 c^2 t\right) \sin \alpha x\right]$$

where $\alpha = k\pi/a$ and $\eta = (-1)^k$.

(a) Notation (16.05a) a, c, E, P = positive constants

$U(x, t)$ = displacement $U(x, 0)$ = initial displacement ($t = 0$)

$U(0, t)$ = left end displacement $U(a, t)$ = right end displacement

(b) Vibration equation (16.05b) in one dimension,

$$\boxed{\frac{\partial^2 U(x, t)}{\partial x^2} - \frac{1}{c^2}\frac{\partial^2 U(x, t)}{\partial t^2} = 0} \qquad (0 \le x \le a \qquad t \ge 0)$$

with boundary conditions,

(1) $U(0, t) = 0$ (2) $U(a, t) = P/E$

and initial conditions,

(3) $U(x, 0) = 0$ (4) $dU(x, 0)/dx = 0$

is transformed with respect to t by (16.03c) as

$$\mathcal{L}\left\{\frac{\partial^2 U(x, t)}{\partial x^2}\right\} = \frac{1}{c^2}\mathcal{L}\left\{\frac{\partial^2 U(x, t)}{\partial t^2}\right\}$$

and becomes $\dfrac{d^2 u(x, s)}{dx^2} = \dfrac{1}{c^2}[s^2 u(x, s) - sU(x, 0) - U'(x, 0)]$

where $u(x, s) = \displaystyle\int_0^\infty e^{-st}U(x, t)\, dt$

is the transform of $U(x, t)$ and by (3) and (4), $U(x, 0) = 0$, $U'(x, 0) = 0$.

(c) Transformed equation is

$$\boxed{\frac{d^2 u(x, s)}{dx^2} - \frac{s^2}{c^2}u(x, s) = 0}$$

and its general solution (Ref. 1.20, p. 180) is

$$u(x, s) = A \cosh (xs/c) + B \sinh (xs/c)$$

(d) Integration constants from (1) and (2) in (b) are

$$A = 0 \qquad B = \frac{cP}{Es^2 \cosh (as/c)}$$

and the final form of the integral of the transformed differential equation is

$$u(x, s) = \frac{P}{E}\frac{c \sinh (xs/c)}{s^2 \cosh (as/c)}$$

(e) Inverse Laplace transform of this equation taken from (16.12b-4) is the solution of the vibration equation in (b).

$$\boxed{U(x, t) = \frac{P}{E}\left[x + \frac{2}{a}\sum_{k=1}^{\infty}\frac{\eta}{\beta^2}\sin \beta x \cos c\beta t\right]}$$

where $\beta = (2k - 1)/2a$ and $\eta = (-1)^k$.

(a) Solution of a cartesian second-order partial differential equation in two dimensions (16.01c) by the finite-difference method requires first the replacement of the region of integration by a rectangular grid of n horizontal lines and m vertical lines whose points of intersection are the *nodal points* (pivotal points, joints) shown in the adjacent figure, where each point i, j is given by

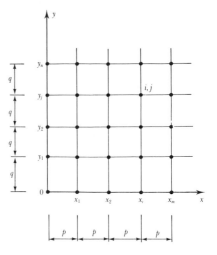

$$x_i = ip \qquad (i = 0, 1, \ldots, m)$$

$$y_j = jq \qquad (j = 0, 1, \ldots, n)$$

and p, q are the spacings of the grid lines.

(b) Partial derivatives of the unknown function

$$\Phi = \Phi(x, y)$$

which is the solution of the partial differential equation in (16.01c), are next expressed by the finite-difference approximations (3.02d) related to a particular point. If the point C shown in the adjacent graph is that point, then the partial derivatives of Φ in difference form become

$$\frac{\partial \Phi}{\partial x} = \Phi_x \cong \frac{\Phi_R - \Phi_L}{2p}$$

$$\frac{\partial \Phi}{\partial y} = \Phi_y \cong \frac{\Phi_T - \Phi_B}{2q}$$

$$\frac{\partial^2 \Phi}{\partial x^2} = \Phi_{xx} \cong \frac{\Phi_L - 2\Phi_C + \Phi_R}{p^2}$$

$$\frac{\partial^2 \Phi}{\partial x\,\partial y} = \Phi_{xy} \cong \frac{\Phi_{LB} - \Phi_{LT} - \Phi_{RB} + \Phi_{RT}}{4pq}$$

$$\frac{\partial^2}{\partial y^2} = \Phi_{yy} \cong \frac{\Phi_B - 2\Phi_C + \Phi_T}{q^2}$$

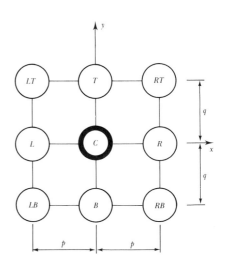

where the subscripts define the location of Φ as L = left point, C = central point, R = right point, B = bottom point, T = top point, LB = left-bottom point, RB = right-bottom point, LT = left-top point, and RT = right-top point. These equations apply to all points of the grid, including the points on its boundary.

(c) Forcing function of the partial differential equation (16.01c),

$$G = G(x, y)$$

at the same point C is taken as the joint value

$$G_C = G(x_C, y_C)$$

or is approximated more accurately by the equivalent joint value in terms of the respective formulas given below.

(d) Singular function P_z located at

$$x = x_C + (1 - a)p \qquad y = y_C + (1 - b)q$$

as shown in the adjacent figure is replaced by

$$G_C = \frac{ab}{pq} P_z$$

where a, b are the dimensionless coordinates of P_z in the rectangle C, R, RT, T.

(e) Singular function Q_y located at the same point as P_z in (d) as shown in the adjacent figure is replaced by

$$G_C = \frac{b}{pq^2} Q_y$$

where a, b are the same as in (d).

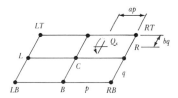

(f) Singular function Q_x located at the same point as P_z in (d) as shown in the adjacent figure is replaced by

$$G_C = \frac{a}{p^2 q} Q_x$$

where a, b are the same as in (d).

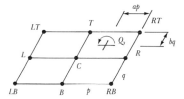

(g) Constant function G represented by the diagram of the adjacent figure is replaced by

$$G_C = G$$

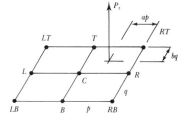

(h) Variable function $G(x, y)$ represented by the diagram of the adjacent figure is replaced by

$$G_C = \frac{1}{36} \begin{pmatrix} G_{LT} + & 4G_T + & G_{RT} \\ + \, 4G_L & + \, 16G_C + & 4G_R \\ + \, G_{LB} & + \, 4G_B + & G_{RB} \end{pmatrix}$$

where $G_{LT}, G_T, \ldots, G_B, G_{RB}$ are the values of G at the respective joints and the terms in braces represent an algebraic sum.

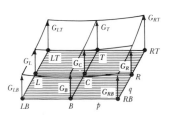

(a) Notation

$U = \Phi(x, y)$ = solution of differential equation in region R with boundary S

k = grid points L, T, C, B, R, depicted in (16.16b)

U_k = value of U at k $\qquad$ K_k = stiffness coefficient at k

A, C, D, E, F = given functions $\qquad$ A_C, C_C, D_C, E_C, F_C = particular functions at C

p, q = spacing of grid lines $\qquad$ $r = p/q$ = grid ratio

G = given forcing function $\qquad$ G_C = value of G at C

(b) General form of the cartesian elliptic equation in two dimensions (16.01d) is

$$\boxed{AU_{xx} + CU_{yy} + DU_x + EU_y + FU = G}$$

where U_{xx}, U_{yy}, U_x, U_y are the partial derivatives of U and $A > 0$, $C > 0$, $F \leq 0$ so that the elliptic condition is satisfied. As such, this equation states always a boundary-value problem, and its boundary conditions must be one of the following types.

(c) First boundary condition called the *Dirichlet condition* prescribes the value of U on the boundaries as

$$\boxed{U^* = \Phi(x_s, y_s)}$$

where x_s, y_s are the coordinates of the boundaries and U^* is continuous on S.

(d) Second boundary condition called the *Neumann condition* prescribes the value of the normal derivatives of U on the boundaries as

$$\boxed{U_n^* = \partial \Phi(x_s, y_s)/\partial n}$$

where x_s, y_s are the same as in (c) and U_n^* is continuous on S.

(e) Third boundary condition called the *mixed condition* prescribes the value

$$\boxed{U_n^* + \alpha U^* = \partial \Phi(x_s, y_s)/\partial n + \alpha(x_s, y_s)\Phi(x_s, y_s)}$$

where the symbols are the same as in (c) and (d) and $\alpha(x_s, y_s) > 0$.

(f) Governing difference equation at C (16.16b) in a molecule form

applies at all points of the grid. The molecule which forms the left side of the equation represents an algebraic sum of products of the stiffness coefficients $K_L, K_C, K_R, \ldots$ defined in (g) on the opposite page and of the respective values of $U_L, U_C, U_R, \ldots$ defined in (a). The right side of the equation includes the joint value G_C calculated by the formulas given in (16.16).

(g) Stiffness factors in (f) are

	$K_T = (r^2 C_C + \frac{1}{2} r p E_C)/p^2$	
$K_L = (A_C - \frac{1}{2} p D_C)/p^2$	$K_C = (-2A_C - 2r^2 C_C + p^2 F_C)/p^2$	$K_R = (A_C + \frac{1}{2} p D_C)/p^2$
	$K_B = (r^2 C_C - \frac{1}{2} r p E_C)/p^2$	

(h) Modified difference equation is required in cases where the grid lines and the boundary of the region R do not intersect at the grid joints. In such cases, depicted by the adjacent graph, the stiffness factors of the molecule (f) become

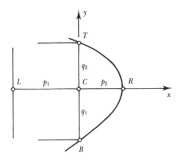

$$K_T^* = \frac{2C_C + q_1 E_C}{q_2(q_1 + q_2)}$$

$$K_L^* = \frac{2A_C - p_2 D_C}{p_1(p_1 + p_2)}$$

$$K_R^* = \frac{2A_c + p_1 D_C}{p_2(p_1 + p_2)}$$

$$K_C^* = F_C - \frac{1}{p_1 p_2}[2A_C + (p_1 - p_2)D_C] - \frac{1}{q_1 q_2}[2C_c + (q_1 - q_2)E_C]$$

$$K_B^* = \frac{2C_C - q_2 E_C}{q_1(q_1 + q_2)}$$

where p_1, q_2, q_1, q_2 are the new spacings shown in the figure.

(i) Modified forcing function adjusted to this new spacing is derived by (16.16*d*) to (16.16*g*), and for the variable function $G(x,y)$ the joint value becomes

$$G_C = \frac{\begin{pmatrix} + p_1 q_2 G_{LT} & + 2(p_1 + p_2)q_2 G_T & + p_2 q_2 G_{RT} \\ + 2(q_1 + q_2)p_1 G_L & + 4(p_1 + p_2)(q_1 + q_2)G_C & + 2(q_1 + q_2)p_2 G_R \\ + p_1 q_1 G_{LB} & + 2(p_1 + p_2)q_1 G_B & + p_2 q_1 G_{RB} \end{pmatrix}}{36(p_1 + p_2)(q_1 + q_2)}$$

where the numerator of the fraction is an algebraic sum in which $G_{LB}, G_G, G_{RB}, \ldots$ have the same meaning as their counterparts in (16.16*h*).

(j) System matrix equation consisting of the difference equations written for all joints of the grid (including the boundary points) involves as many unknown U_k as there are joints in the grid and yields the approximate values of these unknowns as shown in (16.18) and (16.19).

(a) Laplace's equation in cartesian system

$$\boxed{U_{xx} + U_{yy} = 0}$$

in the triangular region shown in the adjacent figure is subject of the Dirichlet conditions

$$U(x_s,y_s) = x_s - y_s,$$

where x_s,y_s are the coordinates of the boundary. The grid of the triangular region is formed by $p = q = 2$, and the values of U on the boundaries are

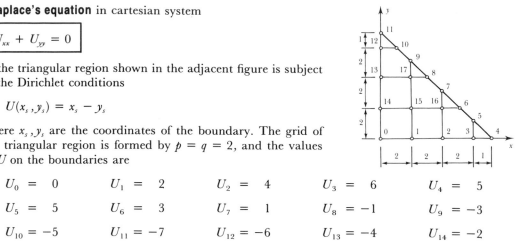

$U_0 = 0$	$U_1 = 2$	$U_2 = 4$	$U_3 = 6$	$U_4 = 5$
$U_5 = 5$	$U_6 = 3$	$U_7 = 1$	$U_8 = -1$	$U_9 = -3$
$U_{10} = -5$	$U_{11} = -7$	$U_{12} = -6$	$U_{13} = -4$	$U_{14} = -2$

(b) Intermediate point 15 is the only regular point. With $A = C = 1$, $D = E = F = G = 0$, $r = p/q = 1$ and $K_L = K_B = K_T = K_R = 1$, $K_C = -4$, the difference equation at 15 by (16.17f) is

$$
\begin{array}{|c|c|c|}
\hline
 & U_{17} & \\
\hline
-2 & -4U_{15} & U_{16} \\
\hline
 & 2 & \\
\hline
\end{array} = 0
$$

(c) Points 16,17 are near to the boundary with points 6 to 9 not being the points of the regular grid. Thus the difference equations at these points must be modified according to (16.17h). At 16, with $p_1 = 2$, $p_2 = 1$, $q_1 = 2$, $q_2 = 1$,

$$K_L^* = K_{15}^* = \tfrac{1}{3} \qquad K_R^* = K_6^* = \tfrac{2}{3} \qquad K_C^* = K_{16}^* = -2 \qquad K_B^* = K_2^* = \tfrac{1}{3} \qquad K_T^* = K_7^* = \tfrac{2}{3}$$

and at 17 with the same p's and q's,

$$K_L^* = K_{13}^* = \tfrac{1}{3} \qquad K_R^* = K_8^* = \tfrac{2}{3} \qquad K_C^* = K_{17}^* = -2 \qquad K_B^* = K_{15}^* = \tfrac{1}{3} \qquad K_T^* = K_9^* = \tfrac{2}{3}$$

which yields two additional difference equations

$$
\begin{array}{|c|c|c|}
\hline
 & \tfrac{2}{3} & \\
\hline
\tfrac{1}{3}U_{14} & -2U_{16} & 2 \\
\hline
 & \tfrac{4}{3} & \\
\hline
\end{array} = 0
\qquad
\begin{array}{|c|c|c|}
\hline
 & -2 & \\
\hline
-\tfrac{4}{3} & -2U_{17} & -\tfrac{2}{3} \\
\hline
 & \tfrac{1}{3}U_{15} & \\
\hline
\end{array} = 0
$$

(d) System matrix equation in terms of (*b*) and (*c*) is

$$
\begin{bmatrix} -4 & 1 & 1 \\ 1 & -6 & 0 \\ 1 & 0 & -6 \end{bmatrix}
\begin{bmatrix} U_{15} \\ U_{16} \\ U_{17} \end{bmatrix}
=
\begin{bmatrix} 0 \\ -12 \\ 12 \end{bmatrix}
$$

from which $U_{15} = 0$, $U_{16} = 2$, $U_{17} = -2$. In this particular case, these values are exact, since $U(x,y) = x - y$ is the exact solution.

(a) Poisson's equation

$$U_{xx} + U_{yy} = -1$$

in the rectangular region shown in the adjacent figure is
the subject of the Dirichlet condition

$$U(x_s, y_s) = 0$$

where x_s, y_s are coordinates of the boundaries. The grid
of the rectangular region is formed by $p = q = 1$, and
the values of G_C at all points equal -1.

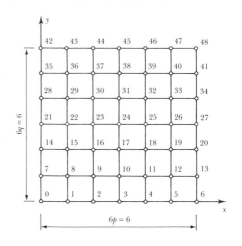

(b) Two-axial symmetry offers five additional conditions

$$U_8 = U_{12} = U_{36} = U_{40} \qquad U_{16} = U_{18} = U_{30} = U_{32}$$

$$U_9 = U_{11} = U_{37} = U_{39} = U_{15} = U_{19} = U_{29} = U_{33}$$

$$U_{10} = U_{22} = U_{26} = U_{38} \qquad U_{17} = U_{23} = U_{25} = U_{31}$$

and only six values of U remain to be calculated.

(c) Intermediate points 8 to 10, 16, 17, 24. With $A = C = 1$, $D = E = F = 0$, $G_C = 1$, $r = p/q = 1$ and $K_L, K_B, K_C, K_T, K_R = 1$, $K_C = -4$, the respective difference equations (16.17f) become

	U_9	
0	$-4U_8$	U_9
	0	

$= -1$

	U_{16}	
U_8	$-4U_9$	U_{10}
	0	

$= -1$

	U_{17}	
U_9	$-4U_{10}$	U_9
	0	

$= -1$

	U_{17}	
U_9	$-4U_{16}$	U_{17}
	U_9	

$= -1$

	U_{24}	
U_{16}	$-4U_{17}$	U_{16}
	U_{10}	

$= -1$

	U_{17}	
U_{17}	$-4U_{24}$	U_{17}
	U_{17}	

$= -1$

(d) System matrix equation in terms of (d) is

$$\begin{bmatrix} -4 & 2 & & & & \\ 1 & -4 & 1 & 1 & & \\ & 2 & -4 & & 1 & \\ & 2 & & -4 & 2 & \\ & & 1 & 2 & -4 & 1 \\ & & & & 4 & -4 \end{bmatrix} \begin{bmatrix} U_8 \\ U_9 \\ U_{10} \\ U_{16} \\ U_{17} \\ U_{24} \end{bmatrix} = \begin{bmatrix} -1 \\ -1 \\ -1 \\ -1 \\ -1 \\ -1 \end{bmatrix}$$

From which $U_8 = 0.9519$ $U_9 = 1.404$ $U_{10} = 1.538$

$U_{16} = 2.125$ $U_{17} = 2.346$ $U_{24} = 2.596$

For better results a finer grid must be used.

(a) **Notation.** $U = \Phi(x, y)$ = solution of differential equation in region R with boundary S. The symbols used below and on the opposite page are those defined in $(16.17a)$.

(b) **General form** of the cartesian hyperbolic equation in two dimensions $(16.01d)$ is

$$AU_{xx} - CU_{yy} + DU_x + EU_y + FU = G$$

where U_{xx}, U_{yy}, U_x, U_y are the partial derivatives of U. The solution of this equation is not restricted to a closed region and frequently is allowed to propagate in one or two directions with given boundary conditions.

(c) **Governing difference equation** at C in molecule form is the same as $(16.17f)$, but the stiffness coefficients must be changed to

$$K_T = (-r^2 C_C + \tfrac{1}{2}rpE_C)/p^2$$

$$K_L = (A_C - \tfrac{1}{2}pD_C)/p^2 \qquad K_C = (-2A_C - 2R^2 C_C + p^2 F_C)/p^2 \qquad K_R = (A_C + \tfrac{1}{2}pD_C)/p^2$$

$$K_B = (-r^2 C_C - \tfrac{1}{2}rpE_C)/p^2$$

where $r = p/q$.

(d) **Modified difference equation** required in cases when the grid lines and the boundary of the region R do not intersect at the grid joints shown in $(16.17h)$ has the same form as in (c) above, but the stiffness factors in the molecule $(16.17f)$ become

$$K_T^* = -\frac{2C_C + q_1 E_C}{q_2(q_1 + q_2)}$$

$$K_L^* = \frac{2A_C - p_2 D_C}{p_1(p_1 + p_2)} \qquad\qquad K_R^* = \frac{2A_c + p_1 D_C}{p_2(p_1 + p_2)}$$

$$K_C^* = F_C - \frac{1}{p_1 p_2}\left[2A_C + (p_1 - p_2)D_C - \frac{1}{q_1 q_2}*2C_c + (q_1 - q_2)E_C \right]$$

$$K_B^* = -\frac{2C_C - q_2 E_C}{q_1(q_1 + q_2)}$$

where p_1, p_2, q_1, q_2 are the new spacings shown in $(16.17h)$.

(e) **Modified load values** adjusted for the new spacing are formally the same as in $(16.17i)$. For G_C on the boundary the following substitutions must be made.

At $x = 0, y = y$: $\quad p_1 = 0$ $\qquad\qquad$ At $x = x, y = 0$: $\quad q_1 = 0$

At $x = mp, y = y$: $\quad p_2 = 0$ $\qquad\qquad$ At $x = x, y = nq$: $\quad q_2 = 0$

(f) **System matrix equation** is constructed and solved as shown in (16.21) on the opposite page. It is important to note that the solution of the system may be obtained to any degree of accuracy since the system matrix equation is a transport matrix equation. The stability of the solution requires K_L, K_R, K_B, K_T to be equal to or less than 1.

(a) Wave equation in cartesian system

$$U_{xx} - \frac{1}{c^2} U_{tt} = 0$$

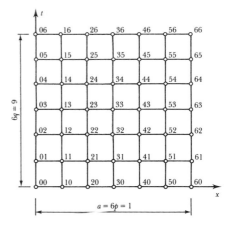

in the rectangular region shown in the adjacent figure $[0 \le x \le a, t \ge 0]$ is subject to the conditions

$$U(0, t) = U(a, t) = 0$$

$$U(x, 0) = \sin \pi x \qquad U_t(x, 0) = 0$$

where t = time (s), a = length (m), and c = constant (m^{-2}). The grid is formed by six vertical and horizontal lines with $p = \frac{1}{6}$ and $q = \frac{3}{2}$.

(b) Governing difference equation in molecule form is

	$-\lambda U_T$	
U_L	$-s U_C$	U_R
	$-\lambda U_B$	

$= 0$

where $\lambda = p^2/(cq)^2$ and $s = 2(1 - \lambda)$.

(c) Boundary conditions yield

$$U_{00} = U_{01} = \cdots = U_{05} = U_{06} = 0 \qquad U_{60} = U_{61} = \cdots = U_{65} = U_{66} = 0$$

$$U_{10} = U_{50} = \sin p\pi \qquad U_{20} = U_{40} = \sin 2p\pi \qquad U_{30} = \sin 3p\pi \qquad \cdots$$

$$U_{11} - U_{1,-1} = 0 \qquad U_{21} - U_{2,-1} = 0 \qquad U_{31} - U_{3,-1} = 0 \qquad \cdots$$

where $U_{1,-1}, \ldots, U_{5,-1}$ are ficticious points below the x axis on verticals $1, 2, \ldots$ at the distance q.

(d) System matrix equation. Since $U_{10}, U_{20}, \ldots, U_{50}$ are known and $U_{1,-1}, U_{2,-1}, \ldots, U_{5,-1}$ are defined by the last boundary condition, the values of U at $11, 21, \ldots, 51$ can be computed directly as

$$\begin{bmatrix} U_{11} \\ U_{21} \\ U_{31} \\ U_{41} \\ U_{51} \end{bmatrix} = \frac{1}{\lambda} \begin{bmatrix} -s & 1 & & & \\ 1 & -s & 1 & & \\ & 1 & -s & 1 & \\ & & 1 & -s & 1 \\ & & & 1 & -s \end{bmatrix} \begin{bmatrix} U_{10} \\ U_{20} \\ U_{30} \\ U_{40} \\ U_{50} \end{bmatrix} - \begin{bmatrix} U_{1,-1} \\ U_{2,-1} \\ U_{3,-1} \\ U_{4,-1} \\ U_{5,-1} \end{bmatrix}$$

or symbolically as $\mathbf{H}_1 = (1/\lambda)\mathbf{K}\mathbf{H}_0 - \mathbf{H}_{-1}$, where $\mathbf{H}_0$, $\mathbf{H}_1$, $\mathbf{H}_{-1}$ are the U values of the 0 row, 1 row, and -1 row, respectively, and $\mathbf{K}$ is the transport matrix. Recurrent use of this formula yields

$$\begin{bmatrix} \mathbf{H}_N \\ \mathbf{H}_{N+1} \end{bmatrix} = \begin{bmatrix} -\mathbf{I} & \mathbf{K} \\ -\mathbf{K} & \mathbf{K}^2 - \mathbf{I} \end{bmatrix}^{N/2} \begin{bmatrix} \mathbf{H}_0 \\ \mathbf{H}_1 \end{bmatrix}$$

where N is an even integer.

(a) **Notation.** $U = \Phi(x, y)$ = solution of differential equation in region R with boundary S. The symbols used below and on the opposite page are those defined in (16.17a).

(b) **General form** of the cartesian parabolic equation in two dimensions (16.01d) is

$$AU_{xx} + DU_x - EU_y + FU = G$$

where U_{xx}, U_x, U_y are the partial derivatives of U. The solution of this equation is not restricted to a closed region and frequently is allowed to propagate in one or two directions with boundary conditions given.

(c) **Governing difference equation** at C in molecule form may be one of the following types:

(1) *Forward difference form*

	$-prE_C U_T$		
$(A_C - \frac{1}{2}pD_C)U_L$	$-(2A_C - prE_C - p^2F_C)U_C$	$(A_C + \frac{1}{2}pD_C)U_R$	$= G_C p^2$

(2) *Backward difference form*

	$-(2A_C + prE_C - p^2F_C)U_C$	$(A_C + \frac{1}{2}pD_C)U_R$	$= G_C p^2$
$(A_C - \frac{1}{2}pD_C)U_L$	$prE_C U_B$		

(3) *Average difference form*

$\frac{1}{2}(A_C - \frac{1}{2}pD_C)U_{LT}$	$-(A_C + prE_C)U_T$	$\frac{1}{2}(A_C + \frac{1}{2}pD_C)U_{RT}$	
$\frac{1}{2}(A_C - \frac{1}{2}pD_C)U_L$	$-(A_C - pRE_C - p^2F_C)U_C$	$\frac{1}{2}(A_C + \frac{1}{2}pD_C)U_R$	$= G_C p^2$

(d) **Reduced equations** for the special case $A = 1$, $D = F = 0$, and $E = c^{-2}$ become in terms of

$$\alpha = c^2/pr \qquad \beta = pr/c^2 \qquad r = p/q$$

(1)	$U_T = \alpha U_L + (1 - 2\alpha)U_C + \alpha U_R - \alpha G_C$
(2)	$U_B = -\alpha U_L + (1 + 2\alpha)U_C - \alpha U_R + \alpha G_C$
(3)	$-\frac{1}{2}U_{LT} + (1 + \beta)U_T - \frac{1}{2}U_{RT} = \frac{1}{2}U_L + (\beta - 1)U_C + \frac{1}{2}U_R - \beta G_C q$

The construction of the system matrix equations based on these relations is shown in (16.23) on the opposite page.

(e) **Load values** in cases (c) and (d) are calculated by formulas given in (16.16).

(a) Diffusion equation in cartesian system

$$\boxed{U_{xx} = U_t}$$

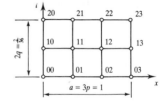

in the grid shown in the adjacent figure is subject to the conditions

$$U(0, t) = U(a, t) = 0$$

$$U(x, 0) = \sin \pi x$$

where t = time (s) and a = length (m). The grid is formed by four vertical and three horizontal lines with $p = \frac{1}{3}$ and $q = \frac{1}{36}$.

(b) System matrix equation of the first type (16.22d-1) with $\alpha = 0.25$ is

$$\begin{bmatrix} U_{11} \\ U_{12} \end{bmatrix} = \begin{bmatrix} 0.5 & 0.25 \\ 0.25 & 0.5 \end{bmatrix} \begin{bmatrix} U_{01} \\ U_{02} \end{bmatrix} = \begin{bmatrix} 0.649\ 52 \\ 0.639\ 52 \end{bmatrix}$$

where $U_{01} = \sin \pi/3$ and $U_{02} = \sin 2\pi/3$. Similarly,

$$\begin{bmatrix} U_{21} \\ U_{22} \end{bmatrix} = \begin{bmatrix} 0.5 & 0.25 \\ 0.25 & 0.5 \end{bmatrix} \begin{bmatrix} U_{11} \\ U_{12} \end{bmatrix} = \begin{bmatrix} 0.487\ 14 \\ 0.487\ 14 \end{bmatrix}$$

where U_{11}, U_{12} are the values calculated by the preceding matrix equation.

(c) System matrix equation of the second type (16.22d-2) with $\alpha = 4$ is

$$\begin{bmatrix} U_{01} \\ U_{02} \end{bmatrix} = \begin{bmatrix} 1.5 & -0.25 \\ -0.25 & 1.5 \end{bmatrix} \begin{bmatrix} U_{11} \\ U_{12} \end{bmatrix} \quad \text{or} \quad \begin{bmatrix} U_{11} \\ U_{12} \end{bmatrix} = \begin{bmatrix} 0.685\ 71 & 0.114\ 29 \\ 0.114\ 29 & 0.687\ 51 \end{bmatrix} \begin{bmatrix} U_{01} \\ U_{02} \end{bmatrix} = \begin{bmatrix} 0.692\ 82 \\ 0.692\ 82 \end{bmatrix}$$

where U_{01}, U_{02} are the same as in (b) above. Similarly,

$$\begin{bmatrix} U_{21} \\ U_{22} \end{bmatrix} = \begin{bmatrix} 0.685\ 71 & 0.114\ 29 \\ 0.114\ 29 & 0.685\ 71 \end{bmatrix} \begin{bmatrix} U_{11} \\ U_{12} \end{bmatrix} = \begin{bmatrix} 0.554\ 26 \\ 0.554\ 26 \end{bmatrix}$$

where U_{11}, U_{12} are the values calculated by the preceding matrix equation.

(d) System matrix equation of the third type (16.22d-3) with $\alpha = 0.25$, $\beta = 4$ is

$$\begin{bmatrix} 5 & -0.5 \\ -0.5 & 5 \end{bmatrix} \begin{bmatrix} U_{11} \\ U_{12} \end{bmatrix} = \begin{bmatrix} 3 & 0.5 \\ 0.5 & 3 \end{bmatrix} \begin{bmatrix} U_{10} \\ U_{20} \end{bmatrix}$$

from which by inversion

$$\begin{bmatrix} U_{11} \\ U_{12} \end{bmatrix} = \begin{bmatrix} 0.616\ 16 & 0.161\ 61 \\ 0.161\ 61 & 0.616\ 16 \end{bmatrix} \begin{bmatrix} U_{01} \\ U_{02} \end{bmatrix} = \begin{bmatrix} 0.673\ 34 \\ 0.673\ 34 \end{bmatrix}$$

where U_{01}, U_{02} are the same as in the preceding cases (b) and (c). By the same procedure $U_{21} = U_{22} = 0.523\ 70$.

(e) Exact solution of this equation is $U(x, t) = \exp(-\pi^2 t) \sin \pi x$ and $U_{11} = U_{12} = 0.658\ 36$, $U_{21} = U_{22} = 0.500\ 50$, which calls for a finer grid in (b) to (d).

(a) Approximate partial derivatives

(a) Approximate partial derivatives of $U = \Phi(x, y)$ at the point C of the adjacent grid can be found by expressing each central difference in terms of the values of U corresponding to the nodes of this grid. The equidistant spacings are the same as before, and only the central differences of the respective orders are considered in the approximations listed in (*b*) below, where the subscripts define the location of U. These relations apply to all nodes of the grid including the points on the boundary. Their utilization in the approximation of the biharmonic equation is shown on the opposite page in (16.25).

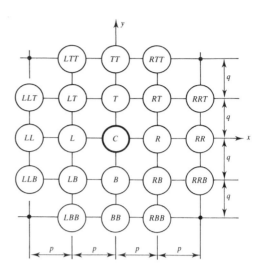

(b) Particular cases

$$\left[\frac{\partial U}{\partial x}\right]_C \cong \frac{U_R - U_L}{2p} \qquad\qquad \left[\frac{\partial U}{\partial x}\right]_C \cong \frac{U_T - U_B}{2q}$$

$$\left[\frac{\partial^2 U}{\partial x^2}\right]_C \cong \frac{U_R - 2U_C + U_L}{p^2} \qquad\qquad \left[\frac{\partial^2 U}{\partial y^2}\right]_C \cong \frac{U_T - 2U_C + U_B}{q^2}$$

$$\left[\frac{\partial^2 U}{\partial x\, \partial y}\right]_C \cong \frac{U_{RT} - U_{LT} - U_{RB} + U_{LB}}{4pq}$$

$$\left[\frac{\partial^3 U}{\partial x^3}\right]_C \cong \frac{U_{RR} - 2U_R + 2U_L - U_{LL}}{2p^3} \qquad\qquad \left[\frac{\partial^3 U}{\partial y^3}\right]_C \cong \frac{U_{TT} - 2U_T + 2U_B - U_{BB}}{2q^3}$$

$$\left[\frac{\partial^3 U}{\partial x^2\, \partial y}\right]_C \cong \frac{U_{RT} - 2U_T + U_{LT} - U_{RB} + 2U_B - U_{LB}}{2p^2 q}$$

$$\left[\frac{\partial^3 U}{\partial x\, \partial y^2}\right]_C \cong \frac{U_{RT} - 2U_R + U_{RB} - U_{LT} + 2U_L - U_{LB}}{2pq^2}$$

$$\left[\frac{\partial^4 U}{\partial x^4}\right]_C \cong \frac{U_{RR} - 4U_R + 6U_C - 4U_L + U_{LL}}{p^4} \qquad\qquad \left[\frac{\partial^4 U}{\partial y^4}\right]_C \cong \frac{U_{TT} - 4U_T + 6U_C - 4U_B + U_{BB}}{q^4}$$

$$\left[\frac{\partial^4 U}{\partial x^3\, \partial y}\right]_C \cong \frac{U_{RRT} - 2U_{RT} - U_{RRB} + 2U_{RB} - U_{LLT} + 2U_{LT} + U_{LLB} - U_{LB}}{4p^3 q}$$

$$\left[\frac{\partial^4 U}{\partial x^2\, \partial y^2}\right]_C \cong \frac{U_{LT} - 2U_T + U_{RT} - 2U_L + 4U_C - 2U_R + U_{LB} - 2U_B + U_{RB}}{p^2 q^2}$$

$$\left[\frac{\partial^4 U}{\partial x\, \partial y^3}\right]_C \cong \frac{U_{RTT} - 2U_{RT} + 2U_{RB} - U_{RBB} - U_{LTT} + 2U_{LT} - 2U_{LB} + U_{LBB}}{4pq^3}$$

(a) Notation. $U = \Phi(x, y)$ = solution of differential equation in region R

U_k = value of U at k K_k = stiffness factor at k

A, B, C, D, E, F, G = given functions $A_C, B_C, C_C, D_C, E_C, F_C$ = particular functions at C

$r = p/q$ = grid ratio parameter G_C = forcing function at C

(b) General form of the cartesian biharmonic equation in two dimensions is

$$AU_{xxxx} + BU_{xxyy} + CU_{yyyy} + DU_{xx} + EU_{yy} + FU = G$$

where $U_{xxxx}, U_{xxyy}, \ldots$ are the partial derivatives of U and $A, B, \ldots, G$ are functions in x, y or constants.

(c) Governing difference equation C in a molecule form

$$
\left\{
\begin{array}{ccccc}
 & & K_{TT}U_{TT} & & \\
 & K_{LT}U_{LT} & K_T U_T & K_{RT}U_{RT} & \\
K_{LL}U_{LL} & K_L U_L & K_C U_C & K_R U_R & K_{RR}U_{RR} \\
 & K_{LB}U_{LN} & K_B U_B & K_{RB}U_{RB} & \\
 & & K_{BB}U_B & &
\end{array}
\right\} = G_C
$$

applies at all points of the grid defined in (16.24*a*). The molecule which forms the left side of this equation represents an algebraic sum of products of the stiffness coefficients $K_{LL}, K_L, \ldots$ defined in (*d*) below and of the respective values of U. The right side is the forcing function at C.

(d) Stiffness factors in (*c*) are

	$K_{TT} = r^4 C_C/p_4$		
$K_{LT} = r^2 B_C/p^4$	$K_T = (-2r^2 B_C - 4r^4 C_C + r^2 E_C)/p^4$	$K_{RT} = r^2 B_C/p^4$
$K_{LL} = A_C/p^4$	$K_L = (-4A_C - 2r^2 B_C + D_C)/p^4 = K_R$	$K_{RR} = A_C/p^4$
$K_C = (6A_C + 4r^2 B_C + 6r^4 C_C - 2D_C - 2r^2 E_C + F)/p^4$		
$K_{LB} = r^2 B_C/p^4$	$K_B = (-2r^2 B_C - 4r^4 C_C + r^2 E_C)/p^4$	$K_{RB} = r^2 B_C/p^4$
$K_{BB} = r^4 C_C/p^4$		

(e) System matrix equation consisting of the difference equations written for all points of the grid involves exactly the same number of equations as there are unknown values of U. The solution of this matrix equation yields the approximate values of U.

(a) Solution of a polar second-order partial differential equation by the finite-difference method requires first the replacement of the region of integration by a polar grid of m radial lines and n concentric circles whose points of intersection are the *nodal points* (pivotal points, joints) shown in the adjacent figure, where each point i,j is given by

$$\theta_i = ip \qquad (i = 0, 1, \ldots, m)$$

$$r_j = jq \qquad (j = 0, 1, \ldots, n)$$

and p, q are the angular and linear spacings of the grid lines and circles, respectively.

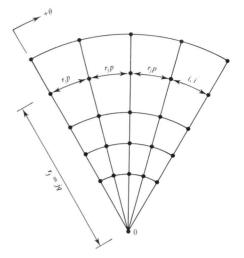

(b) Partial derivatives of the unknown function

$$\Phi = \Phi(r, \theta)$$

which is the solution of the partial differential equation, are next expressed by the finite-difference approximations (3.02d) related to the particular points. If the point C shown in the adjacent graph is that point, then the partial derivatives of Φ in difference form become

$$\frac{\partial \Phi}{\partial \theta} = \Phi_\theta \cong \frac{\Phi_R - \Phi_L}{2p}$$

$$\frac{\partial \Phi}{\partial r} = \Phi_r \cong \frac{\Phi_T - \Phi_B}{2q}$$

$$\frac{\partial^2 \Phi}{\partial \theta^2} = \Phi_{\theta\theta} \cong \frac{\Phi_L - 2\Phi_C + \Phi_R}{p^2}$$

$$\frac{\partial^2 \Phi}{\partial p\,\partial q} = \Phi_{\theta r} \cong \frac{\Phi_{LB} - \Phi_{LT} - \Phi_{RB} + \Phi_{RT}}{4pq}$$

$$\frac{\partial^2 \Phi}{\partial r^2} = \Phi_{rr} \cong \frac{\Phi_B - 2\Phi_C + \Phi_T}{q^2}$$

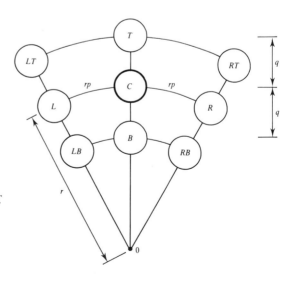

where r is the radius of circle LCR, $r + q$ is the radius of circle $LTTRT$, $r - q$ is the radius of circle $LBBRB$, and rp is the length of the circular arcs LC and CR. The subscripts of Φ have the same meaning as in (16.16b). These equations again apply to all points of the grid, including the points on the boundary.

(a) Notation. $U = \Phi(r, \theta)$ = solution of differential equation in region R. The symbols used below are those defined in $(16.17a)$ with modifications introduced on the opposite page. Also, r is the radius of the circle of the central point C.

(b) General form of Poisson's equation (16.19) in polar coordinates is

$$U_{rr} + \frac{1}{r^2}U_{\theta\theta} + \frac{1}{r}U_r = G$$

where U_{rr}, $U_{\theta\theta}$, U_r are the respective partial derivatives of U and G is the forcing function in r, θ.

(c) Governing difference equation at C in molecule form is formally identical to $(16.17f)$ and may be applied similarly. The stiffness factors used in its molecule are

	$K_T = \dfrac{1 + 1/2j}{q^2}$	
$K_L = \dfrac{(1/jp)^2}{q^2}$	$K_C = \dfrac{2[1 + (1/jp)^2]}{q^2}$	$K_R = \dfrac{(1/jp)^2}{q^2}$
	$K_B = \dfrac{1 - 1/2j}{q^2}$	

where j = number of the ring measured from the center of the grid. The joint value of the forcing function at C may be calculated by the relations given in (16.16).

(d) Polar grid introduced in $(16.26a)$ creates a special condition at the center 0, where the governing difference equation has the following special form (Refs. 16.08, 16.21):

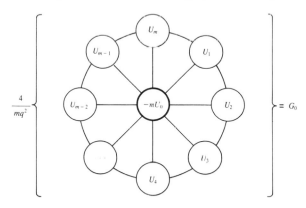

where m = number of radial lines and G_0 is the joint value of G at 0.

(e) Axisymmetrical case. If the forcing function and the boundary conditions are symmetrical with respect to the center 0 of the region, then the stiffness factors in (b) reduce to

$$K_L = K_R = 0 \qquad K_C = 2 \qquad K_B = K_T = \frac{1 + 1/2j}{q^2}$$

and the central difference equation at 0 becomes with $U_1 = U_2 = \cdots = U_m$,

$$\frac{4}{q^2}(U_m - U_0) = G_0$$

(a) Difference method for a well-posed problem must satisfy the conditions of consistency, stability, and convergence.

    (a) *Consistency.* If in the limit $p \to 0$, $q \to 0$ the difference and differential equation become equivalent, then the difference equation is said to be consistent with the differential equation.

    (2) *Stability.* If the errors at one stage of calculations remain finite all the time (do not cause larger errors as the computation continues), then the finite-difference method is said to be stable.

    (3) *Convergence.* If the results of the finite-difference method approach the true values as both p and q approach zero, then the finite-difference method is said to be convergent.

(b) Elliptic equation. Since the finite-difference solution of this equation (16.17) is based on the solution of finite linear system of equations, no stability problem is present. As almost exclusively some iterative method is used, only the convergence of this method is required. The local discretization error is of order $O(p^2 + q^2)$.

(c) Hyperbolic equation. Since the finite-difference solution of this equation (16.20) is based upon the transport chain, only the stability may become a problem. The procedure is stable if λ in (16.21b) is equal to or less than 1. The local discretization error is of order $O(p^2 + q^2)$.

(d) Parabolic equation. Solution of (16.22) by the forward difference method (explicit method) requires for convergence and stability that α given in (16.22d) be equal to or less than $\frac{1}{2}$, which implies small q and necessitates an enormous amount of numerical work. In turn the backward difference method (implicit method) is unconditionally stable and convergent. The only limitation is the size of p and q since the discretization error is of order $O(p^2 + q)$. The desire to eliminate small q's led to the development of the average difference method (Crank–Nicholson method), which is not only unconditionally stable and convergent but since $O(p^2 + q^2)$, it allows q to be of the same magnitude as p.

Appendix A
NUMERICAL TABLES

2	233	547	877	1229	1597	1993	2371	2749	3187	3581
3	239	557	881	1231	1601	1997	2377	2753	3191	3583
5	241	563	883	1237	1607	1999	2381	2767	3203	3593
7	251	569	887	1249	1609	2003	2383	2777	3209	3607
11	257	571	907	1259	1613	2011	2389	2789	3217	3613
13	263	577	911	1277	1619	2017	2393	2791	3221	3617
17	269	587	919	1279	1621	2027	2399	2797	3229	3623
19	271	593	929	1283	1627	2029	2411	2801	3251	3631
23	277	599	937	1289	1637	2039	2417	2803	3253	3637
29	281	601	941	1291	1657	2053	2423	2819	3257	3643
31	283	607	947	1297	1663	2063	2437	2833	3259	3659
37	293	613	953	1301	1667	2069	2441	2837	3271	3671
41	307	617	967	1303	1669	2081	2447	2843	3299	3673
43	311	619	971	1307	1693	2083	2459	2851	3301	3677
47	313	631	977	1319	1697	2087	2467	2857	3307	3691
53	317	641	983	1321	1699	2089	2473	2861	3313	3697
59	331	643	991	1327	1709	2099	2477	2879	3319	3701
61	337	647	997	1361	1721	2111	2503	2887	3323	3709
67	347	653	1009	1367	1723	2113	2521	2897	3329	3719
71	349	659	1013	1373	1733	2129	2531	2903	3331	3727
73	353	661	1019	1381	1741	2131	2539	2909	3343	3733
79	359	673	1021	1399	1747	2137	2543	2917	3347	3739
83	367	677	1031	1409	1753	2141	2549	2927	3359	3761
89	373	683	1033	1423	1759	2143	2551	2939	3361	3767
97	379	691	1039	1427	1777	2153	2557	2953	3371	3769
101	383	701	1049	1429	1783	2161	2579	2957	3373	3779
103	389	709	1051	1433	1787	2179	2591	2963	3389	3793
107	397	719	1061	1439	1789	2203	2593	2969	3391	3797
109	401	727	1063	1447	1801	2207	2609	2971	3407	3803
113	409	733	1069	1451	1811	2213	2617	2999	3413	3821
127	419	739	1087	1453	1823	2221	2621	3001	3433	3823
131	421	743	1091	1459	1831	2237	2633	3011	3449	3833
137	431	751	1093	1471	1847	2239	2647	3019	3457	3847
139	433	757	1097	1481	1861	2243	2657	3023	3461	3851
149	439	761	1103	1483	1867	2251	2659	3037	3463	3853
151	443	769	1109	1487	1871	2267	2663	3041	3467	3863
157	449	773	1117	1489	1873	2269	2671	3049	3469	3877
163	457	787	1123	1493	1877	2273	2677	3061	3491	3881
167	461	797	1129	1499	1879	2281	2683	3067	3499	3889
173	463	809	1151	1511	1889	2287	2687	3079	3511	3907
179	467	811	1153	1523	1901	2293	2689	3083	3517	3911
181	479	821	1163	1531	1907	2297	2693	3089	3527	3917
191	487	823	1171	1543	1913	2309	2699	3109	3529	3919
193	491	827	1181	1549	1931	2311	2707	3119	3533	3923
197	499	829	1187	1553	1933	2333	2711	3121	3539	3929
199	503	839	1193	1559	1949	2339	2713	3137	3541	3931
211	509	853	1201	1567	1951	2341	2719	3163	3547	3943
223	521	857	1213	1571	1973	2347	2729	3167	3557	3947
227	523	859	1217	1579	1979	2351	2731	3169	3559	3967
229	541	863	1223	1583	1987	2357	2741	3181	3571	3989

* Prime number is an integer divisible by 1 and itself only. There is an infinite number of prime numbers, but no general formula.

4001	4421	4861	5281	5701	6143	6577	7001	7507	7927	8389
4003	4423	4871	5297	5711	6151	6581	7013	7517	7933	8419
4007	4441	4877	5303	5717	6163	6599	7019	7523	7937	8423
4013	4447	4889	5309	5737	6173	6607	7027	7529	7949	8429
4019	4451	4903	5323	5741	6197	6619	7039	7537	7951	8431
4021	4457	4909	5333	5743	6199	6637	7043	7541	7963	8443
4027	4463	4919	5347	5749	6203	6653	7057	7547	7993	8447
4049	4481	4931	5351	5779	6211	6659	7069	7549	8009	8461
4051	4483	4933	5381	5783	6217	6661	7079	7559	8011	8467
4057	4493	4937	5387	5791	6221	6673	7103	7561	8017	8501
4073	4507	4943	5393	5801	6229	6679	7109	7573	8039	8513
4079	4513	4951	5399	5807	6247	6689	7121	7577	8053	8521
4091	4517	4957	5407	5813	6257	6691	7127	7583	8059	8527
4093	4519	4967	5413	5821	6263	6701	7129	7589	8069	8537
4099	4523	4969	5417	5827	6269	6703	7151	7591	8081	8539
4111	4547	4973	5419	5839	6271	6709	7159	7603	8087	8543
4127	4549	4987	5431	5843	6277	6719	7177	7607	8089	8563
4129	4561	4993	5437	5849	6287	6733	7187	7621	8093	8573
4133	4567	4999	5441	5851	6299	6737	7193	7639	8101	8581
4139	4583	5003	5443	5857	6301	6761	7207	7643	8111	8597
4153	4591	5009	5449	5861	6311	6763	7211	7649	8117	8599
4157	4597	5011	5471	5867	6317	6779	7213	7669	8123	8609
4159	4603	5021	5477	5869	6323	6781	7219	7673	8147	8623
4177	4621	5023	5479	5879	6329	6791	7229	7681	8161	8627
4201	4637	5039	5483	5881	6337	6793	7237	7687	8167	8629
4211	4639	5051	5501	5897	6343	6803	7243	7691	8171	8641
4217	4643	5059	5503	5903	6353	6823	7247	7699	8179	8647
4219	4649	5077	5507	5923	6359	6827	7253	7703	8191	8663
4229	4651	5081	5519	5927	6361	6829	7283	7717	8209	8669
4231	4657	5087	5521	5939	6367	6833	7297	7723	8219	8677
4241	4663	5099	5527	5953	6373	6841	7307	7727	8221	8681
4243	4673	5101	5531	5981	6379	6857	7309	7741	8231	8689
4253	4679	5107	5557	5987	6389	6863	7321	7753	8233	8693
4259	4691	5113	5563	6007	6397	6869	7331	7757	8237	8699
4261	4703	5119	5569	6011	6421	6871	7333	7759	8243	8707
4271	4721	5147	5573	6029	6427	6883	7349	7789	8263	8713
4273	4723	5153	5581	6037	6449	6899	7351	7793	8269	8719
4283	4729	5167	5591	6043	6451	6907	7369	7817	8273	8731
4289	4733	5171	5623	6047	6469	6911	7393	7823	8287	8737
4297	4751	5179	5639	6053	6473	6917	7411	7829	8291	8741
4327	4759	5189	5641	6067	6481	6947	7417	7841	8293	8747
4337	4783	5197	5647	6073	6491	6949	7433	7853	8297	8753
4339	4787	5209	5651	6079	6521	6959	7451	7867	8311	8761
4349	4789	5227	5653	6089	6529	6961	7457	7873	8317	8779
4357	4793	5231	5657	6091	6547	6967	7459	7877	8329	8783
4363	4799	5233	5659	6101	6551	6971	7477	7879	8353	8803
4373	4801	5237	5669	6113	6553	6977	7481	7883	8363	8807
4391	4813	5261	5683	6121	6563	6983	7487	7901	8369	8819
4397	4817	5273	5689	6131	6569	6991	7489	7907	8377	8821
4409	4831	5279	5693	6133	6571	6997	7499	7919	8387	8831

* For higher prime numbers, see Ref. 1.01, pp. 870–873.

k, m, n = positive integers				$\mathcal{P}_{m,n} = \sum_{k=1}^{n} k^m$ (5.11)

n \\ m	1	2	3	4	5
3	6	14	36	98	276
4	10	30	100	354	1 300
5	15	55	225	979	4 425
6	21	91	441	2 275	12 201
7	28	140	784	4 676	29 008
8	36	204	1 296	8 772	61 776
9	45	285	2 025	15 333	120 825
10	55	385	3 025	25 333	220 825
11	66	506	4 356	39 974	381 876
12	78	650	6 084	60 710	630 708
13	91	819	8 281	89 271	1 002 001
14	105	1 015	11 025	127 687	1 539 825
15	120	1 240	14 400	178 312	2 299 200
16	136	1 496	18 496	243 848	3 347 776
17	153	1 785	23 409	327 369	4 767 633
18	171	2 109	29 241	432 345	6 657 201
19	190	2 470	36 100	562 666	9 133 300
20	210	2 870	44 100	722 666	12 333 300
21	231	3 311	53 361	917 147	16 417 401
22	253	3 795	64 009	1 151 403	21 571 033
23	276	4 324	76 176	1 431 244	28 007 376
24	300	4 900	90 000	1 763 020	35 970 000
25	325	5 525	105 625	2 153 645	45 735 625
26	351	6 201	123 201	2 610 621	57 617 001
27	378	6 930	142 884	3 142 062	71 965 908
28	406	7 714	164 836	3 756 718	89 176 276
29	435	8 555	189 225	4 463 999	109 687 425
30	465	9 455	216 225	5 273 999	133 987 425
31	496	10 416	246 016	6 197 520	162 616 576
32	528	11 440	278 784	7 246 096	196 171 008
33	561	12 529	314 721	8 432 017	235 306 401
34	595	13 685	354 025	9 768 353	280 741 825
35	630	14 910	396 900	11 268 978	333 263 700
36	666	16 206	443 556	12 948 594	393 729 876
37	703	17 575	494 209	14 822 755	463 073 833
38	741	19 019	549 081	16 907 891	542 309 001
39	780	20 540	608 400	19 221 332	632 533 200
40	820	22 140	672 400	21 781 332	734 933 200
41	861	23 821	741 321	24 607 093	850 789 401
42	903	25 585	815 409	27 718 789	981 480 633
43	946	27 434	894 916	31 137 590	1 128 489 076
44	990	29 370	980 100	34 885 686	1 293 405 300
45	1035	31 395	1 071 225	38 986 311	1 477 933 425
46	1081	33 511	1 168 561	43 463 767	1 683 896 401
47	1128	35 720	1 272 384	48 343 448	1 913 241 408
48	1176	38 024	1 382 976	53 651 864	2 168 045 376
49	1225	40 425	1 500 625	59 416 665	2 450 520 625
50	1275	42 925	1 625 625	65 666 665	2 763 020 625

n \ m	1	2	3	4	5
51	1326	45 526	1 758 276	72 431 866	3 108 045 876
52	1378	48 230	1 898 884	79 743 482	3 488 249 908
53	1431	51 039	2 047 761	87 633 963	3 906 445 401
54	1485	53 955	2 205 225	96 137 019	4 365 610 425
55	1540	56 980	2 371 600	105 287 644	4 868 894 800
56	1596	60 116	2 547 216	115 122 140	5 419 626 576
57	1653	63 365	2 732 409	125 678 141	6 021 318 633
58	1711	66 729	2 927 521	136 994 637	6 677 675 401
59	1770	70 210	3 132 900	149 111 998	7 392 599 700
60	1830	73 810	3 348 900	162 071 998	8 170 199 700
61	1891	77 531	3 575 881	175 917 839	9 014 796 001
62	1953	81 375	3 814 209	190 694 175	9 930 928 833
63	2016	85 344	4 064 256	206 447 136	10 923 365 376
64	2080	89 440	4 326 400	223 224 352	11 997 107 200
65	2145	93 665	4 601 025	241 074 977	13 157 397 825
66	2211	98 021	4 888 521	260 049 713	14 409 730 401
67	2278	102 510	5 189 284	280 200 834	15 759 855 508
68	2346	107 134	5 503 716	301 582 210	17 213 789 076
69	2415	111 895	5 832 225	324 249 331	18 777 820 425
70	2485	116 795	6 175 225	348 259 331	20 458 520 425
71	2556	121 836	6 533 136	373 671 012	22 262 749 776
72	2628	127 020	6 906 384	400 544 868	24 197 667 408
73	2701	132 349	7 295 401	428 943 109	26 270 739 001
74	2775	137 825	7 700 625	458 929 685	28 489 745 625
75	2850	143 450	8 122 500	490 570 310	30 862 792 500
76	2926	149 226	8 561 476	523 932 486	33 398 317 876
77	3003	155 155	9 018 009	559 085 527	36 105 102 033
78	3081	161 239	9 492 561	596 100 583	38 992 276 401
79	3160	167 480	9 985 600	635 050 664	42 069 332 800
80	3240	173 880	10 497 600	676 010 664	45 346 132 800
81	3321	180 441	11 029 041	719 057 385	48 832 917 201
82	3403	187 165	11 580 409	764 269 561	52 540 315 633
83	3486	194 054	12 152 196	811 727 882	56 479 356 276
84	3570	201 110	12 744 900	861 515 018	60 661 475 700
85	3655	208 335	13 359 025	913 715 643	65 098 528 825
86	3741	215 731	13 995 081	968 416 459	69 802 799 001
87	3828	223 300	14 653 584	1 025 706 220	74 787 008 208
88	3916	231 044	15 335 056	1 085 675 756	80 064 327 376
89	4005	238 965	16 040 025	1 148 417 997	85 648 386 825
90	4095	247 065	16 769 025	1 214 027 997	91 553 286 825
91	4186	255 346	17 522 596	1 282 602 958	97 793 608 276
92	4278	263 810	18 301 284	1 354 242 254	104 384 423 508
93	4371	272 459	19 105 641	1 429 047 455	111 341 307 201
94	4465	281 295	19 936 225	1 507 122 351	118 680 347 425
95	4560	290 320	20 793 600	1 588 572 976	126 418 156 800
96	4656	299 536	21 678 336	1 673 507 632	134 571 883 776
97	4753	308 945	22 591 009	1 762 036 913	143 159 224 033
98	4851	318 549	23 532 201	1 854 273 729	152 198 432 001
99	4950	328 350	24 502 500	1 950 333 330	161 708 332 500
100	5050	338 350	25 502 500	2 050 333 330	171 708 332 500

| $n! = n$ factorial   (2.09a) | | | $n!! = n$ double factorial   (2.30d) | | |

n	$n!$	$n!!$	n	$n!$	$n!!$
1	1.000 000 000 (00)	1.000 000 000 (00)	51	1.551 118 753 (066)	2.980 227 914 (33)
2	2.000 000 000 (00)	2.000 000 000 (00)	52	8.065 817 517 (067)	2.706 443 182 (34)
3	6.000 000 000 (00)	3.000 000 000 (00)	53	4.274 883 284 (069)	1.579 520 794 (35)
4	2.400 000 000 (01)	8.000 000 000 (00)	54	2.308 436 973 (071)	1.461 479 318 (36)
5	1.200 000 000 (02)	1.500 000 000 (01)	55	1.269 640 335 (073)	8.687 364 368 (36)
6	7.200 000 000 (02)	4.800 000 000 (01)	56	7.109 985 878 (074)	8.184 284 181 (37)
7	5.040 000 000 (03)	1.050 000 000 (02)	57	4.052 691 950 (076)	4.951 797 690 (38)
8	4.032 000 000 (04)	3.840 000 000 (02)	58	2.350 561 331 (078)	4.746 884 825 (39)
9	3.628 800 000 (05)	9.450 000 000 (02)	59	1.386 831 185 (080)	2.921 560 637 (40)
10	3.628 800 000 (06)	3.840 000 000 (03)	60	8.320 987 112 (081)	2.848 130 895 (41)
11	3.991 680 000 (07)	1.039 500 000 (04)	61	5.075 802 139 (083)	1.782 151 989 (42)
12	4.790 016 000 (08)	4.608 000 000 (04)	62	3.146 997 326 (085)	1.765 841 155 (43)
13	6.227 020 800 (09)	1.351 350 000 (05)	63	1.982 608 315 (087)	1.122 755 753 (44)
14	8.717 829 120 (10)	6.451 200 000 (05)	64	1.268 869 322 (089)	1.130 138 339 (45)
15	1.307 674 368 (12)	2.027 025 000 (06)	65	8.247 650 562 (090)	7.297 912 393 (46)
16	2.092 278 989 (13)	1.032 192 000 (07)	66	5.443 449 391 (092)	7.458 913 039 (46)
17	3.556 874 281 (14)	3.445 942 500 (07)	67	3.647 111 092 (094)	4.889 601 304 (47)
18	6.402 373 706 (15)	1.857 945 600 (08)	68	2.480 035 542 (096)	5.072 060 866 (48)
19	1.216 451 004 (17)	6.547 290 750 (08)	69	1.711 224 524 (098)	3.373 824 899 (49)
20	2.432 902 008 (18)	3.715 891 200 (09)	70	1.197 857 167 (100)	3.550 442 606 (50)
21	5.109 094 217 (19)	1.374 931 058 (10)	71	8.504 785 855 (101)	2.395 415 679 (51)
22	1.124 000 728 (21)	8.174 960 640 (10)	72	6.123 445 837 (103)	2.556 318 677 (52)
23	2.585 201 674 (22)	3.162 341 432 (11)	73	4.470 115 461 (105)	1.748 653 445 (53)
24	6.204 484 017 (23)	1.961 990 554 (12)	74	3.307 885 441 (107)	1.891 675 821 (54)
25	1.551 121 004 (25)	7.905 853 581 (12)	75	2.480 914 081 (109)	1.311 490 084 (55)
26	4.032 914 611 (26)	5.101 175 439 (13)	76	1.885 494 702 (111)	1.437 673 624 (56)
27	1.088 886 945 (28)	2.134 580 467 (14)	77	1.451 830 920 (113)	1.009 847 365 (57)
28	3.048 883 446 (29)	1.428 329 123 (15)	78	1.132 428 118 (115)	1.121 385 427 (58)
29	8.841 761 994 (30)	6.190 283 354 (15)	79	8.946 182 130 (116)	7.977 794 181 (58)
30	2.652 528 598 (32)	4.284 987 369 (16)	80	7.156 945 704 (118)	8.971 083 412 (59)
31	8.222 838 654 (33)	1.918 987 840 (17)	81	5.797 126 020 (120)	6.462 013 287 (60)
32	2.631 308 369 (35)	1.371 195 858 (18)	82	4.753 643 337 (122)	7.356 228 389 (61)
33	8.683 317 619 (36)	6.332 659 871 (18)	83	3.945 523 970 (124)	5.363 471 028 (62)
34	2.952 327 990 (38)	4.662 066 258 (19)	84	3.314 240 134 (126)	6.179 282 254 (63)
35	1.033 314 797 (40)	2.216 430 955 (20)	85	2.817 104 114 (128)	4.558 950 374 (64)
36	3.719 933 268 (41)	1.678 343 853 (21)	86	2.422 709 638 (130)	5.314 182 739 (65)
37	1.376 375 309 (43)	8.200 794 533 (21)	87	2.107 757 298 (132)	3.966 286 825 (66)
38	5.230 226 175 (44)	6.377 706 640 (22)	88	1.854 826 422 (134)	4.676 480 810 (67)
39	2.039 788 208 (46)	3.198 309 868 (23)	89	1.650 795 516 (136)	3.529 995 274 (68)
40	8.159 152 832 (47)	2.551 082 656 (24)	90	1.485 715 964 (138)	4.208 832 729 (69)
41	3.345 252 661 (49)	1.311 307 046 (25)	91	1.352 001 528 (140)	3.212 295 700 (70)
42	1.405 006 118 (51)	1.071 454 716 (26)	92	1.243 841 405 (142)	3.872 126 111 (71)
43	5.041 526 306 (52)	5.638 620 297 (26)	93	1.156 772 507 (144)	3.987 435 001 (72)
44	2.658 271 575 (54)	4.714 400 749 (27)	94	1.087 366 157 (146)	3.639 798 544 (73)
45	1.196 222 209 (56)	2.537 379 134 (28)	95	1.032 997 849 (148)	2.838 063 251 (74)
46	5.502 622 160 (57)	2.168 624 344 (29)	96	9.916 779 348 (149)	3.494 206 602 (75)
47	2.586 232 415 (59)	1.192 568 193 (30)	97	9.619 275 968 (151)	2.752 921 353 (76)
48	1.241 391 553 (61)	1.040 939 685 (31)	98	9.426 890 448 (153)	3.424 322 470 (77)
49	6.082 818 640 (62)	5.843 585 145 (31)	99	9.332 621 544 (155)	2.725 393 140 (78)
50	3.041 409 320 (64)	5.204 698 426 (32)	100	9.332 621 544 (157)	3.424 322 470 (79)

| $\Gamma(u+1) = u! =$ gamma function   (2.28d) | | | | $\Pi(u) = u! =$ pi function   (2.31a) | | | |

u	$u!$	u	$u!$	u	$u!$	u	$u!$
0.005	0.997 138 535	0.255	0.905 385 766	0.505	0.886 398 974	0.755	0.920 209 222
0.010	0.994 325 851	0.260	0.904 397 118	0.510	0.886 591 685	0.760	0.921 374 885
0.015	0.991 561 289	0.265	0.903 426 295	0.515	0.886 804 980	0.765	0.922 559 518
0.020	0.988 844 203	0.270	0.902 503 065	0.520	0.887 038 783	0.770	0.923 763 128
0.025	0.986 173 963	0.275	0.901 597 199	0.525	0.887 293 023	0.775	0.924 985 721
0.030	0.983 549 951	0.280	0.900 718 477	0.530	0.887 567 838	0.780	0.926 227 306
0.035	0.980 971 561	0.285	0.899 866 677	0.535	0.887 862 529	0.785	0.927 487 893
0.040	0.978 438 201	0.290	0.899 041 586	0.540	0.888 177 659	0.790	0.928 767 490
0.045	0.975 949 292	0.295	0.898 242 995	0.545	0.888 512 953	0.795	0.930 066 112
0.050	0.973 504 266	0.300	0.897 470 696	0.550	0.888 868 348	0.800	0.931 383 771
0.055	0.971 102 566	0.305	0.896 724 490	0.555	0.889 243 783	0.805	0.932 720 481
0.060	0.968 743 650	0.310	0.896 004 177	0.560	0.889 638 199	0.810	0.934 076 259
0.065	0.966 426 982	0.315	0.895 309 564	0.565	0.890 054 539	0.815	0.935 451 120
0.070	0.964 152 043	0.320	0.894 640 463	0.570	0.890 489 746	0.820	0.936 845 083
0.075	0.961 918 319	0.325	0.893 996 687	0.575	0.890 944 769	0.825	0.938 258 168
0.080	0.959 725 311	0.330	0.893 378 054	0.580	0.891 419 554	0.830	0.939 690 395
0.085	0.957 572 527	0.335	0.892 784 385	0.585	0.891 914 052	0.835	0.941 141 786
0.090	0.955 459 488	0.340	0.892 215 507	0.590	0.892 428 214	0.840	0.942 612 363
0.095	0.953 385 723	0.345	0.891 671 249	0.595	0.892 961 995	0.845	0.944 102 152
0.100	0.951 350 770	0.350	0.891 151 442	0.600	0.893 515 349	0.850	0.945 611 176
0.105	0.949 541 178	0.355	0.890 655 924	0.605	0.894 098 234	0.855	0.947 129 464
0.110	0.947 955 504	0.360	0.890 184 532	0.610	0.894 680 609	0.860	0.948 687 042
0.115	0.945 474 315	0.365	0.889 737 112	0.615	0.895 292 433	0.865	0.950 253 939
0.120	0.943 590 186	0.370	0.889 313 807	0.620	0.895 923 669	0.870	0.951 840 186
0.125	0.941 742 700	0.375	0.888 913 569	0.625	0.896 574 280	0.875	0.953 445 813
0.130	0.939 931 450	0.380	0.888 537 149	0.630	0.897 244 233	0.880	0.955 070 853
0.135	0.939 138 036	0.385	0.888 184 104	0.635	0.897 933 493	0.885	0.956 715 340
0.140	0.936 416 066	0.390	0.887 854 292	0.640	0.898 642 030	0.890	0.958 379 308
0.145	0.934 711 134	0.395	0.887 547 575	0.645	0.899 369 814	0.895	0.960 062 793
0.150	0.933 040 931	0.400	0.887 263 818	0.650	0.900 116 816	0.900	0.961 765 832
0.155	0.931 405 022	0.405	0.887 002 888	0.655	0.900 983 010	0.905	0.963 488 463
0.160	0.929 803 967	0.410	0.886 764 658	0.660	0.901 668 371	0.910	0.965 230 726
0.165	0.928 234 712	0.415	0.886 548 999	0.665	0.902 472 875	0.915	0.966 992 661
0.170	0.926 699 611	0.420	0.886 355 790	0.670	0.903 296 500	0.920	0.968 774 309
0.175	0.925 197 423	0.425	0.886 184 908	0.675	0.904 139 224	0.925	0.970 575 713
0.180	0.923 727 814	0.430	0.886 036 236	0.680	0.905 001 030	0.930	0.972 396 918
0.185	0.922 290 459	0.435	0.885 909 659	0.685	0.905 881 900	0.935	0.974 237 967
0.190	0.920 885 037	0.440	0.885 805 064	0.690	0.906 781 816	0.940	0.976 098 908
0.195	0.919 511 234	0.445	0.885 722 340	0.695	0.907 700 765	0.945	0.977 979 786
0.200	0.918 168 742	0.450	0.885 661 380	0.700	0.908 638 733	0.950	0.979 880 651
0.205	0.916 857 261	0.455	0.885 622 080	0.705	0.909 595 708	0.955	0.981 801 552
0.210	0.915 576 493	0.460	0.885 604 336	0.710	0.910 571 680	0.960	0.983 742 540
0.215	0.914 326 150	0.465	0.885 608 050	0.715	0.911 566 639	0.965	0.985 703 666
0.220	0.913 105 948	0.470	0.885 633 122	0.720	0.912 580 578	0.970	0.987 684 984
0.225	0.911 915 607	0.475	0.885 679 458	0.725	0.913 613 490	0.975	0.989 686 546
0.230	0.910 754 856	0.480	0.885 746 965	0.730	0.914 665 373	0.980	0.991 708 408
0.235	0.909 623 427	0.485	0.885 835 552	0.735	0.915 736 217	0.985	0.993 750 627
0.240	0.908 521 058	0.490	0.885 945 132	0.740	0.916 826 025	0.990	0.995 813 260
0.245	0.907 447 492	0.495	0.886 075 617	0.745	0.917 934 795	0.995	0.997 896 364
0.250	0.906 402 477	0.500	0.886 226 926	0.750	0.919 062 527	1.000	1.000 000 000

*Ref. 1.20, p. 457.

$\psi(x) =$ digamma function (2.32a) $\qquad d\psi(x)/dx =$ trigamma function (2.34a)

x	$\psi(x)$	$d\psi(x)/dx$	x	$\psi(x)$	$d\psi(x)/dx$
0.000	−0.577 215 665	1.644 934 067	0.250	−0.227 453 533	1.197 329 155
0.005	−0.569 020 911	1.632 994 157	0.255	−0.221 483 427	1.190 725 058
0.010	−0.560 885 458	1.621 213 528	0.260	−0.215 546 169	1.184 189 480
0.015	−0.552 808 516	1.609 589 182	0.265	−0.209 641 419	1.177 721 403
0.020	−0.544 789 311	1.598 118 192	0.270	−0.203 768 844	1.171 319 830
0.025	−0.536 827 083	1.586 797 699	0.275	−0.197 928 112	1.164 983 782
0.030	−0.528 921 087	1.575 624 915	0.280	−0.192 118 898	1.158 712 299
0.035	−0.521 070 592	1.564 597 116	0.285	−0.186 340 883	1.152 504 439
0.040	−0.513 274 879	1.553 711 643	0.290	−0.180 593 749	1.146 359 276
0.045	−0.505 533 243	1.542 965 897	0.295	−0.174 877 187	1.140 275 903
0.050	−0.497 844 991	1.532 357 342	0.300	−0.169 190 889	1.134 253 435
0.055	−0.490 209 445	1.521 883 500	0.305	−0.163 534 553	1.128 290 992
0.060	−0.482 625 936	1.511 541 950	0.310	−0.157 907 880	1.122 387 718
0.065	−0.475 093 809	1.501 330 326	0.315	−0.152 310 578	1.116 542 771
0.070	−0.467 612 420	1.491 246 316	0.320	−0.146 742 357	1.110 755 325
0.075	−0.460 181 137	1.481 287 662	0.325	−0.141 202 931	1.105 024 568
0.080	−0.452 799 338	1.471 452 156	0.330	−0.135 692 018	1.099 349 704
0.085	−0.445 466 414	1.461 737 638	0.335	−0.130 209 342	1.093 729 950
0.090	−0.438 181 764	1.452 141 999	0.340	−0.124 754 628	1.088 164 538
0.095	−0.430 944 799	1.442 663 176	0.345	−0.119 327 607	1.082 652 714
0.100	−0.423 754 940	1.433 299 151	0.350	−0.113 928 013	1.077 193 736
0.105	−0.416 611 619	1.424 047 951	0.355	−0.108 555 583	1.071 786 877
0.110	−0.409 514 276	1.414 907 648	0.360	−0.103 210 058	1.066 431 423
0.115	−0.402 462 361	1.405 876 354	0.365	−0.097 891 184	1.061 126 667
0.120	−0.395 455 334	1.396 952 221	0.370	−0.092 598 708	1.055 871 929
0.125	−0.388 492 663	1.388 133 445	0.375	−0.087 332 383	1.050 666 522
0.130	−0.381 573 827	1.379 418 257	0.380	−0.082 091 962	1.045 509 783
0.135	−0.374 698 311	1.370 804 929	0.385	−0.076 877 205	1.040 401 058
0.140	−0.367 865 611	1.362 291 767	0.390	−0.071 687 872	1.035 339 704
0.145	−0.361 075 229	1.353 877 115	0.395	−0.066 523 730	1.030 325 088
0.150	−0.354 326 678	1.345 559 352	0.400	−0.061 384 545	1.025 356 591
0.155	−0.347 619 477	1.337 336 890	0.405	−0.056 270 088	1.020 433 600
0.160	−0.340 953 153	1.329 208 175	0.410	−0.051 180 134	1.015 555 517
0.165	−0.334 327 241	1.321 171 686	0.415	−0.046 114 459	1.010 721 752
0.170	−0.327 741 285	1.313 225 932	0.420	−0.041 072 843	1.005 931 724
0.175	−0.321 194 833	1.305 369 455	0.425	−0.036 055 070	1.001 184 864
0.180	−0.314 687 444	1.297 600 825	0.430	−0.031 060 924	0.996 480 611
0.185	−0.308 218 681	1.289 918 642	0.435	−0.026 090 194	0.991 818 415
0.190	−0.301 788 116	1.282 321 536	0.440	−0.021 142 670	0.987 197 733
0.195	−0.295 395 326	1.274 808 162	0.445	−0.016 218 148	0.982 618 032
0.200	−0.289 039 897	1.267 377 205	0.450	−0.011 316 423	0.978 078 789
0.205	−0.282 721 419	1.260 027 376	0.455	−0.006 437 293	0.973 579 487
0.210	−0.276 439 490	1.252 757 400	0.460	−0.001 580 562	0.969 119 622
0.215	−0.270 193 714	1.245 566 061	0.465	+0.003 253 968	0.964 698 692
0.220	−0.263 983 700	1.238 452 130	0.470	0.008 066 489	0.960 316 209
0.225	−0.257 809 065	1.231 414 426	0.475	0.012 857 193	0.955 971 690
0.230	−0.251 669 431	1.224 451 770	0.480	0.017 626 268	0.951 664 659
0.235	−0.245 564 424	1.217 563 025	0.485	0.022 373 901	0.947 394 651
0.240	−0.239 493 679	1.210 747 071	0.490	0.027 100 276	0.943 161 205
0.245	−0.233 456 834	1.204 002 806	0.495	0.031 805 574	0.938 963 870

$\psi(x)$ = digamma function (2.32a) $d\psi(x)/dx$ = trigamma function (2.34a)

x	$\psi(x)$	$d\psi(x)/dx$	x	$\psi(x)$	$d\psi(x)/dx$
0.500	0.036 489 974	0.934 802 201	0.750	0.247 472 454	0.764 101 870
0.505	0.041 153 654	0.930 675 759	0.755	0.251 285 956	0.761 302 377
0.510	0.045 796 790	0.926 584 114	0.760	0.255 085 510	0.758 522 687
0.515	0.050 419 553	0.922 526 843	0.765	0.258 871 215	0.755 762 595
0.520	0.055 022 115	0.918 503 527	0.770	0.262 643 169	0.753 021 900
0.525	0.059 604 644	0.914 513 755	0.775	0.266 401 466	0.750 300 404
0.530	0.064 167 307	0.910 557 125	0.780	0.270 146 204	0.747 597 911
0.535	0.068 710 270	0.906 633 236	0.785	0.273 877 477	0.744 914 227
0.540	0.073 233 694	0.902 741 698	0.790	0.277 595 378	0.742 249 162
0.545	0.077 737 740	0.898 882 125	0.795	0.281 299 999	0.739 602 527
0.550	0.082 222 568	0.895 054 137	0.800	0.284 991 433	0.736 974 138
0.555	0.086 688 333	0.891 257 360	0.805	0.288 669 771	0.734 363 809
0.560	0.091 135 193	0.887 491 425	0.810	0.292 335 101	0.731 771 362
0.565	0.095 563 298	0.883 755 970	0.815	0.295 987 514	0.729 196 617
0.570	0.099 972 802	0.880 050 638	0.820	0.299 627 097	0.726 639 397
0.575	0.104 363 854	0.876 375 077	0.825	0.303 253 937	0.724 099 530
0.580	0.108 736 602	0.872 728 940	0.830	0.306 868 121	0.721 576 843
0.585	0.113 091 193	0.869 111 887	0.835	0.310 469 734	0.719 071 166
0.590	0.117 427 769	0.865 523 582	0.840	0.314 058 860	0.716 582 333
0.595	0.121 746 475	0.861 963 692	0.845	0.317 635 585	0.714 110 179
0.600	0.126 047 453	0.858 431 893	0.850	0.321 199 999	0.711 654 540
0.605	0.130 330 841	0.854 927 863	0.855	0.324 752 157	0.709 215 255
0.610	0.134 596 777	0.851 451 286	0.860	0.328 292 169	0.706 792 165
0.615	0.138 845 399	0.848 001 849	0.865	0.331 820 106	0.704 385 114
0.620	0.143 076 840	0.844 579 246	0.870	0.335 336 047	0.701 993 946
0.625	0.147 291 235	0.841 183 173	0.875	0.338 840 071	0.699 618 509
0.630	0.151 488 716	0.837 813 333	0.880	0.342 332 258	0.697 258 651
0.635	0.155 669 412	0.834 469 432	0.885	0.345 812 684	0.694 914 224
0.640	0.159 833 453	0.831 151 179	0.890	0.349 281 426	0.692 585 079
0.645	0.163 980 966	0.827 858 290	0.895	0.352 738 560	0.690 271 072
0.650	0.168 112 078	0.824 590 483	0.900	0.356 184 161	0.687 972 058
0.655	0.172 226 912	0.821 347 480	0.905	0.359 618 305	0.685 687 897
0.660	0.176 325 593	0.818 129 009	0.910	0.363 041 065	0.683 418 447
0.665	0.180 408 243	0.814 934 800	0.915	0.366 452 514	0.681 163 570
0.670	0.184 474 981	0.811 764 588	0.920	0.369 852 724	0.678 923 129
0.675	0.188 525 928	0.808 618 109	0.925	0.373 241 769	0.676 696 990
0.680	0.192 561 202	0.805 495 108	0.930	0.376 619 718	0.674 485 019
0.685	0.196 580 918	0.802 395 328	0.935	0.379 986 642	0.672 287 085
0.690	0.200 585 193	0.799 318 520	0.940	0.383 342 612	0.670 103 056
0.695	0.204 574 141	0.796 264 435	0.945	0.386 687 696	0.667 932 804
0.700	0.208 547 875	0.793 232 830	0.950	0.390 021 963	0.665 776 203
0.705	0.212 506 506	0.790 223 464	0.955	0.393 345 481	0.663 633 127
0.710	0.216 450 146	0.787 236 101	0.960	0.396 658 316	0.661 503 451
0.715	0.220 378 904	0.784 270 506	0.965	0.399 960 537	0.659 387 054
0.720	0.224 292 887	0.781 326 449	0.970	0.403 252 209	0.657 283 813
0.725	0.228 192 204	0.778 403 701	0.975	0.406 533 397	0.655 193 610
0.730	0.232 076 959	0.775 502 040	0.980	0.409 804 166	0.653 116 327
0.735	0.235 947 259	0.772 621 242	0.985	0.413 064 582	0.651 051 845
0.740	0.239 803 206	0.769 761 092	0.990	0.416 314 706	0.649 000 051
0.745	0.243 644 904	0.766 921 371	0.995	0.419 554 603	0.646 960 829

n	$\binom{n}{0}$	$\binom{n}{1}$	$\binom{n}{2}$	$\binom{n}{3}$	$\binom{n}{4}$	$\binom{n}{5}$	$\binom{n}{6}$	$\binom{n}{7}$	$\binom{n}{8}$	$\binom{n}{9}$	$\binom{n}{10}$
0	1										
1	1	1									
2	1	2	1								
3	1	3	3	1							
4	1	4	6	4	1						
5	1	5	10	10	5	1					
6	1	6	15	20	15	6	1				
7	1	7	21	35	35	21	7	1			
8	1	8	28	56	70	56	28	8	1		
9	1	9	36	84	126	126	84	36	9	1	
10	1	10	45	120	210	252	210	120	45	10	1
11	1	11	55	165	330	462	462	330	165	55	11
12	1	12	66	220	495	792	924	792	495	220	66
13	1	13	78	286	715	1 287	1 716	1 716	1 287	715	286
14	1	14	91	364	1 001	2 002	3 003	3 432	3 003	2 002	1 001
15	1	15	105	455	1 365	3 003	5 005	6 435	6 435	5 005	3 003
16	1	16	120	560	1 820	4 368	8 008	11 440	12 870	11 440	8 008
17	1	17	136	680	2 380	6 188	12 376	19 448	24 310	24 310	19 448
18	1	18	153	816	3 060	8 568	18 564	31 824	43 758	48 620	43 758
19	1	19	171	969	3 876	11 628	27 132	50 388	75 582	92 378	92 378
20	1	20	190	1 140	4 845	15 504	38 760	77 520	125 970	167 960	184 756
21	1	21	210	1 330	5 985	20 349	54 264	116 280	203 490	293 930	352 716
22	1	22	231	1 540	7 315	26 334	74 613	170 544	319 770	497 420	646 646
23	1	23	253	1 771	8 855	33 649	100 947	245 157	490 314	817 190	1 144 066
24	1	24	276	2 024	10 626	42 504	134 596	346 104	735 471	1 307 504	1 961 256
25	1	25	300	2 300	12 650	53 130	177 100	480 700	1 081 575	2 042 975	3 268 760
26	1	26	325	2 600	14 950	65 780	230 230	657 800	1 562 275	3 124 550	5 311 735
27	1	27	351	2 925	17 550	80 730	296 010	888 030	2 220 075	4 686 825	8 436 285
28	1	28	378	3 276	20 475	98 280	376 740	1 184 040	3 108 105	6 906 900	13 123 110
29	1	29	406	3 654	23 751	118 755	475 020	1 560 780	4 292 145	10 015 005	20 030 010
30	1	30	435	4 060	27 405	142 506	593 775	2 035 800	5 852 925	14 307 150	30 045 015

$\binom{n}{11}$	$\binom{n}{12}$	$\binom{n}{13}$	$\binom{n}{14}$	$\binom{n}{15}$	n
					0
					1
					2
					3
					4
					5
					6
					7
					8
					9
					10
1					11
12	1				12
78	12	1			13
364	91	14	1		14
1 365	455	105	15	1	15
4 368	1 820	560	120	16	16
12 376	6 188	2 380	680	136	17
31 824	18 564	8 568	3 060	816	18
75 582	50 388	27 132	11 628	3 876	19
167 960	123 970	77 520	38 760	15 504	20
352 716	293 930	203 490	116 280	54 263	21
705 432	646 646	497 420	319 770	170 544	22
1 352 078	1 352 078	1 144 066	817 190	490 314	23
2 496 144	2 704 156	2 496 144	1 961 256	1 307 504	24
4 457 400	5 200 300	5 200 300	4 457 400	3 268 760	25
7 726 160	9 657 700	10 400 600	9 657 700	7 726 160	26
13 037 895	17 383 860	20 058 300	20 058 300	17 383 860	27
21 474 180	30 421 755	37 442 160	40 116 600	37 442 160	28
34 597 290	51 895 935	67 863 915	77 558 760	77 558 760	29
54 627 300	86 493 225	119 759 850	145 422 675	155 117 520	30

(Formula box, spanning the upper portion of the table:)

$\binom{n}{k}$ = binomial coefficients (2.10*b*)

$$\binom{n}{k} = \frac{n!}{k!(n-k)!} = \frac{n(n-1)(n-2)\cdots(n-k+1)}{k!}$$

$$\binom{n}{0} = 1 \qquad \binom{n}{1} = n \qquad \binom{n}{n-1} = n \qquad \binom{n}{n} = 1$$

For coefficients not given below, use $\binom{n}{k} = \binom{n}{n-k}$

For example, $\binom{25}{21} = \binom{25}{25-21} = \binom{25}{4} = 12\,650$

$\mathscr{S}_k^{(p)}$ = Stirling number of the first kind (2.11a) $\qquad$ $X_1^{(p)}$ = factorial polynomial (2.11a)

$$X_1^{(p)} = x(x-1)(x-2)(x-3)\cdots(x-p+1) = x\mathscr{S}_1^{(p)} + x^2\mathscr{S}_2^{(p)} + x^3\mathscr{S}_3^{(p)} + \cdots + x^p\mathscr{S}_k^{(p)}$$

For numbers not shown below use the recurrent relation $\mathscr{S}_k^{(m)} = \mathscr{S}_{k-1}^{(m-1)} - (m-1)\mathscr{S}_k^{(m-1)}$

k	$\mathscr{S}_k^{(1)}$	$\mathscr{S}_k^{(2)}$	$\mathscr{S}_k^{(3)}$	$\mathscr{S}_k^{(4)}$	$\mathscr{S}_k^{(5)}$	$\mathscr{S}_k^{(6)}$	$\mathscr{S}_k^{(7)}$	$\mathscr{S}_k^{(8)}$	$\mathscr{S}_k^{(9)}$
1	1	−1	2	−6	24	−120	720	−5 040	40 320
2		1	−3	11	−50	274	−1 764	13 068	−109 584
3			1	−6	35	−225	1 624	−15 132	118 121
4				1	−10	85	−735	6 769	−67 284
5					1	−15	175	−1 960	22 449
6						1	−21	322	−4 536
7							1	−28	546
8								1	−36
9									1

k	$\mathscr{S}_k^{(10)}$	$\mathscr{S}_k^{(11)}$	$\mathscr{S}_k^{(12)}$	$\mathscr{S}_k^{(13)}$	$\mathscr{S}_k^{(14)}$
1	−362 880	3 628 800	−39 916 800	479 001 600	−6 227 020 800
2	1 026 576	−10 628 640	120 543 840	−1 486 442 880	19 802 759 040
3	−1 172 700	12 753 576	−150 917 976	1 931 559 552	−26 596 717 056
4	723 680	−8 409 500	105 258 076	−1 414 014 888	20 313 753 096
5	−269 325	3 416 930	−45 995 730	657 206 836	−9 957 703 756
6	63 273	−902 055	13 339 535	−206 070 150	3 336 118 786
7	−9 450	157 773	−2 637 558	44 990 231	−790 943 153
8	870	−18 150	357 423	−6 926 624	135 036 473
9	−45	1 320	−32 670	749 463	−16 669 653
10	1	−55	1 925	−55 770	1 474 473
11		1	−66	2 717	−91 091
12			1	−78	3 731
13				1	−91
14					1

k	$\mathscr{S}_k^{(15)}$	$\mathscr{S}_k^{(16)}$	$\mathscr{S}_k^{(17)}$	$\mathscr{S}_k^{(18)}$
1	87 178 291 200	−1 307 674 368 000	20 922 789 888 000	−355 687 428 096 000
2	−283 465 647 360	4 339 163 001 600	−70 734 282 393 600	1 223 405 590 579 200
3	392 156 797 824	−6 165 817 614 720	102 992 244 837 120	−1 821 602 444 624 640
4	−310 982 260 400	5 056 995 703 824	−87 077 748 875 904	1 583 313 975 727 489
5	159 721 805 680	−2 706 813 348 600	48 366 009 233 424	−909 299 905 844 112
6	−56 663 366 760	1 009 672 107 080	−18 861 567 058 880	369 012 649 234 384
7	14 409 322 928	−272 803 210 680	5 374 523 477 960	−110 228 466 184 200
8	−2 681 453 775	54 631 129 553	−1 146 901 283 528	24 871 845 297 936
9	368 411 615	−8 207 628 800	185 953 177 553	−4 308 105 301 929
10	−37 312 275	928 095 740	−23 057 159 840	577 924 894 833
11	2 749 747	−78 558 480	2 185 031 420	−60 202 693 980
12	−143 325	4 899 622	−156 952 432	4 853 222 764
12	5 005	−218 400	8 394 022	−299 650 806
13	−105	6 580	−323 680	13 896 582
14	1	−120	8 500	−468 180
16		1	−136	10 812
17			1	−153
18				1

$\bar{\mathscr{S}}_k^{(p)}$ = Stirling number of the second kind (2.11d) x^p = pth power of x (2.11d)

$$x^p = \bar{\mathscr{S}}_1^{(p)}X_1^{(1)} + \bar{\mathscr{S}}_2^{(p)}X_1^{(2)} + \bar{\mathscr{S}}_3^{(p)}X_1^{(3)} + \cdots + \bar{\mathscr{S}}_p^{(p)}X_1^{(p)}$$

For numbers not shown below use the recurrent relation $\bar{\mathscr{S}}_k^{(m)} = k\bar{\mathscr{S}}_k^{(m-1)} + \bar{\mathscr{S}}_{k-1}^{(m-1)}$

k	$\bar{\mathscr{S}}_k^{(1)}$	$\bar{\mathscr{S}}_k^{(2)}$	$\bar{\mathscr{S}}_k^{(3)}$	$\bar{\mathscr{S}}_k^{(4)}$	$\bar{\mathscr{S}}_k^{(5)}$	$\bar{\mathscr{S}}_k^{(6)}$	$\bar{\mathscr{S}}_k^{(7)}$	$\bar{\mathscr{S}}_k^{(8)}$	$\bar{\mathscr{S}}_k^{(9)}$
1	1	1	1	1	1	1	1	1	1
2		1	3	7	15	31	63	127	255
3			1	6	25	90	301	966	3 025
4				1	10	65	350	1 701	7 770
5					1	15	140	1 050	6 950
6						1	21	266	2 646
7							1	28	462
8								1	36
9									1

k	$\bar{\mathscr{S}}_k^{(10)}$	$\bar{\mathscr{S}}_k^{(11)}$	$\bar{\mathscr{S}}_k^{(12)}$	$\bar{\mathscr{S}}_k^{(13)}$	$\bar{\mathscr{S}}_k^{(14)}$
1	1	1	1	1	1
2	511	1 023	2 047	4 095	8 191
3	9 330	28 501	86 526	261 625	288 970
4	34 105	145 750	611 501	2 532 530	10 391 745
5	42 525	246 730	1 379 400	7 508 511	37 795 758
6	22 827	179 487	1 323 652	9 321 312	63 436 373
7	5 880	63 987	627 396	5 715 424	49 329 280
8	750	11 880	159 027	1 899 612	20 912 320
9	45	1 155	22 275	359 502	5 135 130
10	1	55	1 705	39 325	752 752
11		1	66	2 431	66 066
12			1	78	3 367
13				1	91
14					1

k	$\bar{\mathscr{S}}_k^{(15)}$	$\bar{\mathscr{S}}_k^{(16)}$	$\bar{\mathscr{S}}_k^{(17)}$	$\bar{\mathscr{S}}_k^{(18)}$
1	1	1	1	1
2	16 383	32 767	65 535	131 071
3	2 375 101	7 141 686	21 457 825	64 439 010
4	42 355 950	171 798 901	694 337 290	2 798 806 985
5	210 766 920	1 096 190 550	5 652 751 651	28 958 095 545
6	420 693 273	2 934 926 558	17 505 749 898	110 687 251 439
7	408 741 333	3 281 882 604	25 708 104 786	197 462 483 400
8	216 627 840	2 141 764 053	20 415 995 028	189 036 065 010
9	67 128 490	820 784 250	9 528 822 303	106 175 395 755
10	12 662 650	193 754 990	2 758 334 150	37 112 163 803
11	1 479 478	28 936 908	512 060 078	8 391 004 908
12	106 470	2 757 118	62 022 324	1 256 328 866
13	4 550	165 620	4 910 178	125 854 638
14	105	6 020	249 900	8 408 778
15	1	120	7 820	367 200
16		1	136	9 996
17			1	153
18				1

A.09 BERNOULLI POLYNOMIALS

$\bar{B}_r(x)$ = Bernoulli polynomial (2.23a)
$\bar{B}_r$ = Bernoulli number (2.22a, A.10)

$$\bar{B}_r(x) = \sum_{k=0}^{r} \binom{r}{k} x^{r-k} \bar{B}_k = \sum_{k=0}^{r} b_k x^k$$

For example,

$$\bar{B}_3(x) = x^3 \bar{B}_0 + \binom{3}{1} x^2 \bar{B}_1 + \binom{3}{2} x \bar{B}_2 + \binom{3}{3} \bar{B}_3$$
$$= \tfrac{1}{2}x - \tfrac{3}{2}x^2 + x^3$$

$\bar{B}_r(x) \diagdown b_k$	b_0	b_1	b_2	b_3	b_4	b_5	b_6	b_7	b_8	b_9	b_{10}	b_{11}	b_{12}	b_{13}	b_{14}	b_{15}
$\bar{B}_0(x)$	1															
$\bar{B}_1(x)$	$-\frac{1}{2}$	1														
$\bar{B}_2(x)$	$\frac{1}{6}$	-1	1													
$\bar{B}_3(x)$	0	$\frac{1}{2}$	$-\frac{3}{2}$	1												
$\bar{B}_4(x)$	$-\frac{1}{30}$	0	1	-2	1											
$\bar{B}_5(x)$	0	$-\frac{1}{6}$	0	$\frac{5}{3}$	$-\frac{5}{2}$	1										
$\bar{B}_6(x)$	$\frac{1}{42}$	0	$-\frac{1}{2}$	0	$\frac{5}{2}$	-3	1									
$\bar{B}_7(x)$	0	$\frac{1}{6}$	0	$-\frac{7}{6}$	0	$\frac{7}{2}$	$-\frac{7}{2}$	1								
$\bar{B}_8(x)$	$-\frac{1}{30}$	0	$\frac{2}{3}$	0	$-\frac{7}{3}$	0	$\frac{14}{3}$	-4	1							
$\bar{B}_9(x)$	0	$-\frac{3}{10}$	0	2	0	$-\frac{21}{5}$	0	6	$-\frac{9}{2}$	1						
$\bar{B}_{10}(x)$	$\frac{5}{66}$	0	$-\frac{3}{2}$	0	5	0	-7	0	$\frac{15}{2}$	-5	1					
$\bar{B}_{11}(x)$	0	$\frac{5}{6}$	0	$-\frac{11}{2}$	0	11	0	-11	0	$\frac{55}{6}$	$-\frac{11}{2}$	1				
$\bar{B}_{12}(x)$	$-\frac{691}{2730}$	0	5	0	$-\frac{33}{2}$	0	22	0	$-\frac{33}{2}$	0	11	-6	1			
$\bar{B}_{13}(x)$	0	$-\frac{691}{210}$	0	$\frac{65}{3}$	0	$-\frac{429}{10}$	0	$\frac{286}{7}$	0	$-\frac{143}{6}$	0	13	$-\frac{13}{2}$	1		
$\bar{B}_{14}(x)$	$\frac{7}{6}$	0	$-\frac{691}{30}$	0	$\frac{455}{6}$	0	$-\frac{1001}{10}$	0	$\frac{143}{2}$	0	$-\frac{1001}{30}$	0	$\frac{91}{6}$	-7	1	
$\bar{B}_{15}(x)$	0	$\frac{35}{2}$	0	$-\frac{691}{6}$	0	$\frac{455}{2}$	0	$-\frac{429}{2}$	0	$\frac{715}{6}$	0	$-\frac{91}{2}$	0	$\frac{35}{2}$	$-\frac{15}{2}$	1

A.10 BERNOULLI NUMBERS

B_r = auxiliary Bernoulli number (2.22b)		$\bar{B}_r$ = Bernoulli number (2.22a)	$r = 0, 1, 2, \ldots$	$\alpha = (-1)^{r+1}$

B_r	$\bar{B}_{2r}$	Decimals
	$\bar{B}_0$	1.000 000 000 (+00)
	$-\bar{B}_1$	5.000 000 000 (−01)
B_1	$\bar{B}_2$	1.666 666 667 (−01)
B_2	$-\bar{B}_4$	3.333 333 333 (−02)
B_3	$\bar{B}_6$	2.380 952 381 (−02)
B_4	$-\bar{B}_8$	3.333 333 333 (−02)
B_5	$\bar{B}_{10}$	7.575 757 576 (−02)
B_6	$-\bar{B}_{12}$	2.531 135 531 (−01)
B_7	$\bar{B}_{14}$	1.666 666 667 (+00)
B_8	$-\bar{B}_{16}$	7.092 156 863 (+00)
B_9	$\bar{B}_{18}$	5.497 117 794 (+01)
B_{10}	$-\bar{B}_{20}$	5.291 242 424 (+02)
B_{11}	$\bar{B}_{22}$	6.192 123 188 (+03)
B_{12}	$-\bar{B}_{24}$	8.658 025 311 (+04)
B_{13}	$\bar{B}_{26}$	1.425 517 167 (+06)
B_{14}	$-\bar{B}_{28}$	2.729 823 107 (+07)
B_{15}	$\bar{B}_{30}$	6.015 808 739 (+08)

$\bar{B}_3 = \bar{B}_5 = \bar{B}_7 = \cdots = 0$

B_r	$\bar{B}_{2r}$	Decimals
B_{16}	$-\bar{B}_{32}$	1.511 631 577 (+10)
B_{17}	$\bar{B}_{34}$	4.296 146 431 (+11)
B_{18}	$-\bar{B}_{36}$	1.371 165 521 (+13)
B_{19}	$\bar{B}_{38}$	4.883 323 190 (+14)
B_{20}	$-\bar{B}_{40}$	1.929 657 934 (+16)
B_{21}	$\bar{B}_{42}$	8.416 930 476 (+17)
B_{22}	$-\bar{B}_{44}$	4.033 807 185 (+19)
B_{23}	$\bar{B}_{46}$	2.115 074 864 (+21)
B_{24}	$-\bar{B}_{48}$	1.208 662 652 (+23)
B_{25}	$\bar{B}_{50}$	7.500 866 746 (+24)
B_{26}	$-\bar{B}_{52}$	5.038 778 101 (+26)
B_{27}	$\bar{B}_{54}$	3.652 877 638 (+28)
B_{28}	$-\bar{B}_{56}$	2.849 876 930 (+30)
B_{29}	$\bar{B}_{58}$	2.386 542 750 (+32)
B_{30}	$-\bar{B}_{60}$	2.139 994 926 (+34)
B_{31}	$\bar{B}_{62}$	2.017 031 483 (+36)
B_{32}	$-\bar{B}_{64}$	2.060 029 615 (+38)
B_{33}	$\bar{B}_{66}$	2.238 571 753 (+40)

B_r	$\bar{B}_{2r}$	Decimals
B_{34}	$-\bar{B}_{68}$	2.583 419 885 (+42)
B_{35}	$\bar{B}_{70}$	3.160 693 563 (+44)
B_{36}	$-\bar{B}_{72}$	4.092 733 821 (+46)
B_{37}	$\bar{B}_{74}$	5.600 261 977 (+48)
B_{38}	$-\bar{B}_{76}$	8.085 808 705 (+50)
B_{39}	$\bar{B}_{78}$	1.230 124 459 (+53)
B_{40}	$-\bar{B}_{80}$	1.969 275 126 (+55)
B_{41}	$\bar{B}_{82}$	3.313 183 806 (+57)
B_{42}	$-\bar{B}_{84}$	8.851 176 136 (+59)
B_{43}	$\bar{B}_{86}$	1.083 429 888 (+62)
B_{44}	$-\bar{B}_{88}$	2.101 081 990 (+64)
B_{45}	$\bar{B}_{90}$	4.263 004 386 (+66)

Series representation for $r \geq 2$:

$$B_r = \alpha \bar{B}_{2r} = 2 \frac{(2r)!}{(2\pi)^{2r}} \sum_{k=1}^{\infty} \left(\frac{1}{k}\right)^{2r}$$

Recurrence relations for $r \geq 32$:

$$B_r = \alpha \bar{B}_{2r} = \frac{2r(2r-1)}{(2\pi)^2} |\bar{B}_{2r-2}|$$

A.11 EULER POLYNOMIALS

$\bar{E}_r(x)$ = Euler polynomial (2.25a)
$\bar{E}_r$ = Euler number (2.24a), (A.12)

$$\bar{E}_r(x) = \frac{2}{r+1}\sum_{k=1}^{r+1}(1-2^k)\binom{r+1}{k}x^{r+k-1}\bar{B}_k = \sum_{k=0}^{r}e_k x^k$$

For example,

$$\bar{E}_3(x) = \tfrac{1}{2}\left[-1\binom{4}{1}x^3\bar{B}_1 - 3\binom{4}{2}x^2\bar{B}_2 - 7\binom{4}{3}x\bar{B}_3 - 15\binom{4}{4}\bar{B}_4\right]$$
$$= \tfrac{1}{4} - \tfrac{3}{2}x^2 + x^3$$

$E_r(x)\backslash e_k$	e_0	e_1	e_2	e_3	e_4	e_5	e_6	e_7	e_8	e_9	e_{10}	e_{11}	e_{12}	e_{13}	e_{14}	e_{15}
$E_0(x)$	1															
$E_1(x)$	$-\frac{1}{2}$	1														
$E_2(x)$	0	-1	1													
$E_3(x)$	$\frac{1}{4}$	0	$-\frac{3}{2}$	1												
$E_4(x)$	0	1	0	-2	1											
$E_5(x)$	$-\frac{1}{2}$	0	$\frac{5}{2}$	0	$-\frac{5}{2}$	1										
$E_6(x)$	0	-3	0	5	0	-3	1									
$E_7(x)$	$\frac{17}{8}$	0	$-\frac{21}{2}$	0	$\frac{35}{4}$	0	$-\frac{7}{2}$	1								
$E_8(x)$	0	17	0	-28	0	14	0	-4	1							
$E_9(x)$	$-\frac{31}{2}$	0	$\frac{153}{2}$	0	-63	0	21	0	$-\frac{9}{2}$	1						
$E_{10}(x)$	0	-155	0	255	0	-126	0	30	0	-5	1					
$E_{11}(x)$	$\frac{691}{4}$	0	$-\frac{1705}{2}$	0	$\frac{2805}{4}$	0	-231	0	$\frac{165}{4}$	0	$-\frac{11}{2}$	1				
$E_{12}(x)$	0	2073	0	-3410	0	1683	0	-396	0	55	0	-6	1			
$E_{13}(x)$	$-\frac{5461}{2}$	0	$\frac{26949}{2}$	0	$-\frac{22165}{2}$	0	$\frac{7293}{2}$	0	$-\frac{1287}{2}$	0	$\frac{143}{2}$	0	$-\frac{13}{2}$	1		
$E_{14}(x)$	0	-38227	0	62881	0	-31031	0	7293	0	-1001	0	91	0	-7	1	
$E_{15}(x)$	$\frac{929569}{16}$	0	$-\frac{573405}{2}$	0	$\frac{943215}{4}$	0	$-\frac{155155}{2}$	0	$\frac{109395}{8}$	0	$-\frac{3003}{2}$	0	$\frac{455}{4}$	0	$-\frac{15}{2}$	1

A.12 EULER NUMBERS

E_r = auxiliary Euler number (2.24b)	$\bar{E}_r$ = Euler number (2.24a)	$r = 0, 1, 2, \ldots$	$\alpha = (-1)^{r+1}$	$\beta = (-1)^{k+1}$

E_r	$\bar{E}_{2r}$	Decimals	E_r	$\bar{E}_{2r}$	Decimals	E_r	$\bar{E}_{2r}$	Decimals
	$\bar{E}_0$	1.000 000 000 (+00)	E_{16}	$\bar{E}_{32}$	1.775 193 916 (+29)	E_{34}	$\bar{E}_{68}$	1.456 184 439 (+83)
	$\bar{E}_1$	0.000 000 000 (+00)	E_{17}	$-\bar{E}_{34}$	8.072 232 992 (+31)	E_{35}	$-\bar{E}_{70}$	2.850 517 834 (+86)
			E_{18}	$\bar{E}_{36}$	4.122 206 033 (+34)	E_{36}	$\bar{E}_{72}$	5.905 747 212 (+89)
E_1	$-\bar{E}_2$	1.000 000 000 (+00)	E_{19}	$-\bar{E}_{38}$	2.348 958 053 (+37)	E_{37}	$-\bar{E}_{74}$	1.292 973 665 (+93)
E_2	$\bar{E}_4$	5.000 000 000 (+00)	E_{20}	$\bar{E}_{40}$	1.485 115 077 (+40)	E_{38}	$\bar{E}_{76}$	2.968 928 185 (+96)
E_3	$-\bar{E}_6$	6.100 000 000 (+01)	E_{21}	$-\bar{E}_{42}$	1.036 462 273 (+43)	E_{39}	$-\bar{E}_{78}$	7.270 601 719 (+99)
E_4	$\bar{E}_8$	1.385 000 000 (+03)	E_{22}	$\bar{E}_{44}$	7.947 579 423 (+45)	E_{40}	$\bar{E}_{80}$	1.862 291 577 (+104)
E_5	$-\bar{E}_{10}$	5.052 100 000 (+04)	E_{23}	$-\bar{E}_{46}$	6.667 537 517 (+48)	E_{41}	$-\bar{E}_{82}$	5.013 104 944 (+107)
E_6	$\bar{E}_{12}$	2.702 765 000 (+06)	E_{24}	$\bar{E}_{48}$	6.096 278 646 (+51)	E_{42}	$\bar{E}_{84}$	1.416 525 577 (+111)
E_7	$-\bar{E}_{14}$	1.993 609 810 (+08)	E_{25}	$-\bar{E}_{50}$	6.053 285 248 (+54)	E_{43}	$-\bar{E}_{86}$	4.196 643 167 (+114)
E_8	$\bar{E}_{16}$	1.939 151 215 (+10)	E_{26}	$\bar{E}_{52}$	6.506 162 487 (+57)	E_{44}	$\bar{E}_{88}$	1.302 159 592 (+118)
E_9	$-\bar{E}_{18}$	2.404 879 675 (+12)	E_{27}	$-\bar{E}_{54}$	7.546 659 939 (+60)	E_{45}	$-\bar{E}_{90}$	4.227 240 691 (+121)
E_{10}	$\bar{E}_{20}$	3.703 711 882 (+14)	E_{28}	$\bar{E}_{56}$	9.420 321 896 (+63)			
E_{11}	$-\bar{E}_{22}$	6.934 887 439 (+16)	E_{29}	$-\bar{E}_{58}$	1.262 201 925 (+67)			
E_{12}	$\bar{E}_{24}$	1.551 453 416 (+19)	E_{30}	$\bar{E}_{60}$	1.810 891 150 (+70)			
E_{13}	$-\bar{E}_{26}$	4.087 072 509 (+21)	E_{31}	$-\bar{E}_{62}$	2.775 710 171 (+73)			
E_{14}	$\bar{E}_{28}$	1.252 259 641 (+24)	E_{32}	$\bar{E}_{64}$	4.535 810 335 (+76)			
E_{15}	$-\bar{E}_{30}$	4.415 438 932 (+26)	E_{33}	$-\bar{E}_{66}$	7.886 284 211 (+79)			

$\bar{E}_3 = \bar{E}_5 = \bar{E}_7 = \cdots = 0$

Series representation for $r \geq 2$:

$$E_r = -\alpha \bar{E}_{2r} = 2\left(\frac{2}{\pi}\right)^{2r+1}(2r)! \sum_{k=1}^{\infty} \beta \left(\frac{1}{k}\right)^{2r+1}$$

Recurrence relations for $r \geq 32$:

$$E_r = -\alpha \bar{E}_{2r} = \left(\frac{2}{\pi}\right)^2 2r(2r-1)|\bar{E}_{2r-1}|$$

$$\text{Si}(x) = \int_0^x \frac{\sin x}{x}\,dx \qquad (10.01b) \qquad\qquad \text{Ci}(x) = \int_\infty^x \frac{\cos x}{x}\,dx \qquad (10.01b)$$

x	Si(x)	Ci(x)	x	Si(x)	Ci(x)
0.00	0.000 000 000	$-\infty$	0.50	0.493 107 418	−0.177 784 079
0.01*	0.009 999 944	−4.027 979 521	0.51	0.502 687 751	−0.160 453 239
0.02	0.019 999 556	−3.334 907 339	0.52	0.512 251 521	−0.143 553 736
0.03	0.029 998 500	−2.929 567 224	0.53	0.521 798 423	−0.127 070 794
0.04	0.039 996 444	−2.642 060 133	0.54	0.531 328 149	−0.110 990 457
0.05	0.049 993 056	−2.424 760 102	0.55	0.540 840 395	−0.095 299 527
0.06	0.059 988 601	−2.237 094 917	0.56	0.550 334 856	−0.079 985 513
0.07	0.069 980 847	−2.083 269 122	0.57	0.559 811 230	−0.065 036 574
0.08	0.079 971 561	−1.950 112 553	0.58	0.569 269 214	−0.050 441 482
0.09	0.089 959 510	−1.832 754 260	0.59	0.578 708 507	−0.036 189 571
0.10	0.099 944 461	−1.727 868 387	0.60	0.588 128 810	−0.022 270 707
0.11	0.109 926 082	−1.633 082 724	0.61	0.597 529 823	−0.008 675 249
0.12	0.119 904 042	−1.546 645 511	0.62	0.606 911 250	+0.004 605 985
0.13	0.129 878 006	−1.467 227 190	0.63	0.616 272 794	0.017 581 742
0.14	0.139 847 645	−1.393 793 192	0.64	0.625 614 160	0.030 260 369
0.15	0.149 812 627	−1.325 524 049	0.65	0.634 935 054	0.042 649 829
0.16	0.159 772 619	−1.261 758 976	0.66	0.644 235 183	0.054 757 734
0.17	0.169 727 292	−1.201 957 483	0.67	0.653 514 256	0.066 591 359
0.18	0.179 676 315	−1.145 671 936	0.68	0.662 771 982	0.078 157 666
0.19	0.189 619 357	−1.092 526 967	0.69	0.672 008 072	0.089 463 320
0.20	0.199 556 089	−1.042 205 596	0.70	0.681 222 239	0.100 514 707
0.21	0.209 486 180	−0.994 436 844	0.71	0.690 414 197	0.111 317 953
0.22	0.219 409 303	−0.938 987 667	0.72	0.699 583 659	0.121 878 932
0.23	0.229 325 127	−0.905 656 189	0.73	0.708 730 343	0.132 203 288
0.24	0.239 233 326	−0.864 266 174	0.74	0.717 853 966	0.142 296 440
0.25	0.249 133 570	−0.824 663 062	0.75	0.726 954 247	0.152 163 601
0.26	0.259 025 534	−0.786 710 386	0.76	0.736 030 907	0.161 809 783
0.27	0.268 908 891	−0.750 287 386	0.77	0.745 083 666	0.171 239 811
0.28	0.278 783 309	−0.715 286 095	0.78	0.754 112 249	0.180 458 334
0.29	0.288 648 469	−0.681 610 153	0.79	0.763 116 380	0.189 469 829
0.30	0.298 504 044	−0.649 172 933	0.80	0.772 095 786	0.198 278 616
0.31	0.308 402 750	−0.617 896 321	0.81	0.781 050 192	0.206 888 861
0.32	0.318 185 138	−0.587 709 639	0.82	0.789 979 329	0.215 304 586
0.33	0.328 010 011	−0.558 548 724	0.83	0.798 882 928	0.223 529 675
0.34	0.337 827 399	−0.530 355 151	0.84	0.807 760 719	0.231 567 882
0.35	0.347 626 791	−0.503 075 569	0.85	0.816 612 437	0.239 422 837
0.36	0.357 418 056	−0.476 661 125	0.86	0.825 437 817	0.247 098 049
0.37	0.367 197 475	−0.451 066 976	0.87	0.834 236 595	0.254 596 915
0.38	0.376 951 556	−0.426 251 855	0.88	0.843 008 510	0.261 922 726
0.39	0.386 219 499	−0.402 177 704	0.89	0.851 753 309	0.269 078 669
0.40	0.396 461 465	−0.378 809 346	0.90	0.860 470 711	0.276 067 831
0.41	0.406 190 310	−0.356 114 201	0.91	0.869 160 481	0.282 893 207
0.42	0.415 905 717	−0.334 062 035	0.92	0.877 822 356	0.289 557 702
0.43	0.425 582 945	−0.312 624 740	0.93	0.886 456 084	0.296 064 136
0.44	0.435 294 951	−0.291 776 135	0.94	0.895 061 411	0.302 415 246
0.45	0.444 968 145	−0.271 491 799	0.95	0.903 638 088	0.308 613 691
0.46	0.454 626 684	−0.251 748 913	0.96	0.912 185 866	0.314 662 055
0.47	0.464 270 136	−0.232 526 111	0.97	0.920 704 497	0.320 562 850
0.48	0.473 898 301	−0.213 803 372	0.98	0.929 193 737	0.326 318 518
0.49	0.486 878 344	−0.195 561 916	0.99	0.937 653 342	0.331 931 438

* For $x < 0.01$ use the formula in (10.03a).

$$\text{Si}(x) = \int_0^x \frac{\sin x}{x}\, dx \quad (10.01b) \qquad\qquad \text{Ci}(x) = \int_\infty^x \frac{\cos x}{x}\, dx \quad (10.01b)$$

x	$\text{Si}(x)$	$\text{Ci}(x)$	x	$\text{Si}(x)$	$\text{Ci}(x)$
1.00	0.946 083 070	0.337 403 923	5.0	1.549 931 245	−0.190 029 750
1.05	0.987 775 223	0.362 736 681	5.1	1.531 253 205	−0.183 476 263
1.10	1.028 685 219	0.384 873 377	5.2	1.513 670 947	−0.175 253 602
1.15	1.068 784 876	0.404 045 365	5.3	1.497 315 064	−0.165 505 959
1.20	1.108 047 199	0.420 459 183	5.4	1.482 300 083	−0.154 385 926
1.25	1.146 446 416	0.434 300 724	5.5	1.468 724 073	−0.142 052 948
1.30	1.183 958 009	0.445 738 568	5.6	1.456 668 385	−0.128 671 749
1.35	1.220 558 751	0.454 926 675	5.7	1.446 197 529	−0.114 410 781
1.40	1.256 226 733	0.462 006 585	5.8	1.437 359 182	−0.099 440 665
1.45	1.290 941 390	0.467 109 209	5.9	1.430 184 334	−0.083 932 674
1.50	1.324 683 531	0.470 356 317	6.0	1.424 687 551	−0.068 057 244
1.55	1.357 435 538	0.471 861 764	6.1	1.420 867 373	−0.051 982 529
1.60	1.389 180 486	0.471 732 517	6.2	1.418 706 824	−0.035 873 019
1.65	1.419 903 964	0.470 069 510	6.3	1.418 174 035	−0.019 888 221
1.70	1.449 592 290	0.466 968 364	6.4	1.419 222 974	−0.004 181 411
1.75	1.478 233 419	0.462 519 997	6.5	1.421 794 274	+0.011 101 520
1.80	1.505 816 780	0.456 811 129	6.6	1.425 816 149	0.025 823 138
1.85	1.532 333 281	0.449 924 724	6.7	1.431 205 385	0.039 855 440
1.90	1.557 775 314	0.441 940 350	6.8	1.437 868 416	0.053 080 717
1.95	1.582 136 757	0.432 934 494	6.9	1.445 702 443	0.065 392 314
2.0	1.605 412 977	0.422 980 829	7.0	1.454 596 614	0.076 695 279
2.1	1.648 698 636	0.400 511 988	7.1	1.464 433 244	0.086 906 888
2.2	1.687 624 827	0.375 074 599	7.2	1.475 089 055	0.095 957 064
2.3	1.722 207 482	0.347 175 617	7.3	1.486 436 445	0.103 788 666
2.4	1.752 485 501	0.317 291 617	7.4	1.498 344 753	0.110 357 667
2.5	1.778 520 173	0.285 871 196	7.5	1.510 681 531	0.115 633 203
2.6	1.800 394 451	0.253 336 616	7.6	1.523 313 791	0.119 597 529
2.7	1.818 212 077	0.220 084 879	7.7	1.536 109 238	0.122 245 832
2.8	1.832 096 589	0.186 488 390	7.8	1.548 937 458	0.123 585 954
2.9	1.842 190 195	0.152 895 324	7.9	1.561 671 070	0.123 638 007
3.0	1.848 652 528	0.119 629 786	8.0	1.574 186 822	0.122 433 883
3.1	1.851 659 308	0.086 991 831	8.1	1.586 366 622	0.120 016 673
3.2	1.851 400 897	0.055 257 412	8.2	1.598 098 511	0.116 440 006
3.3	1.848 080 783	+0.024 678 285	8.3	1.609 277 542	0.111 767 293
3.4	1.841 913 983	−0.004 518 078	8.4	1.619 806 597	0.106 070 920
3.5	1.833 125 399	−0.032 128 549	8.5	1.629 597 100	0.099 431 359
3.6	1.821 948 116	−0.057 974 352	8.6	1.638 569 645	0.091 936 240
3.7	1.808 621 681	−0.081 901 001	8.7	1.646 654 531	0.083 679 370
3.8	1.793 390 355	−0.103 778 150	8.8	1.653 792 186	0.074 759 720
3.9	1.776 501 360	−0.123 499 349	8.9	1.659 933 505	0.065 280 385
4.0	1.758 203 139	−0.140 981 698	9.0	1.665 040 076	0.055 347 531
4.1	1.738 743 627	−0.156 165 392	9.1	1.669 084 306	0.045 069 333
4.2	1.718 368 564	−0.169 013 157	9.2	1.672 049 448	0.034 554 913
4.3	1.697 319 851	−0.179 509 573	9.3	1.673 929 528	0.023 913 305
4.4	1.675 833 959	−0.187 660 287	9.4	1.674 729 173	0.013 252 419
4.5	1.654 140 414	−0.193 491 122	9.5	1.674 463 342	+0.002 678 059
4.6	1.632 460 353	−0.197 047 080	9.6	1.673 156 980	−0.007 707 036
4.7	1.611 005 172	−0.198 391 247	9.7	1.670 844 570	−0.017 804 098
4.8	1.589 975 278	−0.197 603 613	9.8	1.667 569 617	−0.027 519 181
4.9	1.569 558 938	−0.194 779 806	9.9*	1.663 384 057	−0.036 763 956

* For $x > 9.9$ use the formula in (10.03c).

$$\text{Ei}(x) = \int_{\infty}^{x} \frac{e^{-x}}{x}\,dx \quad (10.01b) \qquad\qquad \bar{\text{Ei}}(x) = \int_{-\infty}^{x} \frac{e^{x}}{x}\,dx \quad (10.01b)$$

x	$\text{Ei}(x)$	$\bar{\text{Ei}}(x)$	x	$\text{Ei}(x)$	$\bar{\text{Ei}}(x)$
0.00	$-\infty$	∞	0.50	$-0.559\,773\,595$	0.454 219 905
0.01*	$-4.037\,929\,576$	$-4.140\,925\,226$	0.51	$-0.547\,822\,352$	0.487 032 167
0.02	$-3.354\,707\,783$	$-3.314\,706\,894$	0.52	$-0.536\,219\,798$	0.519 530 633
0.03	$-2.959\,118\,724$	$-2.899\,115\,724$	0.53	$-0.524\,951\,510$	0.551 730 445
0.04	$-2.681\,263\,689$	$-2.601\,256\,518$	0.54	$-0.514\,003\,886$	0.583 645 931
0.05	$-2.468\,051\,020$	$-2.367\,884\,599$	0.55	$-0.503\,364\,081$	0.615 290 657
0.06	$-2.295\,306\,918$	$-2.125\,282\,916$	0.56	$-0.493\,019\,959$	0.646 677 490
0.07	$-2.150\,838\,180$	$-2.010\,800\,064$	0.57	$-0.482\,960\,034$	0.677 818 642
0.08	$-2.026\,941\,003$	$-1.866\,884\,103$	0.58	$-0.473\,173\,433$	0.708 725 720
0.09	$-1.918\,744\,770$	$-1.738\,663\,750$	0.59	$-0.463\,649\,849$	0.739 409 764
0.10	$-1.822\,923\,958$	$-1.622\,812\,814$	0.60	$-0.454\,379\,503$	0.769 881 290
0.11	$-1.737\,106\,694$	$-1.516\,958\,751$	0.61	$-0.445\,353\,112$	0.800 150 320
0.12	$-1.659\,541\,752$	$-1.419\,349\,669$	0.62	$-0.436\,561\,854$	0.830 226 417
0.13	$-1.588\,899\,305$	$-1.328\,665\,020$	0.63	$-0.427\,997\,338$	0.860 118 716
0.14	$-1.524\,145\,722$	$-1.243\,560\,654$	0.64	$-0.419\,651\,581$	0.889 835 949
0.15	$-1.464\,461\,671$	$-1.164\,086\,417$	0.65	$-0.411\,516\,976$	0.919 386 468
0.16	$-1.409\,186\,699$	$-1.088\,731\,238$	0.66	$-0.403\,586\,275$	0.948 778 277
0.17	$-1.357\,780\,653$	$-1.017\,234\,290$	0.67	$-0.395\,852\,563$	0.978 019 042
0.18	$-1.309\,796\,135$	$-0.949\,147\,505$	0.68	$-0.388\,309\,243$	1.007 116 121
0.19	$-1.264\,858\,424$	$-0.884\,095\,487$	0.69	$-0.380\,950\,010$	1.036 076 576
0.20	$-1.222\,650\,544$	$-0.821\,760\,587$	0.70	$-0.373\,768\,843$	1.064 907 195
0.21	$-1.182\,901\,986$	$-0.761\,871\,623$	0.71	$-0.366\,759\,981$	1.093 614 501
0.22	$-1.145\,600\,055$	$-0.704\,195\,224$	0.72	$-0.359\,917\,914$	1.122 204 777
0.23	$-1.109\,881\,139$	$-0.648\,529\,102$	0.73	$-0.353\,237\,364$	1.150 684 069
0.24	$-1.076\,235\,415$	$-0.594\,696\,758$	0.74	$-0.346\,713\,279$	1.179 058 208
0.25	$-1.044\,282\,634$	$-0.542\,543\,264$	0.75	$-0.340\,340\,813$	1.207 332 816
0.26	$-1.013\,888\,737$	$-0.491\,931\,883$	0.76	$-0.334\,115\,321$	1.235 513 319
0.27	$-0.984\,933\,104$	$-0.442\,941\,312$	0.77	$-0.328\,032\,346$	1.263 604 960
0.28	$-0.957\,308\,300$	$-0.394\,863\,444$	0.78	$-0.322\,087\,610$	1.291 612 805
0.29	$-0.930\,918\,246$	$-0.348\,201\,510$	0.79	$-0.316\,277\,004$	1.319 541 753
0.30	$-0.905\,676\,652$	$-0.302\,668\,539$	0.80	$-0.310\,596\,579$	1.347 396 548
0.31	$-0.881\,505\,746$	$-0.258\,186\,075$	0.81	$-0.305\,042\,539$	1.375 181 783
0.32	$-0.858\,335\,189$	$-0.214\,683\,096$	0.82	$-0.299\,611\,236$	1.402 901 910
0.33	$-0.836\,101\,162$	$-0.172\,095\,092$	0.83	$-0.294\,299\,155$	1.430 561 245
0.34	$-0.814\,745\,580$	$-0.130\,363\,293$	0.84	$-0.289\,102\,918$	1.458 163 978
0.35	$-0.794\,217\,357$	$-0.089\,434\,001$	0.85	$-0.284\,019\,269$	1.485 714 176
0.36	$-0.774\,462\,218$	$-0.049\,258\,017$	0.86	$-0.279\,045\,070$	1.513 215 791
0.37	$-0.755\,441\,428$	$-0.009\,790\,148$	0.87	$-0.274\,177\,301$	1.540 672 664
0.38	$-0.737\,112\,144$	$+0.029\,011\,221$	0.88	$-0.269\,413\,046$	1.568 088 534
0.39	$-0.719\,436\,652$	$0.067\,184\,501$	0.89	$-0.264\,749\,496$	1.595 467 036
0.40	$-0.702\,380\,119$	$0.104\,265\,219$	0.90	$-0.260\,183\,939$	1.622 811 714
0.41	$-0.685\,910\,311$	$0.141\,786\,307$	0.91	$-0.255\,713\,758$	1.650 126 019
0.42	$-0.669\,997\,342$	$0.195\,347\,657$	0.92	$-0.251\,336\,425$	1.677 413 317
0.43	$-0.654\,613\,448$	$0.214\,269\,821$	0.93	$-0.247\,049\,501$	1.704 676 891
0.44	$-0.639\,732\,798$	$0.249\,787\,245$	0.94	$-0.242\,850\,627$	1.731 919 946
0.45	$-0.625\,331\,316$	$0.284\,855\,406$	0.95	$-0.238\,737\,524$	1.759 145 612
0.46	$-0.611\,386\,530$	$0.319\,497\,483$	0.96	$-0.234\,707\,988$	1.786 356 947
0.47	$-0.597\,877\,426$	$0.353\,735\,196$	0.97	$-0.230\,759\,890$	1.813 556 941
0.48	$-0.584\,784\,345$	$0.387\,588\,924$	0.98	$-0.226\,891\,167$	1.840 748 519
0.49	$-0.572\,088\,836$	$0.421\,077\,818$	0.99	$-0.223\,099\,826$	1.867 934 543

* For $x < 0.01$ use the formula in (10.03a).

$$\text{Ei}(x) = \int_{\infty}^{x} \frac{e^{-x}}{x}\, dx \qquad (10.01b) \qquad\qquad \bar{\text{Ei}}(x) = \int_{-\infty}^{x} \frac{e^{x}}{x}\, dx \qquad (10.01b)$$

x	$\text{Ei}(x)$	$\bar{\text{Ei}}(x)$	x	$\text{Ei}(x)$	$\bar{\text{Ei}}(x)$
1.00	−0.219 383 934	1.895 117 816	1.50	−0.100 019 582	3.301 285 449
1.01	−0.215 741 624	1.922 301 085	1.51	−0.098 544 365	3.331 213 449
1.02	−0.212 171 083	1.949 487 042	1.52	−0.097 093 466	3.361 242 701
1.03	−0.208 670 559	1.976 678 325	1.53	−0.095 666 424	3.391 374 858
1.04	−0.205 238 352	2.003 877 525	1.54	−0.094 262 786	3.421 611 576
1.05	−0.201 872 813	2.031 087 184	1.55	−0.092 882 108	3.451 954 503
1.06	−0.198 572 347	2.058 309 800	1.56	−0.091 523 960	3.482 405 289
1.07	−0.195 335 403	2.085 547 825	1.57	−0.090 187 917	3.512 965 580
1.08	−0.192 160 479	2.112 803 672	1.58	−0.088 873 566	3.543 637 024
1.09	−0.189 046 118	2.140 079 712	1.59	−0.087 580 504	3.574 421 266
1.10	−0.185 990 905	2.167 378 280	1.60	−0.086 308 334	3.605 319 949
1.11	−0.182 993 465	2.194 701 672	1.61	−0.085 056 670	3.636 334 719
1.12	−0.180 052 467	2.222 052 152	1.62	−0.083 825 133	3.667 467 221
1.13	−0.177 166 615	2.249 431 949	1.63	−0.082 613 354	3.698 719 099
1.14	−0.174 334 651	2.276 843 260	1.64	−0.081 420 970	3.730 091 999
1.15	−0.171 555 354	2.304 288 252	1.65	−0.080 247 627	3.761 587 569
1.16	−0.168 827 535	2.331 769 062	1.66	−0.079 092 978	3.793 207 456
1.17	−0.166 150 040	2.359 287 800	1.67	−0.077 956 684	3.824 953 310
1.18	−0.163 521 748	2.386 846 549	1.68	−0.076 838 412	3.856 826 783
1.19	−0.160 941 567	2.414 447 367	1.69	−0.075 737 839	3.888 829 528
1.20	−0.158 408 437	2.442 092 285	1.70	−0.074 654 644	3.920 963 201
1.21	−0.155 921 324	2.469 783 315	1.71	−0.073 588 518	3.953 229 462
1.22	−0.153 479 226	2.497 522 442	1.72	−0.072 539 154	3.985 629 972
1.23	−0.151 081 164	2.525 311 634	1.73	−0.071 506 255	4.018 166 395
1.24	−0.148 726 188	2.553 152 836	1.74	−0.070 489 527	4.050 840 400
1.25	−0.146 413 373	2.581 047 974	1.75	−0.069 488 685	4.083 653 659
1.26	−0.144 141 815	2.608 998 956	1.76	−0.068 503 447	4.116 607 847
1.27	−0.141 910 639	2.637 007 673	1.77	−0.067 533 539	4.149 704 645
1.28	−0.139 718 989	2.665 075 997	1.78	−0.066 578 691	4.182 945 736
1.29	−0.137 566 032	2.693 205 785	1.79	−0.065 638 641	4.216 332 809
1.30	−0.135 450 958	2.721 398 880	1.80	−0.064 713 129	4.249 867 557
1.31	−0.133 372 975	2.749 657 110	1.81	−0.063 801 903	4.283 551 681
1.32	−0.131 331 314	2.777 982 287	1.82	−0.062 904 715	4.317 386 883
1.33	−0.129 325 224	2.806 376 214	1.83	−0.062 021 320	4.351 374 872
1.34	−0.127 353 972	2.834 840 677	1.84	−0.061 151 482	4.385 517 364
1.35	−0.125 416 844	2.863 377 453	1.85	−0.060 294 967	4.419 816 080
1.36	−0.123 513 146	2.891 988 308	1.86	−0.059 451 545	4.454 272 746
1.37	−0.121 642 198	2.920 674 997	1.87	−0.058 620 994	4.488 889 097
1.38	−0.119 803 337	2.949 439 263	1.88	−0.057 803 091	4.523 666 872
1.39	−0.117 995 919	2.978 282 844	1.89	−0.056 997 623	4.558 607 817
1.40	−0.116 219 313	3.007 207 464	1.90	−0.056 204 378	4.593 713 687
1.41	−0.114 472 903	3.036 214 843	1.91	−0.055 423 149	4.628 986 242
1.42	−0.112 756 090	3.065 306 691	1.92	−0.054 653 731	4.664 427 249
1.43	−0.111 068 287	3.094 484 712	1.93	−0.053 895 927	4.700 038 485
1.44	−0.109 408 923	3.123 750 601	1.94	−0.053 149 540	4.735 821 734
1.45	−0.107 777 440	3.153 106 049	1.95	−0.052 414 380	4.771 778 785
1.46	−0.106 173 291	3.182 552 741	1.96	−0.051 690 257	4.807 911 438
1.47	−0.104 595 946	3.212 092 355	1.97	−0.050 976 988	4.844 221 501
1.48	−0.103 044 882	3.241 726 566	1.98	−0.050 274 392	4.880 710 791
1.49	−0.101 519 593	3.271 457 042	1.99*	−0.049 582 291	4.917 381 131

* For $x > 1.99$ use the formula in (10.03c).

$$\text{erf}(x) = \frac{2}{\sqrt{\pi}} \int_0^x e^{-x^2}\, dx \qquad (10.01b) \qquad\qquad \frac{d[\text{erf}(x)]}{dx} = \frac{2}{\sqrt{\pi}} e^{-x^2} \qquad (10.01b)$$

x	erf (x)	$d[\text{erf}(x)]/dx$	x	erf (x)	$d[\text{erf}(x)]/dx$
0.00	0.000 000 000	1.128 379 167	0.50	0.520 499 878	0.878 782 579
0.01*	0.011 283 416	1.128 266 335	0.51	0.529 243 620	0.869 951 547
0.02	0.022 564 575	1.127 927 906	0.52	0.537 898 631	0.861 037 034
0.03	0.033 841 222	1.127 364 083	0.53	0.546 464 097	0.852 043 444
0.04	0.045 111 106	1.126 575 204	0.54	0.554 939 251	0.842 975 181
0.05	0.056 371 978	1.125 561 742	0.55	0.563 323 366	0.833 836 647
0.06	0.067 621 594	1.124 324 305	0.56	0.571 615 764	0.824 632 240
0.07	0.078 857 720	1.122 863 633	0.57	0.579 815 806	0.815 366 346
0.08	0.090 078 126	1.121 180 600	0.58	0.587 922 900	0.806 043 343
0.09	0.101 280 594	1.119 276 213	0.59	0.595 936 497	0.796 667 591
0.10	0.112 462 916	1.117 151 607	0.60	0.603 856 091	0.787 243 432
0.11	0.123 622 896	1.114 808 050	0.61	0.611 681 219	0.777 775 185
0.12	0.134 758 352	1.112 246 938	0.62	0.619 411 462	0.768 267 144
0.13	0.145 867 115	1.109 469 793	0.63	0.627 046 443	0.758 723 576
0.14	0.156 947 033	1.106 478 265	0.64	0.634 585 829	0.749 148 716
0.15	0.167 995 971	1.103 274 127	0.65	0.642 029 327	0.739 546 763
0.16	0.179 011 813	1.099 859 273	0.66	0.649 376 680	0.729 921 881
0.17	0.189 992 461	1.096 235 719	0.67	0.656 627 702	0.720 278 193
0.18	0.200 935 839	1.092 405 601	0.68	0.663 782 203	0.710 619 778
0.19	0.211 839 892	1.088 371 168	0.69	0.670 840 062	0.700 950 672
0.20	0.222 702 589	1.084 134 787	0.70	0.677 801 194	0.691 274 860
0.21	0.233 521 923	1.079 698 934	0.71	0.684 665 550	0.681 596 279
0.22	0.244 295 912	1.075 066 196	0.72	0.691 433 123	0.671 918 811
0.23	0.255 022 600	1.070 239 267	0.73	0.698 103 943	0.662 246 284
0.24	0.265 700 059	1.065 220 945	0.74	0.704 678 078	0.652 582 467
0.25	0.276 326 390	1.060 014 129	0.75	0.711 155 634	0.642 931 069
0.26	0.286 899 723	1.054 621 819	0.76	0.717 536 753	0.633 295 740
0.27	0.297 418 212	1.049 047 110	0.77	0.723 821 614	0.623 680 063
0.28	0.307 880 067	1.043 293 189	0.78	0.730 010 431	0.614 087 551
0.29	0.318 283 496	1.037 363 333	0.79	0.736 103 454	0.604 521 670
0.30	0.328 626 760	1.031 260 910	0.80	0.742 100 965	0.594 985 786
0.31	0.338 908 150	1.024 989 366	0.81	0.748 003 281	0.585 483 216
0.32	0.349 125 995	1.018 552 231	0.82	0.753 810 751	0.576 017 197
0.33	0.359 278 655	1.011 953 112	0.83	0.759 523 757	0.566 590 894
0.34	0.369 364 529	1.005 195 689	0.84	0.765 142 712	0.557 207 397
0.35	0.379 382 054	0.998 283 712	0.85	0.770 668 058	0.547 869 717
0.36	0.389 329 701	0.991 221 000	0.86	0.776 100 268	0.538 580 792
0.37	0.399 205 984	0.984 011 434	0.87	0.781 439 846	0.529 343 477
0.38	0.409 009 453	0.976 658 954	0.88	0.786 687 319	0.520 160 557
0.39	0.418 738 700	0.969 167 559	0.89	0.791 843 247	0.511 034 712
0.40	0.428 392 355	0.961 541 299	0.90	0.796 908 212	0.501 968 574
0.41	0.437 969 090	0.953 784 273	0.91	0.801 882 826	0.492 964 674
0.42	0.447 467 618	0.945 900 626	0.92	0.806 767 722	0.484 025 464
0.43	0.456 886 695	0.937 894 544	0.93	0.811 563 559	0.475 153 313
0.44	0.466 225 115	0.929 770 254	0.94	0.816 271 019	0.466 350 509
0.45	0.475 481 720	0.921 532 013	0.95	0.820 890 807	0.457 619 255
0.46	0.484 655 390	0.913 184 112	0.96	0.825 423 650	0.448 961 670
0.47	0.493 745 051	0.904 730 869	0.97	0.829 870 293	0.440 379 791
0.48	0.502 749 671	0.896 176 622	0.98	0.834 231 504	0.431 875 571
0.49	0.511 668 261	0.887 525 734	0.99	0.838 508 070	0.423 450 878

* For $x < 0.01$ use the formula in $(10.03a)$.

$$\text{erf}(x) = \frac{2}{\sqrt{\pi}} \int_0^x e^{-x^2}\, dx \qquad (10.01b) \qquad\qquad \frac{d[\text{erf}(x)]}{dx} = \frac{2}{\sqrt{\pi}} e^{-x^2} \qquad (10.01b)$$

x	$\text{erf}(x)$	$d[\text{erf}(x)]/dx$	x	$\text{erf}(x)$	$d[\text{erf}(x)]/dx$
1.00	0.842 700 793	0.415 107 497	1.50	0.966 105 147	0.118 930 289
1.01	0.846 810 496	0.406 847 132	1.51	0.967 276 748	0.115 403 827
1.02	0.850 838 018	0.398 671 399	1.52	0.968 413 497	0.111 959 536
1.03	0.854 784 212	0.390 581 837	1.53	0.969 516 209	0.108 596 320
1.04	0.858 649 947	0.382 579 899	1.54	0.970 585 690	0.105 313 068
1.05	0.862 436 106	0.374 666 957	1.55	0.971 622 733	0.102 108 658
1.06	0.866 143 587	0.366 844 303	1.56	0.972 628 122	0.098 981 951
1.07	0.869 773 297	0.359 113 149	1.57	0.973 602 628	0.095 931 800
1.08	0.873 326 158	0.351 474 625	1.58	0.974 547 009	0.092 957 046
1.09	0.876 803 102	0.343 929 783	1.59	0.975 462 016	0.090 056 524
1.10	0.880 205 070	0.336 479 598	1.60	0.976 348 383	0.087 229 059
1.11	0.883 533 012	0.329 124 967	1.61	0.977 206 837	0.084 473 470
1.12	0.886 787 890	0.321 866 710	1.62	0.978 038 088	0.081 788 571
1.13	0.889 970 670	0.314 705 574	1.63	0.978 842 840	0.079 173 173
1.14	0.893 082 328	0.307 642 230	1.64	0.979 621 780	0.076 626 082
1.15	0.896 123 843	0.300 677 276	1.65	0.980 375 585	0.074 146 103
1.16	0.899 096 203	0.293 811 239	1.66	0.981 104 921	0.071 732 040
1.17	0.902 000 399	0.287 044 575	1.67	0.981 810 442	0.069 382 697
1.18	0.904 837 427	0.280 377 670	1.68	0.982 492 787	0.067 096 878
1.19	0.907 608 286	0.273 810 844	1.69	0.983 152 587	0.064 873 390
1.20	0.910 313 978	0.267 344 347	1.70	0.983 790 459	0.062 711 041
1.21	0.912 955 508	0.260 978 366	1.71	0.984 407 008	0.060 608 644
1.22	0.915 533 881	0.254 713 024	1.72	0.985 002 827	0.058 565 016
1.23	0.918 050 104	0.248 548 381	1.73	0.985 578 500	0.056 578 979
1.24	0.920 505 184	0.242 484 434	1.74	0.986 134 595	0.054 649 361
1.25	0.922 900 128	0.236 521 122	1.75	0.986 671 671	0.052 774 996
1.26	0.925 235 942	0.230 658 328	1.76	0.987 190 275	0.050 954 726
1.27	0.927 513 629	0.224 895 875	1.77	0.987 690 942	0.049 187 401
1.28	0.929 734 193	0.219 233 532	1.78	0.988 174 196	0.047 471 879
1.29	0.931 898 633	0.213 671 015	1.79	0.988 640 549	0.045 807 027
1.30	0.934 007 945	0.208 207 987	1.80	0.989 090 502	0.044 191 723
1.31	0.936 063 123	0.202 844 062	1.81	0.989 524 545	0.042 624 854
1.32	0.938 065 155	0.197 578 805	1.82	0.989 943 157	0.041 105 319
1.33	0.940 015 026	0.192 411 733	1.83	0.990 346 805	0.039 632 026
1.34	0.941 913 715	0.187 342 317	1.84	0.990 735 948	0.038 203 897
1.35	0.943 762 196	0.182 369 907	1.85	0.991 111 030	0.036 819 865
1.36	0.945 561 437	0.177 494 126	1.86	0.991 472 488	0.035 478 877
1.37	0.947 312 398	0.172 714 081	1.87	0.991 820 748	0.034 179 892
1.38	0.949 016 035	0.168 029 157	1.88	0.992 156 223	0.032 921 881
1.39	0.950 673 296	0.163 438 622	1.89	0.992 479 318	0.031 703 831
1.40	0.952 285 120	0.158 941 708	1.90	0.992 790 429	0.030 524 740
1.41	0.953 852 439	0.154 537 613	1.91	0.993 089 940	0.029 383 624
1.42	0.955 376 179	0.150 225 503	1.92	0.993 378 225	0.028 279 510
1.43	0.956 857 253	0.146 004 511	1.93	0.993 655 650	0.027 211 441
1.44	0.958 296 570	0.141 873 741	1.94	0.993 922 571	0.026 178 475
1.45	0.959 695 026	0.137 832 271	1.95	0.994 179 334	0.025 179 685
1.46	0.961 053 510	0.133 879 149	1.96	0.994 426 276	0.024 214 158
1.47	0.962 372 900	0.130 013 399	1.97	0.994 663 725	0.023 280 999
1.48	0.963 654 065	0.126 234 024	1.98	0.994 892 000	0.022 379 324
1.49	0.964 897 865	0.122 540 001	1.99*	0.995 111 413	0.021 508 270

* For $x > 1.99$ use the formula in (10.03c).

$J_r(x)$ = ordinary Bessel function of the first kind* (10.07b)

x	$J_0(x)$	$J_1(x)$	x	$J_0(x)$	$J_1(x)$
0.0	1.000 000 000	0.000 000 000	5.0	−0.177 596 771	−0.327 579 138
0.1	0.997 501 562	0.049 937 526	5.1	−0.144 334 747	−0.337 097 202
0.2	0.990 024 972	0.099 500 833	5.2	−0.110 290 440	−0.343 223 006
0.3	0.977 626 246	0.148 318 816	5.3	−0.075 803 112	−0.345 960 834
0.4	0.960 398 227	0.196 026 578	5.4	−0.041 210 101	−0.345 344 791
0.5	0.938 469 807	0.242 268 458	5.5	−0.006 843 869	−0.341 438 215
0.6	0.912 004 863	0.286 700 988	5.6	+0.026 970 885	−0.334 332 836
0.7	0.881 200 889	0.328 995 742	5.7	0.059 920 010	−0.324 147 680
0.8	0.846 287 353	0.368 842 046	5.8	0.091 702 568	−0.311 027 744
0.9	0.807 523 798	0.405 949 546	5.9	0.122 033 355	−0.295 142 445
1.0	0.765 197 687	0.440 050 586	6.0	0.150 645 257	−0.276 683 858
1.1	0.719 622 019	0.470 902 395	6.1	0.177 291 422	−0.255 864 773
1.2	0.671 132 744	0.498 289 058	6.2	0.201 747 223	−0.232 916 567
1.3	0.620 085 960	0.522 023 247	6.3	0.223 812 006	−0.208 086 940
1.4	0.566 855 120	0.541 947 714	6.4	0.243 310 605	−0.181 637 509
1.5	0.511 827 672	0.557 936 501	6.5	0.260 094 606	−0.153 841 301
1.6	0.455 402 168	0.569 895 935	6.6	0.274 043 361	−0.124 980 165
1.7	0.397 984 859	0.577 765 232	6.7	0.285 064 738	−0.095 342 118
1.8	0.339 986 411	0.581 516 952	6.8	0.293 095 603	−0.065 218 663
1.9	0.281 818 559	0.581 157 073	6.9	0.298 102 035	−0.034 902 096
2.0	0.223 890 779	0.576 724 801	7.0	0.300 079 271	−0.004 682 824
2.1	0.166 606 980	0.568 292 136	7.1	0.299 051 381	+0.025 153 274
2.2	0.110 362 267	0.555 963 050	7.2	0.295 070 691	0.054 327 420
2.3	0.055 539 784	0.539 872 533	7.3	0.288 216 948	0.082 570 431
2.4	+0.002 507 683	0.520 185 268	7.4	0.278 596 233	0.109 625 095
2.5	−0.048 383 776	0.497 094 103	7.5	0.266 339 658	0.135 248 428
2.6	−0.096 804 954	0.470 818 267	7.6	0.251 601 834	0.159 213 768
2.7	−0.142 449 370	0.441 601 379	7.7	0.234 559 140	0.181 312 715
2.8	−0.185 036 033	0.409 709 247	7.8	0.215 407 801	0.201 356 873
2.9	−0.224 311 546	0.375 427 482	7.9	0.194 361 845	0.219 179 400
3.0	−0.260 051 955	0.339 058 959	8.0	0.171 650 807	0.234 636 347
3.1	−0.292 064 348	0.300 921 133	8.1	0.147 517 454	0.247 607 767
3.2	−0.320 188 170	0.261 343 249	8.2	0.122 215 302	0.257 998 598
3.3	−0.344 296 260	0.220 663 453	8.3	0.096 006 101	0.265 739 302
3.4	−0.364 295 597	0.179 225 852	8.4	0.069 157 262	0.270 786 268
3.5	−0.380 127 740	0.137 377 527	8.5	0.041 939 252	0.273 121 964
3.6	−0.391 768 984	0.095 465 547	8.6	+0.014 622 991	0.272 754 845
3.7	−0.399 230 203	0.053 833 988	8.7	−0.012 522 732	0.269 719 024
3.8	−0.402 556 410	+0.012 821 003	8.8	−0.039 233 803	0.264 073 703
3.9	−0.401 826 015	−0.027 244 040	8.9	−0.065 253 247	0.255 902 371
4.0	−0.397 149 810	−0.066 043 328	9.0	−0.090 333 611	0.245 311 789
4.1	−0.388 669 680	−0.103 273 258	9.1	−0.114 239 233	0.232 430 745
4.2	−0.376 557 054	−0.138 646 942	9.2	−0.136 748 371	0.217 408 655
4.3	−0.361 011 117	−0.171 896 560	9.3	−0.157 655 190	0.200 413 928
4.4	−0.342 256 790	−0.202 775 522	9.4	−0.176 771 575	0.181 632 204
4.5	−0.320 542 509	−0.231 060 432	9.5	−0.193 928 748	0.161 264 431
4.6	−0.296 137 816	−0.256 552 836	9.6	−0.208 978 718	0.139 524 812
4.7	−0.269 330 789	−0.279 080 734	9.7	−0.221 795 482	0.116 638 648
4.8	−0.240 425 327	−0.298 499 858	9.8	−0.232 276 028	0.092 840 091
4.9	−0.209 738 328	−0.314 694 671	9.9	−0.240 341 106	0.068 369 832

* For $r = 2, 3, \ldots$ use the recurrent relations in (10.07d).

$Y_r(x)$ = ordinary Bessel function of the second kind* (10.07c)

x	$Y_0(x)$	$Y_1(x)$	x	$Y_0(x)$	$Y_1(x)$
0.0	$-\infty$	$-\infty$	5.0	−0.308 517 625	0.147 863 143
0.1	−1.534 238 651	−6.458 951 095	5.1	−0.321 602 449	0.113 736 442
0.2	−1.081 105 322	−3.323 824 988	5.2	−0.331 250 935	0.079 190 343
0.3	−0.807 273 578	−2.293 105 138	5.3	−0.337 437 301	0.044 547 619
0.4	−0.606 024 568	−1.780 872 044	5.4	−0.340 167 878	+0.010 127 267
0.5	−0.444 518 734	−1.471 472 393	5.5	−0.339 480 593	−0.023 758 239
0.6	−0.308 509 870	−1.260 391 347	5.6	−0.335 444 181	−0.056 805 614
0.7	−0.190 664 929	−1.103 249 872	5.7	−0.328 157 141	−0.088 723 341
0.8	−0.086 802 280	−0.978 144 177	5.8	−0.317 746 430	−0.119 234 114
0.9	+0.005 628 307	−0.873 126 583	5.9	−0.304 365 930	−0.148 077 153
1.0	0.088 256 964	−0.781 212 821	6.0	−0.288 194 684	−0.175 010 344
1.1	0.162 163 203	−0.698 119 560	6.1	−0.269 434 930	−0.199 812 205
1.2	0.228 083 503	−0.621 136 380	6.2	−0.248 309 951	−0.222 283 641
1.3	0.286 535 357	−0.548 519 730	6.3	−0.225 061 750	−0.242 249 501
1.4	0.337 895 130	−0.479 146 974	6.4	−0.199 948 595	−0.259 559 893
1.5	0.382 448 924	−0.412 308 627	6.5	−0.173 242 435	−0.274 091 274
1.6	0.420 426 896	−0.347 578 008	6.6	−0.145 226 217	−0.285 747 279
1.7	0.452 027 000	−0.284 726 245	6.7	−0.116 191 143	−0.294 459 313
1.8	0.477 431 715	−0.223 664 868	6.8	−0.086 433 868	−0.300 186 876
1.9	0.496 819 971	−0.164 405 772	6.9	−0.056 253 692	−0.302 917 634
2.0	0.510 375 673	−0.107 032 432	7.0	−0.025 949 744	−0.302 667 237
2.1	0.518 293 738	−0.051 678 612	7.1	+0.004 181 793	−0.299 478 875
2.2	0.520 784 285	+0.001 487 789	7.2	0.033 850 405	−0.293 422 594
2.3	0.518 075 396	0.052 277 316	7.3	0.062 773 886	−0.284 594 372
2.4	0.510 414 749	0.100 488 938	7.4	0.090 680 880	−0.273 114 960
2.5	0.498 070 360	0.145 918 138	7.5	0.117 313 286	−0.259 128 511
2.6	0.481 330 591	0.188 363 544	7.6	0.142 428 525	−0.242 801 002
2.7	0.460 503 549	0.227 632 446	7.7	0.165 801 632	−0.224 318 474
2.8	0.435 915 986	0.263 545 391	7.8	0.187 227 173	0.203 885 095
2.9	0.407 911 769	0.295 940 055	7.9	0.206 520 948	−0.181 721 077
3.0	0.376 850 010	0.324 674 425	8.0	0.223 521 489	−0.158 060 462
3.1	0.343 102 889	0.349 629 482	8.1	0.238 091 329	−0.133 148 796
3.2	0.307 053 250	0.370 711 338	8.2	0.250 118 028	−0.107 240 722
3.3	0.269 091 995	0.387 852 931	8.3	0.259 514 964	−0.080 597 504
3.4	0.229 615 337	0.401 015 292	8.4	0.266 221 867	−0.053 484 508
3.5	0.189 021 944	0.410 188 418	8.5	0.270 205 105	−0.026 168 679
3.6	0.147 710 013	0.415 391 762	8.6	0.271 457 712	+0.001 083 992
3.7	0.106 074 315	0.416 674 373	8.7	0.269 999 170	0.028 010 959
3.8	0.064 503 247	0.414 114 689	8.8	0.265 874 942	0.054 355 563
3.9	+0.023 375 908	0.407 820 019	8.9	0.259 155 762	0.079 869 397
4.0	−0.016 940 739	0.397 925 711	9.0	0.249 936 698	0.104 314 575
4.1	−0.056 094 627	0.384 594 035	9.1	0.238 335 992	0.127 465 882
4.2	−0.093 751 201	0.368 012 808	9.2	0.224 493 687	0.149 112 788
4.3	−0.129 595 903	0.348 393 758	9.3	0.208 570 068	0.169 061 307
4.4	−0.163 336 463	0.325 970 671	9.4	0.190 743 919	0.187 135 685
4.5	−0.194 705 009	0.300 997 323	9.5	0.171 210 626	0.203 179 899
4.6	−0.223 459 953	0.273 745 242	9.6	0.150 180 135	0.217 058 966
4.7	−0.249 387 647	0.244 501 297	9.7	0.127 874 792	0.228 660 030
4.8	−0.272 303 795	0.213 565 167	9.8	0.104 527 088	0.237 893 242
4.9	−0.292 054 594	0.181 246 692	9.9	0.080 377 305	0.244 692 477

* For $r = 2, 3, \ldots$ use the recurrent relations in (10.07d).

$I_r(x) =$ modified Bessel function of the first kind* (10.13b)

x	$e^{-x}I_0(x)$	$e^{-x}I_1(x)$	x	$e^{-x}I_0(x)$	$e^{-x}I_1(x)$
0.0	1.000 000 000	0.000 000 000	5.0	0.183 540 813	0.163 972 267
0.1	0.907 100 926	0.045 298 447	5.1	0.181 615 102	0.162 663 855
0.2	0.826 938 552	0.082 283 124	5.2	0.179 749 488	0.161 383 285
0.3	0.757 580 625	0.112 377 561	5.3	0.177 940 865	0.160 129 791
0.4	0.697 402 171	0.136 763 224	5.4	0.176 186 348	0.158 902 615
0.5	0.645 035 271	0.156 420 803	5.5	0.174 483 256	0.157 701 009
0.6	0.599 327 203	0.172 164 420	5.6	0.172 829 095	0.156 524 241
0.7	0.559 305 527	0.184 669 983	5.7	0.171 221 536	0.155 371 592
0.8	0.524 148 942	0.194 498 693	5.8	0.169 658 406	0.154 242 364
0.9	0.493 162 966	0.202 116 531	5.9	0.168 137 673	0.153 135 874
1.0	0.465 759 608	0.207 910 415	6.0	0.166 657 433	0.152 051 459
1.1	0.441 440 378	0.212 201 613	6.1	0.165 215 902	0.150 988 475
1.2	0.419 782 079	0.215 256 859	6.2	0.163 811 406	0.149 946 298
1.3	0.400 424 913	0.217 297 588	6.3	0.162 442 372	0.148 924 321
1.4	0.383 062 515	0.218 507 592	6.4	0.161 107 318	0.147 921 960
1.5	0.367 433 609	0.219 039 387	6.5	0.159 804 849	0.146 938 646
1.6	0.353 314 998	0.219 019 490	6.6	0.158 533 651	0.145 973 831
1.7	0.340 515 688	0.218 552 807	6.7	0.157 292 483	0.145 026 987
1.8	0.328 871 950	0.217 726 279	6.8	0.156 080 172	0.144 097 599
1.9	0.318 243 163	0.216 611 911	6.9	0.154 895 609	0.143 185 175
2.0	0.308 508 323	0.215 269 289	7.0	0.153 737 745	0.142 289 235
2.1	0.299 563 095	0.213 747 672	7.1	0.152 605 584	0.141 409 319
2.2	0.291 317 333	0.212 087 733	7.2	0.151 498 186	0.140 544 981
2.3	0.283 692 986	0.210 323 005	7.3	0.150 414 653	0.139 695 792
2.4	0.276 622 323	0.208 481 089	7.4	0.149 354 137	0.138 861 335
2.5	0.270 046 442	0.206 584 650	7.5	0.148 315 830	0.138 041 212
2.6	0.263 913 996	0.204 652 254	7.6	0.147 298 964	0.137 235 033
2.7	0.258 180 124	0.202 699 061	7.7	0.146 302 806	0.136 442 427
2.8	0.252 805 534	0.200 737 411	7.8	0.145 326 661	0.135 663 032
2.9	0.247 755 730	0.198 777 282	7.9	0.144 369 864	0.134 896 500
3.0	0.243 000 354	0.196 826 712	8.0	0.143 431 782	0.134 142 493
3.1	0.238 512 619	0.194 892 131	8.1	0.142 511 810	0.133 400 688
3.2	0.234 268 832	0.192 978 623	8.2	0.141 609 369	0.132 670 771
3.3	0.230 247 981	0.191 090 173	8.3	0.140 723 910	0.131 952 436
3.4	0.226 431 401	0.189 229 851	8.4	0.139 854 903	0.131 245 392
3.5	0.222 802 438	0.187 399 977	8.5	0.139 001 843	0.130 549 355
3.6	0.219 346 225	0.185 602 248	8.6	0.138 164 247	0.129 864 051
3.7	0.216 049 442	0.183 837 858	8.7	0.137 341 653	0.129 189 213
3.8	0.212 900 131	0.182 107 581	8.8	0.136 533 615	0.128 524 587
3.9	0.209 887 528	0.180 411 854	8.9	0.135 739 708	0.127 869 924
4.0	0.207 001 921	0.178 750 839	9.0	0.134 959 525	0.127 224 984
4.1	0.204 234 527	0.177 124 476	9.1	0.134 192 672	0.126 589 534
4.2	0.201 577 384	0.175 532 526	9.2	0.133 438 774	0.125 963 350
4.3	0.199 023 257	0.173 974 609	9.3	0.132 697 469	0.125 346 214
4.4	0.196 565 559	0.172 450 234	9.4	0.131 968 409	0.124 737 915
4.5	0.194 198 278	0.170 958 822	9.5	0.131 251 261	0.124 138 248
4.6	0.191 915 915	0.169 499 731	9.6	0.130 545 702	0.123 547 015
4.7	0.189 713 433	0.168 072 268	9.7	0.129 851 422	0.122 964 026
4.8	0.187 586 204	0.166 675 706	9.8	0.129 168 125	0.122 389 093
4.9	0.185 529 972	0.165 309 294	9.9	0.128 495 522	0.121 822 036

* For $r = 2, 3, \ldots$ use the recurrent relations in (10.13d).

$K_r(x) =$ modified Bessel function of the second kind* (10.13b)

x	$e^x K_0(x)$	$e^x K_1(x)$	x	$e^x K_0(x)$	$e^x K_1(x)$
0.0	∞	∞	5.0	0.547 807 564	0.600 273 859
0.1	2.682 326 102	10.890 182 683	5.1	0.542 635 352	0.593 625 046
0.2	2.140 757 323	5.833 386 037	5.2	0.537 607 354	0.587 188 606
0.3	1.852 627 301	4.125 157 762	5.3	0.532 716 974	0.580 953 609
0.4	1.662 682 089	3.258 673 880	5.4	0.527 958 033	0.574 909 887
0.5	1.524 109 386	2.731 009 708	5.5	0.523 324 732	0.569 047 974
0.6	1.416 737 621	2.373 920 038	5.6	0.518 811 625	0.563 359 039
0.7	1.330 123 656	2.115 011 313	5.7	0.514 413 594	0.557 834 835
0.8	1.258 203 122	1.917 930 299	5.8	0.510 125 818	0.552 467 650
0.9	1.197 163 380	1.762 388 220	5.9	0.505 943 758	0.547 250 264
1.0	1.144 463 080	1.636 153 486	6.0	0.501 863 131	0.542 175 910
1.1	1.098 330 283	1.531 403 754	6.1	0.497 879 893	0.537 238 239
1.2	1.057 484 532	1.442 897 552	6.2	0.493 990 224	0.532 431 283
1.3	1.020 973 161	1.366 987 284	6.3	0.490 190 509	0.527 749 434
1.4	0.988 069 996	1.301 053 740	6.4	0.486 477 329	0.523 187 410
1.5	0.958 210 053	1.243 165 874	6.5	0.482 847 441	0.518 740 234
1.6	0.930 945 981	1.191 867 565	6.6	0.479 297 773	0.514 403 211
1.7	0.905 918 139	1.146 039 246	6.7	0.475 825 407	0.510 171 910
1.8	0.882 833 527	1.104 805 373	6.8	0.472 427 572	0.506 042 142
1.9	0.861 450 617	1.067 470 930	6.9	0.469 101 637	0.502 009 947
2.0	0.841 568 215	1.033 476 847	7.0	0.465 845 096	0.498 071 575
2.1	0.823 017 153	1.002 368 053	7.1	0.462 655 566	0.494 223 474
2.2	0.805 653 981	0.973 770 168	7.2	0.459 530 776	0.490 462 276
2.3	0.789 356 131	0.947 372 225	7.3	0.156 468 562	0.486 784 784
2.4	0.774 018 141	0.922 913 665	7.4	0.453 466 859	0.483 187 965
2.5	0.759 548 690	0.900 174 424	7.5	0.450 523 699	0.479 668 934
2.6	0.745 868 243	0.878 967 281	7.6	0.447 637 200	0.476 224 949
2.7	0.732 907 152	0.859 131 887	7.7	0.444 805 564	0.472 853 400
2.8	0.720 604 125	0.840 530 060	7.8	0.442 027 072	0.469 551 801
2.9	0.708 904 977	0.823 042 040	7.9	0.439 300 082	0.466 317 785
3.0	0.697 761 598	0.806 563 480	8.0	0.436 623 019	0.463 149 093
3.1	0.687 131 101	0.791 003 016	8.1	0.433 994 375	0.460 043 571
3.2	0.676 975 114	0.776 280 282	8.2	0.431 412 708	0.456 999 162
3.3	0.667 259 183	0.762 324 286	8.3	0.428 876 633	0.454 013 900
3.4	0.657 952 273	0.749 072 061	8.4	0.426 384 821	0.451 085 909
3.5	0.649 026 338	0.736 467 548	8.5	0.423 935 999	0.448 213 392
3.6	0.640 455 965	0.724 460 661	8.6	0.421 528 943	0.445 394 630
3.7	0.632 218 059	0.713 006 501	8.7	0.419 162 478	0.442 627 978
3.8	0.624 291 581	0.702 064 693	8.8	0.416 835 474	0.439 911 860
3.9	0.616 657 315	0.691 598 821	8.9	0.414 546 846	0.437 244 765
4.0	0.609 297 669	0.681 575 945	9.0	0.412 295 550	0.434 625 245
4.1	0.602 196 506	0.671 966 195	9.1	0.410 080 578	0.432 051 912
4.2	0.595 338 989	0.662 742 411	9.2	0.407 900 966	0.429 523 430
4.3	0.588 711 449	0.653 879 840	9.3	0.405 755 781	0.427 038 520
4.4	0.582 301 270	0.645 355 869	9.4	0.403 644 125	0.424 595 952
4.5	0.576 096 790	0.637 149 800	9.5	0.401 565 132	0.422 194 543
4.6	0.570 087 202	0.629 242 638	9.6	0.399 517 969	0.419 833 157
4.7	0.564 262 484	0.621 616 931	9.7	0.397 501 831	0.417 510 699
4.8	0.558 613 319	0.614 256 600	9.8	0.395 515 942	0.415 226 118
4.9	0.553 131 040	0.607 146 813	9.9	0.393 559 551	0.412 978 400

* For $r = 2, 3, \ldots$ use the recurrent relations in (10.13d).

$$K = \int_0^{\pi/2} (1 - \sin^2 \omega \sin^2 \theta)^{-1/2} \, d\theta \qquad E = \int_0^{\pi/2} (1 - \sin^2 \omega \sin^2 \theta)^{1/2} \, d\theta \qquad (10.05b)$$

$\omega,°$	K	E	$\omega,°$	K	E
0	1.570 796 327	1.570 796 327	45	1.854 074 468	1.350 643 881
1	1.570 915 958	1.570 676 709	46	1.869 147 546	1.341 806 058
2	1.571 274 952	1.570 317 920	47	1.884 808 657	1.332 869 954
3	1.572 873 611	1.569 720 150	48	1.901 083 033	1.323 842 184
4	1.572 712 435	1.568 883 380	49	1.917 997 546	1.314 729 560
5	1.573 792 131	1.567 809 074	50	1.935 581 096	1.305 539 094
6	1.573 113 608	1.566 496 788	51	1.953 864 809	1.296 278 008
7	1.576 677 982	1.564 947 563	52	1.972 882 266	1.286 953 739
8	1.578 486 578	1.563 162 230	53	1.992 669 756	1.277 573 948
9	1.580 540 934	1.561 141 745	54	2.013 266 565	1.268 146 531
10	1.582 842 804	1.558 887 197	55	2.034 715 312	1.258 679 625
11	1.585 394 164	1.556 399 798	56	2.057 062 323	1.249 181 621
12	1.588 197 213	1.553 680 892	57	2.080 358 067	1.239 661 175
13	1.591 254 382	1.550 731 951	58	2.104 657 658	1.230 127 224
14	1.594 568 341	1.547 554 576	59	2.130 021 438	1.220 588 996
15	1.598 142 002	1.544 150 500	60	2.156 515 647	1.211 056 028
16	1.601 978 530	1.540 521 574	61	2.184 213 217	1.201 538 184
17	1.606 081 349	1.536 669 798	62	2.213 194 695	1.192 045 677
18	1.610 454 154	1.532 597 288	63	2.243 549 342	1.182 589 085
19	1.615 100 916	1.528 306 296	64	2.275 376 430	1.173 179 383
20	1.620 025 899	1.523 799 205	65	2.308 785 798	1.163 827 964
21	1.625 233 668	1.519 078 530	66	2.343 904 724	1.154 546 678
22	1.630 729 102	1.514 146 917	67	2.380 870 191	1.145 347 857
23	1.636 517 409	1.509 007 148	68	2.419 841 654	1.136 244 365
24	1.642 604 144	1.503 662 135	69	2.460 999 458	1.127 249 638
25	1.648 995 218	1.498 114 928	70	2.504 550 079	1.118 377 738
26	1.655 696 926	1.492 368 711	71	2.550 731 450	1.109 643 414
27	1.662 715 958	1.486 426 804	72	2.599 819 730	1.101 062 169
28	1.670 059 426	1.480 292 664	73	2.652 138 005	1.092 650 346
29	1.677 734 884	1.473 969 887	74	2.708 067 615	1.084 425 219
30	1.685 750 355	1.467 462 209	75	2.768 063 145	1.076 405 113
31	1.694 114 357	1.460 773 506	76	2.832 672 583	1.068 609 533
32	1.702 835 936	1.453 907 796	77	2.902 564 941	1.061 059 334
33	1.711 924 695	1.446 869 241	78	2.978 568 951	1.053 776 920
34	1.721 390 831	1.439 662 147	79	3.061 728 612	1.046 786 499
35	1.731 245 176	1.432 290 969	89	3.153 385 252	1.040 114 396
36	1.741 499 234	1.424 760 310	81	3.255 302 942	1.033 789 462
37	1.752 165 236	1.417 074 923	82	3.369 868 927	1.027 843 620
38	1.763 256 184	1.409 239 716	83	3.500 422 499	1.022 312 588
39	1.774 785 909	1.401 259 751	84	3.651 855 948	1.017 236 918
40	1.786 769 135	1.393 140 249	85	3.831 742 000	1.012 663 506
41	1.799 221 544	1.384 886 591	86	4.052 758 170	1.008 647 960
42	1.812 159 854	1.576 504 326	87	4.338 653 977	1.005 258 587
43	1.825 601 898	1.367 999 166	88	4.742 717 266	1.002 584 086
44	1.839 566 721	1.359 376 997	89	5.434 909 830	1.000 751 578
			90	∞	1.000 000 000

Appendix B
GLOSSARY
OF SYMBOLS

= or : :	Equals	$\neq$ or $\neq$	Does not equal
>	Greater than	<	Less than
$\geq$	Greater than or equal	$\leq$	Less than or equal
$\equiv$	Identical	$\approx$	Approximately equal

B.2 NUMERICAL CONSTANTS

π	Archimedes constant (2.17a)	G	Catalan constant (2.27b)
e	Euler's constant (2.18a)	γ	Euler's constant (2.19a)
$\bar{B}_r$	Bernoulli number (2.22a)	B_r	Auxilliary Bernoulli number (2.22b)
$\bar{E}_r$	Euler number (2.24a)	E_r	Auxilliary Euler number (2.24b)
$\mathcal{S}_k^{(p)}$	Stirling number of the first kind (2.11a)	α, β	Golden mean numbers (2.21b)
$\bar{\mathcal{S}}_k^{(p)}$	Stirling number of the second kind (2.11d)		
F_r	Fibonacci number (2.21b)		

B.3 ALGEBRA

+	Plus or positive	$-$	Minus or negative
$\pm\}$	Plus or minus / Positive or negative	$\mp\}$	Minus or plus / Negative or positive
$\times$	Multiplied by	$\div$ or :	Divided by
a^n	nth power of a	$\sqrt[n]{a}$	nth root of a
log / $\log_{10}\}$	Common logarithm or Brigg's logarithm	ln / $\log_e\}$	Natural logarithm or Napier's logarithm
( )	Parentheses [ ] Brackets	{ } Braces	$|-a|$ Positive a

$i = \sqrt{-1}$	Unit imaginary number	$z = x + iy$	Complex variable
$|z|$	Absolute value of z	$\bar{z} = x - iy$	Conjugate of z
$\begin{vmatrix} a_1 & a_2 & \cdots \\ b_1 & b_2 & \cdots \\ \cdots \end{vmatrix}$	Determinant	$\begin{bmatrix} a_1 & a_2 & \cdots \\ b_1 & b_2 & \cdots \\ \cdots \end{bmatrix}$	Matrix
$\mathbf{I}$	Unit matrix	Adj	Adjoint matrix
$\mathbf{A}^{-1}$	Inverse of the $\mathbf{A}$ matrix	$\mathbf{A}^T$	Transpose of the $\mathbf{A}$ matrix
$n!$	n factorial (2.09a)	$\binom{n}{k}$	Binomial coefficient (2.10d)
$n!!$	n double factorial (2.30d)	$X_h^{(p)}$	Factorial polynomial (2.11a)

B.4 NUMERICAL APPROXIMATIONS

ϵ	Absolute error	$\bar{\epsilon}$	Relative error
ϵ_T	Truncation eror	$\bar{y}$	Substitute function
Δy_n	Forward difference	∇y_n	Backward difference
δy_n	Central difference	Δx_n	Divided difference

(a, b)	Bounded open interval	$[a, b]$	Bounded closed interval
$f(x), F(x)$	Function of x	$f^{-1}(x), F^{-1}(x)$	Inverse function of x
$\displaystyle\sum_{i=1}^{n} u_i$	Sum of n terms	$\displaystyle\prod_{i=1}^{n} u_i$	Product of n terms

$$\bigwedge_{k=1}^{n} [1 \pm a_k s] = (1 \pm a_1 s(1 \pm a_2 s(1 \pm a_3 s(1 \pm \cdots \pm a_n s)))) \quad \text{Nested sum of } n + 1 \text{ terms}$$

Δu	Increment of u	du	Differential of u

$\left.\begin{array}{l} \dfrac{dy}{dx}, y', D_x y \\[2ex] \dfrac{df(x)}{dx}, f'(x), D_x f(x) \end{array}\right\}$ First-order derivative of $y = f(x)$ with respect to x

$\left.\begin{array}{l} \dfrac{d^n y}{dx^n}, y^{(n)}, D_x^{(n)} y \\[2ex] \dfrac{d^n f(x)}{dx^n}, f^{(n)}(x), D_x^{(n)} f(x) \end{array}\right\}$ nth order derivative of $y = f(x)$ with respect to x

$\left.\begin{array}{l} \dfrac{\partial w}{\partial x}, w_x, D_x w \\[2ex] \dfrac{\partial f}{\partial x}, f_x, F_x \end{array}\right\}$ First-order partial derivative of $w = f(x, y, \ldots)$ with respect to x

$\left.\begin{array}{l} \dfrac{\partial w}{\partial y}, w_y, D_y w \\[2ex] \dfrac{\partial f}{\partial y}, f_y, F_y \end{array}\right\}$ First-order partial derivative of $w = f(x, y, \ldots)$ with respect to y

$\left.\begin{array}{l} \dfrac{\partial^2 w}{\partial x^2}, w_{xx}, D_{xx} w \\[2ex] \dfrac{\partial^2 f}{\partial x^2}, f_{xx}, F_{xx} \end{array}\right\}$ Second-order partial derivative of $w = f(x, y, \ldots)$ with respect to x

$\left.\begin{array}{l} \dfrac{\partial^2 w}{\partial y^2}, w_{yy}, D_{yy} w \\[2ex] \dfrac{\partial^2 f}{\partial y^2}, f_{yy}, F_{yy} \end{array}\right\}$ Second-order partial derivative of $w = f(x, y, \ldots)$ with respect to y

$\left.\begin{array}{l} \dfrac{\partial^2 w}{\partial x \, \partial y}, w_{xy}, D_{xy} w \\[2ex] \dfrac{\partial^2 w}{\partial x \, \partial y}, f_{xy}, F_{xy} \end{array}\right\}$ Second-order partial derivative of $w = f(x, y, \ldots)$ with respect to x and then with respect to y

$\displaystyle\int f(x)\, dx$ Indefinite integral of $y = f(x)$ $\qquad$ $\displaystyle\int_a^b f(x)\, dx$ Definite integral of $y = f(x)$ in $[a, b]$

$u = bx$ Variable	$v = bx \ln a$ Variable

$e^u, e^{-u}, a^u, a^{-u}, e^v, e^{-v}$	Exponential functions (9.05a)
$\ln x, \log x$	Logarithmic functions (9.06a)
$\sin x$	Circular sine (9.07a)
$\cos x$	Circular cosine (9.07a)
$\sinh x$	Hyperbolic sine (9.08a)
$\cosh x$	Hyperbolic cosine (9.08a)
$\tan x$	Circular tangent (9.09a)
$\tanh x$	Hyperbolic tangent (9.10a)
$\sin^{-1} x, \arcsin x$	Inverse circular sine (9.11a)
$\cos^{-1} x, \arccos x$	Inverse circular cosine (9.11a)
$\sinh^{-1} x, \text{arcsinh } x$	Inverse hyperbolic sine (9.12a)
$\cosh^{-1} x, \text{arccosh } x$	Inverse hyperbolic cosine (9.12a)
$\tan^{-1} x, \arctan x$	Inverse circular tangent (9.13a)
$\tanh^{-1} x, \text{arctanh } x$	Inverse hyperbolic tangent (9.13a)

B.7   SPECIAL FUNCTIONS

$\Gamma(x)$	Gamma function (2.28a)	$\Pi(x)$	Pi function (2.31a)
$B(x)$	Beta function (2.31b)	$\text{erf}(x)$	Error function (10.01b)
$\Psi(x)$	Digamma function (2.32a)	$\Psi^{(m)}(x)$	Polygamma function (2.34a)
$Z(x)$	Zeta function (2.26a)	$\bar{Z}(x)$	Complementary zeta function (2.27a)

$Si(x), \bar{Si}(x)$	Circular and hyperbolic sine integrals (10.01b)
$Ci(x), \bar{Ci}(x)$	Circular and hyperbolic cosine integrals (10.01b)
$Ei(x), \bar{Ei}(x)$	Alternating and monotonic exponential integrals (10.01b)
$Li(x)$	Logarithmic integral (10.01b)
$S(x), C(x)$	Fresnel integrals (10.04a)
$F(k, x)$	Incomplete elliptic integral of the first kind (10.05b)
$E(k, x)$	Incomplete elliptic integral of the second kind (10.05b)
$\Pi(k, x)$	Incomplete elliptic integral of the third kind (10.05b)
$K, F(k, 1)$	Complete elliptic integral of the first kind (10.05b)
$E, E(k, 1)$	Complete elliptic integral of the second kind (10.05b)
$\Pi, \Pi(k, 1)$	Complete elliptic integral of the third kind (10.05b)
$\left.\begin{array}{l} S_1(x), S_2(x) \\ C_1(x), C_2(x) \end{array}\right\}$	Related Fresnel integrals (10.04b)

$J_r(x)$ Ordinary Bessel function of the first kind of order r (10.07b)

$Y_r(x)$ Ordinary Bessel function of the second kind of order r (10.07c)

$j_r(x)$ Spherical Bessel function of the first kind of order r (10.12a)

$y_r(x)$ Spherical Bessel function of the second kind of order r (10.12a)

$I_r(x)$ Modified Bessel function of the first kind of order r (10.13b)

$K_r(x)$ Modified Bessel function of the second kind of order r (10.13c)

$i_r(x)$ Modified spherical Bessel function of the first kind of order r (10.17a)

$k_r(x)$ Modified spherical Bessel functions of the second kind of order r (10.17a)

B.9   ORTHOGONAL POLYNOMIALS

$\bar{P}_r(x)$ Legendre function of the first kind of order r (11.09a)

$\bar{Q}_r(x)$ Legendre function of the second kind of order r (11.09a)

$P_r(x)$ Legendre polynomial of the first kind of order r (11.09a)

$Q_r(x)$ Legendre polynomial of the second kind of order r (11.09a)

$\bar{T}_r(x)$ Chebyshev function of the first kind of order r (11.14a)

$\bar{U}_r(x)$ Chebyshev function of the second kind of order r (11.14a)

$T_r(x)$ Chebyshev polynomial of the first kind of order r (11.14a)

$U_r(x)$ Chebyshev polynomial of the second kind of order r (11.14a)

$\bar{L}_r(x)$ Laguerre function of the first kind of order r (11.16a)

$\bar{L}_r^*(x)$ Laguerre function of the second kind of order r (11.16a)

$L_r(x)$ Laguerre polynomial of the first kind of order r (11.16a)

$L_r^*(x)$ Laguerre polynomial of the second kind of order r (11.16a)

$\bar{H}_r(x)$ Hermite function of the first kind of order r (11.16a)

$\bar{H}_r^*(x)$ Hermite function of the second kind of order r (11.16a)

$H_r(x)$ Hermite polynomial of the first kind of order r (11.16a)

$H_r^*(x)$ Hermite polynomial of the second kind of order r (11.16a)

B.10   GREEK ALPHABET

A	α	Alpha	H	η	Eta	N	ν	Nu	T	τ	Tau
B	β	Beta	Θ	θ	Theta	Ξ	ξ	Xi	Y	υ	Upsilon
Γ	γ	Gamma	I	ι	Iota	O	o	Omicron	Φ	ϕ	Phi
Δ	δ	Delta	K	κ	Kappa	Π	π	Pi	X	χ	Chi
E	ϵ	Epsilon	Λ	λ	Lambda	P	ρ	Rho	Ψ	ψ	Psi
Z	ζ	Zeta	M	μ	Mu	Σ	σ	Sigma	Ω	ω	Omega

Symbol	Name	Quantity	Relations
cd	candela	luminous intensity	basic unit
kg	kilogram	mass	basic unit
lm	lumen	luminous flux	$cd \cdot sr$
lx	lux	illumination	$lm \cdot m^{-2}$
m	meter	length	basic unit
mol	mole	amount of substance	basic unit
rad	radian	plane angle	supplementary unit
s	second	time	basic unit
sr	steradian	solid angle	supplementary unit
A	ampere	electric current	basic unit
C	coulomb	electric charge	$A \cdot s$
F	farad	electric capacitance	$A \cdot s \cdot V^{-1}$
H	henry	electric inductance	$V \cdot s \cdot A^{-1}$
Hz	hertz	frequency	s^{-1}
J	joule	work, energy	$N \cdot m$
K	kelvin	temperature degree	basic unit
N	newton	force	$kg \cdot m \cdot s^{-2}$
Pa	pascal	pressure, stress	$N \cdot m^{-2}$
S	siemens	electric conductance	Ω^{-1}
T	tesla	magnetic flux density	$Wb \cdot m^{-2}$
V	volt	voltage, electromotive force	$W \cdot A^{-1}$
W	watt	power	$J \cdot s^{-1}$
Wb	weber	magnetic flux	$V \cdot s$
Ω	ohm	electric resistance	$V \cdot A^{-1}$

B.12 DECIMAL MULTIPLES AND FRACTIONS OF UNITS

Factor	Prefix	Symbol	Factor	Prefix	Symbol
10^1	deka	D*	10^{-1}	deci	d
10^2	hecto	h	10^{-2}	centi	c
10^3	kilo	k	10^{-3}	milli	m
10^6	mega	M	10^{-6}	micro	μ
10^9	giga	G	10^{-9}	nano	n
10^{12}	tera	T	10^{-12}	pico	p
10^{15}	peta	P	10^{-15}	femto	f
10^{18}	exa	E	10^{-18}	atto	a

* In some literature, the symbol "da" is used for deka

Symbol*	Name	Quantity	Relations
deg	degree	plane angle	supplementary unit
ft	foot	length	basic unit
g	standard gravity	acceleration	$32.174 \, \text{ft} \cdot \text{sec}^{-2}$
hp	horsepower	power	$550 \, \text{lbf} \cdot \text{ft} \cdot \text{sec}^{-1}$
lb	pound-mass	mass	$\text{lbf} \cdot g^{-1}$
lbf	pound-force	force	basic unit
pd	poundal	force	$\text{lb} \cdot \text{ft} \cdot \text{sec}^{-2}$
sec	second	time	basic unit
sl	slug	mass	$\text{lbf} \cdot \text{ft}^{-1} \cdot \text{sec}^2$
Btu	British thermal unit	work, energy	$778.128 \, \text{ft} \cdot \text{lbf}$
°F	degree Fahrenheit	temperature	basic unit

* For A, C, F, . . . refer to (B.11).

B.14 METRIC SYSTEM OF UNITS (MKS SYSTEM)

Symbol*	Name	Quantity	Relations
cal	calorie	work, energy	$0.426 \, 65 \, \text{m} \cdot \text{kgf}$
deg	degree	plane angle	supplementary unit
dyn	dyne	force	$10^{-6} \, \text{kgf}$
erg	erg	work, energy	$10^{-8} \, \text{m} \cdot \text{kgf}$
g	standard gravity	acceleration	$9.806 \, 65 \, \text{m} \cdot \text{sec}^{-2}$
hp	horsepower	power	$75 \, \text{kgf} \cdot \text{m} \cdot \text{sec}^{-1}$
kg	kilogram-mass	mass	$\text{kgf} \cdot g^{-1}$
kgf	kilogram-force	force	basic unit
m	meter	length	basic unit
sec	second	time	basic unit
°C	degree Celsius	temperature	basic unit

* For A, C, F, . . . refer to (B.11).

Selected unit	SI units		USCS units	
Acre	4.046 86	(+03) m^2	4.356 00	(+04) ft^2
Angstrom (Å)	1.000 00	(−10) m	3.937 01	(−09) in
Atmosphere (atm, phs.)	1.013 25	(+05) Pa	1.469 39	(+01) lb · in^{-2}
Atmosphere (at, tec.)	9.806 65	(+04) Pa	1.422 33	(+01) lb · in^{-2}
Astronomical unit	1.495 98	(+11) m	4.908 81	(+11) ft
Bar (b)	1.000 00	(+05) Pa	1.450 38	(+01) lbf · in^{-2}
British thermal unit (Btu, IST)	1.055 06	(+03) J	7.781 69	(+02) ft · lbf
British thermal unit (Btu, tec.)	1.055 00	(+03) J	7.781 28	(+02) ft · lbf
British thermal unit (Btu, thm.)	1.054 35	(+03) J	7.776 49	(+02) ft · lbf
Bushel (U.S.)	3.523 91	(−02) m^3	1.244 46	(+00) ft^3
Calorie (cal, IST)	4.186 80	(+00) J	3.088 03	(+00) ft · lbf
Calorie (cal, tec.)	4.186 00	(+00) J	3.087 44	(+00) ft · lbf
Calorie (cal, thm.)	4.184 00	(+00) J	3.085 96	(+00) ft · lbf
Circular mil	5.067 07	(−10) m^2	7.854 00	(−07) in^2
Dyne	1.000 00	(−05) N	2.248 09	(−06) lbf
Erg	1.000 00	(−07) J	7.375 62	(−08) ft · lbf
Foot (ft)	3.048 00	(−01) m	1.000 00	(+00) ft
Gallon (U.S., liq.)	3.785 41	(−03) m^3	1.336 81	(−01) ft^3
Grain	6.479 89	(−05) kg	1.428 57	(−04) lb
Horsepower (hp, FPS)	7.456 99	(+02) W	5.500 00	(+02) ft · lbf · s^{-1}
Horsepower (hp, MKS)	7.354 99	(+02) W	9.863 20	(−01) hp (FPS)
Inch (in.)	2.540 00	(−02) m	8.333 33	(−02) ft
Kilogram-force (kgf)	9.806 65	(+00) N	2.204 62	(+00) lbf
Knot (int.)	5.144 44	(−01) m · s^{-1}	1.687 81	(+00) ft · s^{-1}
Light-year	9.460 55	(+15) m	3.103 86	(+16) ft
Liter (l)	1.000 00	(−03) m^3	3.531 47	(−02) ft^3
Mile (U.S., nat.)	1.852 00	(+03) m	6.076 12	(+03) ft
Mile (U.S., stu.)	1.609 34	(+03) m	5.280 00	(+03) ft
Ounce (U.S., avd.)	2.834 95	(−02) kg	6.250 00	(−02) lb
Ounce (U.S., liq.)	2.957 353	(−05) m^3	1.044 38	(−03) ft^3
Pint (U.S., liq.)	4.731 76	(−04) m^3	1.671 01	(−02) ft^3
Poise	1.000 00	(−01) N · s · m^{-2}	2.088 54	(−03) lbf · s · ft^{-2}
Pound (avd.)	4.535 92	(−01) kg	1.000 00	(+00) lb
Pound-force (lbf)	4.448 22	(+00) N	1.000 00	(+00) lbf
Poundal (pd)	1.382 55	(−01) N	3.108 10	(−02) lbf
Psi (lbf/in^2)	6.894 76	(+03) Pa	1.440 00	(+02) lbf · ft^2
Quart (U.S., dry)	1.101 22	(−03) m^3	3.888 93	(−02) ft^3
Quart (U.S., liq.)	9.463 53	(−04) m^3	3.342 01	(−02) ft^3
Slug (sl)	1.459 39	(+01) kg	3.217 40	(+01) lb
Stoke	1.000 00	(−04) m^2 · s^{-1}	1.076 39	(−03) ft^2 · s^{-1}
Torr (0°C)	1.333 22	(+02) Pa	2.784 50	(+00) lbf · ft^{-2}
Yard (yd)	9.144 40	(−01) m	3.000 00	(+00) ft

*Abbreviations:

	nat. = nautical	tec. = technical	MKS = metric system
avd. = avoirdupois	phs. = physical	thm. = thermochemical	FPS = U.S. Customary System
liq. = liquid	stu. = statute	IST = international steam tables	(USCS)

Appendix C
REFERENCES AND BIBLIOGRAPHY

C.1 NUMERICAL CALCULATIONS

1.01   Abramowitz, M., and I. A. Stegun: "Handbook of Mathematical Functions," National Bureau of Standards, Washington, D.C., 1964.

1.02   Acton, F. S.: "Numerical Methods That Work," Harper & Row, New York, 1970.

1.03   Atkinson, K. E.: "An Introduction to Numerical Analysis," Wiley, New York, 1978.

1.04   Burdin, R. L., and J. D. Faires: "Numerical Analysis," 3rd ed, Prindle, Weber & Schmidt, Boston, 1981.

1.05   Chapra, S. C., and R. P. Canale: "Numerical Methods for Engineers," 2d ed., McGraw-Hill, New York, 1988.

1.06   Collatz, L.: "Functional Analysis and Numerical Mathematics," Academic, New York, 1966.

1.07   Demidovich, B. P., and I. A. Maron; "Computational Mathematics," Mir, Moscow, 1981.

1.08   Fröburg, C. E.: "Numerical Mathematics," Benjamin-Cummings, Reading, Mass., 1985.

1.09   Hamming, R.: "Numerical Methods for Scientists and Engineers," 2d ed., McGraw-Hill, New York, 1973.

1.10   Hildebrand, F. B.: "Introduction to Numerical Analysis," McGraw-Hill, New York, 1956.

1.11   Jain, M. K., S. R. K. Inyengar, and R. K. Jain: "Numerical Methods for Scientific and Engineering Computations," Halstead, New York, 1982.

1.12   Johnson, L. W., and R. D. Riess: "Numerical Analysis," 2d ed., Addison-Wesley, New York, 1982.

1.13   Korn, G. A., and T. M. Korn: "Mathematical Handbook for Scientists and Engineers," 2d ed., McGraw-Hill, New York, 1968.

1.14   Marchuk, G. I.: "Methods of Numerical Mathematics," Springer, New York, 1975.

1.15   Pachner, J.: "Handbook of Numerical Analysis Applications," McGraw-Hill, New York, 1984.

1.16   Ralston, A., and P. Rabinowitz: "A First Course in Numerical Analysis," McGraw-Hill, New York, 1978.

1.17   Scarborough, J. B.: "Numerical Mathematical Analysis," 3d ed., Johns Hopkins, Baltimore, 1955.

1.18   Scheid, F.: "Schaum's Outline of Numerical Analysis," McGraw-Hill, New York, 1968.

1.19   Tod, J.: "Survey of Numerical Analysis," McGraw-Hill, New York, 1962.

1.20   Tuma, J. J.: "Engineering Mathematics Handbook," 3d ed., McGraw-Hill, New York, 1987.

1.21   Tuma, J. J.: "Handbook of Physical Calculations," 2d ed., McGraw-Hill, New York, 1983.

1.22   Tuma, J. J.: "Technology Mathematics Handbook," McGraw-Hill, New York, 1975.

1.23   Young, D., and R. T. Gregory: "A Survey of Numerical Mathematics," vols. I, II, Addison-Wesley, Reading, Mass., 1972.

C.2 NUMERICAL CONSTANTS

2.01   Artin, E.: "The Gamma Function," Holt, New York, 1964.

2.02   Beckmann, P.: "A History of π," 3d ed., St. Martin's New York, 1974.

2.03   Dwight, H. B.: "Tables of Integrals and Other Mathematical Data," 4th ed., Macmillan, New York, 1961.

2.04   Jahnke, E., and F. Emden: "Tables of Functions," Dover, New York, 1945.

2.05   Lösch, F., and F. Schoblik: "Die Fakutät und Vervandte Funktionen," Teubner, Leipzig, 1962.

2.06   Nielsen, N.: "Handbuch der Theorie der Gammafunktion," Teubner, Leipzig, 1906.

2.07   Perron, O.: "Die Lehre von Kettenbrüchen," vols. I, II, Teubner, Stuttgart, 1957.

2.08   Sibagaki, W.: "Theory and Applications of the Gamma Function," Iwanami Syoten, Tokyo, 1952.

2.09   Titchmarsh, E. C.: "The Zeta Function of Riemann," Cambridge, London, 1930.

2.10   Wahl, H. S.: "Analytical Theory of Continued Fractions", Van Nostrand, New York, 1948.

C.3 NUMERICAL DIFFERENCES

3.01   Boole, G.: "Finite Differences," 3d ed., Stechert, New York, 1931.

3.02   Fort, T.: "Finite Differences and Difference Equations," Oxford, New York, 1948.

3.03   Jordan, C.: "Calculus of Finite Differences," Chelsea, New York, 1947.

3.04   Kunzmann, J.: "Méthodes Numériques," Dunod, Paris, 1959.

3.05   Milne, W. E.: "Numerical Calculus," Princeton, Princeton, N.J., 1949.

3.06   Milne-Thompson, L. N.: "Calculus of Finite Differences," Macmillan, London, 1951.

C.3 NUMERICAL DIFFERENCES (*Continued*)

3.07 Nörlund, N. E.: "Vorlesungen über Differnzenrechnung," Chelsea, New York, 1954.

3.08 Sauer, R., and I. Szabó: "Mathematishe Hilfsmittel des Ingenieurs," vol. III, Springer, New York, 1968.

3.09 Schulz, G.: "Formmelsammlung zur Praktischen Mathematik," De Gruyter, Berlin, 1945.

3.10 Steffensen, J. P.: "Interpolation," Baillière, London, 1937.

3.11 Zurmühl, R.: "Praktische Mathematik," 5th ed., Springer, Berlin, 1965.

C.4 NUMERICAL INTEGRALS

4.01 Davis, S., and P. Rabinowitz: "Methods of Numerical Integration," Academic, New York, 1975.

4.02 Gradshteyn, I. S., and I. M. Ryzhik: "Tables of Integrals, Series and Products," 4th ed., Academic, New York, 1965.

4.03 Gröbner, W., and N. Hofreiter: "Integraltafeln," 4th ed., vols. I, II, Springer, Vienna, 1965.

4.04 Krylov, V. I.: "Approximate Calculation of Integrals," Macmillan, New York, 1962.

4.05 Nikosky, S. M.: "Quadrature Formulas," Fizmatgiz, Moscow, 1958.

4.06 Spiegel, M. R.: "Mathematical Handbook," McGraw-Hill, New York, 1968.

4.07 Spiegel, M. R.: "Schaum's Outline of Finite Differences and Difference Equations," McGraw-Hill, New York, 1971.

4.08 Stroud, A.: "Approximate Calculation of Multiple Integrals," Prentice-Hall, Englewood Cliffs, N.J., 1971.

4.09 Stroud, A., and D. Secrest: "Gaussian Quadrature Formulas," Prentice-Hall, Englewood Cliffs, N.J., 1966.

C.5 SERIES AND PRODUCTS OF CONSTANTS

5.01 Adams, E. P.: "Smithsonian Mathematical Formulas," The Smithsonian Institute, Washington, 1922.

5.02 Bromwich, T. J.: "Introduction of the Theory of Infinite Series," Macmillan, London, 1926.

5.03 Erdélyi, A.: "Asymptotic Expansions," Dover, New York, 1956.

5.04 Fort, T.: "Infinite Series," Oxford, New York, 1930.

5.05 Jolley, L. B.: "Summation of Series," 2d ed., Dover, New York, 1948.

5.06 Knopp, K.: "Theory and Application of Infinite Series," Blackie, London, 1963.

6.07 Smail, L. L.: "Elements of the Theory of Infinite Processes," McGraw-Hill, New York, 1923.

C.6 ALGEBRAIC AND TRANSCENDENTAL EQUATIONS

6.01 Acton, F. C.: "Numerical Methods That Work,", Harper & Row, New York, 1970.

6.02 Bairstow, L.: "Investigations Relating to the Stability of the Aeroplane," Report and Memoranda No. 154, AEA, Washington, D.C., 1914.

6.03 Dickson, L. E.: "Algebraic Theories," Dover, New York, 1959.

6.04 Housholder, A. S.: "Principles of Numerical Analysis," McGraw-Hill, New York, 1953.

6.05 Kopal, Z.: "Numerical Analysis," Wiley, New York, 1955.

6.06 Kunz, K. S.: "Numerical Analysis," McGraw-Hill, New York, 1957.

6.07 Oliver, F. W. J.: The Evaluation of Zeros of High Degree Polynomials, *Philos. Trans. R. Soc. London, Ser. A.*, vol. 244, 1952.

6.08 Upensky, J. V.: "Theory of Equations," McGraw-Hill, New York, 1948.

6.09 Zurmühl, R.: "Praktische Mathematik," 5th ed., Springer, Berlin, 1965.

C.7 MATRIX EQUATIONS

7.01 Havner, K. S., and J. J. Tuma: Influence Lines for Continuous Beams, *Okla., State Univ. Agri. Appl. Sci., Eng. Exp. Stn., Publ. 106*, 1959.

7.02 Forsythe, G., and C. Moler: "Computational Solution of Linear Algebraic Systems," Prentice-Hall, Englewood Cliffs, N.J., 1967.

7.03 Gantmacher, F. R.: "Matrix Theory," vols. I, II, Chelsea, New York, 1950.

7.04 Ortega, J., and W. Reinboldt: "Iterative Solution of Equations of Several Variables," Academic, New York, 1970.

C.7 MATRIX EQUATIONS (*Continued*)

7.05   Pipes, L. A.: "Matrix Methods for Engineers," Prentice-Hall, Englewood Cliffs, N.J., 1963.

7.06   Smirnov, V. I.: "Linear Algebra and Group Theory," McGraw-Hill, New York, 1961.

7.07   Stephenson, G.: "An Introduction to Matrices, Sets and Groups for Science Students," Dover, New York, 1986.

7.08   Tuma, J. J.: Analysis of Continuous Beams by Carryover Moments, *ASCE J. Struct. Div.* vol. 84, 1958.

7.09   Varga, R. S.: "Iterative Analysis," Prentice-Hall, Englewood Cliffs, N.J., 1962.

7.10   Westlake, J. R.: "Handbook of Numerical Matrix Inversions and Solutions of Linear Equations," Wiley, New York, 1968.

7.11   Young, D.: "Iterative Methods for Solution of Large Linear Systems," Academic, New York, 1971.

7.12   Zurmühl, R.: "Matrizen," 3d ed., Springer, Berlin, 1961.

C.8 EIGENVALUE EQUATIONS

8.01   Collatz, L.: "Eigenweteaufgaben mit Technischen Anwendungen," Springer, Berlin, 1959.

8.02   Faddeev, D. K., and V. N. Faddeeva: "Computational Methods of Linear Algebra," Freeman, San Francisco, 1963.

8.03   Gourley, A. R., and G. A. Watson: "Computational Methods for Matrix Eigenproblems," Wiley, New York, 1973.

8.04   National Bureau of Standards: "Simultaneous Linear Equations and the Determination of Eigenvalues," Applied Mathematics Series, no. 29, Washington, 1953.

8.05   National Bureau of Standards: "Simultaneous Linear Equations and the Determination of Eigenvalues," Applied Mathematics Series, no. 38, Washington, 1954.

8.06   Wilkinson, J.: "The Algebraic Eigenvalue Problems,", Oxford, London, 1965.

C.9 SERIES OF FUNCTIONS

9.01   Calson, B., and M. Goldstein: "Rational Approximation of Functions," Los Alamos Scientific Laboratory, LA-1943, Los Alamos, N. Mex., 1955.

9.02   Clenshaw, C. W.: Polynomial Approximations to Elementary Functions, *Math. Tables Aids Comp.*, vol. 8, 1954.

9.03   Hastings, C.: "Approximations for Digital Computors," Princeton Univ. Press, Princeton, N.J., 1955.

9.04   Mangulis, V.: "Handbook of Series," Academic, New York, 1965.

9.05   Meinardus, G.: "Approximations of Functions," Springer, New York, 1967.

9.06   Ralston, A., and H. F. Wilf (eds.): "Mathematical Methods for Digital Computers," vol. I, Wiley, New York, 1960.

9.07   Rice, J. R.: "Approximations of Functions," Addison-Wesley, Reading, Mass., 1964.

9.08   Schwartz, I. J.: "Introduction to the Operations with Series," Chelsea, New York, 1924.

C.10 SPECIAL FUNCTIONS

10.01   Andrews, L. C.: "Special Functions," Macmillan, New York, 1985.

10.02   Allen, E. E.: Analytical Approximations, *Math. Tables Aids Comp.*, vol. 8, 1954.

10.03   Allen, E. E.: "Polynomial Approximations to Some Modified Bessel Functions," *Math. Tables Aids Comp.*, vol. 10, 1956.

10.04   Bell, W. W.: "Special Functions," Van Nostrand, London, 1968.

10.05   Carlson, B. C.: "Special Functions of Applied Mathematics," Academic, New York, 1977.

10.06   Erdélyi et al.: "Higher Transcendental Functions," vols. I to III, McGraw-Hill, New York, 1955.

10.07   Klein, F.: "Vorlesungen über die Hypergeometrische Funktion," Teubner, Berlin, 1933.

10.08   Lebedev, N. N.: "Special Functions and Their Applications," Dover, New York, 1972.

10.09   Magnus, W., F. Oberhettinger, and R. P. Soni: "Formulas and Theorems of the Special Functions of Mathematical Physics," Springer, New York, 1966.

10.10   McLachlan, N. W.: "Bessel Functions for Engineers," Oxford, London, 1955.

10.11   Oberhettinger, F., and W. Magnus: "Anwendung der Elliptischen Funktionen in Physik und Technik," Springer, Berlin, 1949.

C.10 SPECIAL FUNCTIONS (*Continued*)

10.12 Rainville, E. D.: "Special Functions," Macmillan, New York, 1960.

10.13 Sanson, G.: "Orthogonal Functions," Wiley, New York, 1959.

10.14 Spanier, J., and K. B. Oldham: "An Atlas of Functions," Hemisphere, New York, 1987.

10.15 Whittaker, E. T., and G. N. Watson: "A Course of Modern Analysis," 4th ed., Cambridge, London, 1952.

C.11 ORTHOGONAL POLYNOMIALS

11.01 Askey, R. A. (ed.): "Theory and Applications of Special Functions," Academic, New York, 1975.

11.02 Johnson, D. E., and J. R. Johnson: "Mathematical Methods in Engineering and Physics," Prentice-Hall, Englewood Cliffs, N.J., 1982.

11.03 Madelung, E.: "Die Mathematischen Hilfsmittel des Physikers," 7th ed., Springer, Berlin, 1964.

11.04 Szegö, G.: "Orthogonal Polynomials," American Mathematical Society, New York, 1959.

11.05 Sneddon, I. N.: "Special Functions of Mathematical Physics and Chemistry," Oliver & Boyd, Edinburgh, 1961.

11.06 Shotat, J. A., E. Hille, and J. L. Walsh: Bibliography on Orthogonal Polynomials, *Bull. Natl. Res. Counc.*, no. 103, 1940.

11.07 Tricomi, F. G.: "Vorlesungen über Orthogonalreihen," Springer, Berlin, 1953.

C.12 LEAST-SQUARES APPROXIMATIONS

12.01 Clenshaw, C. W.: "A Note on the Summation of Chebyshev Series," *Math. Tables Aids Comp.*, vol. 9, 1955.

12.02 Davis, P. J.: "Interpolation and Approximation," Dover, New York, 1975.

12.03 Fox, L., and I. B. Parker: "Chebyshev Polynomials in Numerical Analysis," Oxford, London, 1968.

12.04 Lanczos, C.: "Applied Analysis," Prentice-Hall, Englewood Cliffs, N.J., 1956.

12.05 Lawson, C. L., and R. J. Hanson: "Solving Least-Squares Problems," Prentice-Hall, Englewood Cliffs, N.J., 1974.

12.06 Powell, M. J.: "Approximation Theory and Methods," Cambridge, London, 1981.

12.07 Rivlin, T. K.: "An Introduction to the Approximation of Functions," Dover, New York, 1981.

C.13 FOURIER APPROXIMATIONS

13.01 Carslow, H. S.: "Theory of Fourier Series and Integrals," Dover, New York, 1930.

13.02 Churchill, R. V.: "Fourier Series and Boundary Value Problems," 2d ed., McGraw-Hill, New York, 1963.

13.03 Franklin, P.: "Introduction to Fourier Methods and the Laplace Transforms," Dover, New York, 1958.

13.04 Jackson, D.: Fourier Series and Orthogonal Polynomials, *Carus Math. Monogr.* no. 6, Mathematical Association of America, 1941.

13.05 Rogosinski, W.: "Fourier Series," Chelsea, New York, 1950.

13.06 Tolstov, G. P.: "Fourier Series," Dover, New York, 1976.

13.07 Zygmund, A.: "Trigonometric Series," Dover, New York, 1955.

C.14 LAPLACE TRANSFORMS

14.01 Churchill, R. V.: "Operational Mathematics," 2d ed., McGraw-Hill, New York, 1958.

14.02 DiStefano III, J. J., A. R. Stubberud, and I. J. Williams: "Schaum's Outline of Feedback and Control Systems," McGraw-Hill, New York, 1967.

14.03 Doetsch, G.: "Theorie und Anwendung der Laplace-Transformation," Springer, Berlin, 1937.

14.04 Doetsch, G., H. Kniess, and D. Voelker: "Taballen zur Laplace-Transformation und Anleitung zum Gebrauch," Springer, Berlin, 1947.

14.05 Erdeleyi, A., W. Magnus, F. Oberhettingen, and F. Tricomi: "Tables of Integral Transforms," vols. I, II, McGraw-Hill, New York, 1954.

14.06 Spiegel, M. R.: "Schaum's Outline of Laplace Transforms," McGraw-Hill, New York, 1965.

14.07 Widder, D. V.: "The Laplace Transform," Princeton Univ. Press, Princeton, N.J., 1941.

C.15 ORDINARY DIFFERENTIAL EQUATIONS

15.01   Allen, D. N.: "Relaxation Methods in Engineering and Science," McGraw-Hill, New York, 1954.

15.02   Babuška, I., M. Práger, and E. Vitásek: "Numerical Processes in Differential Equations," Interscience, New York, 1966.

15.03   Collatz, L.: "Numerical Treatment of Differential Equations," 3d ed., Springer, Berlin, 1960.

15.04   Fox, L. (ed.): "Numerical Solution of Ordinary and Partial Differential Equations," Pergamon, London, 1962.

15.05   Frank, R., und R. V. Mises: "Die Differential und Integral Gleichungen der Mechanik und Physik," vols. I, II, Dover, New York, 1961.

15.06   Goldberg, S.: "Introduction to Difference Equations," Dover, New York, 1986.

15.07   Grinter, L. E. (ed.): "Numerical Methods of Analysis in Engineering," Macmillan, New York, 1949.

15.08   Henrici, P.: "Discrete Variable Methods in Ordinary Differential Equations," Wiley, New York, 1962.

15.09   Hildebrand, F. B.: "Finite-difference Equations and Simulations," Prentice-Hall, Englewood Cliffs, N.J., 1968.

15.10   Ince, E. L.: "Ordinary Differential Equations," Dover, New York, 1956.

15.11   Kamke, E.: "Differential Gleichungen," 3d ed., Chelsea, New York, 1948.

15.12   Levy, H., and E. A. Baggolt: "Numerical Solutions of Differential Equations," Dover, New York, 1950.

15.13   Milne, W. E.: "Numerical Solution of Differential Equations," Dover, New York, 1970.

C.16 PARTIAL DIFFERENTIAL EQUATIONS

16.01   Ames, W. F.: "Numerical Methods for Partial Differential Equations," Academic, New York, 1977.

16.02   Bateman, H.: "Partial Differential Equations of Mathematical Physics," Cambridge, London, 1959.

16.03   Bender, L., and S. Orzag: "Advanced Mathematics for Engineers," McGraw-Hill, New York, 1978.

16.04   Carrier, G. F., and C. E. Pearson: "Partial Differential Equations," Academic, New York, 1976.

16.05   Duffy, D. G.: "Solution of Partial Differential Equations," TAB, Blue Ridge Summit, Pa., 1986.

16.06   Forsythe, G. E., and W. R. Watson: "Finite Difference Methods for Partial Differential Equations," Wiley, New York, 1960.

16.07   Gladwell, I., and R. Wait: "Survey of Numerical Metods for Partial Differential Equations," Oxford, New York, 1979.

16.08   Havner, K. S.: "Algebriac Carry-Over in Two-Dimensional Systems," dissertaiton, Oklahoma State University Library, Stillwater, 1959.

16.09   Jain, M. K.: "Numerical Solution of Differential Equations," Halstead, New York, 1984.

16.10   Kardestuncer, H., and D. H. Norrie (eds.): "Finite Element Handbook," McGraw-Hill, New York, 1987.

16.11   Ketter, R. L., and S. P. Prawel, Jr.: "Modern Methods of Engineering Computation," McGraw-Hill, New York, 1969.

16.12   Miller, K. S.: "Partial Differential Equations in Engineering Problems," Prentice-Hall, Englewood Cliffs, N.J., 1953.

16.13   Oden, J. T.: "Finite Elements in Nonlinear Continua," McGraw-Hill, New York, 1972.

16.14   Panov, D. J.: "Formulas for Numerical Solution of Partial Differential Equations by the Method of Differences," Ungar, New York, 1963.

16.15   Salvadori, M. G., and M. L. Baron: "Numerical Methods in Engineering," Prentice-Hall, Englewood Cliffs, N.J., 1961.

16.16   Shaw, F. S.: "An Introduction to Relaxation Methods," Dover, New York, 1953.

16.17   Soare, M.: "Application of Finite Difference Equations to Shell Analysis," Pergamon, London, 1967.

16.18   Sommerfeld, A.: "Partial Differential Equations of Physics," Academic, New York, 1949.

16.19   Southwell, R. V.: "Relaxation Methods in Engineering Science," Oxford, New York, 1940.

16.20   Tikhovov, A. N., and A. A. Samarski: "Equations of Mathematical Physics," Macmillan, New York, 1963.

16.21   Tuma, J. J.: "Handbook of Structural and Mechanical Matrices," McGraw-Hill, New York, 1988.

16.22   Vladimirov, V. S.: "Equations of Mathematical Physics," Dekker, New York, 1971.

16.23   Zienkiewicz, O. C.: "The Finite Element Method," 3d ed., McGraw-Hill, New York, 1977.

INDEX

References are made to section numbers for the text material. Numers preceded by the letters A and B refer to the Appendixes. In the designation of units and systems of units the following abbreviations are used:

avd = avoirdupois   thm = thermochemical
liq = liquid   try = troy(apothecary)
nat = nautical   FPS = English system
phs = physical   IST = international steam tables
stu = statute   MKS = metric system
tec = technical   SI = international system

Fundamental Physical Constants

The values of constants given in this table are those recommended by the CODATA Task Group on Fundamental Constants, E. R. Cohen, Chairman.

Quantity	Symbol	Value	Units* SI	Units* CGS
Speed of light in vacuum	c	2.997 924 580	$10^8 \, \mathrm{m \cdot s^{-1}}$	$10^{10} \, \mathrm{cm \cdot s^{-1}}$
Fine-structure constant,	α	7.297 350 6	10^{-3}	10^{-3}
$[\mu_0 c^2/4\pi](e^2/\hbar c)$	α^{-1}	137.036 04		
Electron charge	e	1.602 189 2	$10^{-19} \, \mathrm{C}$	$10^{-20} \, \mathrm{emu}$
		4.803 242		$10^{-10} \, \mathrm{esu}$
Planck constant	h	6.626 176	$10^{-34} \, \mathrm{J \cdot s}$	$10^{-27} \, \mathrm{erg \cdot s}$
	$\hbar = h/2\pi$	1.054 588 7	$10^{-34} \, \mathrm{J \cdot s}$	$10^{-27} \, \mathrm{erg \cdot s}$
Avogadro constant	N	6.022 045	$10^{26} \, \mathrm{kmol^{-1}}$	$10^{23} \, \mathrm{mol^{-1}}$
Atomic mass unit	u	1.660 565 5	$10^{-27} \, \mathrm{kg}$	$10^{-24} \, \mathrm{g}$
Electron rest mass	m_e	9.109 534	$10^{-31} \, \mathrm{kg}$	$10^{-28} \, \mathrm{g}$
$Nm_e = M_e$	M_e	5.485 802 6	$10^{-4} \, \mathrm{u}$	$10^{-4} \, \mathrm{u}$
Proton rest mass	m_p	1.672 648 5	$10^{-27} \, \mathrm{kg}$	$10^{-24} \, \mathrm{g}$
$Nm_p = M_p$	M_p	1.007 276 471	u	u
Ratio of proton mass to electron mass	m_p/m_e	1836.151 52		
Neutron rest mass	m_n	1.674 954 3	$10^{-27} \, \mathrm{kg}$	$10^{-24} \, \mathrm{g}$
$Nm_m = M_n$	M_n	1.008 665 0	u	u
Electron charge to mass ratio	e/m_e	1.758 804 7	$10^{11} \, \mathrm{C \cdot kg^{-1}}$	$10^7 \, \mathrm{emu \cdot g^{-1}}$
		5.272 764		$10^{17} \, \mathrm{esu \cdot g^{-1}}$
Magnetic flux quantum,	Φ_0	2.067 850 6	$10^{-15} \, \mathrm{Wb}$	$10^{-7} \, \mathrm{G \cdot cm^2}$
$[c]^{-1} (hc/2e)$	h/e	4.135 701	$10^{-15} \, \mathrm{J \cdot s \cdot C^{-1}}$	$10^{-7} \, \mathrm{erg \cdot s \cdot emu^{-1}}$
		1.379 521 5		$10^{-17} \, \mathrm{erg \cdot s \cdot esu^{-1}}$
Josephson frequency-voltage ratio	$2e/h$	4.835 939	$10^{14} \, \mathrm{Hz \cdot V^{-1}}$	
Quantum of circulation	$h/2m_e$	3.636 945 5	$10^{-4} \, \mathrm{J \cdot s \cdot kg^{-1}}$	$\mathrm{erg \cdot s \cdot g^{-1}}$
	h/m_e	7.273 891	$10^{-4} \, \mathrm{J \cdot s \cdot kg^{-1}}$	$\mathrm{erg \cdot s \cdot g^{-1}}$
Faraday constant	$\mathscr{F}$	9.648 456	$10^7 \, \mathrm{C \cdot kmol^{-1}}$	$10^4 \, \mathrm{C \cdot mol^{-1}}$
		2.892 534 2		$10^{14} \, \mathrm{esu \cdot mol^{-1}}$
Rydberg constant, $[\mu_0 c^2/4\pi]^2 (m_e e^4/4\pi\hbar^3 c)$	R_∞	1.097 373 177	$10^7 \, \mathrm{m^{-1}}$	$10^5 \, \mathrm{cm^{-1}}$
Bohr radius, $[\mu_0 c^2/4\pi]^{-1}(\hbar^2/m_e e^2) = \alpha/4\pi R_\infty$	a_0	5.291 770 6	$10^{-11} \, \mathrm{m}$	$10^{-9} \, \mathrm{cm}$
Classical electron radius, $[\mu_0 c^2/4\pi](e^2/m_e c^2) = \alpha^3/4\pi R_\infty$	r_0	2.817 938 0	$10^{-15} \, \mathrm{m}$	$10^{-13} \, \mathrm{cm}$
Free electron g-factor, or electron magnetic moment in Bohr magnetons	$g_{j/2} = \mu_e/\mu_B$	1.001 159 656 7		
Free muon g-factor, or muon magnetic moment in units of $[c] (e\hbar/2m_\mu c)$	$g_{\mu/2}$	1.001 166 16		